ELECTRIC VEHICLE MACHINES AND DRIVES

ELECTRIC VEHICLE MACHINES AND DRIVES

DESIGN, ANALYSIS AND APPLICATION

K. T. Chau

International Research Centre for Electric Vehicles
The University of Hong Kong

IEEE

WILEY

This edition first published 2015

Registered office
John Wiley & Sons Singapore Pte. Ltd., 1 Fusionopolis Walk, #07-01 Solaris South Tower, Singapore 138628.

For details of our global editorial offices, for customer services and for information about how to apply for permission to reuse the copyright material in this book please see our website at www.wiley.com.

Library of Congress Cataloging-in-Publication Data applied for.

ISBN 978-1-118-75252-4

Typeset in 9.5/11.5pt TimesLTStd by SPi Global, Chennai, India

1 2015

To my parents, family, colleagues, and friends worldwide

Contents

Preface xiii

Organization of This Book xv

Acknowledgments xvii

About the Author xix

1 Introduction **1**
1.1 What Is an Electric Vehicle? 1
1.2 Overview of EV Challenges 3
1.2.1 Pure Electric Vehicle 4
1.2.2 Hybrid Electric Vehicle 4
1.2.3 Gridable Hybrid Electric Vehicle 5
1.2.4 Fuel-Cell Electric Vehicle 5
1.3 Overview of EV Technologies 6
1.3.1 Motor Drive Technology 6
1.3.2 Energy Source Technology 9
1.3.3 Battery Charging Technology 11
1.3.4 Vehicle-to-Grid Technology 14
References 16

2 DC Motor Drives **19**
2.1 System Configurations 19
2.2 DC Machines 20
2.2.1 Structure of DC Machines 20
2.2.2 Principle of DC Machines 22
2.2.3 Modeling of DC Machines 22
2.3 DC–DC Converters 24
2.3.1 DC–DC Converter Topologies 24
2.3.2 Soft-Switching DC–DC Converter Topologies 27
2.4 DC Motor Control 28
2.4.1 Speed Control 29
2.4.2 Regenerative Braking 31
2.5 Design Criteria of DC Motor Drives for EVs 33
2.6 Design Example for EVs 34

2.7 Application Examples of DC Motor Drives in EVs 35
2.8 Fading Technology for EVs? 38
References 38

3 Induction Motor Drives 39
3.1 System Configurations 39
3.2 Induction Machines 40
3.2.1 Structure of Induction Machines 41
3.2.2 Principle of Induction Machines 42
3.2.3 Modeling of Induction Machines 44
3.3 Inverters for Induction Motors 46
3.3.1 PWM Switching Inverters 47
3.3.2 Soft-Switching Inverters 50
3.4 Induction Motor Control 51
3.4.1 Variable-Voltage Variable-Frequency Control 51
3.4.2 Field-Oriented Control 53
3.4.3 Direct Torque Control 57
3.5 Design Criteria of Induction Motor Drives for EVs 61
3.6 Design Example of Induction Motor Drives for EVs 64
3.7 Application Examples of Induction Motor Drives in EVs 67
3.8 Matured Technology for EVs? 67
References 68

4 Permanent Magnet Brushless Motor Drives 69
4.1 PM Materials 69
4.2 System Configurations 70
4.3 PM Brushless Machines 72
4.3.1 Structure of PM Brushless Machines 72
4.3.2 Principle of PM Brushless Machines 75
4.3.3 Modeling of PM Brushless Machines 78
4.4 Inverters for PM Brushless Motors 82
4.4.1 Inverter Requirements 82
4.4.2 Switching Schemes for Brushless AC Operation 83
4.4.3 Switching Schemes for Brushless DC Operation 84
4.5 PM Brushless Motor Control 87
4.5.1 PM Synchronous Motor Control 87
4.5.2 PM Brushless DC Motor Control 91
4.6 Design Criteria of PM Brushless Motor Drives for EVs 93
4.7 Design Examples of PM Brushless Motor Drives for EVs 96
4.7.1 Planetary-Geared PM Synchronous Motor Drive 96
4.7.2 Outer-Rotor PM Brushless DC Motor Drive 100
4.8 Application Examples of PM Brushless Motor Drives in EVs 103
4.9 Preferred Technology for EVs? 104
References 106

5 Switched Reluctance Motor Drives 108
5.1 System Configurations 108
5.2 SR Machines 110

	5.2.1 Structure of SR Machines	110
	5.2.2 Principle of SR Machines	112
	5.2.3 Modeling of SR Machines	115
5.3	SR Converters	117
	5.3.1 SR Converter Topologies	118
	5.3.2 Soft-Switching SR Converter Topologies	119
	5.3.3 Comparison of SR Converters for EVs	123
5.4	SR Motor Control	124
	5.4.1 Speed Control	124
	5.4.2 Torque-Ripple Minimization Control	126
	5.4.3 Position Sensorless Control	128
5.5	Design Criteria of SR Motor Drives for EVs	131
	5.5.1 Machine Initialization	132
	5.5.2 Suppression of Acoustic Noise	136
5.6	Examples of SR Motor Drives for EVs	137
	5.6.1 Planetary-Geared SR Motor Drive	137
	5.6.2 Outer-Rotor In-Wheel SR Motor Drive	141
5.7	Application Examples of SR Motor Drives in EVs	144
5.8	Maturing Technology for EVs?	144
	References	145
6	**Stator-Permanent Magnet Motor Drives**	**147**
6.1	Stator-PM versus Rotor-PM	147
6.2	System Configurations	148
6.3	Doubly-Salient PM Motor Drives	149
6.4	Flux-Reversal PM Motor Drives	157
6.5	Flux-Switching PM Motor Drives	160
6.6	Hybrid-Excited PM Motor Drives	161
6.7	Flux-Mnemonic PM Motor Drives	165
6.8	Design Criteria of Stator-PM Motor Drives for EVs	173
6.9	Design Examples of Stator-PM Motor Drives for EVs	177
	6.9.1 Outer-Rotor Hybrid-Excited DSPM Motor Drive	177
	6.9.2 Outer-Rotor Flux-Mnemonic DSPM Motor Drive	181
6.10	Potential Applications of Stator-PM Motor Drives in EVs	192
	References	194
7	**Magnetic-Geared Motor Drives**	**195**
7.1	System Configurations	195
7.2	Magnetic Gears	197
	7.2.1 Converted Magnetic Gears	198
	7.2.2 Field-Modulated Magnetic Gears	200
7.3	MG Machines	203
	7.3.1 Principle of MG Machines	205
	7.3.2 Modeling of MG Machines	211
7.4	Inverters for MG Motors	211
7.5	MG Motor Control	212
7.6	Design Criteria of MG Motor Drives for EVs	213

7.7 Design Examples of MG Motor Drives for EVs 215
7.7.1 MG PM Brushless DC In-Wheel Motor Drive 215
7.7.2 MG PM Brushless AC In-Wheel Motor Drive 218
7.8 Potential Applications of MG Motor Drives in EVs 224
References 225

8 Vernier Permanent Magnet Motor Drives 227
8.1 System Configurations 227
8.2 Vernier PM Machines 228
8.2.1 Vernier PM versus Magnetic-Geared PM 228
8.2.2 Structure of Vernier PM Machines 229
8.2.3 Principle of Vernier PM Machines 234
8.2.4 Modeling of Vernier PM Machines 237
8.3 Inverters for Vernier PM Motors 238
8.4 Vernier PM Motor Control 239
8.5 Design Criteria of Vernier PM Motor Drives for EVs 240
8.6 Design Examples of Vernier PM Motor Drives for EVs 242
8.6.1 Outer-Rotor Vernier PM Motor Drive 242
8.6.2 Outer-Rotor Flux-Controllable Vernier PM Motor Drive 245
8.7 Potential Applications of Vernier PM Motor Drives in EVs 251
References 251

9 Advanced Magnetless Motor Drives 253
9.1 What Is Advanced Magnetless? 253
9.2 System Configurations 254
9.3 Synchronous Reluctance Motor Drives 255
9.4 Doubly-Salient DC Motor Drives 257
9.5 Flux-Switching DC Motor Drives 260
9.6 Vernier Reluctance Motor Drives 264
9.7 Doubly-Fed Vernier Reluctance Motor Drives 266
9.8 Axial-Flux Magnetless Motor Drives 269
9.9 Design Criteria of Advanced Magnetless Motor Drives for EVs 272
9.10 Design Examples of Advanced Magnetless Motor Drives for EVs 272
9.10.1 Multi-tooth Doubly-Salient DC Motor Drive 272
9.10.2 Multi-tooth Flux-Switching DC Motor Drive 274
9.10.3 Axial-Flux Doubly-Salient DC Motor Drive 276
9.10.4 Axial-Flux Flux-Switching DC Motor Drive 283
9.11 Potential Applications of Advanced Magnetless Motor Drives in EVs 288
References 289

10 Integrated-Starter-Generator Systems 291
10.1 Classification of HEVs 291
10.2 ISG System Configurations 295
10.3 ISG Machines 296
10.4 ISG Operations 298
10.4.1 Cranking 298
10.4.2 Electricity Generation 298
10.4.3 Idle Stop-Start 298
10.4.4 Regenerative Braking 299
10.4.5 Power Assistance 300

10.5 Design Criteria of ISG Systems 300
10.6 Design Examples of ISG Systems 301
10.6.1 Double-Stator PM Synchronous Machine-Based ISG System 301
10.6.2 Hybrid-Excited Doubly-Salient PM Machine-Based ISG System 303
10.7 Application Examples of ISG Systems in HEVs 312
10.8 Matured Technology for HEVs? 313
References 313

11 Planetary-Geared Electric Variable Transmission Systems 315
11.1 System Configurations 315
11.2 Planetary Gears 316
11.3 Input-Split PG EVT Systems 319
11.3.1 Toyota Hybrid System 319
11.3.2 Ford Hybrid System 324
11.4 Compound-Split PG EVT Systems 326
11.4.1 GM Two-Mode Hybrid System 327
11.4.2 Renault Hybrid System 331
11.4.3 Timken Hybrid System 332
11.5 Design Criteria of PG EVT Systems 333
11.6 Design Example of PG EVT Systems 334
11.6.1 PM Synchronous PG EVT System Configuration 334
11.6.2 PM Synchronous Machine Design 335
11.6.3 PM Synchronous Machine Analysis 336
11.7 Application Examples of PG EVT Systems in HEVs 339
11.8 Matured Technology for HEVs? 341
References 342

12 Double-Rotor Electric Variable Transmission Systems 343
12.1 System Configurations 343
12.2 Double-Rotor Machines 345
12.2.1 Multi-port Machine Concept 345
12.2.2 DR Machine Structure 346
12.3 Basic Double-Rotor EVT Systems 347
12.3.1 DR EVT Structure 347
12.3.2 DR EVT Modeling 349
12.3.3 DR EVT Operation 350
12.4 Advanced Double-Rotor EVT Systems 351
12.4.1 PM DR EVT System 353
12.4.2 SR DR EVT System 354
12.4.3 Axial-Flux DR EVT System 356
12.4.4 Advanced Magnetless DR EVT System 357
12.5 Design Criteria of DR EVT Systems 359
12.6 Design Example of DR EVT Systems 359
12.6.1 DSDC DR EVT System Configuration 360
12.6.2 DSDC DR Machine Design 360
12.6.3 DSDC DR Machine Analysis 360
12.7 Potential Applications of DR EVT Systems in HEVs 364
References 365

13 Magnetic-Geared Electric Variable Transmission Systems 367
13.1 System Configurations 367
13.2 Multi-port Magnetic Gears 369
13.2.1 Magnetic Planetary Gears 369
13.2.2 Magnetic Concentric Gears 371
13.3 Magnetic Planetary-Geared EVT System 373
13.4 Magnetic Concentric-Geared EVT System 375
13.5 Design Criteria of MG EVT Systems 380
13.6 Design Example of MG EVT Systems 382
13.6.1 MCG EVT System Configuration 382
13.6.2 Integrated MCG Machine Design 384
13.6.3 Integrated MCG Machine Analysis 386
13.7 Potential Applications of MG EVT Systems in HEVs 392
References 392

Index 393

Preface

Born in a metropolitan city, I have fully witnessed the decolorization of our blue sky over the past few decades. Hoping to contribute an effort to recover our blue sky, I have devoted myself to researching electric vehicle (EV) technologies for over 20 years – starting from my undergraduate final year project in EV battery monitoring in 1987 to my latest research project in EV in-wheel motor drives in 2014. In fact, EVs have been globally identified to be the greenest road transportation, and the use of EVs will be the most practical way to combat smog and soot from road vehicles in the foreseeable future.

Over the years, there have been many books dealing with EV technologies, mainly briefing various technologies in pure EVs, hybrid EVs, or fuel-cell EVs. Because of the multidisciplinary nature of EV technologies, it is hardly possible to embrace all of them in a single book without sacrificing the latest developments or in-depth discussions. In recent years, some books have begun to deal with specific EV technologies, such as electric propulsion, hybrid propulsion, batteries, and fuel cells. Although many EV machines and drives have been widely published as papers in learned journals, a book comprehensively discussing them is highly desirable. In fact, such advancement of machines and drives is beneficial to the whole EV trio – namely the pure, hybrid, and fuel-cell EVs.

The purpose of this book is to provide a comprehensive discussion on machines and drives for pure electric, hybrid, and fuel-cell vehicles, including both electric propulsion and hybrid propulsion. The corresponding motor drives for electric propulsion range from the existing types, namely the DC, induction, permanent magnet (PM) brushless, and switched reluctance motor drives, to the advanced types, namely the stator-PM, magnetic-geared, vernier PM, and advanced magnetless motor drives. The correspondin machine systems for hybrid propulsion cover the existing types, namely the integrated-starter-generat and planetary-geared electric variable transmission (EVT) systems, and the advanced types, namely double-rotor EVT and magnetic-geared EVT systems. Meanwhile, emphasis is given to the design c ria, performance analyses, and application examples or potentials of various motor drives and ma systems. It is anticipated that various EVs will adopt different machines and drives, and this book a key reference for researchers, engineers, and administrators who need to make such decisions.

While EVs are the driving force for a better environment, my family is the driving power for on EVs. Especially, I would like to take this chance to express my heartfelt gratitude to my Man-Ho, and my wife, Joan Wai-Yi, for their hearty support all the way.

Organization of This Book

This book provides a comprehensive knowledge of electric vehicle (EV) machines and drives, including latest developments and in-depth discussions. It is written for a wide coverage of readers including students, researchers, engineers, administrators, and general readers, and is organized into two themes:

- The first theme is the knowledge of various electric motor drives for EVs, including pure electric and fuel-cell vehicles. It is composed of eight chapters in which Chapters 2–5 deal with those available motor drives for existing EVs, while Chapters 6–9 discuss those advanced motor drives for future EVs.
- The second theme is the knowledge of various electric machine systems for hybrid electric vehicles (HEVs). It consists of four chapters in which Chapters 10 and 11 describe those available machine systems for existing HEVs, while Chapters 12 and 13 elaborate those advanced machine systems for future HEVs.

In this book, there are in total 13 chapters. Each chapter has different numbers of sections and subsections. In order to facilitate selection of reading, all chapters are outlined below:

- Chapter 1 gives an introduction of EVs, including the classification of EVs, overview of EV challenges, and overview of various technologies developed for EVs.
- Chapter 2 is devoted to discussing DC motor drives for EVs, including their system configurations, DC machines, DC-DC converters, and control strategies. The corresponding design criteria, design examples, and application examples are also mentioned.
- Chapter 3 is devoted to discussing induction motor drives for EVs, including their system configurations, induction machines, power inverters, and control strategies. The corresponding design criteria, design examples, and application examples are also discussed.
- Chapter 4 is devoted to discussing permanent magnet (PM) brushless motor drives for EVs, covering both the PM synchronous and PM brushless DC types. Their PM materials, system configurations, PM brushless machines, power inverters, and control strategies are described. The corresponding design criteria, design examples, and application examples are also discussed.
- Chapter 5 is devoted to discussing switched reluctance (SR) motor drives for EVs, including their system configurations, SR machines, SR converters, and control strategies. The corresponding design criteria, design examples, and application examples are also discussed.
- Chapter 6 discusses various stator-PM motor drives for EVs, embracing the doubly-salient PM, flux-reversal PM, flux-switching PM, hybrid-excited PM, and flux-mnemonic PM types. The corresponding design criteria, design examples, and potential applications are also given.
- Chapter 7 discusses magnetic-geared (MG) motor drives for EVs, including their system configurations, magnetic gears, MG machines, power inverters, and control strategies. The corresponding design criteria, design examples, and potential applications are also given.

- Chapter 8 discusses vernier PM motor drives for EVs, including their system configurations, vernier PM machines, power inverters, and control strategies. The corresponding design criteria, design examples, and potential applications are also given.
- Chapter 9 discusses various advanced magnetless motor drives for EVs, covering the synchronous reluctance, doubly-salient DC, flux-switching DC, vernier reluctance, doubly-fed vernier reluctance, and axial-flux magnetless types. The corresponding design criteria, design examples, and potential applications are also given.
- Chapter 10 describes integrated-starter-generator systems for HEVs, including their system configurations, machine structures, and operation modes. The corresponding design criteria, design examples, and application examples are also discussed.
- Chapter 11 describes planetary-geared electric variable transmission systems for HEVs, including the system configurations and planetary gears as well as the input-split and compound-split planetary-geared types. The corresponding design criteria, design example, and application examples are also mentioned.
- Chapter 12 describes double-rotor electric variable transmission systems for HEVs, including the system configurations and double-rotor machines as well as the basic and advanced double-rotor types. The corresponding design criteria, design example, and potential applications are also mentioned.
- Chapter 13 describes MG electric variable transmission systems for HEVs, including the system configurations and multi-port magnetic gears as well as the magnetic planetary-geared and magnetic concentric-geared types. The corresponding design criteria, design example, and potential applications are also briefed.

Readers have the flexibility of reading those chapters that are most interesting to them. The suggestion for reading is as follows:

- Undergraduate students taking a course dedicated to EV technologies may be particularly interested in Chapters 1–5 as well as Chapters 10 and 11.
- Postgraduate students taking a course dedicated to advanced EV technologies may be interested in all chapters.
- Researchers in the area of EV machines and drives may be interested in all chapters. Particularly, they may have special interest in Chapters 6–9 as well as Chapters 12 and 13, which involve newly explored research topics.
- Practicing engineers for product design and development may be more interested in Chapters 6–9 as well as Chapters 12 and 13 in which new ideas can be triggered and commercial products can be derived.
- Administrators and general readers may be interested in all chapters. They are advised to read the book from the beginning to the end, page by page, which is most enjoyable.

Acknowledgments

Material presented in this book is a collection of many years of research and development by the author in the International Research Centre for Electric Vehicles and the Department of Electrical and Electronic Engineering at The University of Hong Kong.

I am grateful to all members of my Electric Vehicle Technologies research group, especially Dr. Wenlong Li, Mr. Christopher Ho-Tin Lee, Mr. Mu Chen, Miss Fei Lin, Mr. Zhen Zhang, Mr. Chun Qiu, Dr. Yubin Wang, Dr. Chunhua Liu, Mr. Fuhua Li, Mr. Xianglin Li, and Mr. Feng Yu for their help in the preparation of this book. I want also to express my sincere gratitude to my PhD graduates and postdoctoral fellows, especially Dr. Herman Tsz-Wood Ching, Prof. Ming Cheng, Dr. Ying Fan, Dr. Xiaoyong Zhu, Dr. Wenxiang Zhao, Dr. Shuangxia Niu, Dr. Linni Jian, Dr. Chuang Yu, Dr. Jiangui Li, Dr. Xiaodong Zhang, Dr. Shuang Gao, and Dr. Diyun Wu for their excellent research outcomes, which are essential ingredients of this book.

I am deeply indebted to my colleagues and friends worldwide for their continuous support and encouragement over the years. I also appreciate the reviewers of this book for their thoughtful and constructive comments and the Editors of John Wiley and Sons for their patience and effective support.

Last but not least, I thank my family for the unconditional support and absolute understanding during the writing of this book.

About the Author

K. T. Chau received his B.Sc. (Eng.) degree with First Class Honors, M.Phil., and PhD, all in Electrical and Electronic Engineering from The University of Hong Kong. He joined the alma mater in 1995 and currently serves as Professor in the Department of Electrical and Electronic Engineering, Director of International Research Centre for Electric Vehicles, and Director of B.Eng. degree in Electrical Engineering.

He is Fellow of the Institute of Electrical and Electronics Engineers (IEEE) for Contributions to Energy Systems for Electric and Hybrid Vehicles. He is also a Chartered Engineer, Fellow of the Institution of Engineering and Technology (IET), and Fellow of the Hong Kong Institution of Engineers (HKIE). He has served as editors and editorial board members of various international journals as well as chairs and organizing committee members of many international conferences, especially in the area of Electric Vehicles. He is also an international consultant for electric vehicle technologies.

His teaching and research interests are electric and hybrid vehicles, electric machines, and drives, as well as clean and renewable energies. In these areas, he has published over 400 refereed technical papers, and many industrial reports. His teaching and research philosophy follows the Confucian classic "Li Ji" – Teaching and learning (research) are mutually motivating.

He has received many awards, including the Chang Jiang Chair Professorship; the Environmental Excellence in Transportation Award for Education, Training, and Public Awareness; the Award for Innovative Excellence in Teaching, Learning, and Technology; and the University Teaching Fellow Award.

1

Introduction

With ever-increasing concern on energy diversification, energy efficiency, and environmental protection, electric vehicles (EVs), including pure electric vehicle (PEV), hybrid electric vehicle (HEV), and fuel-cell electric vehicle (FEV) are becoming attractive for road transportation. Although some of them have become commercially available, there are many challenges and hence opportunities for EV research and development.

In this chapter, the classification of EVs is discussed. Then, an overview of EV challenges is given. Consequently, an overview of various technologies developed for EVs is brought forward.

1.1 What Is an Electric Vehicle?

EVs are nothing new; they were invented 178 years ago but lost the competition for dominance to internal combustion engine vehicles (ICEVs). Actually, the first EV was a battery-powered tricycle built by Thomas Davenport in 1834 (Wakefield, 1994). In 1900, among an annual sale of 4200 automobiles in the US, 38% were EVs, 22% ICEVs, and 40% steam-powered vehicles. At that time, EVs were the preferred road transportation among the wealthy elite. Their cost was equivalent to a Rolls Royce of today. A man with an idea that finished off the EVs for good was Ford. His mass-produced Ford Model T could offer a range double or triple that of the EVs but at only a fraction of their cost. By the 1930s, the EVs almost vanished from the scene. The rekindling of interests in EVs started at the outbreak of the energy crisis and oil shortage in the 1970s. Owing to the growing concern over air quality and the possible consequences of the greenhouse effect in the 1980s, the pace of EV development was accelerated.

In general, EVs are classified as the PEV, HEV, and FEV types on the basis of their energy sources and the propulsion devices (Chan and Chau, 2001; Chau, 2010, 2014). In essence, the PEV is purely fed from electricity, while the propulsion is solely driven by the electric motor; the HEV is sourced from both electricity and gasoline/diesel, while the propulsion involves both the electric motor and engine; and the FEV is directly or indirectly sourced from hydrogen, while the propulsion is solely driven by the electric motor. Moreover, in order to distinguish the refueling means, the HEV can be further categorized into the conventional HEV and the gridable HEV. The conventional one is solely refueled with gasoline/diesel in filling stations, whereas the gridable one can be recharged by electricity via charging ports. On the basis of the hybridization level and the operation feature between the electric motor and engine, the conventional HEV can be further split into the micro HEV, mild HEV, and full HEV. Meanwhile, on the basis of the coordination between the electric motor and engine, the gridable HEV can be further split into the plug-in hybrid electric vehicle (PHEV) and range-extended electric vehicle (REV). This classification is depicted in Figure 1.1.

Electric Vehicle Machines and Drives: Design, Analysis and Application, First Edition. K. T. Chau.

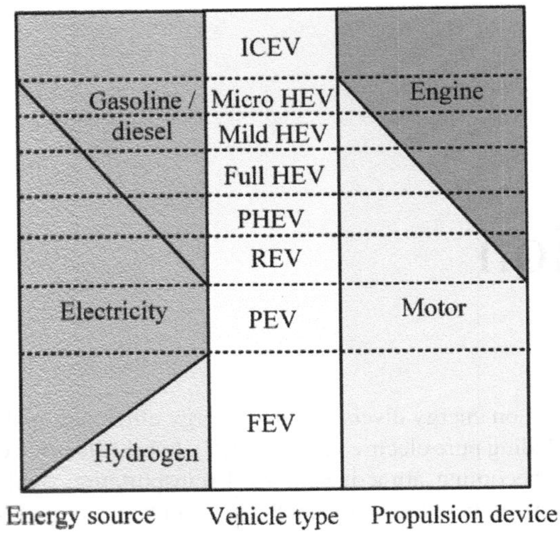

Figure 1.1 Classification of EVs

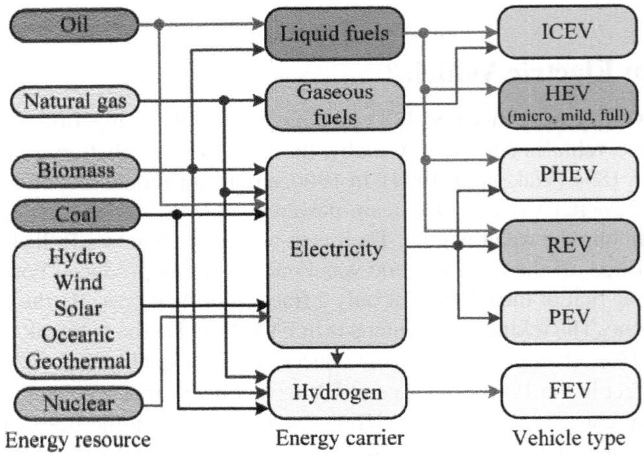

Figure 1.2 Energy diversification of EVs

Deriving from crude oil, the gasoline and diesel are the major liquid fuels for ICEVs. EVs are an excellent solution to rectify this unhealthy dependence because electricity can be generated by almost all kinds of energy resources. Figure 1.2 illustrates the merit of energy diversification due to the use of EVs in which electricity can be produced by thermal power (oil, natural gas, and coal), nuclear power, hydropower, wind power, solar power, oceanic power, geothermal power, and biomass power. In order to compare the overall energy efficiency of EVs with that of ICEVs, their energy conversion processes from crude oil to road load are depicted in Figure 1.3, indicating that EVs are more energy efficient than ICEVs. Moreover, EVs can recover the kinetic energy during braking and utilize it for battery recharging, whereas ICEVs wastefully dissipate this kinetic energy as heat in the brake discs. With this regenerative braking technology, the energy efficiency of EVs can be further boosted by up to 10%.

In many metropolises, ICEVs are responsible for more than 50% of harmful air pollutants and smog-forming compounds. To reduce air pollution from road transportation, the use of EVs is the

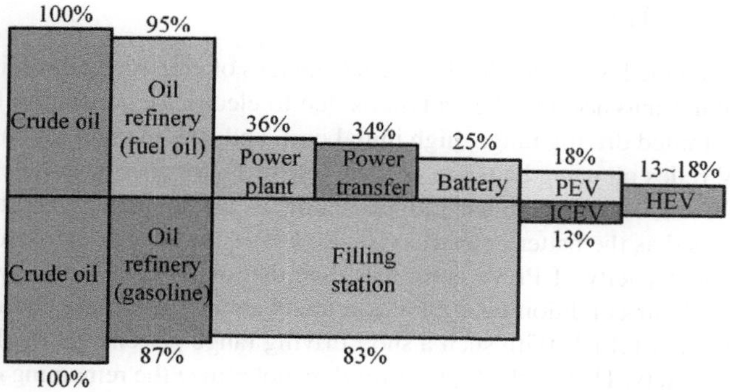

Figure 1.3 Energy efficiency of EVs

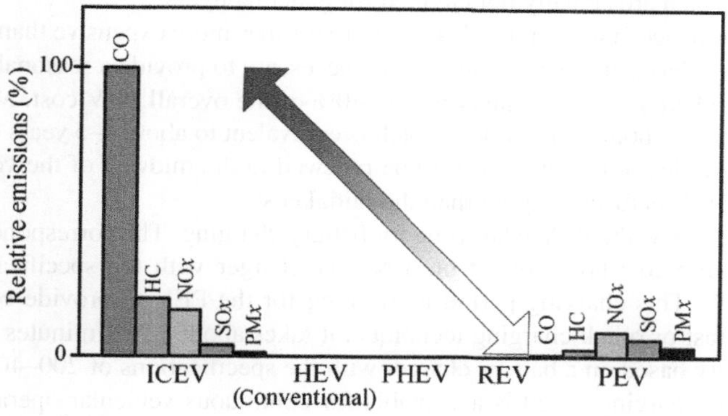

Figure 1.4 Overall harmful emissions of EVs

most viable choice. Definitely, most EVs offer zero roadside emissions. Even taking into account the emissions from refineries to produce gasoline for ICEVs and the emissions from power plants to generate electricity for EVs, the overall harmful emissions of EVs are still much lower than those of ICEVs as indicated in Figure 1.4, where carbon monoxide (CO), hydrocarbons (HC), nitrogen oxides (NO_x), sulfur oxides (SO_x), and particulate matters (PM_x) are taken into account (Chau, 2010). It should be noted that the overall carbon dioxide (CO_2) emission can also be reduced by about 5% with the use of EVs and energy-efficient power plants. This improvement may be further increased when incorporating with higher percentages of clean or renewable power generation, but may even be negative when adopting inefficient coal-fired power plants.

Currently, the conventional HEV has been commercially available and widely accepted as an energy-efficient and environment-friendly vehicle, while the PEV is becoming commercially available and tagged with a zero-emission label. Nevertheless, there are many challenges and opportunities for EV research and development.

1.2 Overview of EV Challenges

There are different types of EVs, including the PEV, conventional HEV, PHEV, REV, and FEV. These EVs can be grouped into the PEV, HEV, gridable HEV, and FEV for discussion, with emphasis on their challenges (Chau, 2012).

1.2.1 *Pure Electric Vehicle*

The PEV, loosely termed the EV, offers the definite advantages of zero roadside emissions and minimum overall emissions (taking into account the emissions due to electricity generation by power plants). Its major challenges are limited driving range, high initial cost, and lack of charging infrastructure.

Currently, the PEV relies on using batteries as their sole or major energy storage device to store electricity even though there is an option to use ultracapacitors as the sole energy storage device. Thus, the PEV is sometimes named as the battery electric vehicle (BEV). At the present status of battery technology, the energy storage capacity of PEVs is far less than that of ICEVs. Typically, for a passenger car under urban driving with air-conditioning, a PEV can travel about 120 km per charge, whereas an ICEV can offer about 500 km per refuel. With such a short driving range per charge, the PEV will suffer from the problem of range anxiety. That is, the PEV driver dare not utilize the remaining capacity such as 20% to travel a trip of 24 km. It should be noted that some PEV models purposely install three to four times the battery capacity to enable their driving range comparable with that of ICEVs, hence solving the short range and range anxiety problems. Of course, these PEV models will be two to three times more expensive than general PEVs, which are actually not targeted for general buyers.

With similar performance, a general PEV is two to four times more expensive than an ICEV. Such high initial cost is because a large number of batteries are necessary to provide a reasonable driving range per charge. Typically, the battery cost accounts for 30–40% of the overall PEV cost. Moreover, the battery life can generally last for about 1500 cycles, which is equivalent to about 4–5 years of vehicle operation, indicating that all batteries of the PEV need to be renewed in the midway of the vehicle life. Thus, the effective cost of the PEV is further higher than the initial cost.

Differing from the ICEV, the PEV takes time for battery charging. The corresponding charging period normally ranges from 5 to 8 hours based on a battery charger with the specifications of 110–240 V, 13–40 A, and 2–4 kW. This charging period is too long for the PEV to provide continuous operation. When adopting the fast or quick charging technique, it takes about 20–30 minutes to charge the batteries up to 80% capacity based on a battery charger with the specifications of 200–400 V, 100–200 A, and 50 kW. Although this charging speed is acceptable for continuous vehicular operation, the installation cost and establishment cost of these fast charging stations are very high. Since the power demand for fast charging is high, the fast charging process inevitably causes burden to our existing power system, which violates the merit of using the PEV for load leveling or demand side management. In case the PEV allows for battery swapping, namely replacing the discharged batteries with the fully charged ones using mechanical means, it takes only a few minutes to mechanically charge up the batteries. Although the time required for battery swapping is comparable to that for gas refueling, the necessary space for each swapping station is much larger. Practically, it involves two implementation challenges: the battery size and location inside the PEV have to be standardized and the single ownership of all batteries needs a new business model.

1.2.2 *Hybrid Electric Vehicle*

The HEV, loosely termed the hybrid vehicle, refers to the conventional or nongridable version (Chau and Wong, 2002). For the micro HEV, the conventional starter motor is eliminated, while the conventional generator is replaced by an integrated starter-generator (ISG). Instead of propelling the vehicle, the ISG offers two important hybrid features. One feature is to shut down the engine whenever the vehicle is at rest, the so-called idle stop-start feature, hence improving the fuel economy for urban driving. Another feature is to recharge the battery primarily during vehicle deceleration or braking, thus offering a mild amount of regenerative braking. For the mild HEV, the ISG is generally placed between the engine and the transmission. This ISG not only provides the hybrid features of idle stop-start and regenerative braking but also assists the engine to propel the vehicle, thus allowing for a downsized engine (Liu, Chau, and Jiang, 2010a). However, since the engine and the ISG share the same shaft, it cannot offer electric launch (initial acceleration under electric power only). For the full HEV, the key technology is the electric variable transmission (EVT) system, which mainly functions to perform power splitting. This EVT can offer all hybrid features, including the electric launch, idle stop-start, regenerative braking, and engine downsizing.

Compared with the PEV, the HEV can offer a comparable driving range of the ICEV and use the existing refueling infrastructure of the ICEV, but sacrificing the merits of zero roadside emissions and energy diversification. Its key challenges are how to reduce the system complexity that involves both an electric motor and an engine for propulsion and how to coordinate these two propulsion devices to achieve optimal efficiency operation (Chau and Wong, 2002). The turning point of HEV development was the advent of Toyota Prius in 1997 (Hermance and Sasaki, 1998), which initially adopted the EVT system. The key is to employ a planetary gear for power splitting of the engine output power, one via the ring gear to the driveline shaft while one via the sun gear to the generator, then back-to-back converters, motor, and finally the driveline shaft. Hence, under varying road load, the engine can always operate at its most energy-efficient or optimal operation line (OOL), resulting in a considerable reduction in fuel consumption. However, this EVT system suffers from the reliance on planetary gearing, which involves transmission loss, gear noise, and regular lubrication. In addition, the overall system is relatively heavy and bulky.

1.2.3 Gridable Hybrid Electric Vehicle

The term "gridable" means that the vehicle can be directly connected to the power grid. Therefore, the gridable HEV refers to the vehicles that have gridable capability and HEV features, namely the PHEV and REV. The PHEV is extended from the conventional HEV by incorporating the additional feature of plug-in rechargeable. Since it incorporates a larger bank of batteries that can be recharged by plugging into an external charging port, it can offer a longer electric drive range and hence reduce the requirement of refueling from gas stations. On the other hand, the REV is extended from the PEV by incorporating a small engine coupled with a generator to recharge the battery bank. This avoids the range anxiety problem that is always associated with the PEV. Therefore, it can offer energy-efficient operation throughout its electric drive range and hence significantly reduce refueling from gas stations. Although the PHEV and REV are both a HEV and have similar electric motor and battery ratings, they have different nominal operations. The PHEV generally operates in the blended mode in which the electric motor and the engine are coordinated to work together in such a way that the engine can maintain efficient operation, hence achieving high fuel economy. If necessary, it can operate in the pure-electric mode. In contrast, the REV generally operates in the pure-electric mode all the way, regardless of the driving range or profile. Until the battery pack is depleted to the threshold, it can operate in the extended mode that the engine is turned on, which then drives the generator to produce the desired electricity.

The key challenges of the gridable HEV are the system complexity and high initial cost. Its system complexity is similar to that of the conventional HEV, mainly because of the use of both the electric motor and engine. Differing from the conventional HEV, it needs to install the on-board charger to plug in the power grid for battery charging. Its initial cost is much higher than that of the conventional HEV because of the use of a large number of batteries for the pure-electric mode. Of course, when the PHEV operates at the blended mode or the REV operates at the extended mode, they lose the merit of zero roadside emissions.

1.2.4 Fuel-Cell Electric Vehicle

The FEV, loosely termed the fuel-cell vehicle, offers the same advantages as the PEV, namely, zero roadside emissions and minimum overall emissions (taking into account the emissions due to hydrogen production by chemical plants or an on-board reformer). In addition, it can offer a driving range comparable to that of the ICEV. Its major challenges are the high initial cost and lack of hydrogen refueling infrastructure. The high initial cost is because of the use of expensive fuel cells. Hydrogen refueling infrastructure is generally absent in our society, and the establishment of such an infrastructure involves a huge investment cost. There are three practical ways to store hydrogen in the FEV: the compressed hydrogen gas (CHG), liquid hydrogen (LH), and metal hydride (MH). When adopting the CHG (a pressure of about 350–700 bar) for the FEV, the infrastructure is similar to that of compressed natural gas (a pressure of about 200–248 bar) for some alternative fuel vehicles. When adopting the LH, the infrastructure is very demanding since the hydrogen needs to be cooled to about −253 °C while still pressurized. This requires cryogenic storage technology, which is even more severe than liquid oxygen. When adopting the MH, it needs to have a similar infrastructure as battery swapping to mechanically replace the discharged MH with the fully charged MH. In addition, it requires more energy for providing necessary temperatures (120–200 °C) to discharge the hydrogen and necessary pressure (over 700 bar) to recharge the hydrogen. Both the CHG and LH enjoy the merit of high specific energy (good energy density by weight), which is desirable for the FEV, but also face the same safety concern, which can be an explosion hazard. Meanwhile, the MH takes the merit of safety, which is essential for the FEV, but suffers from the problem of low specific energy, which deteriorates the driving range.

In the coming future, the commercialization of the FEV depends on whether there will be a breakthrough in fuel-cell technology in terms of cost per kilowatt and whether there will be a mandate or energy policy to establish the hydrogen refueling infrastructure.

1.3 Overview of EV Technologies

An overview of key technologies for various types of EVs is presented, with emphasis on their emerging research activities. Among them, the motor drive technology is most actively developed in recent years where there are many innovations and advancements in the design, analysis, and control of motor drives. The energy source technology is also actively developed in recent years. Nevertheless, there is no real breakthrough in the battery technology, especially aspiring the simultaneous possession of low initial cost and high specific energy. Rather than waiting for a breakthrough in battery technology, the battery charging technology is being actively developed. Particularly, the concept of move-and-charge (MAC) for battery charging using wireless power transfer (WPT) is promising to fundamentally solve the long-term shortcomings of EVs. Moreover, in order to promote the advantage of EVs to justify their high initial cost, the vehicle-to-grid (V2G) technology is being actively researched, which can expand the role or function of EVs to increase their cost effectiveness.

1.3.1 Motor Drive Technology

Motor drives are the core technology for EVs that convert the on-board electrical energy to the desired mechanical motion. Meanwhile, electric machines are the key element of motor drive technology. The requirements of electric machines for EVs are much more demanding than that for industrial applications. These requirements are summarized as follows (Zhu and Howe, 2007; Chau, 2009):

- High torque density and high power density
- Wide speed range, covering low-speed creeping and high-speed cruising
- High efficiency over wide torque and speed ranges
- Wide constant-power operating capability
- High torque capability for electric launch and hill climbing
- High intermittent overload capability for overtaking
- High reliability and robustness for vehicular environment
- Low acoustic noise
- Reasonable cost

When the electric machine needs to work with the engine for various HEVs, there are some additional requirements:

- High-efficiency power generation over a wide speed range
- Good voltage regulation over wide speed generation
- Capable of being integrated with the engine

Figure 1.5 shows the classification of electric machines for EVs in which the bold types are those that have been applied to EVs, including the series DC, shunt DC, separately excited DC, permanent magnet (PM) DC, cage-rotor induction, PM brushless AC (BLAC), PM brushless DC (BLDC), and switched reluctance (SR) machines. Basically, EV machines are classified into two main groups: commutator and commutatorless. The former simply denotes that they have a commutator and carbon brushes, while the latter have neither commutator nor carbon brushes. It should be noted that the trend is focused on developing new types of commutatorless or brushless machines (Chau, Chan, and Liu, 2008), especially the class of doubly salient machines and the class of vernier machines.

The key feature of doubly salient machines is the presence of salient poles in both the stator and rotor. The SR machine is a kind of doubly salient machines having the simplest structure. When incorporating PMs in the stator of doubly salient machines, a new class of PM brushless machines is resulted – the stator-PM machine (Liu *et al.*, 2008). Since the rotor has neither PMs nor windings, this class of machines is mechanically simple and robust, hence very suitable for vehicular operation. According to the location of the PMs, it can be split into doubly salient permanent magnet (DSPM), flux-reversal permanent magnet (FRPM), and flux-switching permanent magnet (FSPM) machines. Additionally, with the inclusion of independent field windings in the stator for flux control, the class can be further split into the flux-controllable (FC) types – the FC-DSPM, FC-FRPM, and FC-FSPM. Furthermore, when the PM poles are replaced with DC field windings aiming to get rid of those expensive PM materials and provide flexible flux control, the resulting doubly salient DC (DSDC), flux-reversal DC (FRDC), and flux-switching DC (FSDC) machines are emerging types of advanced magnetless machines.

The key feature of vernier machines is the use of vernier effect to amplify the output torque while stepping down the speed, leading to be a class of brushless machines dedicated to low-speed high-torque direct-drive application. There are two main classes of vernier machines: vernier permanent magnet (VPM) and vernier reluctance (VR). There are three types of VPM machines depending on the location

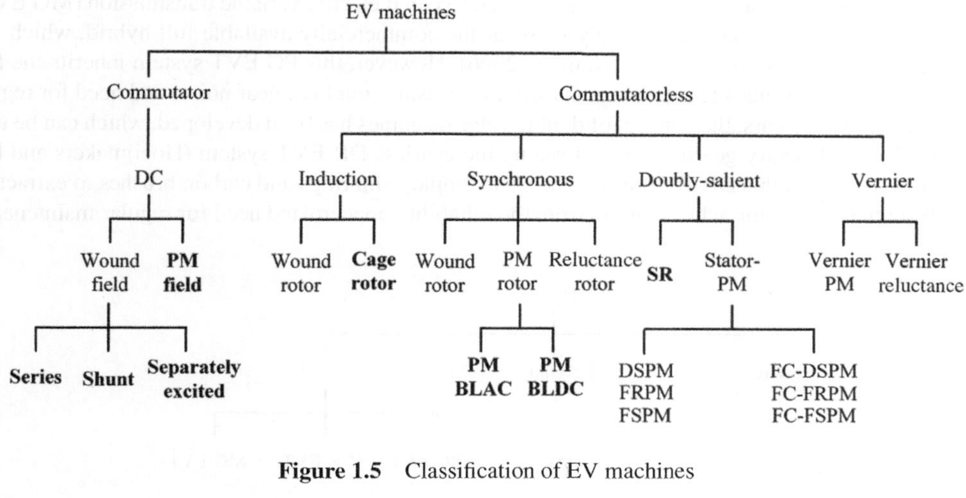

Figure 1.5 Classification of EV machines

of PMs: the rotor-PM type with all PMs mounted on the rotor, the stator-PM type with all PMs mounted on the stator, and the all-PM type with PMs mounted on both the rotor and stator. As the rotor-PM VPM machine is most mature, it is loosely called as the VPM machine (Li, Chau, and Li, 2011). The stator-PM VPM machine is commonly termed the vernier hybrid machine (Spooner and Haydock, 2003). On the other hand, the VR machine is structurally similar to the SR machine, but they operate differently. In essence, the VR machine is fed by three-phase sinusoidal currents to produce the rotating magnetic field, and the rotor runs synchronously at a fraction of the speed of this rotating field (Lee, 1963). Because of its inherently low power factor, an additional supply can be incorporated to feed an additional field winding in the stator of the VR machine, thus resulting in the doubly fed vernier reluctance (DFVR) machine (Taibi, Tounzi, and Piriou, 2006). This additional field winding can be fed by AC or DC current, leading to further create the vernier reluctance AC (VRAC) and vernier reluctance DC (VRDC) machine. The VR and DFVR machines are also classified as emerging types of advanced magnetless machines.

All EV machine topologies developed for the conventional radial-flux morphology can readily be extended to other morphologies such as the axial-flux morphology (Lee, Liu, and Chau, 2014), linear-flux morphology (Du *et al.*, 2011), and transverse-flux morphology (Wang *et al.*, 2008). The axial-flux morphology takes the advantages of higher power density and higher torque density than its radial-flux counterpart, but suffers from the problem of large axial force exerted on the stator by the rotor and limited to pancake shape. As the linear-flux morphology functions to provide linear motion, it is less attractive for EV propulsion. Although the transverse-flux morphology can offer the highest torque density, the corresponding machine structure is very complicated, which limits its manufacturability and practicality for EVs.

Apart from developing EV motor drives for pure electric propulsion, namely for the PEV and FEV, the technological development of EV machines has been extended to hybrid propulsion for HEVs. As depicted in Figure 1.6, there are two main machine systems for hybrid propulsion: the ISG system for the micro and mild hybrids and the EVT system for the full hybrid (Chau and Chan, 2007). The ISG system needs to offer not only the conventional features of engine cranking and electricity generation, but also the hybrid features of idle stop-start, regenerative braking, and power assistance. Therefore, the corresponding machine design, analysis, and control are very demanding (Liu, Chau, and Jiang, 2010a). The EVT functions to offer electrically controllable power transfer from the engine to the wheels with continuously variable transmission, hence providing all hybrid features including the electric launch, idle stop-start, regenerative braking, and power assistance, as well as achieving the highest fuel economy. There are three main types of EVT systems: the planetary-geared electric variable transmission (PG EVT), double-rotor electric variable transmission (DR EVT), and magnetic-geared electric variable transmission (MG EVT). The PG EVT system is almost exclusively used for the commercially available full hybrid, which was first developed by Toyota for its Prius (Kamiya, 2006). However, this PG EVT system inherits the fundamental drawbacks of planetary gearing, namely the transmission loss, gear noise, and need for regular lubrication. In recent years, the concept of double-rotor machines has been developed, which can be used to supersede the planetary gearing, hence forming the gearless DR EVT system (Hoeijmakers and Ferreira, 2006). However, this DR EVT system needs to employ slip rings and carbon brushes to extract the energy from the inner rotor, which suffers from the reliability concern and need for regular maintenance.

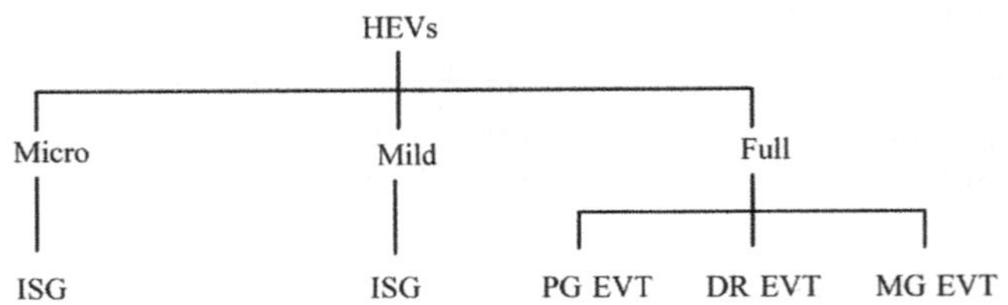

Figure 1.6 EV machine systems for HEVs

Meanwhile, by replacing the planetary gearing with magnetic gearing, the resulting MG EVT system can inherit the distinct advantages of magnetic gearing, namely, the high transmission efficiency, silent operation, and no maintenance, while avoiding the use of slip rings and carbon brushes (Jian and Chau, 2009, 2010). Nevertheless, this pseudo-gearless and brushless EVT system exhibits a complicated structure, and the required precision for manufacture is demanding.

1.3.2 Energy Source Technology

Energy sources are another core technology that provides on-board electrical energy for EVs. Currently, over the last two decades, there are four viable EV energy sources: electrochemical batteries (normally termed batteries), ultracapacitors (also called supercapacitors), ultrahigh-speed flywheels, and fuel cells (Chau, Wong, and Chan, 1999). The batteries are electrochemical devices that store electrical energy during charging and produce electricity during discharging. The ultracapacitors are essentially capacitors with ultrahigh capacitances that store and produce electrical energy by electrostatic means. The ultrahigh-speed flywheels are essentially electric machines spinning at ultrahigh speeds that store and produce electrical energy by electromechanical means, that is, they work as motors during charging and serve as generators during discharging. The fuel cells are electrochemical devices that directly convert chemical fuels into electricity. None of them can simultaneously offer high specific energy and high specific power, analogous to none of athletes can simultaneously be suitable for marathon running and 100-m sprint. Thus, a compromise between these two parameters or a hybridization of two energy sources (one with high specific energy and another with high specific power) is necessary for the PEV or FEV (Chau and Wong, 2001).

In the foreseeable future, batteries are still the major energy source for EVs. Table 1.1 lists the major types of batteries that have been developed for EVs over the past two decades, including valve-regulated lead acid (VRLA), nickel-cadmium (Ni-Cd), nickel-metal hydride (Ni-MH), zinc/air (Zn/air), sodium/sulfur (Na/S), and lithium-ion (Li-ion). Among them, VRLA is accepted for low-cost low-end EVs, Ni-MH is preferred for well-performed EVs, and Li-ion is attractive for high-performance EVs. At the present status of battery technology, the PEV can only offer acceptable driving range with affordable price. In order to enable the PEV offering comparable price and driving range as the ICEV, the battery specific energy and cycle life need to be greatly increased, whereas the battery initial cost needs to be significantly reduced. Currently, the main research on battery technology is being focused on the development of various Li-ion batteries, such as the use of lithium nickel manganese cobalt (NMC) for the positive electrode to improve the specific energy and safety (Omar *et al.*, 2012), and the use of lithium titanate ($Li_4Ti_5O_{12}$) for the negative electrode to improve the cycle life and charging time (Giuliano, Advani, and Prasad, 2011). Meanwhile, another key research direction is to develop lithium/air (Li/air) battery (Christensen *et al.*, 2012) and lithium/sulfur (Li/S) battery (Zhang, 2013) to significantly improve the specific energy.

The ultracapacitor technology is promising for EVs since it offers exceptionally high specific power and practically unlimited cycle life. Nevertheless, the ultracapacitor needs significant improvement before

Table 1.1 Major batteries developed for EVs

	Specific energy (Wh/kg)	Specific power (W/kg)	Cycle life (cycles)	Cost (USD/kWh)
VRLA	30–45	200–300	400–600	150
Ni-Cd	40–60	150–350	600–1200	300
Ni-MH	60–120	150–400	600–1200	200–350
Zn/air	230	105	NA	90–120
Na/S	100	200	800	250–450
Li-ion	90–160	250–450	1200–2000	600–1000

practically applicable as the sole energy source for EVs – its specific energy (5–6 Wh/kg) needs to be greatly increased while its initial cost (2400–6000 USD/kWh) has to be greatly reduced. Current research on ultracapacitor technology is being focused on the improvement of its specific energy, such as the use of graphene (Liu *et al.*, 2010b) and carbon nanotubes (Du and Pan, 2006) to increase the usable surface area and hence the energy storage capacity.

The ultrahigh-speed flywheel technology exhibits potentiality for EVs. By providing vacuum environment to remove air friction and magnetic bearings to eliminate bearing loss, the flywheel can spin up to 60 000 rpm so as to achieve high specific energy and high round-trip efficiency. However, it suffers from the problem of safety concern. When the tensile strength of a flywheel is exceeded or the flywheel is accidentally damaged, the flywheel shatters and instantaneously releases all of its stored energy – the so-called flywheel explosion, which is very dangerous. Current research on ultrahigh-speed flywheel technology is focused on improving its safety precaution such as the use of composite materials, which can disintegrate into tiny powder rather than large chunks, or extending its application to energy storage for EV charging stations where the whole flywheel is embedded in the ground (Strasik *et al.*, 2007).

The fuel-cell technology is one of the most active research areas in recent years. Table 1.2 lists the leading types of fuel cells, including the direct methanol fuel cell (DMFC), alkaline fuel cell (AFC), proton exchange membrane fuel cell (PEMFC), phosphate acid fuel cell (PAFC), molten carbonate fuel cell (MCFC), and solid oxide fuel cell (SOFC). Among them, the PEMFC, which is also called the solid polymer fuel cell (SPFC), is a natural choice for the FEV because of its solid electrolyte nature, low-temperature operation, quick start-up, proper power level, high power density, and good system efficiency. In order to enable the FEV offering affordable price, the fuel-cell initial cost (about 4800 USD/kW) has to be dramatically reduced. Current research on fuel-cell technology is being focused on the reduction of platinum usage for the PEMFC, which requires such noble metals as the electrocatalyst (Martin, Garcia-Ybarra, and Castillo, 2010), and the reduction of operating temperature for the SOFC, which does not desire noble metals as the electrocatalyst (Wang *et al.*, 2011).

Recently, the concept of on-board renewable energy sources has been attractive for EVs. Since the fuel efficiency of the engine for various HEVs is only around 25% and about 40% is lost in the form of waste heat of exhaust gas, a thermoelectric generator (TEG) can be mounted at the exhaust pipe to recover the waste heat energy and help charge the batteries (Yu and Chau, 2009). On the other hand, by mounting a solar panel on the roof of EVs, the photovoltaic generator (PVG) can readily collect renewable solar energy and utilize it to charge the batteries. In general, the TEG and PVG are separately operated, even though they are installed in the same HEV, resulting in higher cost, heavier weight, and larger volume. Therefore, the thermoelectric-photovoltaic (TE-PV) hybrid energy system is promising for application to HEVs. Figure 1.7 shows the system configuration of this TE-PV hybrid energy system, which is composed of the TEG, PVG, maximum power point tracking (MPPT) controller, multiple-input converter (MIC), and battery. The MIC can be a SEPIC–SEPIC converter (Zhang and Chau, 2011a) or a Ćuk–Ćuk converter (Zhang and Chau, 2011b). The SEPIC–SEPIC MIC operating in discontinuous capacitor voltage mode is shown in Figure 1.8. The MPPT controller measures the output voltages and currents of the TEG and

Table 1.2 Viable fuel cells

	Power level (MW)	Power density (W/cm^2)	Operating temperature (°C)	System efficiency (%)
DMFC	<0.001	0.04–0.23	90–120	10–20
AFC	<0.1	0.2–0.3	60–100	62
PEMFC	<0.5	0.35–0.6	50–120	30–50
PAFC	<10	0.2–0.25	150–200	40
MCFC	<100	0.1–0.2	600–650	47
SOFC	<100	0.24–0.3	500–1100	55–60

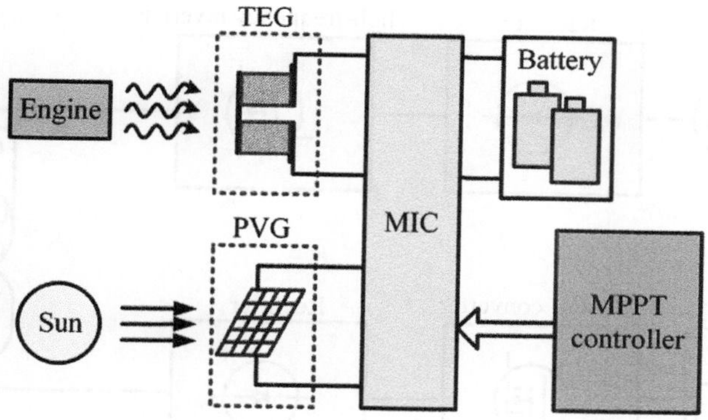

Figure 1.7 Thermoelectric-photovoltaic hybrid energy system

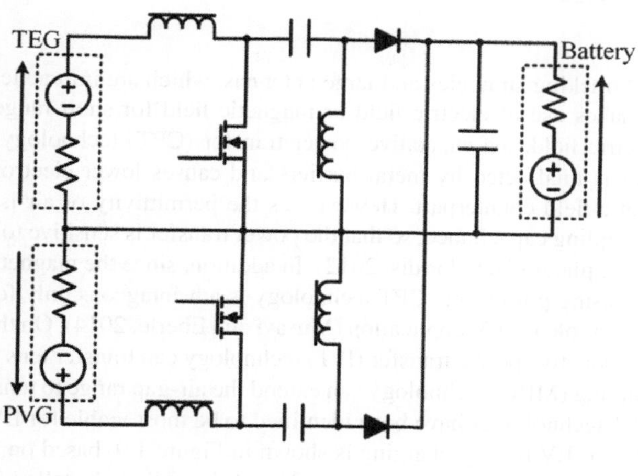

Figure 1.8 SEPIC-SEPIC multiple-input converter

PVG, which exhibit nonlinear characteristics at different temperatures and irradiances, respectively, and then generates proper switching signals to the MIC in such a way that the total output power can be maximized.

1.3.3 Battery Charging Technology

In recent years, many researchers have proposed various methods to alleviate the short driving range per charge of the PEV, focusing on the development of more convenient chargers. Rather than simply building more charging stations and adopting faster battery chargers, which are analogous to gas stations and powerful pumps, the use of WPT for battery charging can greatly facilitate the charging process. Increasingly, because of the absence of metallic contacts, possible electrocution during the charging process can be totally eliminated, which can enable EVs outperforming ICEVs in terms of user safety for self-service recharging or refueling.

There are two main categories of WPT: the far-field and near-field (Qiu *et al.*, 2014a). The far-field WPT makes use of microwave radiation or laser to transfer high power over long distances. However,

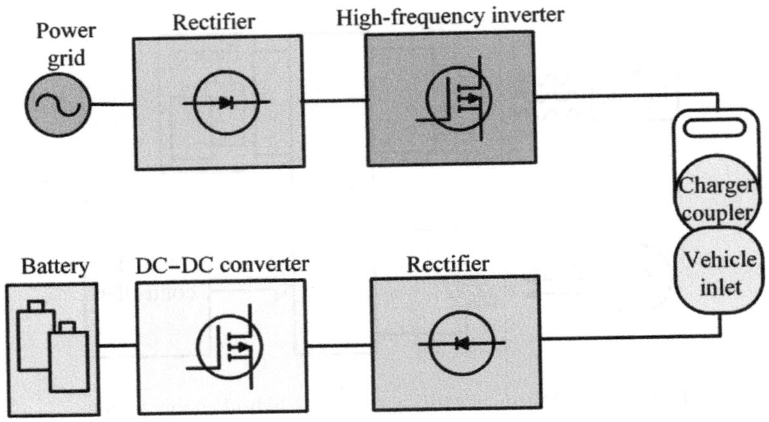

Figure 1.9 Inductive power transfer-based EV charging

it requires complicated tracking strategies and large antennas, which are impractical for EV application. The near-field WPT makes use of electric field or magnetic field for short-range to mid-range power transfer. By using electric field, the capacitive power transfer (CPT) technology takes the advantages that the power transfer is unaffected by metal barriers and causes lower electromagnetic interference (EMI) than the magnetic-field counterpart. However, as the permittivity of air is intrinsically small, it results in inadequate coupling capacitance, so that the power transfer is sensitive to the air-gap length and displacement of coupling plates (Theodoridis, 2012). In addition, since the magnetics involved cannot be scaled down with decreasing power, this CPT technology is advantageous only for low-cost low-power application, and not preferable for EV application (Musavi and Eberle, 2014). On the other hand, by using the magnetic field, the inductive power transfer (IPT) technology can transfer tens of kilowatts, while the magnetic resonant coupling (MRC) technology can extend the air-gap range to tens of centimeters. Thus, the magnetic-field WPT technologies have been identified to be most viable for EV battery charging.

The principle of IPT for EV battery charging is shown in Figure 1.9, based on the magnetic coupling between two coils of a high-frequency transformer. One of the coils is installed in the charger coupler, while the other is embedded in the vehicle inlet. Firstly, the main AC supply with a frequency of 50 or 60 Hz is rectified and converted to a high-frequency AC power of about 80 kHz within the charger module, then the high-frequency AC power is transferred to the PEV side by induction, and finally this high-frequency AC power is converted into DC power for battery charging. This IPT can operate over a wide frequency range and can readily be scaled up to meet various power levels for EV charging. However, the corresponding core losses and EMI are of concern. For instance, a well-known but obsolete IPT-based EV charger, Magne Charge, delivered 6.6 kW at an efficiency of 86%.

In order to facilitate the park-and-charge (PAC) process for EVs, the magnetic-field WPT technology is extended to be plugless, in which the primary coil is installed on the floor of a garage or in a parking lot and the secondary coil is installed on the vehicle. The driver needs no bothering about those cumbersome and dangerous charging cables. This system is very easy to use and the charging process takes place automatically once the driver parks the EV correctly. This plugless PAC not only increases user convenience, but also offers a means of overcoming the standardization of charging plugs. Owing to the existence of a large air-gap or clearance between the primary and secondary coils, the IPT technology is ill-suited. On the basis of MRC, the primary and secondary resonant coils having the same resonant frequency can wirelessly transfer power efficiently with high power density, while dissipating relatively little energy in nonresonant objects such as vehicle bodies or drivers. Increasingly, by incorporating one or more resonant coils between the primary and secondary coils as depicted in Figure 1.10, a strongly

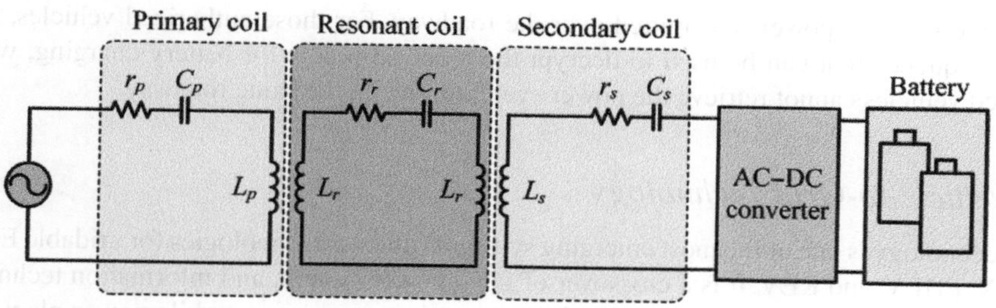

Figure 1.10 Magnetic resonant coupling-based EV charging

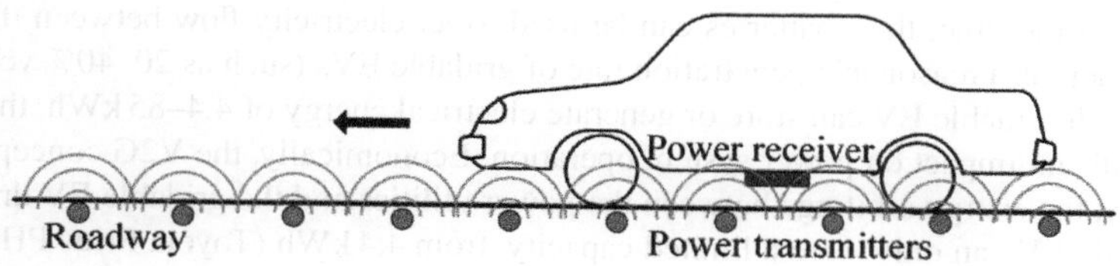

Figure 1.11 EV move-and-charge

coupled magnetic resonance is resulted, which can enable efficient power transfer from the primary coil to the secondary coil with a large air-gap between the floor and vehicle (Cannon *et al.*, 2009; Imura and Hori, 2011). For instance, an on-line electric bus system was demonstrated to transfer 100 kW through an air-gap of 20 cm with an average system efficiency of 75% (Kim *et al.*, 2013). Latest research and development of MRC-based WPT are active and diversified, such as improving the system efficiency, combating the EMI, or compensating the misalignment between magnetic couplers (Qiu *et al.*, 2014b).

Rather than stopping or parking, EVs prefer to be wirelessly charged during moving. An array of power transmitters is embedded beneath the roadway (the so-called charging zone or lane) while a receiver is mounted at the bottom of an EV as depicted in Figure 1.11. This MAC technology has high potentiality to fundamentally solve the long-term problems of the PEV – high initial cost and short driving range. There is no need to install so many batteries in the PEV, hence dramatically cutting its initial cost; and the PEV can be conveniently charged at the charging zone during driving, hence automatically extending the driving range. However, there are some technical problems to be solved before the realistic application of MAC to EVs. First, the efficiency of WPT heavily depends on the distance between the transmitter of the array and the receiver of the vehicle. Since this distance is inevitably time varying and significantly affected by the road condition and vehicle payload, the resonant frequency of WPT is not constant, termed the resonance shifting. Thus, the power converter that excites the power transmitter needs to be dynamically tuned to maintain high-efficiency power transfer. Second, the effectiveness of MAC operation heavily depends on the coverage of WPT as well as the position and speed of vehicles running on the charging zone. The location of power transmitters needs to be optimized in such a way that the electromagnetic field intensities at different locations over the charging zone are uniform. Third, most importantly, as there are many EVs running on the roadway, the MAC operation needs to distinguish which EVs are authorized to retrieve wireless power or to prevent unauthorized vehicles from stealing the energy.

Rather than avoiding the occurrence of chaos, electric drives have positively utilized the chaotic phenomena for various industrial applications (Chau and Wang, 2011) such as industrial mixing (Ye and Chau, 2007) and industrial compaction (Wang and Chau, 2009). Extending from the concept of chaotic modulation (Zhang *et al.*, 2011), the chaotic energy encryption has been developed for EV WPT (Zhang *et al.*, 2014). The key is to employ chaotic sequence to modulate the resonant frequency for WPT, hence encrypting the wireless power transmitted over the roadway. For those authorized vehicles, they know the chaotic sequence that can be used to decrypt the received power for battery charging, whereas the unauthorized vehicles cannot retrieve the power even running on the same roadway.

1.3.4 Vehicle-to-Grid Technology

The V2G technology is one of the most emerging system-crossover technologies for gridable EVs, including the PEV, PHEV, and REV. It is a crossover of EVs, power system, and information technology. The gridable EV is no longer a simple transportation means, but serves as a mobile power plant generating electrical energy to the power grid whenever necessary (Liu *et al.*, 2013). The V2G concept describes a system in which EVs communicate with the power grid to sell services by delivering electricity into the grid or by controlling the charging rate for gridable EVs. Since most gridable EVs are parked with an average of 95% of the time, their batteries can be used to let electricity flow between the vehicles and the grid. When there is a reasonable penetration rate of gridable EVs (such as 20–40% vehicles are gridable EVs) and each gridable EV can store or generate electrical energy of 4.4–85 kWh, the V2G concept will have a significant impact on power system operation. Economically, the V2G concept will be a new business, namely, the energy arbitrage between the power utilities and the gridable EV drivers.

Since a gridable EV can only store a limited capacity, from 4.4 kWh (Toyota Prius PHEV) to 85 kWh (Tesla Model S), an individual V2G operation of each gridable EV with the power grid is ineffective and inefficient. Therefore, an aggregator is introduced, which is responsible for gathering a number of gridable EVs and communicating with the power grid. On the basis of the willingness of drivers and the battery capacity of gridable EVs, the aggregator controls proper gridable EVs to achieve smart charging and discharging (Guille and Gross, 2009). Moreover, the corresponding energy arbitrage can be performed internally by the aggregator, the so-called vehicle-to-vehicle (V2V) operation. For instance, those PEVs preferably perform fast charging for instant usage while those PHEVs preferably sell electricity for profit-making, the aggregator can realize internal V2V operation for energy arbitrage. This dual-grid framework is depicted in Figure 1.12 in which the energy service provider (ESP) sells power directly to homes and businesses, the independent system operator (ISO) oversees the operations of a particular section of the power grid, the regional transmission organization (RTO) integrates the ISOs into larger operations, and the aggregator functions to aggregate the gridable EVs to deal with the ESP and the ISO/RTO (Wu, Chau, and Gao, 2010b). Firstly, the aggregator coordinates the intragrid power flow, minimizes the total power demand and total power loss, optimizes the voltage deviation and total harmonic distortion, and calculates prices to maximize the profit of intragrid operation. Secondly, the aggregator coordinates the intergrid power flow, deals with the ISO/RTO to sell power and energy, deals with the ESP to buy power and energy, and calculates prices to maximize the profit of intergrid operation.

The V2G operation has been identified to have various potential applications: charging coordination, peak shaving, active regulation, spinning reserve, motor starting, reactive regulation, and renewable transients (Ehsani, Falahi, and Lotfifard, 2012). Since the power generation capacity has to match with the load demand, a large fluctuation of load demand will significantly increase the capital cost and operating cost of the power system. As shown in Figure 1.13, the V2G operation preferably performs EV battery charging to absorb or buy electrical energy from the grid during the off-peak period (the so-called coordinated charging), whereas to generate or sell electrical energy to the grid during peak period (the so-called peak shaving). In addition, the corresponding charging and discharging processes of EVs are much faster than the shutoff and start-up processes of standby generators (Wu, Chau, and Gao, 2010a). However, at the present status of battery technology, the EV batteries still suffer from a limited cycle life. Deep discharging of EV batteries for peak shaving will inevitably degrade their life for normal vehicular operation (Gao *et al.*, 2011).

Since renewable power generations such as wind power and solar power are intermittent in nature, the use of standby generators to back up the intermittent power outage is expensive, inefficient, and sluggish. Although the battery energy storage system can provide the desired efficient and fast backup, it is too

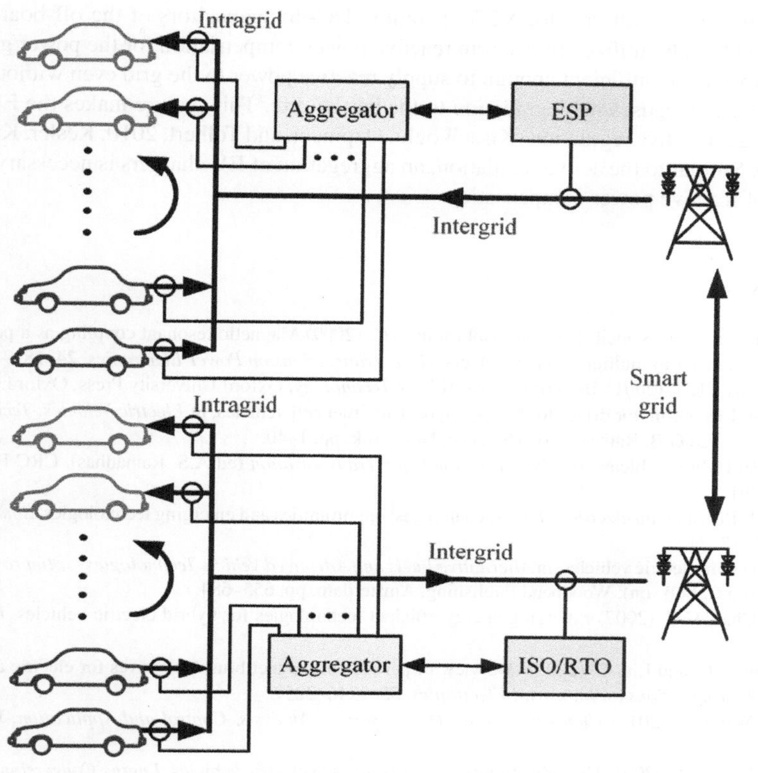

Figure 1.12 Dual-grid framework for vehicle-to-grid operation

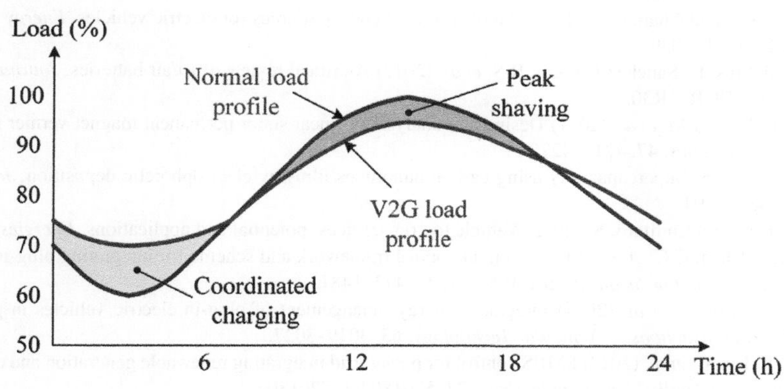

Figure 1.13 Coordinated charging and peak shaving

expensive and bulky. The V2G operation can fully utilize the EV batteries to complement the intermittent outage of power grid with renewables (Gao *et al.*, 2014). Differing from load shaving, the renewable transients will not cause deep discharging of EV batteries or noticeable degradation of battery life. Moreover, by incorporating superconducting magnetic energy storage (SMES) into V2G operation, the system dynamic performance for renewable transients can be further improved (Gao *et al.*, 2012).

Apart from using EV batteries for V2G operation, DC-link capacitors of the off-board bidirectional EV chargers can be fully utilized to perform reactive power compensation for the power grid. In general, the DC-link capacitor is sufficient enough to supply reactive power to the grid even without engaging the EV batteries. Thus, it causes no degradation to the battery life. This feature makes the EV charger very attractive for V2G reactive regulation (Kisacikoglu, Ozpineci, and Tolbert, 2010; Kesler, Kisacikoglu, and Tolbert, 2014). Similar to the active regulation, an aggregation of EV chargers is necessary to provide the desired level of reactive power compensation.

References

Cannon, B.L., Hoburg, J.F., Stancil, D.D. and Goldstein, S.C. (2009) Magnetic resonant coupling as a potential means for wireless power transfer to multiple small receivers. *IEEE Transactions on Power Electronics*, **24**, 1819–1825.

Chan, C.C. and Chau, K.T. (2001) *Modern Electric Vehicle Technology*, Oxford University Press, Oxford.

Chau, K.T. (2009) Electric motor drives for battery, hybrid and fuel cell vehicles, in *Electric Vehicles: Technology, Research and Development* (ed G.B. Raines), Nova Science, New York, pp. 1–40.

Chau, K.T. (2010) Hybrid vehicles, in *Alternative Fuels for Transportation* (ed A.S. Ramadhas), CRC Press, Boca Raton, FL, pp. 361–391.

Chau, K.T. (2012) Research on electric vehicles: challenges, opportunities and emerging technologies. *Studies in Science and Technology*, **1**, 13–24.

Chau, K.T. (2014) Pure electric vehicles, in *Alternative Fuels and Advanced Vehicle Technologies for Improved Environmental Performance* (ed R. Folkson), Woodhead Publishing, Amsterdam, pp. 655–684.

Chau, K.T. and Chan, C.C. (2007) Emerging energy-efficient technologies for hybrid electric vehicles. *Proceedings of the IEEE*, **95**, 821–835.

Chau, K.T., Chan, C.C. and Liu, C. (2008) Overview of permanent-magnet brushless drives for electric and hybrid electric vehicles. *IEEE Transactions on Industrial Electronics*, **55**, 2246–2257.

Chau, K.T. and Wang, Z. (2011) *Chaos in Electric Drive Systems–Analysis, Control and Application*, Wiley-IEEE Press, Singapore.

Chau, K.T. and Wong, Y.S. (2001) Hybridization of energy sources in electric vehicles. *Energy Conversion and Management*, **42**, 1059–1069.

Chau, K.T. and Wong, Y.S. (2002) Overview of power management in hybrid electric vehicles. *Energy Conversion and Management*, **43**, 1953–1968.

Chau, K.T., Wong, Y.S. and Chan, C.C. (1999) An overview of energy sources for electric vehicles. *Energy Conversion and Management*, **40**, 1021–1039.

Christensen, J., Albertus, P., Sanchez-Carrera, R.S. *et al.* (2012) A critical review of Li/air batteries. *Journal of the Electrochemical Society*, **159**, R1–R30.

Du, Y., Chau, K.T., Cheng, M. *et al.* (2011) Design and analysis of linear stator permanent magnet vernier machines. *IEEE Transactions on Magnetics*, **47**, 4219–4222.

Du, C. and Pan, N. (2006) Supercapacitors using carbon nanotubes films by electrophoretic deposition. *Journal of Power Sources*, **160**, 1487–1494.

Ehsani, M., Falahi, M. and Lotfifard, S. (2012) Vehicle to grid services: potential and applications. *Energies*, **5**, 4076–4090.

Gao, S., Chau, K.T., Chan, C.C. *et al.* (2011) Optimal control framework and scheme for integrating plug-in hybrid electric vehicles into grid. *Journal of Asian Electric Vehicles*, **9**, 1473–1481.

Gao, S., Chau, K.T., Liu, C. *et al.* (2014) Integrated energy management of plug-in electric vehicles in power grid with renewables. *IEEE Transactions on Vehicular Technology*, **63**, 3019–3027.

Gao, S., Chau, K.T., Liu, C. *et al.* (2012) SMES control for power grid integrating renewable generation and electric vehicles. *IEEE Transactions on Applied Superconductivity*, **22**, 5701804-1–5701804-4.

Giuliano, M.R., Advani, S.G. and Prasad, A.K. (2011) Thermal analysis and management of lithium-titanate batteries. *Journal of Power Sources*, **196**, 6517–6524.

Guille, C. and Gross, G. (2009) A conceptual framework for the vehicle-to-grid (V2G) implementation. *Energy Policy*, **37**, 4379–4390.

Hermance, D. and Sasaki, S. (1998) Hybrid electric vehicles take to the streets. *IEEE Spectrum*, **35**, 48–52.

Hoeijmakers, M. and Ferreira, J. (2006) The electric variable transmission. *IEEE Transactions on Industry Applications*, **42**, 1092–1100.

Imura, T. and Hori, Y. (2011) Maximizing air gap and efficiency of magnetic resonant coupling for wireless power transfer using equivalent circuit and Neumann formula. *IEEE Transactions on Industrial Electronics*, **58**, 4746–4752.

Jian, L. and Chau, K.T. (2009) A novel electronic-continuously variable transmission propulsion system using coaxial magnetic gearing for hybrid electric vehicles. *Journal of Asian Electric Vehicles*, **7**, 1291–1296.

Jian, L. and Chau, K.T. (2010) Design and analysis of a magnetic-geared electronic-continuously variable transmission system using finite element method. *Progress in Electromagnetics Research*, **107**, 47–61.

Kamiya, M. (2006) Development of traction drive motors for the Toyota hybrid system. *IEEE Transactions on Industry Applications*, **126**, 473–479.

Kesler, M., Kisacikoglu, M.C. and Tolbert, L.M. (2014) Vehicle-to-grid reactive power operation using plug-in electric vehicle bidirectional off-board charger. *IEEE Transactions on Industrial Electronics*, **61**, 6778–6784.

Kim, J., Kim, J., Kong, S. *et al.* (2013) Coil design and shielding methods for a magnetic resonant wireless power transfer system. *Proceedings of the IEEE*, **101**, 1332–1342.

Kisacikoglu, M.C., Ozpineci, B., and Tolbert, L.M. (2010) Examination of a PHEV bidirectional charger for V2G reactive power compensation. Proceedings of IEEE Applied Power Electronics Conference and Exposition, pp. 458–465.

Lee, C.H. (1963) Vernier motor and its design. Proceedings of IEEE Winter General Meeting, pp. 343–349.

Lee, C.H.T., Liu, C. and Chau, K.T. (2014) A magnetless axial-flux machine for range-extended electric vehicle. *Energies*, **7**, 1483–1499.

Li, J., Chau, K.T. and Li, W. (2011) Harmonic analysis and comparison of permanent magnet vernier and magnetic-geared machines. *IEEE Transactions on Magnetics*, **47**, 3649–3652.

Liu, C., Chau, K.T. and Jiang, J.Z. (2010a) A permanent-magnet hybrid brushless integrated starter-generator for hybrid electric vehicles. *IEEE Transactions on Industrial Electronics*, **57**, 4055–4064.

Liu, C., Yu, Z., Neff, D. *et al.* (2010b) Graphene-based supercapacitor with an ultrahigh energy density. *Nano Letters*, **10**, 4863–4868.

Liu, C., Chau, K.T., Jiang, J.Z. and Niu, S. (2008) Comparison of stator-permanent-magnet brushless machines. *IEEE Transactions on Magnetics*, **44**, 4405–4408.

Liu, C., Chau, K.T., Wu, D. and Gao, S. (2013) Opportunities and challenges of vehicle-to-home, vehicle-to-vehicle, and vehicle-to-grid technologies. *Proceedings of the IEEE*, **101**, 2409–2427.

Martin, S., Garcia-Ybarra, P.L. and Castillo, J.L. (2010) Electrospray deposition of catalyst layers with ultra-low Pt loadings for PEM fuel cells cathodes. *Journal of Power Sources*, **195**, 2443–2449.

Musavi, F. and Eberle, W. (2014) Overview of wireless power transfer technologies for electric vehicle battery charging. *IET Power Electronics*, **7**, 60–66.

Omar, N., Daowd, M., Hegazy, O. *et al.* (2012) Standardization work for BEV and HEV applications: critical appraisal of recent traction battery documents. *Energies*, **5**, 138–156.

Qiu, C., Chau, K.T., Ching, T.W. and Liu, C. (2014a) Overview of wireless charging technologies for electric vehicles. *Journal of Asian Electric Vehicles*, **12**, 1–7.

Qiu, C., Chau, K.T., Liu, C. *et al.* (2014b) Quantitative comparison of dynamic flux distribution of magnetic couplers for roadway electric vehicle wireless charging system. *Journal of Applied Physics*, **115**, 17A334-1–17A334-3.

Spooner, E. and Haydock, L. (2003) Vernier hybrid machines. *IEE Proceedings – Electric Power Applications*, **150**, 655–662.

Strasik, M., Johnson, P.E., Day, A.C. *et al.* (2007) Design, fabrication, and test of a 5-kWh/100-kW flywheel energy storage utilizing a high-temperature superconducting bearing. *IEEE Transactions on Applied Superconductivity*, **17**, 2133–2137.

Taibi, S., Tounzi, A. and Piriou, F. (2006) Study of a stator current excited vernier reluctance machine. *IEEE Transactions on Energy Conversion*, **21**, 823–831.

Theodoridis, M.P. (2012) Effective capacitive power transfer. *IEEE Transactions on Power Electronics*, **27**, 4906–4913.

Wakefield, E.H. (1994) *History of the Electric Automobile: Battery-Only Powered Cars*, Society of Automotive Engineers, Warrendale, PA.

Wang, Z. and Chau, K.T. (2009) Design, analysis, and experimentation of chaotic permanent magnet DC motor drives for electric compaction. *IEEE Transactions on Circuits and Systems II*, **56**, 245–249.

Wang, J., Chau, K.T., Jiang, J.Z. and Yu, C. (2008) Design and analysis of a transverse flux permanent magnet machine using three dimensional scalar magnetic potential finite element method. *Journal of Applied Physics*, **103**, 7F107-1–7F107-3.

Wang, X., Ma, Y., Li, S. *et al.* (2011) Ceria-based nanocomposite with simultaneous proton and oxygen ion conductivity for low-temperature solid oxide fuel cells. *Journal of Power Sources*, **196**, 2754–2758.

Wu, D., Chau, K.T. and Gao, S. (2010a) Cost-emission analysis of vehicle-to-grid system. *World Electric Vehicle Journal*, **4**, 767–773.

Wu, D., Chau, K.T., and Gao, S. (2010b) Multilayer framework for vehicle-to-grid operation. Proceedings of IEEE Vehicle Power and Propulsion Conference, pp. 1–6.

Ye, S. and Chau, K.T. (2007) Chaoization of DC motors for industrial mixing. *IEEE Transactions on Industrial Electronics*, **54**, 2024–2032.

Yu, C. and Chau, K.T. (2009) Thermoelectric automotive waste heat energy recovery using maximum power point tracking. *Energy Conversion and Management*, **50**, 1506–1512.

Zhang, S.S. (2013) Liquid electrolyte lithium/sulfur battery: fundamental chemistry, problems, and solutions. *Journal of Power Sources*, **231**, 153–162.
Zhang, X. and Chau, K.T. (2011a) Design and implementation of a new thermoelectric-photovoltaic hybrid energy system for hybrid electric vehicles. *Electric Power Components and Systems*, **39**, 511–525.
Zhang, X. and Chau, K.T. (2011b) An automotive thermoelectric-photovoltaic hybrid energy system using maximum power point tracking. *Energy Conversion and Management*, **52**, 641–647.
Zhang, Z., Chau, K.T., Liu, C. *et al.* (2014) An efficient wireless power transfer system with security considerations for electric vehicle applications. *Journal of Applied Physics*, **115**, 17A328-1–17A328-3.
Zhang, Z., Chau, K.T., Wang, Z. and Li, W. (2011) Improvement of electromagnetic compatibility of motor drives using hybrid chaotic pulse width modulation. *IEEE Transactions on Magnetics*, **47**, 4018–4021.
Zhu, Z.Q. and Howe, D. (2007) Electrical machines and drives for electric, hybrid and fuel cell vehicles. *Proceedings of the IEEE*, **95**, 746–765.

2

DC Motor Drives

DC motor drives were widely used for electric vehicle (EV) propulsion. Various DC motor drives were applied to different EVs because of their merits such as technological maturity and control simplicity. However, they generally suffer from the problems such as lower efficiency and lower power density than their AC counterparts, and the need for regular maintenance of their carbon brushes and commutator. Consequently, besides those small or off-road EVs, they are superseded by AC or brushless DC motor drives for EV propulsion.

In this chapter, various DC motor drives including their system configurations, DC machines, DC–DC converters, and control strategies are described. The corresponding design criteria, design example, and application examples for EV propulsion are also discussed.

2.1 System Configurations

The earliest system configuration of DC motor drives for electric propulsion consisted of a mesh of resistors connected in series and/or in parallel with the DC motor. The motor voltage is equal to the battery voltage minus the voltage drop across the resistors, and can be varied by using electromechanical contactors to open or short those resistors. Although the rheostatic control is simple and low cost, it suffers from poor efficiency because a large amount of energy is dissipated as heat in the resistors. In addition, this rheostatic control cannot offer smooth operation, resulting in discomfort and jerkiness.

With the rapid development of power electronics, the rheostatic control of DC motor drives is obsolete. The DC chopper control has been used extensively for DC motor drives because of the advantages such as smaller size, lighter weight, higher efficiency, and better controllability. Figure 2.1 shows the basic configuration of DC motor drives in which the DC–DC converter functions to control the armature current and hence the output torque of the DC machine. In general, the feedback control variable is only the motor speed, while the armature current feedback is mainly for protection purpose.

For EV propulsion, the DC machine of the DC motor drive usually adopts the high-speed machine design to achieve high power density. Typically, it spins up to 5000 rpm and hence a fixed gear (FG) is incorporated to step it down to 1000 rpm for driving the vehicle wheels. In order to provide backward motoring, such as for parking, the DC machine needs to offer reverse rotation; otherwise, a reverse gear is incorporated. Of course, reverse rotation of the DC machine is preferable to reverse gearing because the reverse gear is bulky, inefficient, and complicated. Moreover, there are various choices of DC–DC converters and DC machines, depending on the desired operation and performance.

Electric Vehicle Machines and Drives: Design, Analysis and Application, First Edition. K. T. Chau.

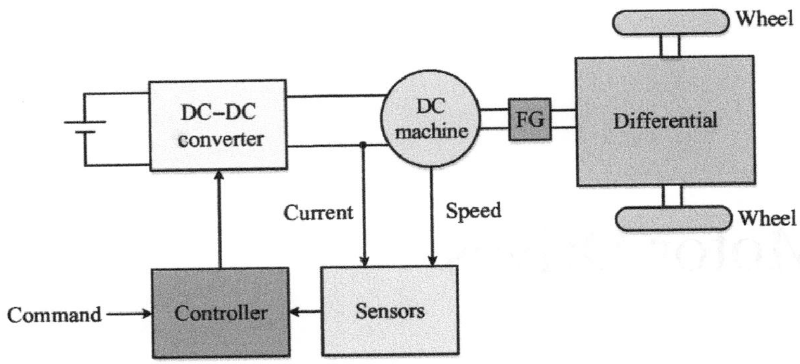

Figure 2.1 Basic configuration of DC motor drives

2.2 DC Machines

DC machines were invented by the British scientist William Sturgeon in 1832, and have been in service for more than a century. Since the advent of AC machines, the role of DC machines for electric propulsion is phasing out. Nevertheless, some low-end or off-road EVs still adopt DC machines mainly for the sake of simplicity.

2.2.1 Structure of DC Machines

Figure 2.2 shows the basic structure of a DC machine which mainly consists of the stator, rotor, and commutator. The stator is the field circuit that incorporates the field winding or permanent magnets (PMs) to produce magnetic field excitation, whereas the rotor is the armature circuit that installs the armature winding where the armature current is bidirectional and switched by the commutator via carbon brushes.

Different arrangements of the field and armature circuits create different types of DC machines, hence providing different torque-speed characteristics (Dubey, 1989). As shown in Figure 2.3, a DC machine can be classified as separately excited, series, shunt, cumulative compound, differential compound, and PM types. In the separately excited DC machine, the field and armature circuits are separately fed by different voltage sources so that their currents are controlled independently. In the shunt DC machine,

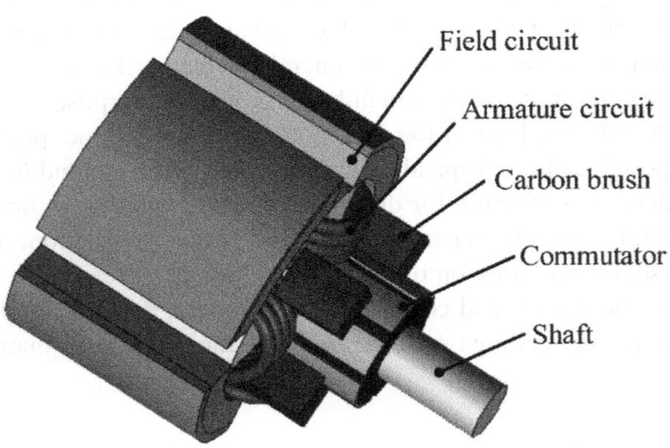

Figure 2.2 DC machine exploded diagram

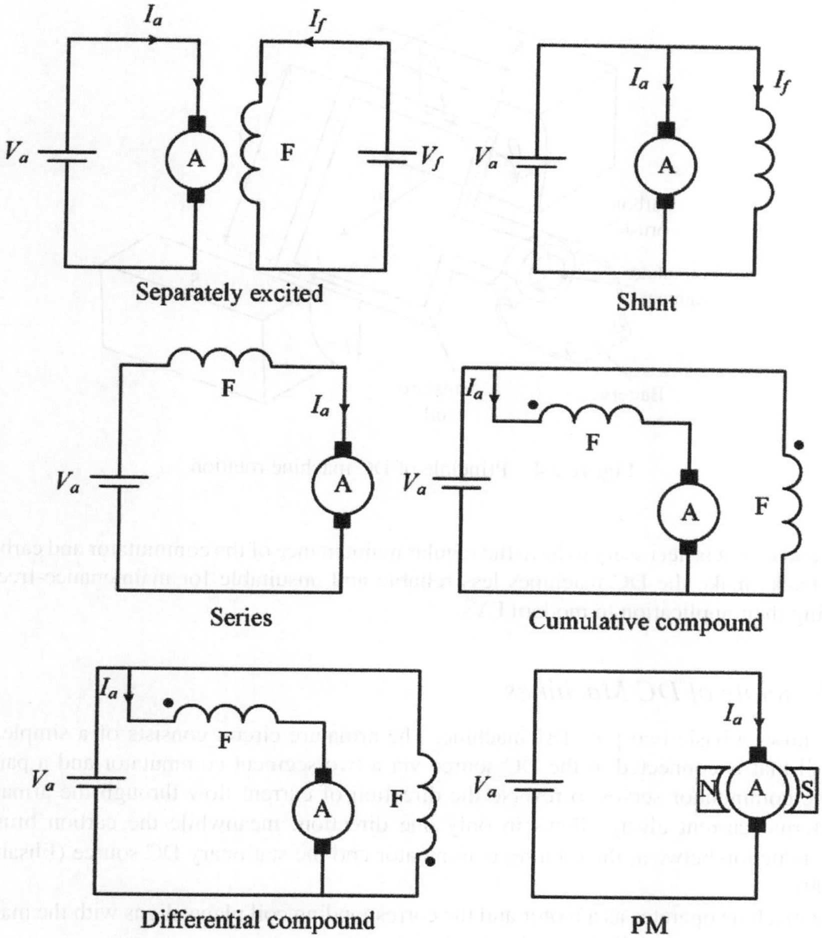

Figure 2.3 DC machine structures

the field and armature circuits are connected in parallel and fed by the same source so that their currents are controlled simultaneously. In the series DC machine, the field circuit is connected in series with the armature circuit so that the field and armature currents are the same, and controlled simultaneously. In the cumulative compound DC machine, there are two field circuits, where one connected in series and the other connected in shunt with the armature circuit, while the flux of the series field is in the same direction as that of the shunt field. Differing from the cumulative compound one, the differential compound DC machine has the flux of the series field opposing the flux of the shunt field. By replacing the field winding and pole structure with PMs, the PM DC machine has relatively higher power density and higher efficiency than the above wound-field types because of the space-saving benefit by PMs and the absence of field losses. However, since the field excitation in the PM DC machine is uncontrollable, it cannot attain the operating characteristics with flux control.

Both the wound-field and PM DC machines suffer from the same fundamental problem because of the use of commutator and carbon brushes. The commutator causes torque ripples and limits the operating speed, while the carbon brushes are responsible for friction and radio-frequency interference. Moreover,

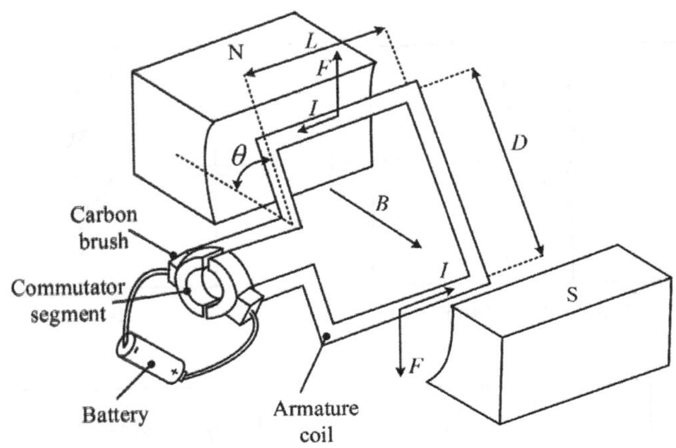

Figure 2.4 Principle of DC machine rotation

due to wear and tear, it is necessary to have the regular maintenance of the commutator and carbon brushes. These drawbacks make the DC machines less reliable and unsuitable for maintenance-free operation, hence limiting their application to modern EVs.

2.2.2 *Principle of DC Machines*

Figure 2.4 shows a basic two-pole DC machine. The armature circuit consists of a simple single-turn armature coil that is connected to the DC source via a two-segment commutator and a pair of carbon brushes. The commutator serves to reverse the direction of current flow through the armature coil so that the external current always flows in only one direction; meanwhile the carbon brushes enable electrical conduction between the rotating commutator and the stationary DC source (Ehsani, Gao, and Emadi, 2009).

When the machine operates as a motor and the corresponding coil plane aligns with the magnetic field, the force F created on each coil side is simply given by

$$F = BIL \tag{2.1}$$

where B is the magnetic flux density produced by the field circuit in the stator, I is the magnitude of armature current, and L is the axial length of the armature coil in the rotor. Hence, the developed torque can be expressed as

$$T = FD\cos\theta \tag{2.2}$$

where D is the diameter of the armature coil or equivalently the rotor and θ is the angle between the coil plane and the magnetic field. In order to produce the maximum torque, the armature coil should be energized at the position $\theta = 0°$. In practice, more commutator segments and pairs of carbon brushes are installed to produce continuous and steady torque.

2.2.3 *Modeling of DC Machines*

The modeling of DC machines is simple. Basically, all types of DC machines can be represented by an equivalent circuit of the armature circuit, and an equivalent circuit of the field circuit or simply constant

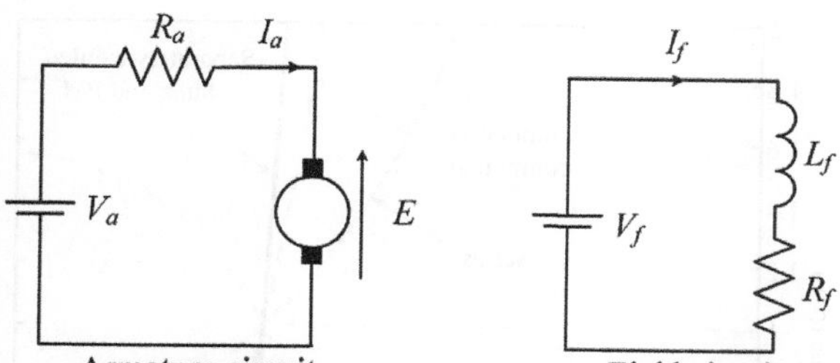

Figure 2.5 Equivalent circuit of separately excited DC machine

flux per pole when the field is excited by PMs. On the basis of the equivalent circuit of the separately excited DC machine as shown in Figure 2.5, the DC machine operation is governed by

$$E = K_e \, \phi \, \omega_m \tag{2.3}$$

$$V_a = E + R_a I_a \tag{2.4}$$

$$T = K_e \, \phi \, I_a \tag{2.5}$$

$$V_f = R_f I_f \tag{2.6}$$

$$\phi = K_f I_f \tag{2.7}$$

where E is the back electromotive force (EMF), K_e is the back EMF constant, ϕ is the air-gap flux per pole, ω_m is the mechanical speed, V_a is the armature voltage, R_a is the armature circuit resistance, I_a is the armature current, T is the developed torque, V_f is the field voltage, R_f is the field circuit resistance, I_f is the field current, and K_f is the slope of magnetization characteristic, which is a constant in the unsaturated region. For those wound-field DC machines, the flux is generally proportional to the field current which may be independently controlled by the field voltage, dependent on the armature voltage, dependent on the armature current, or dependent on both the armature voltage and armature current, respectively, for the separately excited, shunt, series, or compound types. In contrast, the flux is essentially uncontrollable for the PM DC machine.

Without external control, the torque-speed characteristics of DC machines can readily be deduced from Eqs. (2.3)–(2.7) as depicted in Figure 2.6. For the separately excited DC machine, the armature and field voltages are kept at the rated value. It is actually equivalent to the shunt DC machine and PM DC machine at the rated flux. By using Eqs. (2.3)–(2.5), the corresponding torque-speed characteristic can be derived as

$$T = \frac{K_e \phi V_a}{R_a} - \frac{K_e^2 \phi^2 \omega_m}{R_a} \tag{2.8}$$

which indicates that the speed decreases linearly as the torque increases, and the speed regulation depends on the armature circuit resistance. This characteristic is particularly suitable for applications requiring good speed regulation and proper adjustable speed. In case of the series DC machine, by using Eqs. (2.3)–(2.5) and (2.7), the corresponding torque-speed characteristic can be expressed as

$$T = \frac{K_e K_f V_a^2}{(R_a + K_e K_f \omega_m)^2} \tag{2.9}$$

which indicates that the speed inversely relates to the torque. This characteristic is particularly suitable for applications requiring high starting torque and heavy overload. However, it suffers from the problem

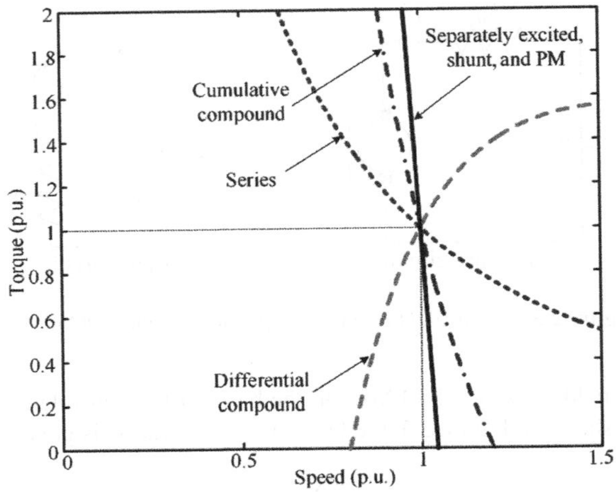

Figure 2.6 Torque-speed characteristics of DC machines

of over-speed when operating at the rated voltage under no load or light load. In case of the cumulative compound DC machine, the torque-speed characteristic lies between those of shunt and series DC machines, depending on the relative strength of the shunt and series fields. The corresponding no-load behavior mainly depends on the shunt field while the drooping behavior is dictated by the series field. This characteristic is attractive for applications requiring high starting torque, and at the same time, the no-load speed can be limited to a safe value. For the differential compound DC machine, the series field reduces the air-gap flux when the torque increases, causing the speed to rise in such a way that it can be used to compensate the speed drop due to the armature resistance. However, on overload, this effect will cause the speed to rise to instability. Thus, this machine is seldom applied to motor drives.

2.3 DC–DC Converters

At fixed supply voltage, the DC machines can only offer the natural torque-speed characteristics as shown in Figure 2.6. In order to provide proper speed control for EV operation, the use of DC–DC converters is almost mandatory.

2.3.1 DC–DC Converter Topologies

When DC–DC converters adopt the chopping mode of operation, they are usually termed as DC choppers and are extensively used for voltage control of DC motor drives. These DC choppers are generally classified as the first-, second-, two-, and four-quadrant versions. The first-quadrant DC chopper, as shown in Figure 2.7, is suitable for motoring and the power flow is from the source to the machine. The second-quadrant DC chopper, which is seldom to be individually applied to DC motor drives, is solely for regenerative braking and the power flow is from the machine to the source. As regenerative braking is essential for EVs, which can significantly extend the vehicle driving range, the two-quadrant DC chopper shown in Figure 2.8 is preferred because it can offer both motoring and regenerative braking. Moreover, instead of using electromechanical contactors to achieve EV backward operation, the four-quadrant DC chopper shown in Figure 2.9 can be employed so that motoring and regenerative braking in both EV forward and backward operations are controlled electronically.

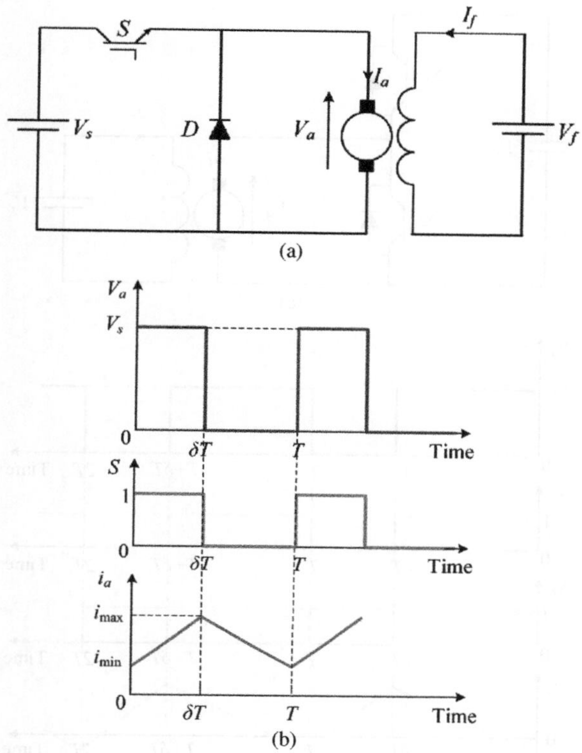

Figure 2.7 First-quadrant DC chopper: (a) circuit and (b) waveforms

In general, pulse-width modulation (PWM) is adopted to control the armature voltage of DC motor drives for EV propulsion. Under the continuous conduction mode, the control is based on the variation of duty cycle δ of the power switch:

$$V_a = \frac{1}{T}\int_0^T v_a\,dt = \frac{1}{T}\int_0^{\delta T} V_s\,dt = \delta V_s \tag{2.10}$$

$$I_a = \frac{V_a - E}{R_a} \tag{2.11}$$

where V_s is the DC source voltage. For the first-quadrant DC chopper, the duty cycle varies between zero and one so that the armature voltage can be controlled from zero to the DC source voltage. Of course, it can only provide a positive armature voltage and a positive armature current for operation in the first quadrant, namely the positive torque and positive speed. On the other hand, for the two-quadrant DC chopper, the power switches are closed alternately and the duty cycle is defined by the upper switch. As the armature current is flowing all the time, the discontinuous conduction mode does not occur. Hence, the motoring operation with $I_a > 0$ occurs when $\delta > (E/V_s)$, whereas regenerative braking with $I_a < 0$ occurs when $\delta < (E/V_s)$. Furthermore, by properly combining two two-quadrant DC choppers, the resulting four-quadrant DC chopper can offer the positive armature voltage with bidirectional armature current and also the negative armature voltage with bidirectional armature current, hence achieving forward motoring and forward regenerative braking as well as backward motoring and backward regenerative braking, respectively.

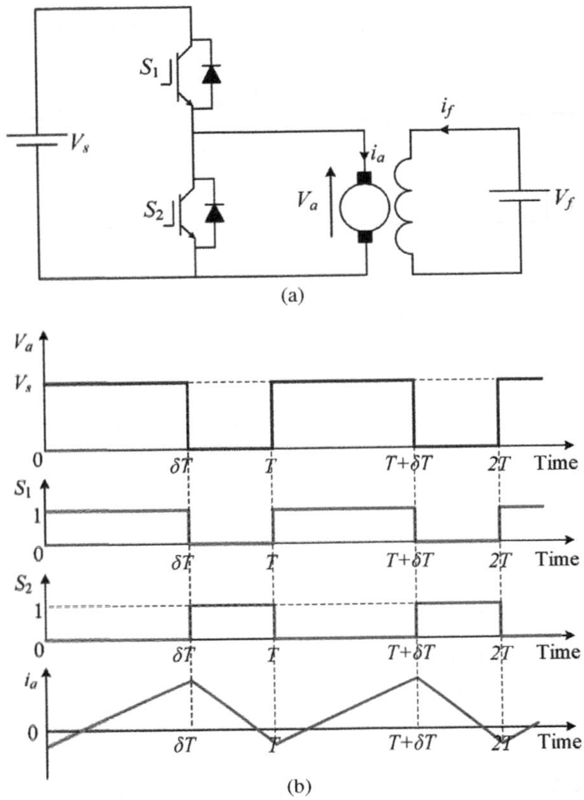

Figure 2.8 Two-quadrant DC chopper: (a) circuit and (b) waveforms

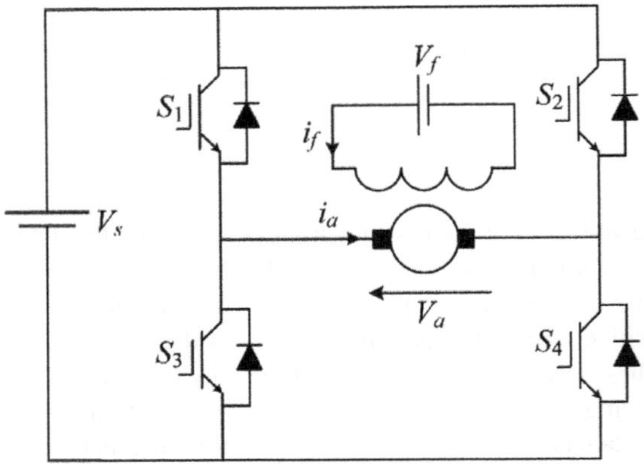

Figure 2.9 Four-quadrant DC chopper

2.3.2 Soft-Switching DC–DC Converter Topologies

Rather than adopting hard or stressed switching, DC–DC converters can employ soft or relaxed switching. The key of soft-switching techniques is to employ a resonant circuit to shape the voltage or current waveform in such a way that the power device switches at the zero-voltage or zero-current condition. In general, the use of soft-switching DC–DC converters possesses the following advantages:

- Because of the zero-voltage switching (ZVS) or zero-current switching (ZCS) condition, the power device switching loss is practically zero, thus offering higher efficiency.
- Owing to practically zero switching loss, the heat-sinking requirement is significantly reduced, thus providing higher power density.
- Because of the minimum switching stress during soft switching, the reliability of power devices can be improved.
- Because of relatively low dv/dt, the effect of electromagnetic interference (EMI) is less severe and the machine insulation is less stressed.
- Because of high-frequency operation, the acoustic noise can be effectively suppressed.

There is no free lunch for the above-mentioned advantages: namely, the use of soft-switching techniques for DC–DC converters has the drawbacks of additional cost of the resonant circuit, increased control complexity, and limited operating ranges.

In the past two decades, many soft-switching DC–DC converters have been successfully developed for switched-mode power supplies, leading to achieve high power density and high efficiency (Hua and Lee, 1995). They are generally classified as the resonant converter, quasi-resonant converter, multi-resonant converter, and zero-transition converter. Most of them initially operate at variable switching frequency and then extend to constant-frequency operation (Chan and Chau, 1993; Chau, 1994). The constant-frequency operation takes the definite advantages over the variable-frequency operation because it favors the optimization of reactive components and closed-loop bandwidth as well as the filtering of EMI and noise. Meanwhile, the control of constant-frequency soft-switching DC–DC converters is generally accomplished by means of PWM techniques (Ching and Chan, 2008).

Soft-switching DC–DC converters are seldom used for DC motor drives. The major reason is due to the fact that the corresponding development has been much less than that for switched-mode power supplies. In addition, most available soft-switching DC–DC converters cannot handle backward power flow during regenerative braking. Nevertheless, some soft-switching DC–DC converters have been developed for DC motor drives. Among them, the zero-transition DC–DC converters are most attractive.

Figure 2.10 shows the two-quadrant zero-voltage transition (2Q-ZVT) converter for DC motor drives (Chau, Ching, and Chan, 1999). This converter topology is created by adding a resonant inductor, a resonant capacitor, and two auxiliary switches to a conventional two-quadrant DC–DC converter. It should be noted that the ZVS condition is particularly desirable for power metal-oxide-semiconductor field-effect transistor (MOSFET)-based power conversion, because the power MOSFET device generally suffers from severe capacitive voltage turn-on losses. This 2Q-ZVT DC–DC converter offers some definite advantages over its counterparts: namely, the ZVS for all main switches and diodes, the capability of bidirectional power flow for both motoring and regenerative braking, the unity device voltage and current stresses during both the motoring and regenerating modes of operation, the use of the same resonant tank for both forward and backward power flows, the utilization of stray capacitances as part of the resonant components, as well as the utilization of all built-in diodes of the power switches to minimize the overall hardware count and cost.

Figure 2.11 shows the two-quadrant zero-current transition (2Q-ZCT) converter for DC motor drives (Ching and Chau, 2001). This converter topology is also created by adding a resonant inductor, a resonant capacitor, and two auxiliary switches to a conventional two-quadrant DC–DC converter. Differing from the previous 2Q-ZVT, this converter is particularly useful for the insulated-gate bipolar transistor

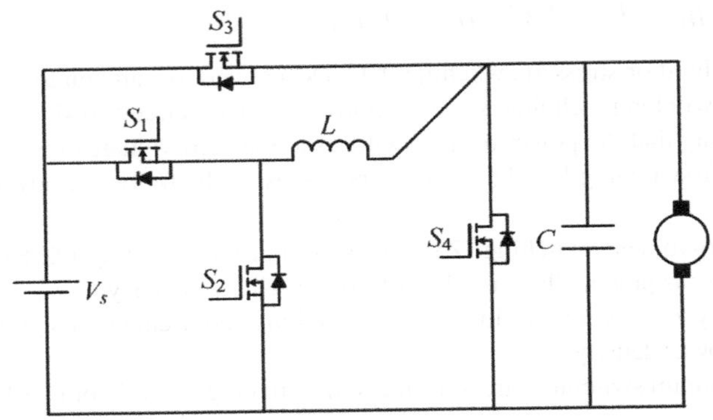

Figure 2.10 Two-quadrant zero-voltage transition DC–DC converter

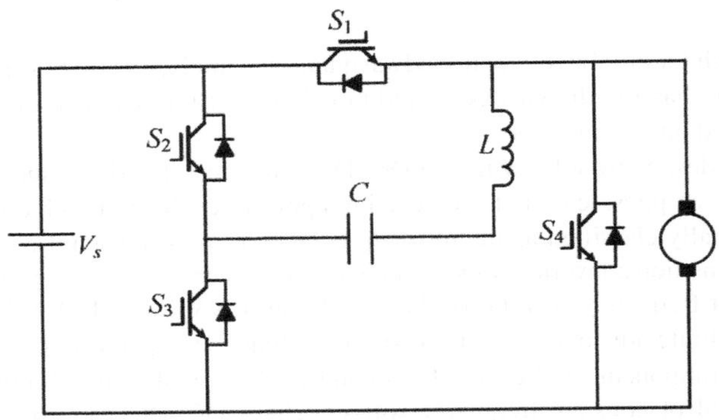

Figure 2.11 Two-quadrant zero-current transition DC–DC converter

(IGBT)-based power conversion, because the IGBT device generally suffers from severe inductive turn-off switching losses. The major advantages of this 2Q-ZCT DC–DC converter are the ZCS for all switches and rectifiers, the bidirectional power flow for both motoring and regenerative braking, the same resonant tank for both forward and backward power flows, and the applicability of high-power DC motor drives employing the IGBT as power devices.

On the basis of the 2Q-ZVT and 2Q-ZCT converter topologies, the four-quadrant (4Q) versions, namely the 4Q-ZVT and 4Q-ZCT converters, can readily be deduced for DC motor drives (Ching, 2005; 2006) as shown in Figures 2.12 and 2.13, respectively. While possessing the feature of 4Q operation, they can directly inherit the merits of their 2Q counterparts.

At present, most low-cost electric scooters and mini EVs utilize hard-switching DC–DC converters for their DC motor drives. They are widely accepted in densely populated cities for example, cities in China and India. Actually, the development of soft-switching DC–DC converters for EV propulsion has a niche market where the power level is about 5 kW.

2.4 DC Motor Control

As mentioned earlier, the DC motor drives take the definite advantage of control simplicity. Among all kinds of DC motors, the separately excited DC motor offers the highest controllability because the armature and field currents can be independently controlled, whereas the series DC motor is relatively inflexible

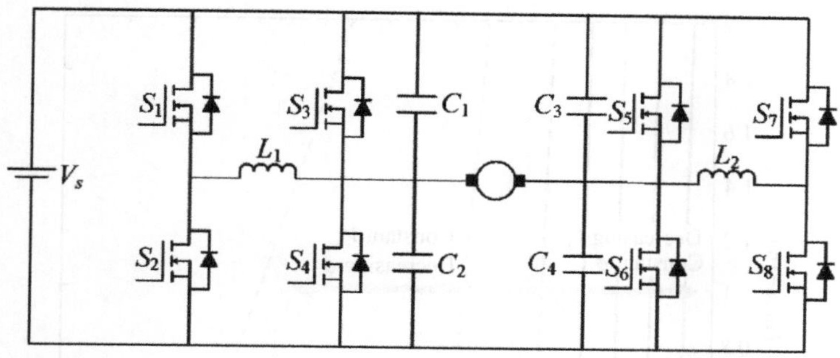

Figure 2.12 Four-quadrant zero-voltage transition DC–DC converter

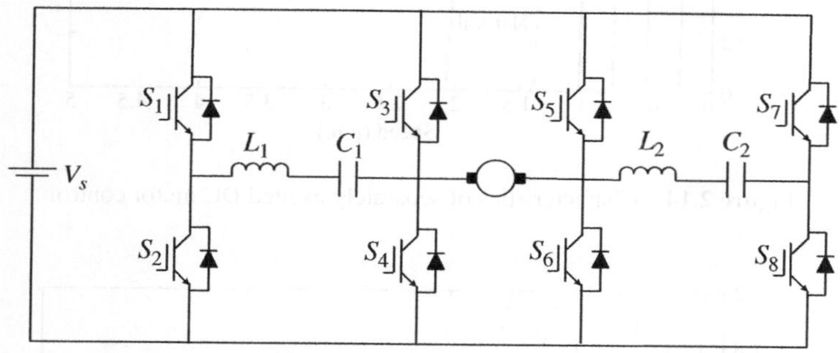

Figure 2.13 Four-quadrant zero-current transition DC–DC converter

since the armature and field currents are forced to be the same. So, in the following discussion, the control characteristics of the separately excited and series DC motors are emphasized.

2.4.1 *Speed Control*

In general, speed control of DC motor drives can be accomplished by two methods: armature voltage control and flux-weakening control. When the armature voltage of the DC motor is reduced, the armature current and hence the motor torque decrease, thereby decreasing the motor speed. In contrast, when the armature voltage is increased, the motor torque increases, thereby increasing the motor speed. Under this armature voltage control, the torque-speed characteristics of the separately excited and series DC motors are shown by solid lines in Figures 2.14 and 2.15, respectively. It can be observed that the motors can operate for any torque-speed characteristics below the natural characteristics. Also, the slope of these characteristics does not change with the speed. Since the maximum allowable armature current remains constant, this armature voltage control has the advantage of retaining the maximum torque capability at all speeds. However, since the armature voltage cannot be further increased beyond the rated value, this control is used only when the DC motor drive operates below its base speed.

When the field voltage of the DC motor is weakened while the armature voltage is fixed, the motor back EMF decreases. Because of low armature resistance, the armature current will increase by an amount much larger than the decrease in the field. Thus, the additional motor torque causes the motor speed to increase. Under this flux-weakening control, the torque-speed characteristics of the separately excited and series DC motors are shown by dotted lines in Figures 2.14 and 2.15, respectively. It can be observed that the slope of these characteristics is no longer kept constant, but is affected by the flux.

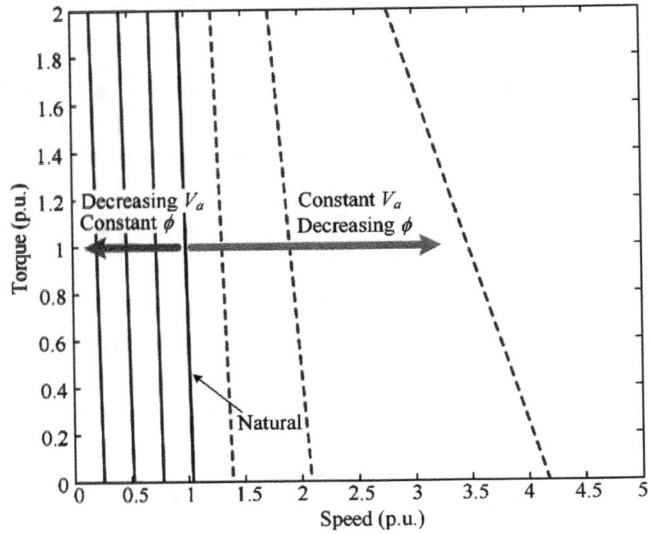

Figure 2.14 Characteristics of separately excited DC motor control

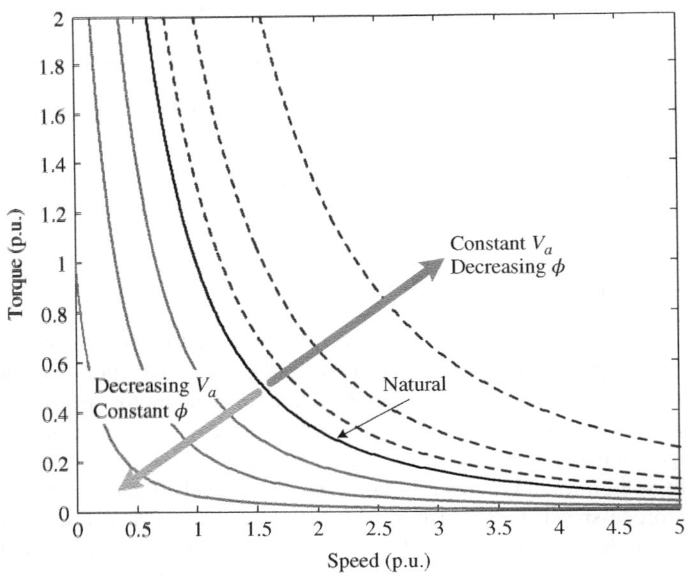

Figure 2.15 Characteristics of series DC motor control

In order to achieve a wide range of speed control, the armature voltage control and flux-weakening control should be independent, which can only be applied to the separately excited DC motor drive. Below the base speed, the armature voltage is varied, whereas the flux is kept at the rated value. Since the maximum allowable armature current is constant, the torque capability under armature voltage control can be kept constant, the so-called constant-torque region. Above the base speed, the flux is weakened, whereas the

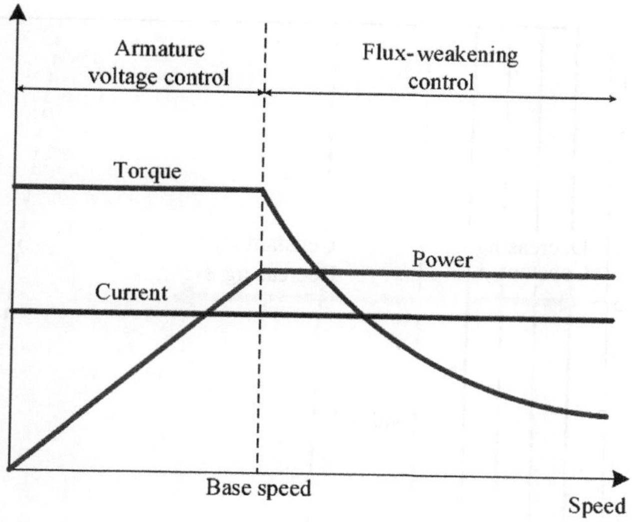

Figure 2.16 Operating capabilities of separately excited DC motor drive under combined armature and flux control

armature voltage is kept at the rated value. Since the armature voltage is fixed and the maximum allowable armature current is constant, the back EMF remains constant for all speeds. Hence, the power capability is constant, the so-called constant-power region, and the torque capability varies inversely with the motor speed. Figure 2.16 depicts these torque and power capabilities.

2.4.2 *Regenerative Braking*

There are various electric braking techniques for DC motor drives, namely the regenerative braking, rheostatic braking, and plugging. The regenerative braking is almost exclusively adopted by EVs because the corresponding battery pack has the ability to absorb and store the mechanical braking energy, whereas the rheostatic braking and plugging are seldom used in EVs since the corresponding braking energy is wasted in resistances.

If the DC motor is fed by a fixed voltage source V_s, the regenerative braking can be performed only for speeds above the rated speed. With the use of DC chopper control, it is possible to obtain regenerative braking down to almost the zero speed. That is, based on the two-quadrant DC chopper, the armature terminal voltage can be controlled by tuning the duty cycle δ to be lower than the back EMF E so that the armature current becomes negative, hence creating a negative torque for regenerative braking. For the separately excited DC motor drive, the torque expression for regenerative braking is the same as that for motoring, but with a negative value of armature current when $\delta < (E/V_s)$. Also, when desiring regenerative braking above the rated speed, the flux needs to be weakened accordingly. The corresponding torque-speed characteristics are shown in Figure 2.17.

For the series DC motor drive, the regenerative braking can be performed based on the same principle as that for the separately excited DC motor drive, but the field needs to be reversed with respect to the armature as compared with the connection for motoring. That is, by connecting the field through a bridge rectifier, the field current can be kept in the same direction under both motoring and regenerative braking, while the back EMF can be maintained to be positive. The corresponding torque-speed characteristics are shown in Figure 2.18.

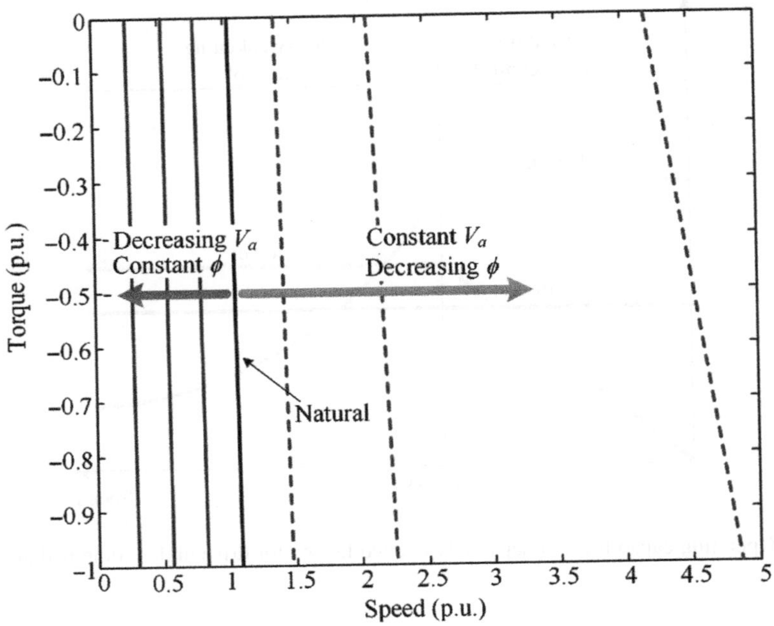

Figure 2.17 Regenerative braking of separately excited DC motor drive

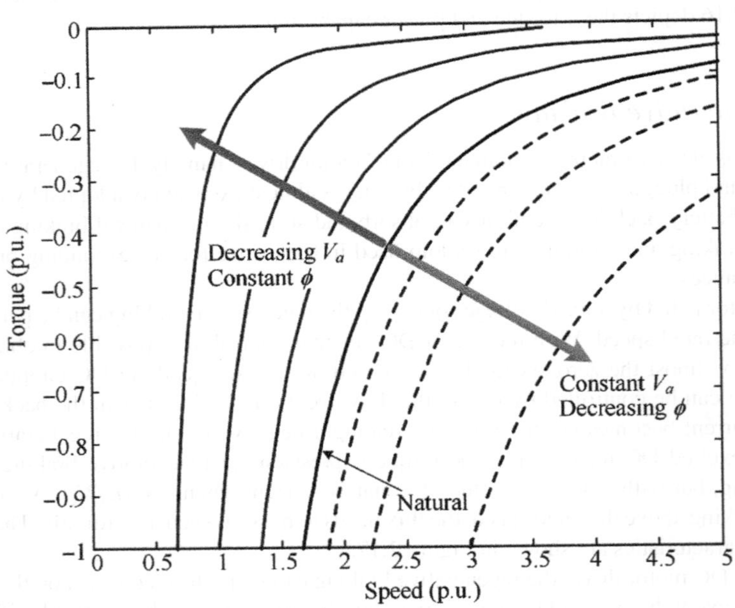

Figure 2.18 Regenerative braking of series DC motor drive

2.5 Design Criteria of DC Motor Drives for EVs

As the DC motor was the first electric machine and has been developed for over a century, the corresponding design criteria and procedure are actually very mature. Nevertheless, the use of modern design approaches, such as the finite element method (FEM), can help optimize the design and hence the performance. The modern design process of DC machines involves an iterative loop. Initially, the machine geometry and dimensions are determined by using analytical equations. Afterward, the machine performances, including the electromagnetic field distribution and operating characteristics, are determined by using electromagnetic FEM analysis. Consequently, the machine dimensions are iteratively tuned to optimize the machine performances for the desired EV propulsion.

As the DC motor technology is very mature, there are many well-written reference books detailing the design criteria and procedure (Say and Taylor, 1986; Chen, 1982).

Firstly, the physical size of the DC motor needs to be estimated. The key component to govern the overall size is the armature, which can be directly related to the converted power P by

$$P = \pi^2 D^2 LABn \tag{2.12}$$

where D is the armature or rotor diameter, L is the core length, A is the electric loading, B is the air-gap flux density or called magnetic loading, and n is the rated speed. Both A and B are experiential parameters: the former is governed by the electric losses in the windings and the latter is governed by the magnetic losses in the teeth and yoke. Thus, the selections of A and B are significantly influenced by the cooling method to dissipate the electric and magnetic losses. Meanwhile, the higher the rated speed is selected, the smaller the machine size is resulted. But, the high-speed rating will cause higher magnetic losses, thus affecting the selection of B. Therefore, based on the desired P and n as well as the selected A and B, D^2L can be deduced. Once D is chosen, L can be determined. Then, the air-gap length δ is selected according to an empirical relationship as given by

$$\delta = 0.008\,D \tag{2.13}$$

Then, the pole arc b_p and pole height h_p of the stator can be estimated by

$$b_p = \alpha_p \tau - 2\delta \tag{2.14}$$

$$h_p = (0.7 \sim 0.9)\,\alpha_p \tau \tag{2.15}$$

where α_p is the pole-arc coefficient in the range of 0.63–0.68 and τ is the pole pitch. Then, the cross-sectional area of the stator yoke S_y can be expressed as follows:

$$S_y = \frac{1}{2}\frac{\sigma\phi}{B_y} \tag{2.16}$$

where ϕ is the rated flux per pole, σ is the flux leakage coefficient (typically in the range of 1.15–1.2), and B_y is the desired stator yoke flux density (typically in the range of 0.95–1.5 T). Meanwhile, ϕ can be calculated by

$$\phi = \alpha_p B \tau L \tag{2.17}$$

The stator yoke thickness h_y can be deduced by

$$h_y = S_y / L_y \tag{2.18}$$

where L_y is the length of the stator yoke, which is equal to or slightly longer than L. Consequently, the stator diameter D_s can be calculated by

$$D_s = D + 2\delta + 2h_p + 2h_y \tag{2.19}$$

Of course, the stator diameter and core length are physically constrained by the available space in the EV.

Secondly, the machine structure, including the field circuit and armature circuit, needs to be devised. Differing from the AC machines, there is no rigid rule on the selection of the number of poles in the field circuit of the DC machine. In general, when more poles are adopted in the field circuit, both iron and copper materials can be reduced; however, the excitation complexity and iron loss will be increased. As a rule of thumb, the number of poles can be estimated simply based on the armature diameter D: for instance, two poles for $D \leq 120$ mm, four poles for 120 mm $< D \leq$ 600 mm, and six poles for 600 mm $<$ $D \leq$ 1200 mm. Concerning the armature circuit, the number of slots and type of winding are the two main factors. In general, the more the number of armature slots, the better the performance is resulted, but accompanied with higher manufacturing cost and complexity. As a rule of thumb, the number of slots can be also estimated based on the armature diameter D: for instance, the maximum number of slots per pole is 12 for $D \leq$ 300 mm, the maximum number is 14 for 300 mm $< D \leq$ 450 mm, and the maximum number is 20 for 450 mm $< D \leq$ 650 mm. Meanwhile, there are two main types of armature winding: lap winding and wave winding. The main difference is that lap winding is preferable for high-current application, whereas wave winding is preferable for high-voltage application. The lap and wave windings are further divided into simplex and duplex types. In addition, the armature can adopt either a single-layer winding or a double-layer winding. The double-layer winding takes the definite advantage of greater flexibility than the single-layer winding as the coil span can be easily selected. Consequently, the required number of armature conductors Z_a can be estimated by

$$Z_a = \frac{2a\pi DA}{I_a} \tag{2.20}$$

where $2a$ is the number of parallel paths and I_a is the armature current. The number of parallel paths depends on the type of armature winding: $a = p$ for the lap simplex, $a = mp$ for the lap duplex, $a = 1$ for the wave simplex, and $a = m$ for wave duplex, where p is the number of pole-pairs and m is the multiplicity of the winding.

Thirdly, two unique components of the DC motor, namely the carbon brushes and commutator, need to be designed. When adopting the lap winding, the number of brushes is equal to the number of poles for the simplex type or twice the number of poles for the duplex type. Meanwhile, when adopting the wave winding, it requires only two brushes or can use as many brushes as poles to improve commutation. The cross-sectional area of the brushes should ensure that the current density lies between 10 and 15 A/cm^2. The commutator diameter D_c is typically equal to (0.6–0.8)D, which can also be estimated by

$$D_c \leq D - 2h_s \tag{2.21}$$

where h_s is the height of armature slot. Hence, the tangential velocity of the commutator is given by

$$v_c = \frac{\pi D_c n}{60} \tag{2.22}$$

which should be less than 35 m/s. The number of commutator segments N_c is equal to the number of active armature coils. Hence, the commutator segment pitch is given by

$$\tau_c = \frac{\pi D_c}{N_c} \tag{2.23}$$

which should be more than 3 mm for $D_c <$ 200 mm and more than 4.5 mm for $D_c >$ 200 mm in order to ensure sufficient mechanical strength.

2.6 Design Example for EVs

As discussed earlier, the use of DC motor drives for EV propulsion is limited to those low-cost mini EVs. For a typical mini EV, the propulsion power is in the range of 3–7 kW and the battery voltage is in the

Table 2.1 Specifications of DC motor drive

DC voltage	72 V
Rated power	5 kW
Rated speed	2000 rpm
Rated torque	23.9 Nm
Constant-torque operation	0–2000 rpm
Constant-power operation	2000–4000 rpm
Gear ratio	5 : 1

Table 2.2 Initialization of separately excited DC motor drive

Stator diameter	267 mm
Rotor diameter	138 mm
Commutator diameter	108 mm
Air-gap length	1 mm
Core length	100 mm
Stator pole arc	67 mm
Stator pole height	50 mm
Stator yoke thickness	27 mm
Number of poles	4
Number of armature slots	14
Number of carbon brushes	4
Number of commutator segments	14
Number of armature conductors	84
Lamination material	50A230

range of 48–96 V. The DC motor drive preferably adopts the high-speed machine design to minimize the overall size, even taking into account the reduction gear to match the motor speed with the wheel speed. Hence, the specifications of the desired DC motor drive are listed in Table 2.1.

In order to provide the desired constant-power operation with flux-weakening control, the separately excited DC motor is selected. On the basis of the specifications, the geometric dimensions and parameters of this DC motor can be initialized by using the aforementioned design equations, and are listed in Table 2.2.

By using the electromagnetic FEM analysis, the magnetic field distribution of the machine under the rated condition is calculated as shown in Figure 2.19. It can be observed that the magnetic saturation is not serious, and most regions of the iron core are with the magnetic flux density less than 1.5 T. The corresponding torque waveform is simulated as shown in Figure 2.20, which confirms that it can offer the desired rated torque of 24 Nm in average. Finally, the torque-speed capability of the motor drive covering both the constant-torque operation and constant-power operation is simulated as plotted in Figure 2.21. It verifies that the motor drive can achieve the desired constant-torque operation from 0 to 2000 rpm and constant-power operation from 2000 to 4000 rpm with field weakening from 5000 A-turn to 2330 A-turn.

2.7 Application Examples of DC Motor Drives in EVs

Almost all types of DC motor drives, namely based on the separately excited DC, series DC, shunt DC, and PM DC machines, have ever been applied to EVs. Among them, the separately excited DC and series DC motor drives were widely adopted for EV propulsion.

Figure 2.19 Magnetic field distribution of DC motor

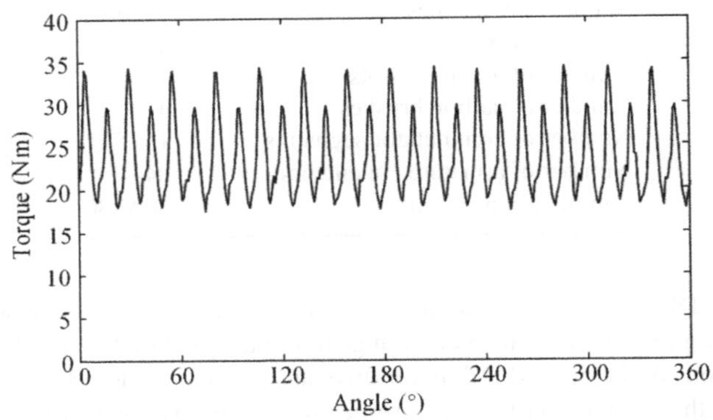

Figure 2.20 Torque waveform of DC motor

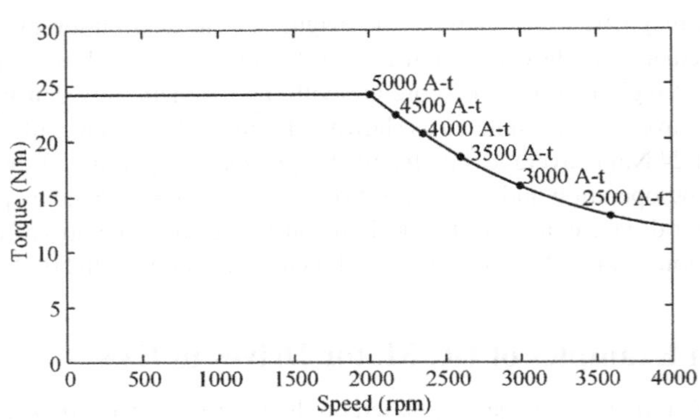

Figure 2.21 Torque-speed capability of DC motor drive

The GMC Conceptor Electric G-Van is a representative EV that adopted the separately excited DC motor drive for electric propulsion. It was the first certified EV complying with Federal Motor Vehicle Safety Standards for line production in North America commenced in 1990. This Electric G-Van adopted the Nelco N200ML012 separately excited DC motor, which could provide the rated power of 45 kW and torque of 323 Nm (Chancey, 2010). The transmission was fixed gearing, forward and reverse, having a ratio of 1.85 : 1. The vehicle top speed was 88 km/h. Installed with the SAFT nickel-cadmium battery pack of about 40 kWh, it could run 96 km at the weight of 3538 kg under the J227a-C driving cycle. The battery pack was charged by an off-board Chloride charger, which took about 8 hours for full charge.

The Fiat Panda Elettra is a representative EV that used the series DC motor drive for electric propulsion. It was introduced in 1990 and was one of the first mass-produced EV models. It adopted the Thrige-Titan TTL180C series DC motor, which could provide the rated power of 9.2 kW at 2500 rpm (Chancey, 2014). It was equipped with a four-speed transmission, which could offer a top speed of 70 km/h. In order to accommodate the lead-acid battery pack of about 9 kWh, it only had two seats and the range per charge was only about 77 km. The battery pack was 350 kg with a charging time of 8 hours.

An important milestone in the use of DC motor drives for EVs was the lunar roving vehicle (LRV) as shown in Figure 2.22, which was designed to operate at the low-gravity vacuum environment of the Moon for the Apollo lunar surface missions (Williams, 2005). The LRV had a mass of 210 kg with a payload of 490 kg. The wheels were separately driven by four series DC motors, which were operated by a nominal input voltage of 36 V to output the rated power of 745 W and speed up to 10 000 rpm via individual 80 : 1 harmonic drives. The corresponding speed control was handled by using the PWM DC chopper. Installed with two 36-V silver-zinc potassium hydroxide nonrechargeable batteries (each with a capacity of 121 Ah), it could offer a driving range of 92 km. By using front and rear steering motors, which were also a series DC motor rated at 75 W each, the LRV could offer excellent maneuverability. The overall cost of four LRVs was astronomically high – US$ 38 million. It was first used for the Apollo 15 mission in 1971.

With the use of power electronics, the separately excited DC motor can perform flexible armature voltage control and flux-weakening control, hence achieving wide-range speed control, including both the constant-torque operation and constant-power operation. Thus, the separately excited DC motor drive is dominant over other DC counterparts for modern mini EVs, especially those neighborhood electric

Figure 2.22 Lunar roving vehicle using series DC motor drive (*Source:* Courtesy Wikimedia Commons, http://en.wikipedia.org/wiki/File:Apollo15LunarRover2.jpg)

vehicles (NEVs) which are a denomination of EVs in the United States that can legally run on public roads with a speed limit of 56 km/h and a loaded weight limit of 1400 kg (Abuelsamid, 2009). The Global Electric Motorcars (GEM) car, which is a major model of NEVs, typically installs a 72-V battery system to feed the separately excited DC motor drive, hence offering a rated power of 5.6 kW at 4950 rpm and a driving range of 48 km per charge.

2.8 Fading Technology for EVs?

There is no doubt that the DC motor drive technology is very mature and well developed. The use of DC motor drives for EV propulsion takes the definite advantages of technological maturity and control simplicity. However, with the ever-increasing demand on high efficiency, high power density, and maintenance-free operation, DC motor drives are no longer attractive for EV propulsion. Actually, the corresponding application to EVs is diminishing and fading.

Currently, some DC motor drives, such as the separately excited DC motor drive, are still adopted by modern mini EVs. Nevertheless, with the continual improvement of AC motor drive technology, it is anticipated that the DC motor drives will be totally superseded by AC or brushless DC motor drives for EV propulsion in the near future.

References

Abuelsamid, S. (2009) What is a Neighborhood Electric Vehicle (NEV)? AutoblogGreen, http://green.autoblog.com/2009/02/06/greenlings-what-is-a-neighborhood-electric-vehicle-nev/ (accessed September 2014).

Chan, C.C. and Chau, K.T. (1993) A new zero-voltage-switching dc/dc boost converter. *IEEE Transactions on Aerospace and Electronic Systems*, **29**, 125–134.

Chancey, M. (2010) 1989 GMC Electric G-Van, EV Album, http://www.evalbum.com/1413 (accessed September 2014).

Chancey, M. (2014) 1990 Fiat Panda Elettra, EV Album, http://www.evalbum.com/3141 (accessed September 2014).

Chau, K.T. (1994) A new class of pulsewidth-modulated multi-resonant converters using resonant inductor freewheeling. *International Journal of Electronics*, **77**, 703–714.

Chau, K.T., Ching, T.W. and Chan, C.C. (1999) A new two-quadrant zero-voltage transition converter for DC motor drives. *International Journal of Electronics*, **86**, 217–231.

Chen, S. (1982) *Electric Machine Design (Chinese)*, China Machine Press, Beijing.

Ching, T.W. (2005) Four-quadrant zero-voltage-transition converter-fed DC motor drives for electric propulsion. *Journal of Asian Electric Vehicles*, **3**, 651–656.

Ching, T.W. (2006) Four-quadrant zero-current-transition converter-fed DC motor drives for electric propulsion. *Journal of Asian Electric Vehicles*, **4**, 911–917.

Ching, T.W. and Chan, K.U. (2008) Review of soft-switching techniques for high-frequency switched-mode power converters. Proceedings of IEEE Vehicle Power and Propulsion Conference, pp. 1–6.

Ching, T.W. and Chau, K.T. (2001) A new two-quadrant zero-current transition converter for DC motor drives. *International Journal of Electronics*, **88**, 719–735.

Dubey, G.K. (1989) *Power Semiconductor Controlled Drives*, Prentice-Hall, Englewood Cliffs, NJ.

Ehsani, M., Gao, Y. and Emadi, A. (2009) *Modern Electric, Hybrid Electric, and Fuel Cell Vehicles: Fundamentals, Theory, and Design*, 2nd edn, CRC Press, Boca Raton, FL.

Hua, G. and Lee, F.C. (1995) Soft-switching techniques in PWM converters. *IEEE Transactions on Industrial Electronics*, **42**, 595–603.

Say, M.G. and Taylor, E.O. (1986) *Direct Current Machines*, Pitman, London.

Williams, D.R. (2005) The Apollo Lunar Roving Vehicle, NASA, http://nssdc.gsfc.nasa.gov/planetary/lunar/apollo_lrv.html (accessed September 2014).

3

Induction Motor Drives

As mentioned in the previous chapters, commutatorless motor drives offer a number of advantages over conventional DC motor drives for electric propulsion. At present, induction motor drives are the most mature technology among various commutatorless motor drives. There are two types of induction machines (IMs): the wound-rotor and squirrel-cage. Because of high cost, need for maintenance, and lack of sturdiness, the wound-rotor induction motor is less attractive than the squirrel-cage counterpart, especially for electric propulsion in electric vehicles (EVs). Hence, the squirrel-cage induction motor is loosely named as the induction motor for EV propulsion. In addition to the common advantages of commutatorless motor drives, the induction motor drives possess the definite advantages of low cost and ruggedness. These advantages can generally outweigh their major disadvantage of control complexity and facilitate this technology being widely accepted for EVs.

In this chapter, various induction motor drives, including their system configurations, IMs, power inverters, and control strategies, are discussed. The corresponding design criteria, design example, and application examples of EV propulsion are also discussed.

3.1 System Configurations

The basic configuration of an induction motor drive is shown in Figure 3.1. It consists of a three-phase squirrel-cage induction motor, a three-phase voltage-fed pulse-width modulated (PWM) inverter, an electronic controller, and some sensors.

For EV propulsion, the system configuration can be single-motor or multiple-motor. The single-motor configuration of the induction motor drive has been widely adopted for commercial EVs. As shown in Figure 3.2, the single-motor configuration uses only one IM and one PWM inverter, which can minimize the corresponding size, weight, and cost. However, it needs a differential to adjust the relative speeds of the driving wheels for cornering. In addition, it generally employs a fixed gear (FG) to reduce the motor speed to match with the wheel speed. It should be noted that the high-speed design of IMs is widely adopted for EV propulsion because this design favors the reduction of machine size and weight, which are crucial factors for EVs.

On the other hand, the multiple-motor configuration uses multiple motors to independently drive individual wheels. As shown in Figure 3.3, the dual-motor configuration of the induction motor drive consists of two IMs, two PWM inverters, and two optional FGs depending on whether using direct-drive or not. Since the two IMs are independently controlled, the differential action can be electronically achieved, thus eliminating the bulky and heavy differential. Whether the IM adopting the high-speed design or low-speed design depends on the trade-off on using the FG. The major concern in adopting the multiple-motor

Electric Vehicle Machines and Drives: Design, Analysis and Application, First Edition. K. T. Chau.

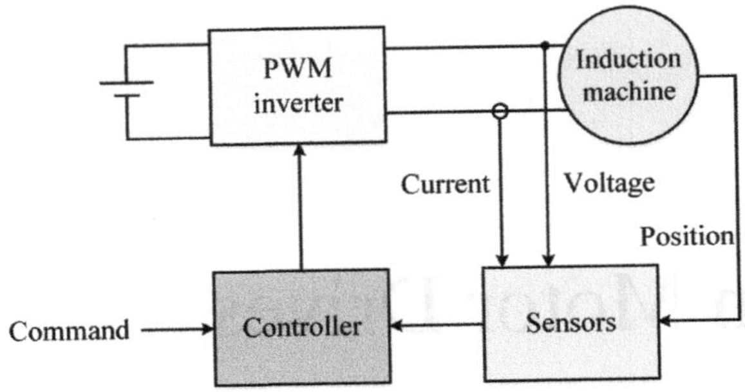

Figure 3.1 Basic configuration of induction motor drive

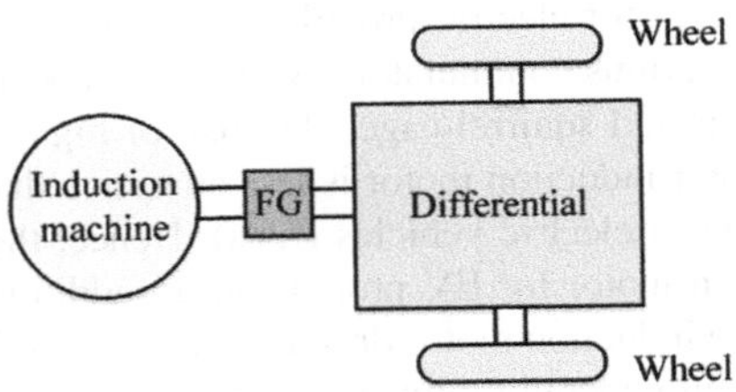

Figure 3.2 Single-motor configuration of induction motor drive for EVs

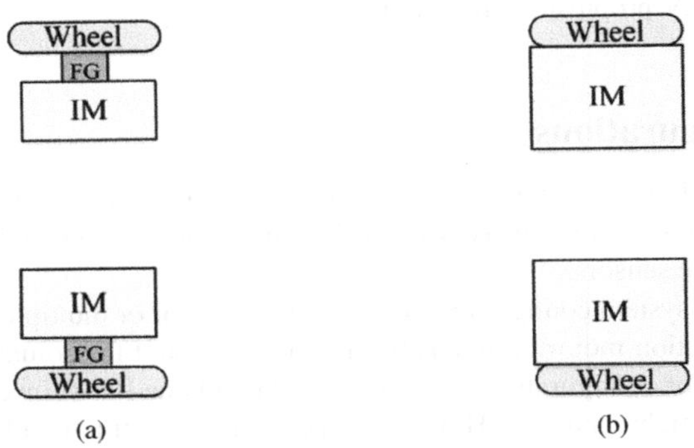

Figure 3.3 Dual-motor configurations of induction motor drive for EVs: (a) high-speed geared structure and (b) low-speed gearless structure

configuration for commercial EVs is the safety issue. If the speed commands for the left and right IMs are wrongly calculated or accidentally corrupted, the cornering action will become disastrous. Additional control monitoring and fault tolerance are inevitable.

3.2 Induction Machines

Among the two types of IMs, the squirrel-cage IM has been exclusively used for electric propulsion. Thus, all future discussions will be based on the squirrel-cage IM, and the wound-rotor IM will not be discussed.

3.2.1 *Structure of Induction Machines*

The most common type of IMs is the squirrel-cage IM. As depicted in Figure 3.4, it consists of a stator incorporated with three-phase armature windings, a rotor incorporated with cage bars that are short-circuited by two end-rings, which can also be molded with fan blades to help cool the machine during rotation, two end-bearings to support the rotor, and a machine frame with two end-bells to house the machine.

Figure 3.5 shows the cross section of a basic three-phase two-pole IM in which the stator is installed with three-phase windings A-X, B-Y, and C-Z, and the rotor is inserted with squirrel-cage bars (loosely called cage bars). The stator windings are distributed in such a way that the winding current produces a sinusoidally distributed air-gap flux density around the periphery. All cage bars are short-circuited together by the front and rear end-rings so that the rotor circuit can be induced to produce the same number of poles as the stator.

Figure 3.6 shows the cross section of an uncommon three-phase 16-pole outer-rotor IM. The principle of operation is the same as the conventional inner-rotor counterpart. This outer-rotor topology preferably adopts the pancake-shaped design, namely a large radial diameter and a short axial length, so that it can readily accommodate a large number of poles, hence offering direct in-wheel drive. Also, the pancake shaft facilitates to implant the machine into the wheel.

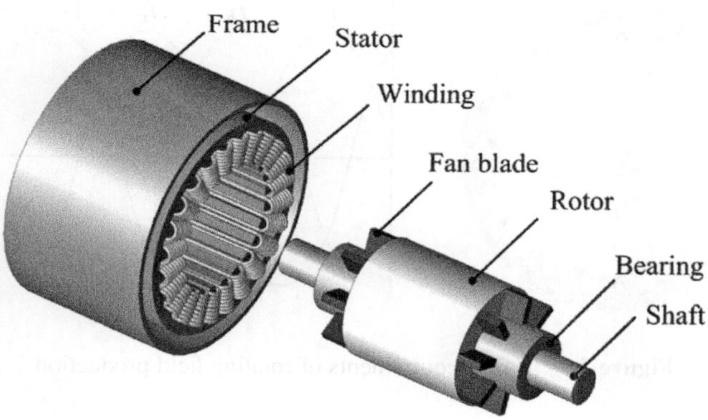

Figure 3.4 Induction machine exploded diagram

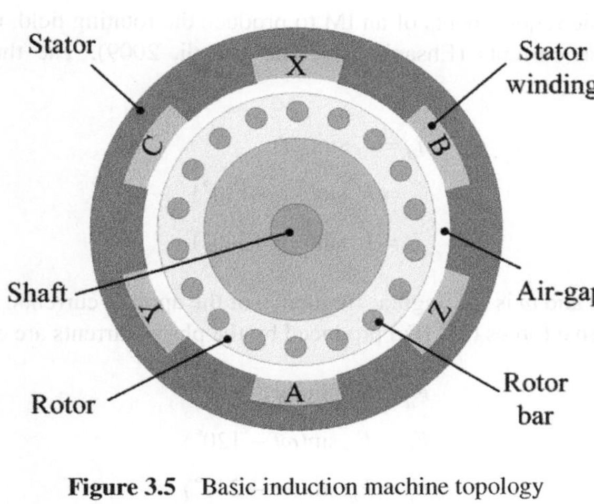

Figure 3.5 Basic induction machine topology

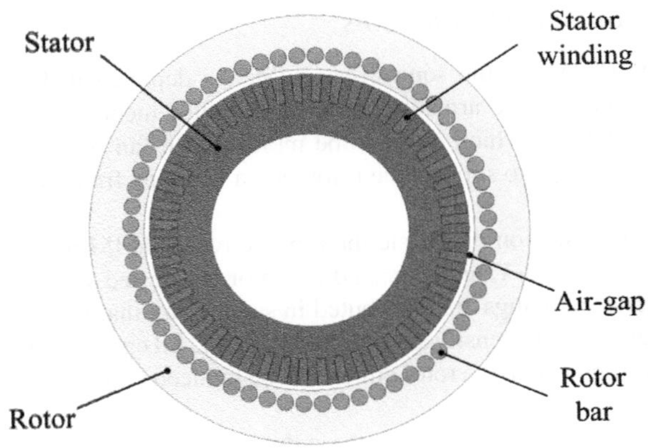

Figure 3.6 Outer-rotor induction machine topology

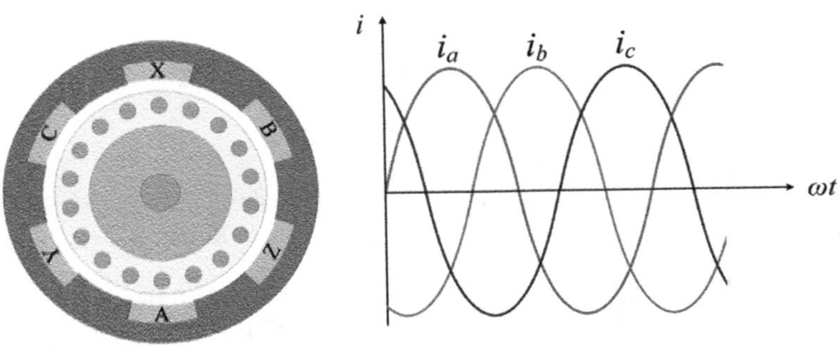

Figure 3.7 Basic requirements of rotating field production

3.2.2 Principle of Induction Machines

Figure 3.7 shows two basic requirements of an IM to produce the rotating field, namely the three-phase windings and three-phase currents (Ehsani, Gao, and Emadi, 2009). The three-phase currents are given by

$$i_a = I_m \sin(\omega t) \tag{3.1}$$

$$i_b = I_m \sin(\omega t - 120^\circ) \tag{3.2}$$

$$i_c = I_m \sin(\omega t - 240^\circ) \tag{3.3}$$

where I_m is the amplitude and ω is the angular frequency of the applied currents.

The stator magnetomotive forces (MMFs) produced by the phase currents are expressed as

$$F_a = F_m \sin(\omega t) \tag{3.4}$$

$$F_b = F_m \sin(\omega t - 120^\circ) \tag{3.5}$$

$$F_c = F_m \sin(\omega t - 240^\circ) \tag{3.6}$$

where $F_m = I_m N$ and N is the number of turns per phase. Hence, the resultant stator MMF vector can be expressed as

$$F = F_a e^{j0°} + F_b e^{j120°} + F_c e^{j240°} \tag{3.7}$$

It can be rewritten as

$$F = \frac{3}{2} F_m e^{j(\omega t - 90°)} \tag{3.8}$$

which is an MMF vector rotating circumferentially with the angular velocity ω, and is 90° delay with respect to the MMF of phase A. Figure 3.8 graphically depicts the stator MMF vectors at $\omega t = 0°$ and $\omega t = 90°$, confirming that the resultant MMF vector rotates by 90°. It is equivalent to the magnetic poles of the stator rotating circumferentially with the angular speed equal to the angular frequency of the phase current. In case the machine has more than two poles, the rotating field speed differs from the angular frequency, which can be expressed as

$$\omega_s = \frac{\omega}{p} \quad \text{or} \quad n_s = \frac{f}{p} \tag{3.9}$$

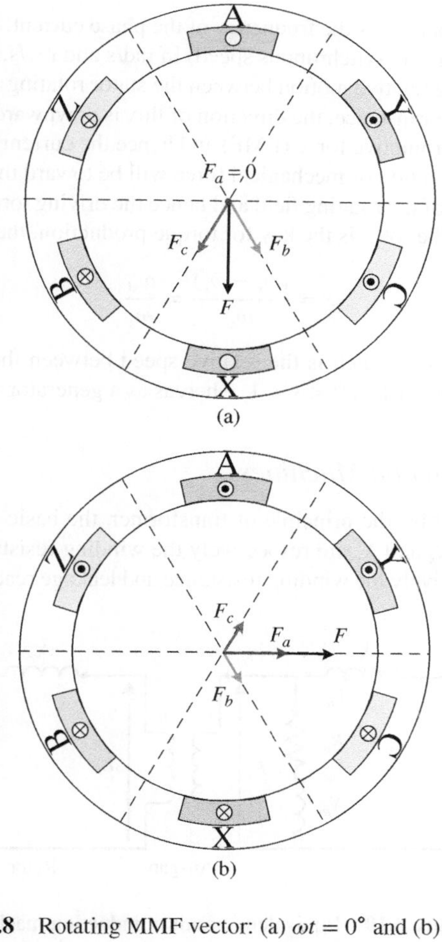

Figure 3.8 Rotating MMF vector: (a) $\omega t = 0°$ and (b) $\omega t = 90°$

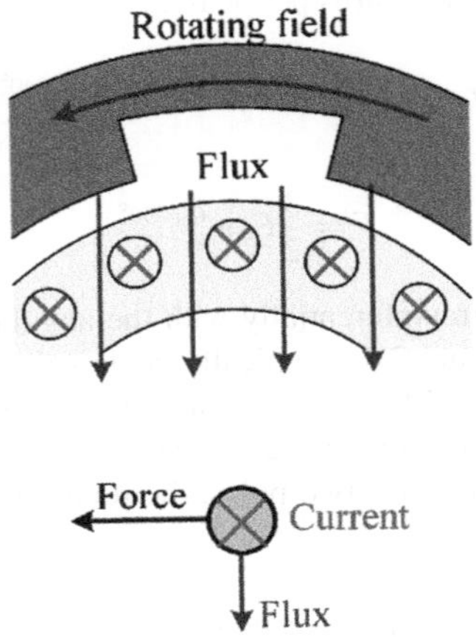

Figure 3.9 Principle of torque production

where p is the number of pole-pairs, f is the frequency of the phase current, and ω_s and n_s are the rotating field velocity (usually dubbed as the synchronous speed) in rad/s and rev/s, respectively.

As depicted in Figure 3.9, the relative motion between the stator rotating field and the rotor conductors induces a voltage in the rotor. For instance, the direction of flux is downward and the field is moving from right to left; the generated electromotive force (EMF) and hence the current in the conductors will be into the paper. It follows that the direction of mechanical force will be toward the left, causing the conductors to follow the moving direction of the rotating field and hence the driving torque. Since the relative motion between the rotating field and the rotor is the key for torque production, the slip s is defined as

$$s = \frac{(\omega_s - \omega_r)}{\omega_s} = \frac{\omega_{\rm sl}}{\omega_s} \tag{3.10}$$

where $\omega_{\rm sl}$ is called the slip speed, which is the relative speed between the stator rotating field and the rotor. The IM operates as a motor when $0 < s < 1$, whereas as a generator when $s < 0$.

3.2.3 *Modeling of Induction Machines*

As the principle of IMs resembles the principle of transformer, the basic equivalent circuit of an IM is shown in Figure 3.10, where R_s and X_s are respectively the winding resistance and leakage reactance in the stator, R_r and X_r are respectively the winding resistance and leakage reactance in the rotor at standstill,

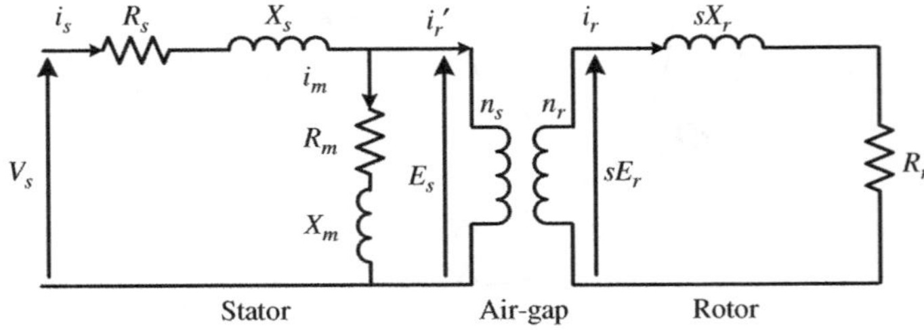

Figure 3.10 Equivalent circuit of induction machine

E_s and E_r are respectively the induced EMF in the stator and induced EMF in the rotor at standstill, R_m and X_m are respectively the core-loss resistance and magnetizing reactance in the stator, and k is the ratio of the number of turns of stator winding to the number of turns of rotor winding. Differing from that, in a transformer, the stator frequency f is not equal to the rotor frequency f_r due to the presence of relative motion. Therefore, the rotor frequency depends on the slip as given by $f_r = sf$. Consequently, the rotor EMF and rotor reactance are all slip-dependent and are expressed as sE_r and sX_r, respectively.

In order to simplify the equivalent circuit of the IM, all rotor quantities are divided by s and then referred to the stator side. The simplified equivalent circuit is shown in Figure 3.11, where the referred rotor quantities are given by $R'_r = k^2R_r$, $X'_r = k^2X_r$, $E'_r = k\,E_r$, and $I'_r = I_r/k$. The stator, magnetizing branch, and referred rotor impedances can be expressed as

$$Z_s = R_s + jX_s \tag{3.11}$$

$$Z_m = R_m + jX_m \tag{3.12}$$

$$Z'_r = \frac{R'_r}{s} + jX'_r \tag{3.13}$$

Hence, the stator current and referred rotor current can be calculated as

$$I_s = \frac{V_s}{(Z_s + Z_m \parallel Z'_r)} \tag{3.14}$$

$$I'_r = \frac{Z_m}{(Z_m + Z'_r)} I_s \tag{3.15}$$

The electromagnetic air-gap power transferred from the stator to the rotor is given by

$$P_g = m\,I'^2_r \frac{R'_r}{s} \tag{3.16}$$

where m is the phase number. After subtracting the rotor copper loss $mI'^2_rR'_r$, the gross mechanical output power can be written as

$$P_m = m\,I'^2_r \frac{R'_r}{s}(1 - s) \tag{3.17}$$

Hence, the output torque can be obtained as

$$T = \frac{P_m}{\omega_r} \tag{3.18}$$

Since $\omega_r = (1 - s)\omega_s$, the output torque can be rewritten as

$$T = \frac{P_g}{\omega_s} \tag{3.19}$$

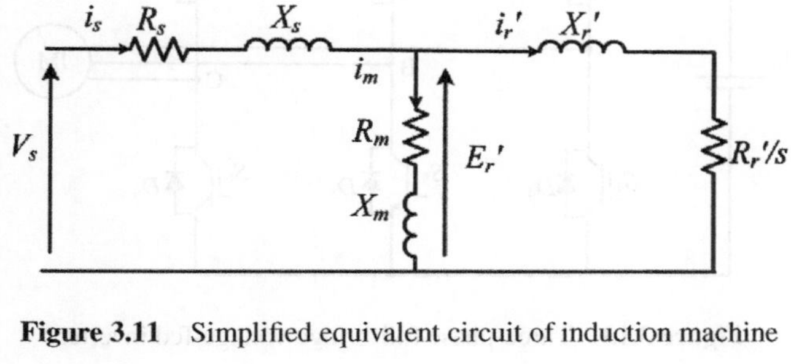

Figure 3.11 Simplified equivalent circuit of induction machine

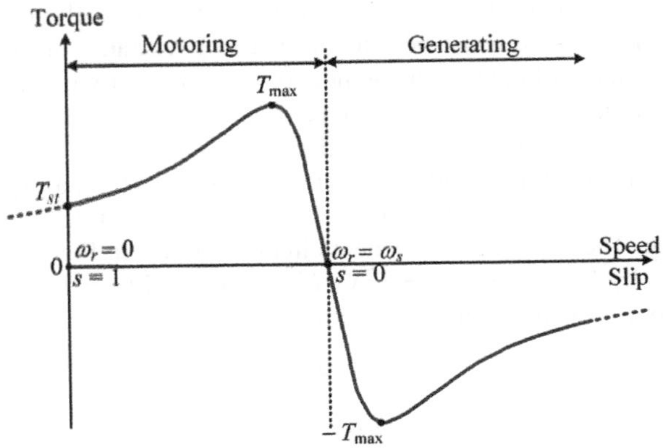

Figure 3.12 Torque-speed characteristic of induction machine

When neglecting the magnetizing branch, the torque equation can be simplified as

$$T = \frac{1}{\omega_s} \frac{mV_s^2}{(R_s + R'_r/s)^2 + (X_s + X'_r)^2} \frac{R'_r}{s} \tag{3.20}$$

Figure 3.12 shows the torque-speed characteristic of the IM under fixed voltage and frequency, in which T_{st} is the starting torque and T_{max} is the maximum torque. It is obvious that this torque-speed characteristic is not good enough for EV propulsion due to the low starting torque and limited speed range. Therefore, the incorporation of power electronics into the induction motor drive is almost mandatory for EV application.

3.3 Inverters for Induction Motors

Inverters for induction motors are generally classified into voltage-fed and current-fed types. Because of the need of a large series inductance to emulate a current source, the current-fed inverters are seldom used for EV propulsion. In fact, the voltage-fed inverters are almost exclusively used because they are very simple and can have power flow in either direction. A typical three-phase full-bridge voltage-fed inverter is shown in Figure 3.13. Its output waveform may be six-step or PWM, depending on the switching strategy for different applications. The six-step inverter takes the advantages of simplicity and low switching loss. However, it is becoming obsolete because its amplitude cannot be directly controlled

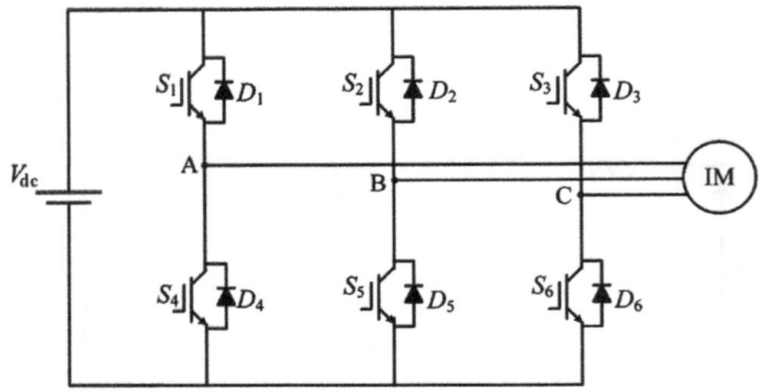

Figure 3.13 Three-phase full-bridge voltage-fed inverter

and its lower order harmonics are rich. On the other hand, the PWM inverter is widely adopted because its waveform is harmonically optimal, and its fundamental magnitude and frequency can be smoothly varied for speed control.

The inverter design highly depends on the technology of power devices. At present, the insulated-gate bipolar transistor (IGBT)-based inverter is most attractive and has been widely accepted by modern EVs. The selection of power devices is based on the criteria that the voltage rating is at least twice the nominal battery voltage because of the voltage surge during switching, the current rating is large enough so that there is no need to connect several power devices in parallel, and the switching speed is sufficiently high to suppress motor harmonics and acoustic noise. Moreover, the power device module is usually a two-in-one or six-in-one type, which aims at minimizing the wiring and stray impedance.

3.3.1 PWM Switching Inverters

For the last two decades, numerous PWM switching schemes have been developed for voltage-fed inverters based on the criteria that the magnitude and frequency of the fundamental component of the output waveform can be smoothly varied, the harmonic distortion of the output waveform is minimum, the switching algorithm can be real-time implemented with minimum hardware and compact software, and the fluctuation of battery voltage can be tolerated. These PWM switching schemes can be voltage-controlled or current-controlled. The use of current control is preferable for high-performance induction motor drives because the motor torque and flux are directly related to the controlled current.

The state-of-the-art PWM switching schemes for the voltage-fed inverters are as follows:

- Sinusoidal PWM
- Regular PWM
- Optimal PWM
- Delta PWM
- Random PWM
- Chaotic PWM
- Hysteresis-current PWM
- Space-vector PWM

Among them, the hysteresis-current PWM and space-vector PWM schemes have been widely adopted by induction motor drives for EV propulsion.

Figure 3.14 shows the principle of hysteresis-current PWM (Bose, 1992). The actual current is online measured and then directly compared with the reference current in such a way that when the actual current exceeds the upper hysteresis band of the reference current, the upper switch is turned off and the current starts to decay; and when the actual current crosses the lower hysteresis band, the upper switch is turned on and the current starts to rise. Thus, the actual current is forced to track the reference current within the hysteresis band, irrespective of the DC voltage fluctuation. This scheme takes the definite advantages of direct current control and fast response. However, the PWM frequency varies with the load so that the current harmonics are varying and not optimal. Also, the fundamental current suffers from phase-lag.

Figure 3.15 shows the principle of space-vector PWM (Bose, 2006), which is actually an indirect current control scheme based on the concept of voltage and current space vectors. It is evident from Figure 3.13 that the three-phase voltage-fed inverter has eight switching states and each state corresponds to a voltage space vector as summarized in Table 3.1. Graphically, the six active vectors $\overline{V}_1(100)$, $\overline{V}_2(110)$, $\overline{V}_3(010)$, $\overline{V}_4(011)$, $\overline{V}_5(001)$, and $\overline{V}_6(101)$ are 60° apart and describe the vertices of a hexagon with a magnitude of $2V_{dc}/3$, while the two zero vectors, $\overline{V}_0(000)$ and $\overline{V}_7(111)$, are at the origin.

The set of phase voltages for each switching state can be combined to derive a voltage space vector as given by

$$\overline{V} = \frac{2}{3}(v_a + v_b e^{j120^\circ} + v_c e^{j240^\circ}) \tag{3.21}$$

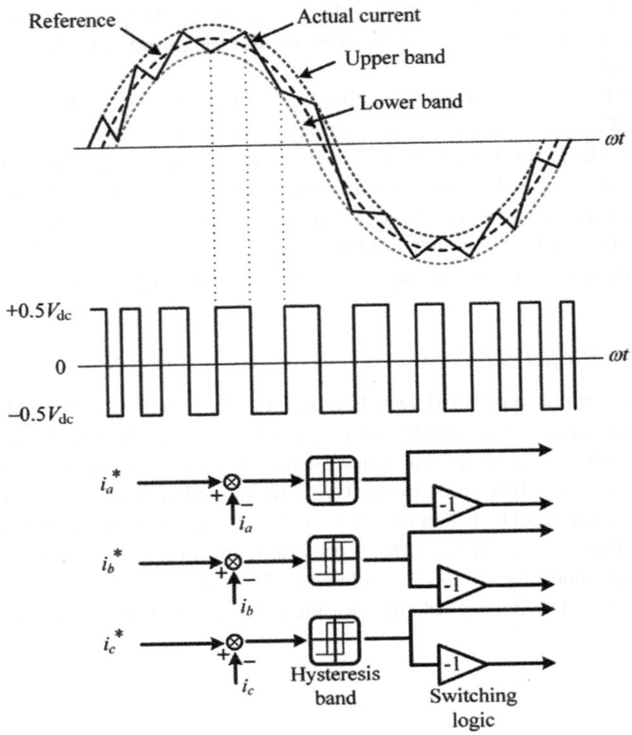

Figure 3.14 Principle of hysteresis-current PWM

Similarly, the current space vector can be expressed as

$$\overline{I} = \frac{2}{3}(i_a + i_b e^{j120°} + i_c e^{j240°}) \tag{3.22}$$

After sampling the phase currents at every period T, the current space vector is calculated. Then, the desired voltage space vector can be deduced by the current error vector. Since the inverter can only operate at one of the eight switching states, this voltage space vector is resolved into any two adjacent active vectors. For instance, the voltage space vector in Figure 3.15 can be resolved into two components V_a and V_b along the directions of $\overline{V}_1(100)$ and $\overline{V}_2(110)$, respectively:

$$V_a = V\cos\alpha - \frac{1}{2}V_b \tag{3.23}$$

$$V_b = \frac{2}{\sqrt{3}}V\sin\alpha \tag{3.24}$$

The durations of the corresponding active states $\overline{V}_1(100)$ and $\overline{V}_2(110)$ are given by

$$t_a = \frac{V_a}{(2V_{dc}/3)}T \tag{3.25}$$

$$t_b = \frac{V_b}{(2V_{dc}/3)}T \tag{3.26}$$

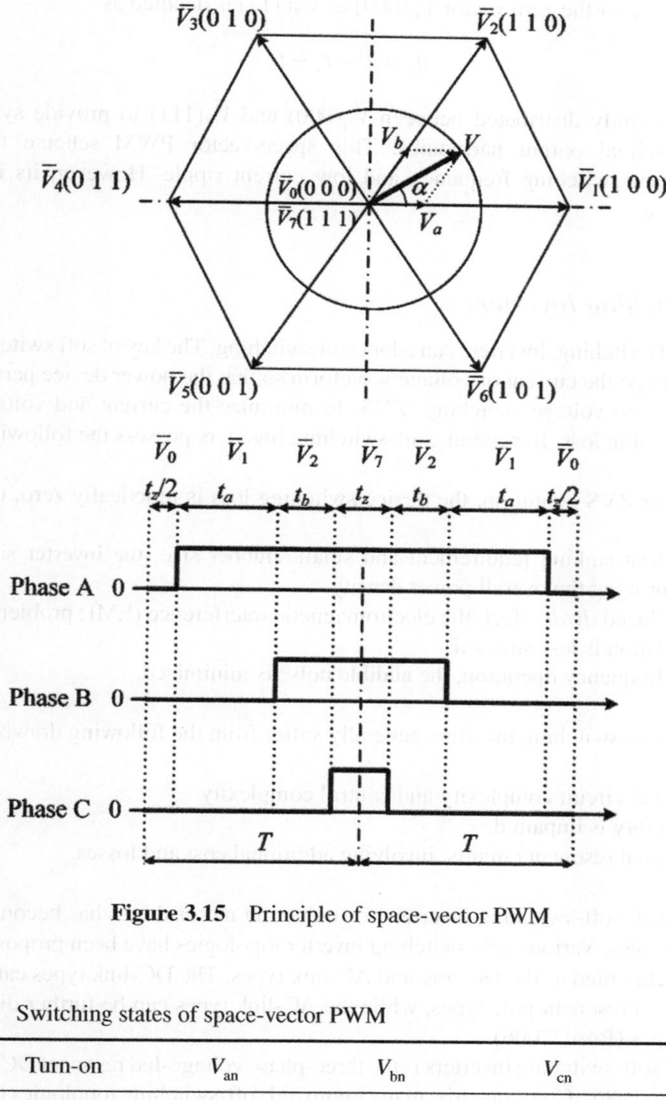

Figure 3.15 Principle of space-vector PWM

Table 3.1 Switching states of space-vector PWM

State	Turn-on	V_{an}	V_{bn}	V_{cn}	Vector
0	$S_4S_5S_6$	0	0	0	$\overline{V}_0(000)$
1	$S_1S_5S_6$	$2V_{dc}/3$	$-V_{dc}/3$	$-V_{dc}/3$	$\overline{V}_1(100)$
2	$S_1S_2S_6$	$V_{dc}/3$	$V_{dc}/3$	$-2V_{dc}/3$	$\overline{V}_2(110)$
3	$S_4S_2S_6$	$-V_{dc}/3$	$2V_{dc}/3$	$-V_{dc}/3$	$\overline{V}_3(010)$
4	$S_4S_2S_3$	$-2V_{dc}/3$	$V_{dc}/3$	$V_{dc}/3$	$\overline{V}_4(011)$
5	$S_4S_5S_3$	$-V_{dc}/3$	$-V_{dc}/3$	$2V_{dc}/3$	$\overline{V}_5(001)$
6	$S_1S_5S_3$	$V_{dc}/3$	$-2V_{dc}/3$	$V_{dc}/3$	$\overline{V}_6(101)$
7	$S_1S_2S_3$	0	0	0	$\overline{V}_7(111)$

The remaining duration for the zero vector $\overline{V}_0(000)$ or $\overline{V}_7(111)$ is denoted as

$$t_z = T - t_a - t_b \tag{3.27}$$

which can be conveniently distributed between $\overline{V}_0(000)$ and $\overline{V}_7(111)$ to provide symmetrical pulses, hence achieving minimal output harmonics. This space-vector PWM scheme takes the definite advantages of constant switching frequency and low current ripple. However, its implementation is computation-intensive.

3.3.2 *Soft-Switching Inverters*

Instead of using hard switching, inverters can adopt soft switching. The key of soft switching is to employ a resonant circuit to shape the current or voltage waveform so that the power device performs zero-current switching (ZCS) or zero-voltage switching (ZVS) to minimize the current and voltage overlap, hence minimizing the switching loss. In general, soft-switching inverters possess the following advantages:

- Because of ZCS or ZVS condition, the device switching loss is practically zero, thus improving the overall efficiency.
- Because of low heat-sinking requirement and small snubber size, the inverter size and weight are reduced, thus improving the overall power density.
- Because of the reduced *dv/dt* effect, the electromagnetic interference (EMI) problem is less severe and the machine insulation is less stressed.
- Because of high-frequency operation, the audible noise is minimized.

On the other hand, soft-switching inverters generally suffer from the following drawbacks:

- There are additional circuit complexity and control complexity.
- The system reliability is impaired.
- They need additional resonant circuits, involving additional cost and losses.

The development of soft-switching inverters for induction motor drives has become a research direction in power electronics. Various soft-switching inverter topologies have been proposed in the literature. Basically, they are classified as the DC-link and AC-link types. The DC-link types can be further divided into resonant link and resonant pole types, while the AC-link types can be further divided into resonant and nonresonant types (Bose, 2006).

The milestone of soft-switching inverters is the three-phase voltage-fed resonant DC-link inverter developed in 1986 (Divan, 1986). Consequently, many improved soft-switching topologies have been proposed, such as the quasi-resonant DC link, series resonant DC link, parallel resonant DC link, synchronized resonant DC link, resonant transition, auxiliary resonant commutated pole, and auxiliary resonant snubber (ARS) inverters. Meanwhile, a number of development goals of soft-switching inverters for EV propulsion have been identified, namely efficiency over 95%, power density over 3.5 W/cm^3, switching frequency over 10 kHz, *dv/dt* below 1000 V/μs, zero EMI, zero failure before the end of the vehicle life, and redundant with "limp-home" mode. The ARS inverter has been actively developed for EV propulsion (Lai, 1997), which has satisfied most of these goals, and has been demonstrated to achieve an output power of 100 kW.

As shown in Figure 3.16, by using auxiliary switches and resonant inductors along with resonant snubber capacitors to achieve the soft-switching condition, the ARS inverter can offer the merits that all main power devices can operate at the ZVS condition, while all auxiliary power switches can operate at the ZCS condition. Moreover, the corresponding parasitic inductance and stray capacitance are utilized as a

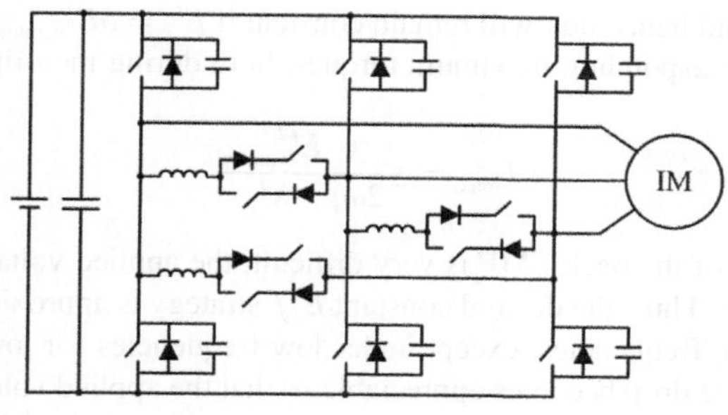

Figure 3.16 Three-phase auxiliary resonant snubber inverter

part of the resonant components, while there are less overvoltage or overcurrent penalty in the main power switches. Although this ARS inverter has promising application to EV propulsion, it still needs continual improvement before practically applying to EVs. Particularly, the corresponding control complexity should be alleviated, while the corresponding PWM switching scheme needs to be modified to enable effective control of induction motor drives.

As the complexity and reliability are the major concerns for practical application, the commercialized EVs have not yet adopted soft-switching inverters for electric propulsion.

3.4 Induction Motor Control

There are three main types of control strategies for induction motor drives: the variable-voltage variable-frequency (VVVF) control, field-oriented control (FOC), and direct torque control (DTC). Hereby, they are described in detail.

3.4.1 *Variable-Voltage Variable-Frequency Control*

VVVF control has been widely adopted for speed control of induction drives. It is based on constant volts/hertz control for frequencies below the rated frequency, and variable-frequency control with constant rated voltage for frequencies beyond the rated frequency. For very low frequencies, voltage boosting is applied to compensate the difference between the applied voltage and induced EMF (Dubey, 1989).

Referring to the equivalent circuit as shown in Figure 3.11 in which R_m is neglected, the magnetizing current I_m at the rated speed can be expressed as

$$I_m = \frac{E'_{r\_\text{rated}}}{X_m} = \frac{E'_{r\_\text{rated}}}{f_\text{rated}} \frac{1}{2\pi L_m} \tag{3.28}$$

where E'_{r_rated} is the back EMF at the rated frequency and L_m is the magnetizing inductance. Below the rated speed, the operating frequency f can be normalized as a frequency ratio a, which is defined as

$$a = f/f_\text{rated} \tag{3.29}$$

Hence, the corresponding magnetizing current can be expressed as

$$I_m = \frac{E'_r}{aX_m} = \frac{E'_r}{af_\text{rated}} \frac{1}{2\pi L_m} \tag{3.30}$$

The magnetizing current and hence flux will remain constant if $E'_r = aE'_{r\_rated}$ or simply called constant E/f. Consequently, the corresponding maximum torques, both during motoring and generating, remain constant as given by

$$T_{max} = \pm\frac{3}{2\omega_s}\frac{E'^2_{r\_rated}}{X'_r} \tag{3.31}$$

Because the measurement of the back EMF is very difficult, the applied voltage is generally adopted to approximate the back EMF. Thus, the desired constant E/f strategy is approximated by the constant V/f strategy for most operating frequencies, except under low frequencies for low-speed operation. At low speeds, the stator impedance drop becomes appreciable so that the applied voltage can no longer be valid to approximate the back EMF. Thus, a boosting voltage is normally required to compensate for the stator impedance drop for low-speed operation using the constant V/f strategy.

Above the rated speed, the applied voltage is kept at the rated voltage V_{rated}, while the operating frequency is increased above the rated frequency. Hence, the maximum torque decreases with the increase in frequency ($a > 1$) as given by

$$T_{max} = \frac{3}{2a\omega_s}\left[\frac{V^2_{rated}}{R_s \pm \sqrt{R_s^2 + a^2(X_s + X'_r)^2}}\right] \tag{3.32}$$

Since the slip is small, the rotor current is almost in phase with the back EMF. When neglecting the rotor resistance loss and stator impedance drop, the output power is equal to the product of rated voltage and rotor current. Thus, under the maximum permissible stator current, this variable-frequency control offers constant-power operation.

Beyond the critical speed, the motor is operated at the slip for the maximum torque. Both the current and power of the motor are allowed to decrease inversely with speed, and the torque decreases inversely as the speed squared.

In summary, the torque-speed characteristics of the induction motor drive using VVVF control are shown in Figure 3.17, and the corresponding torque-speed capabilities are shown in Figure 3.18. It can be observed that there are three operating regions. The first region is called the constant-torque region in which the motor can deliver its rated torque for speeds below the rated speed (normally called the base speed ω_b). In the second region, called the constant-power region, the slip is increased gradually

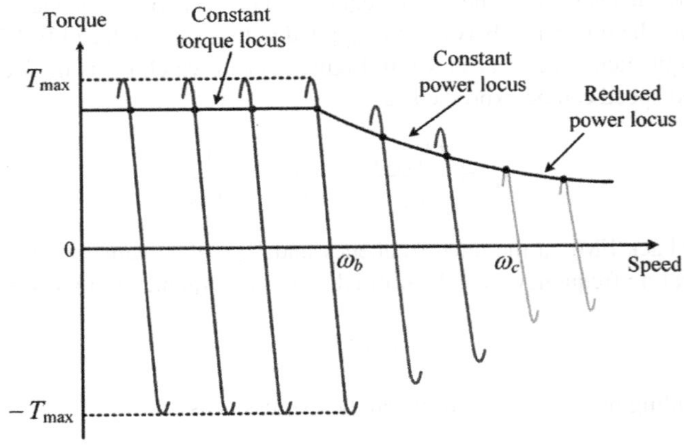

Figure 3.17 Torque-speed characteristics of VVVF-controlled induction motor drive

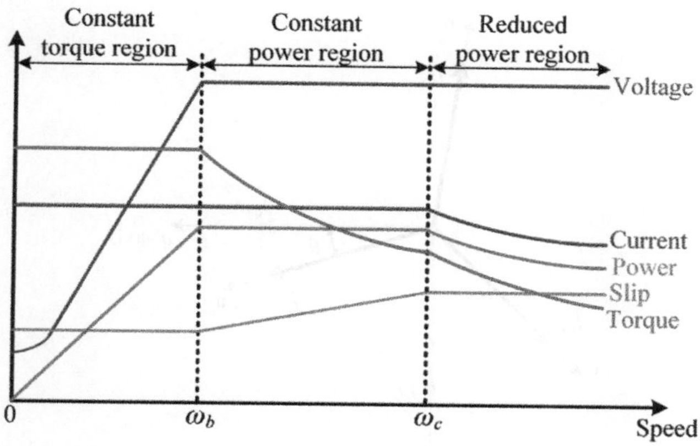

Figure 3.18 Operating capabilities of VVVF-controlled induction motor drive

to the maximum value so that the stator current remains constant and the motor can maintain its rated power capability. When the speed is above the critical speed ω_c, the slip remains constant while the stator current decreases. Thus, the torque capability declines with the square of speed, the so-called reduced power region.

It should be noted that both the torque and air-gap flux under the VVVF control are functions of voltage and frequency. This coupling effect is actually responsible for the sluggish response. That is, the corresponding torque control is not fast and accurate enough for application to high-performance EVs.

3.4.2 Field-Oriented Control

In order to improve the dynamic performance of induction motor drives, FOC is preferred to VVVF control. By using FOC, the mathematical model of the induction motor is first transformed from the stationary *a–b–c* frame to the stationary d^s–q^s frame, then to the synchronously rotating d^e–q^e frame with speed ω_e as shown in Figure 3.19. Thus, at steady state, all sinusoidal variables in the stationary frame such as the stator voltage v_s, stator current i_s, stator flux linkage λ_s, rotor voltage v_r, rotor current i_r, and rotor flux linkage λ_r can be represented by DC quantities in a synchronously rotating frame (Novotny and Lipo, 1996; Sul, 2011).

Firstly, taking θ as the angle between the q^s-axis of the d^s–q^s frame and the *a*-axis of the *a–b–c* frame, the transformed variables in the d^s–q^s frame can be expressed as

$$\begin{bmatrix} f_{ds}^s \\ f_{qs}^s \end{bmatrix} = \frac{2}{3}\begin{bmatrix} \sin\theta & \sin\left(\theta - 120^\circ\right) & \sin(\theta + 120^\circ) \\ \cos\theta & \cos(\theta - 120^\circ) & \cos(\theta + 120^\circ) \end{bmatrix}\begin{bmatrix} f_{as} \\ f_{bs} \\ f_{cs} \end{bmatrix} \tag{3.33}$$

where f can be any circuit variable such as voltage, current, and flux linkage, and the superscript *s* refers to the variable that is on the stationary d^s–q^s frame. Normally, it is convenient to set $\theta = 0$ so that the q^s-axis is aligned with the *a*-axis.

Secondly, as the d^e–q^e frame rotates synchronously at speed ω_e, the angle between the d^e–q^e frame and the d^s–q^s frame is $\theta_e = \omega_e t$. Thus, the transformed variables in the d^e–q^e frame can be obtained as

$$\begin{bmatrix} f_{ds}^e \\ f_{qs}^e \end{bmatrix} = \begin{bmatrix} \cos\theta_e & \sin\theta_e \\ -\sin\theta_e & \cos\theta_e \end{bmatrix}\begin{bmatrix} f_{ds}^s \\ f_{qs}^s \end{bmatrix} \tag{3.34}$$

where the superscript *e* refers to the variable that is on the synchronously rotating d^e–q^e frame.

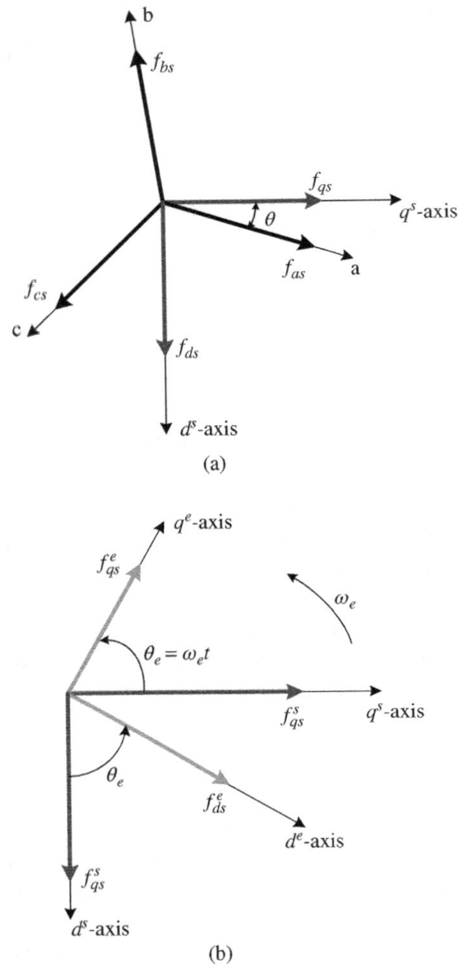

Figure 3.19 Coordinate transformations for FOC: (a) stationary *a–b–c* frame to stationary $d^s - q^s$ frame and (b) stationary $d^s - q^s$ frame to synchronously rotating $d^e - q^e$ frame

After performing the above two transformations, all sinusoidal variables in the stationary *a–b–c* frame become DC quantities in the d^e–q^e frame. Consequently, the voltage equations of the machine can be expressed as

$$v_{ds}^e = R_s i_{ds}^e + p\lambda_{ds}^e - \omega_e \lambda_{qs}^e \tag{3.35}$$

$$v_{qs}^e = R_s i_{qs}^e + p\lambda_{qs}^e + \omega_e \lambda_{ds}^e \tag{3.36}$$

$$v_{dr}^e = R_r i_{dr}^e + p\lambda_{dr}^e - (\omega_e - \omega_r)\lambda_{qr}^e \tag{3.37}$$

$$v_{qr}^e = R_r i_{qr}^e + p\lambda_{qr}^e + (\omega_e - \omega_r)\lambda_{dr}^e \tag{3.38}$$

where R_s is the stator resistance per phase, R_r is the referred rotor resistance per phase, ω_r is the rotor speed, and p is the differential operator. Normally, the rotor circuit of the IM is shorted so that both v_{dr}^e and v_{qr}^e are zero. The stator and rotor flux linkages can be expressed as

$$\lambda_{ds}^e = L_s i_{ds}^e + L_m i_{dr}^e \tag{3.39}$$

$$\lambda_{qs}^e = L_s i_{qs}^e + L_m i_{qr}^e \tag{3.40}$$

$$\lambda_{dr}^e = L_r i_{dr}^e + L_m i_{ds}^e \tag{3.41}$$

$$\lambda_{qr}^e = L_r i_{qr}^e + L_m i_{qs}^e \tag{3.42}$$

where L_m is the mutual inductance per phase, L_s is the stator inductance per phase, and L_r is the rotor inductance per phase. Hence, the output torque T_e can be derived as

$$T_e = \frac{3}{2}\frac{P}{2}\frac{L_m}{L_r}(\lambda_{dr}^e i_{qs}^e - \lambda_{qr}^e i_{ds}^e) \tag{3.43}$$

where P is the number of poles of the machine.

By purposely aligning the d^e-axis with the rotor flux linkage, the rotor flux linkage in the q^e-axis λ_{qr}^e is always zero. Then, the torque expression in Eq. (3.43) can be rewritten as

$$T_e = \frac{3}{2}\frac{P}{2}\frac{L_m}{L_r}\lambda_{dr}^e i_{qs}^e = K_e \lambda_{dr}^e i_{qs}^e \tag{3.44}$$

where K_e is the torque constant. It is analogous to the torque expression of a separately excited DC motor as given by

$$T_e = K_e \phi I_a \tag{3.45}$$

where the flux ϕ in the DC motor (analogous to the d^e-axis rotor flux linkage λ_{dr}^e in the induction motor) and the armature current I_a in the DC motor (analogous to the q^e-axis stator current i_{qs}^e in the induction motor) can be independently controlled. Meanwhile, by substituting $\lambda_{qr}^e = 0$ in Eqs. (3.37) and (3.41), the d^e-axis rotor flux linkage can be expressed as

$$\lambda_{dr}^e = \left(\frac{L_m}{\tau_r p + 1}\right) i_{ds}^e \tag{3.46}$$

where $\tau_r = L_r/R_r$ is the rotor time constant. Thus, the d^e-axis rotor flux linkage can be adjusted by controlling the d^e-axis stator current, which resembles the field current I_f of the separately excited DC motor. In particular, when i_{ds}^e is kept constant, λ_{dr}^e can be approximated by $L_m i_{ds}^e$ so that the torque expression in Eq. (3.44) can be further simplified as

$$T_e = \frac{3}{2}\frac{P}{2}\frac{L_m^2}{L_r} i_{ds}^e i_{qs}^e \tag{3.47}$$

where i_{ds}^e is the so-called field component of the stator current and i_{qs}^e is the so-called torque component of the stator current. Thus, by means of this FOC, the motor torque can be effectively controlled by adjusting the torque component as long as the field component remains constant. Therefore, the induction motor drive can offer the desired fast transient response similar to that of the separately excited DC motor drive. Consequently, the control principles, namely the increase of torque and the flux weakening, are depicted in Figure 3.20.

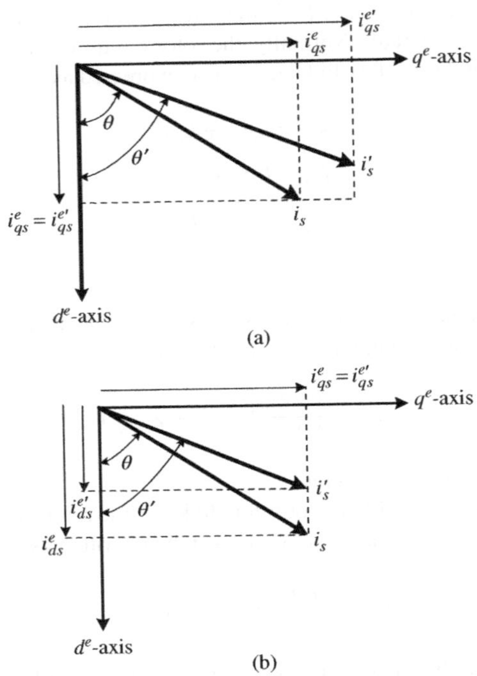

Figure 3.20 Principles of FOC: (a) torque control and (b) flux control

There are two ways to implement the FOC for the induction motor drive: direct and indirect approaches. The direct FOC, also dubbed as the direct vector control, is to instantaneously identify the rotor flux linkage by measurement of air-gap flux or flux estimation from the measured stator voltage and current. The air-gap flux measurement can be achieved by using the Hall sensors or sensing coils. However, mounting of Hall sensors in the air-gap suffers from the problems of mechanical vibration and temperature variations, which is impractical under the harsh operating environment of EVs. In addition, sensing coils suffer from the problems of low sensed voltage and poor signal-to-noise ratio, which make it hardly possible to deduce the air-gap flux during low-speed operation. On the other hand, the rotor flux linkage can be estimated by using the stator voltage equation, which is based on the measured terminal voltage and current as well as the machine parameters R_s, L_m, L_s, and L_r. However, at low speeds, the measured voltage signals are very low, which are ill-suited for the estimation. Meanwhile, the rotor flux linkage can also be estimated by using the rotor voltage equation, which is based on the measured current and speed as well as the machine parameters R_r, L_m, and L_r. Although this technique takes the advantage that the estimation can be down to zero speed, it requires an accurate speed encoder. Both of these flux estimation techniques heavily depend on the machine parameters, proper compensation for parameter variations is desirable.

The indirect FOC, also dubbed as indirect vector control, is widely adopted by the induction motor drive for EV propulsion. This technique does not need to identify the rotor flux linkage. The key is to calculate the slip speed that is required to deduce the instantaneous rotor flux position θ_e for correct field orientation:

$$\theta_e = \int_0^t \omega_e dt = \int_0^t (\omega_{sl} + \omega_r)dt = \int_0^t \omega_{sl} dt + \theta_r \tag{3.48}$$

where ω_e is the synchronous speed, ω_{sl} is the slip speed, ω_r is the rotor speed, and θ_r is the rotor position that is real-time measured by using a position encoder. By using Eqs. (3.38) and (3.42), the slip speed can be expressed as

$$\omega_{sl} = \frac{L_m}{\tau_r} \frac{i_{qs}^e}{\lambda_{dr}^e} \tag{3.49}$$

which is an important criterion to deduce the instantaneous rotor flux position, hence dubbed as the decoupling condition for the indirect FOC. On the basis of Eqs. (3.44), (3.46), (3.48), and (3.49), the indirect FOC control block diagram under the torque command T_e^* and flux command λ_{dr}^{e*} is shown in Figure 3.21. It can be observed that the torque and flux are controlled in a feed-forward manner.

Although the indirect FOC has been widely used for high-performance induction motor drives, it still suffers from some drawbacks. Particularly, the rotor time constant τ_r (which has a dominant effect on the decoupling condition) changes severely with the operating temperature and magnetic saturation, leading to deteriorate the desired indirect FOC. Currently, there are two convenient ways to solve this problem. One way is to perform online identification of the rotor time constant and accordingly to update the parameters used in the indirect FOC controller. Another way is to adopt a sophisticated control algorithm to enable the indirect FOC controller insensitive to motor parameter variations. The model-reference adaptive control (MRAC) scheme has been widely used for indirect FOC of the induction motor drive. Essentially, an adaptive mechanism is adopted that forces the motor drive to follow the reference model even under system parameter variations such as the change of rotor time constant due to prolonged operation. The definite advantage of this MRAC scheme is that there is no need to carry out explicit parameter identification.

3.4.3 *Direct Torque Control*

The DTC is an advanced scalar control scheme that can offer comparable performance as the FOC for the induction motor drive. This scheme is to directly control the stator flux linkage and the torque by properly selecting the switching modes of the voltage-fed PWM inverter. The selection is made to restrict the torque and flux errors within the respective torque and flux hysteresis bands, hence to achieve faster torque response and flexible control (Vas, 1998; Bose, 2001).

In the IM, the electromagnetic torque can be expressed in the form of cross product of the stator flux linkage vector $\bar{\lambda}_s$ and the stator current vector $\bar{i}_s$ as given by

$$T_e = \frac{3}{2}\frac{P}{2}(\lambda_{ds}^s i_{qs}^s - \lambda_{qs}^s i_{ds}^s) = \frac{3}{2}\frac{P}{2}(\bar{\lambda}_s \times \bar{i}_s) \tag{3.50}$$

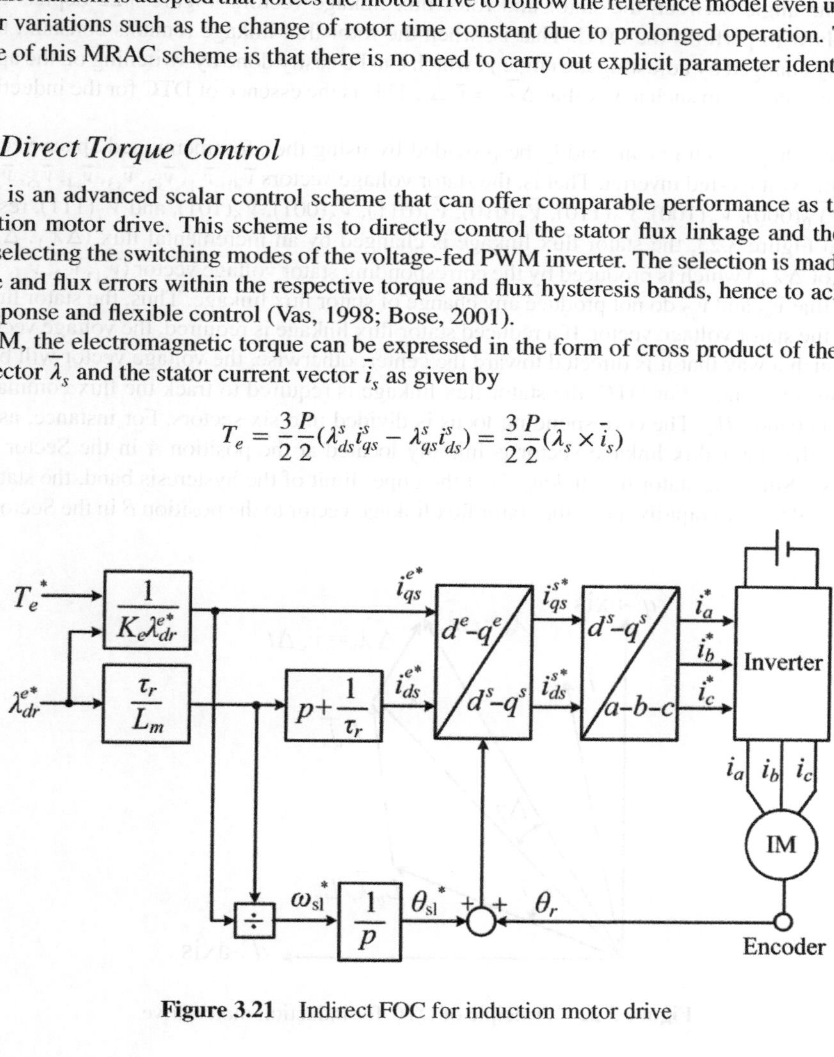

Figure 3.21 Indirect FOC for induction motor drive

The stator flux linkage vector $\overline{\lambda}_s$ and rotor flux linkage vector $\overline{\lambda}_r$ can be expressed in terms of the stator current vector $\overline{i}_s$ and rotor current vector $\overline{i}_r$, which are expressed as

$$\overline{\lambda}_s = L_s\overline{i}_s + L_m\overline{i}_r \tag{3.51}$$

$$\overline{\lambda}_r = L_r\overline{i}_r + L_m\overline{i}_s \tag{3.52}$$

By eliminating $\overline{i}_r$ in Eqs. (3.51) and (3.52), the stator flux linkage vector can be written as

$$\overline{\lambda}_s = \frac{L_m}{L_r}\overline{\lambda}_r + L_x\overline{i}_s \tag{3.53}$$

where $L_x = L_sL_r - L_m^2$. Hence, the torque expression in Eq. (3.50) can be rewritten as

$$T_e = \frac{3}{2}\frac{P}{2}\frac{L_m}{L_xL_r}(\overline{\lambda}_r \times \overline{\lambda}_s) = \frac{3}{2}\frac{P}{2}\frac{L_m}{L_xL_r}|\overline{\lambda}_r||\overline{\lambda}_s| \sin\gamma \tag{3.54}$$

where γ is the angle between the stator and rotor flux linkage vectors. Figure 3.22 depicts the vector diagram on how to perform the DTC. That is, when the rotor flux linkage remains constant, the torque can be rapidly changed by adjusting the angle γ, which can be easily done by switching on the appropriate stator voltage vector $\overline{v}_s$ in such a way that $\Delta\overline{\lambda}_s = \overline{v}_s\Delta t$. This is the essence of DTC for the induction motor drive.

The stator voltage vectors can readily be provided by using the aforementioned space-vector PWM scheme for the voltage-fed inverter. That is, the stator voltage vectors $\overline{v}_{s0}$, $\overline{v}_{s1}$, $\overline{v}_{s2}$, $\overline{v}_{s3}$, $\overline{v}_{s4}$, $\overline{v}_{s5}$, $\overline{v}_{s6}$, and $\overline{v}_{s7}$ are equal to $\overline{V}_0(000)$, $\overline{V}_1(100)$, $\overline{V}_2(110)$, $\overline{V}_3(010)$, $\overline{V}_4(011)$, $\overline{V}_5(001)$, $\overline{V}_6(101)$, and $\overline{V}_7(111)$, respectively. As shown in Figure 3.23, the stator flux linkage is changed by an incremental flux ($\Delta\overline{\lambda}_{s1}$, $\Delta\overline{\lambda}_{s2}$, $\Delta\overline{\lambda}_{s3}$, $\Delta\overline{\lambda}_{s4}$, $\Delta\overline{\lambda}_{s5}$, or $\Delta\overline{\lambda}_{s6}$) which is produced by the corresponding stator voltage vector ($\overline{v}_{s1}$, $\overline{v}_{s2}$, $\overline{v}_{s3}$, $\overline{v}_{s4}$, $\overline{v}_{s5}$, or $\overline{v}_{s6}$). Notice that $\overline{v}_{s0}$ and $\overline{v}_{s7}$ do not produce any change of stator flux linkage. Thus, the stator flux linkage moves with the stator voltage vector. If a reduced stator flux linkage is required, the voltage vector will be selected in such a way that it is directed toward the center; otherwise, the voltage vector will be selected to be outward the center. For DTC, the stator flux linkage is required to track the flux command within the hysteresis band $2H_\lambda$. The corresponding locus is divided into six sectors. For instance, as shown in Figure 3.24, the stator flux linkage vector is initially located at the position A in the Sector 1 rotating anticlockwise. Since the stator flux linkage is at the upper limit of the hysteresis band, the stator voltage vector, $\overline{v}_{s3}$ is selected to rapidly move the stator flux linkage vector to the position B in the Sector 2. As the

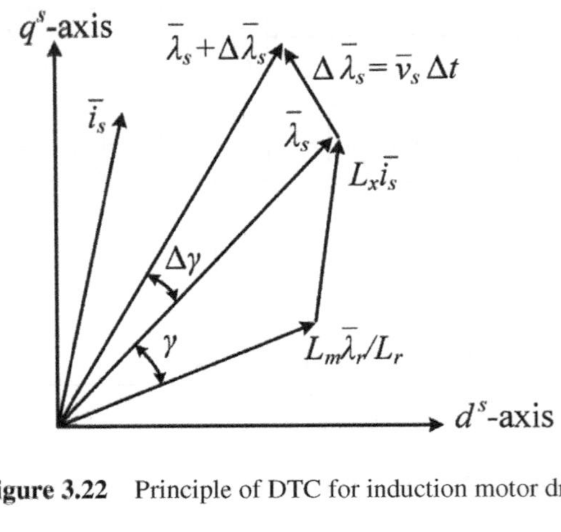

Figure 3.22 Principle of DTC for induction motor drive

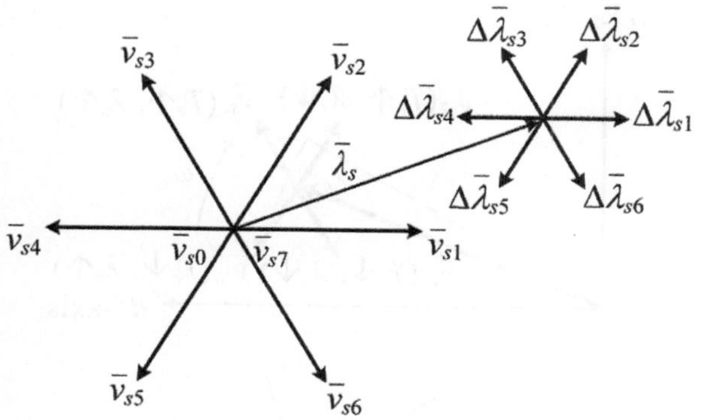

Figure 3.23 Control of stator flux vector for DTC

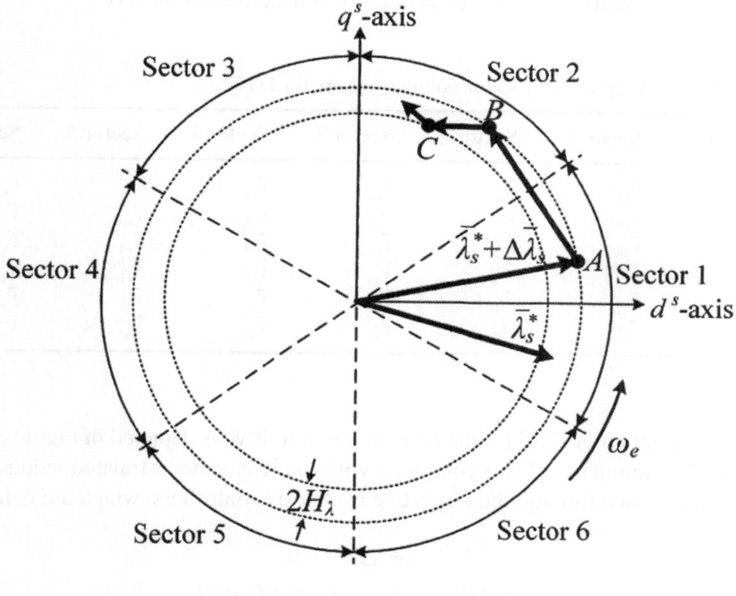

Figure 3.24 Trajectory of stator flux vector in DTC

stator flux linkage again reaches its upper limit, the stator voltage vector $\bar{v}_{s4}$ is applied to move it to the position C. It should be noted that the stator flux linkage vector changes quickly by $\bar{v}_s\Delta t$, whereas the change of rotor flux linkage vector is very sluggish because of its large time constant.

Figure 3.25 illustrates how to select various stator voltage vectors to control the torque and flux (torque increase, torque decrease, flux strengthening, and flux weakening) when the stator flux linkage vector is in the Sector 1. In general, if an increase of the torque is required, the voltage vector will be applied to advance the stator flux linkage vector in the direction of rotation; otherwise, the voltage vector will be applied opposite to the direction of rotation. If zero torque is required, the zero-voltage vector ($\bar{v}_{s0}$ or $\bar{v}_{s7}$) will be applied, depending on which vector can minimize the number of switching. Consequently, the torque demand is reduced to a choice of increase ($\delta T_e = 1$), decrease ($\delta T_e = -1$), or zero ($\delta T_e = 0$), while the stator flux linkage demand is a choice of strengthening ($\delta\lambda_s = 1$) or weakening ($\delta\lambda_s = -1$). The corresponding look-up table of stator voltage vectors for DTC is shown in Table 3.2.

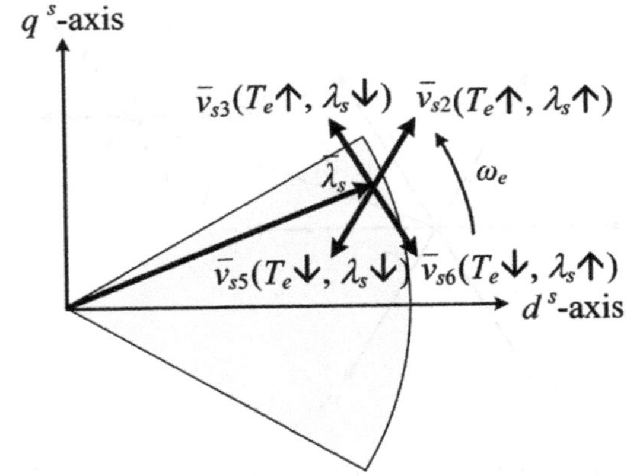

Figure 3.25 Selection of stator voltage vectors for DTC

Table 3.2 Look-up table of stator voltage vectors for DTC

δT_e	$\delta\lambda_s$	Sector 1	Sector 2	Sector 3	Sector 4	Sector 5	Sector 6
1	1	$\bar{v}_{s2}$	$\bar{v}_{s3}$	$\bar{v}_{s4}$	$\bar{v}_{s5}$	$\bar{v}_{s6}$	$\bar{v}_{s1}$
0	1	$\bar{v}_{s0}$	$\bar{v}_{s7}$	$\bar{v}_{s0}$	$\bar{v}_{s7}$	$\bar{v}_{s0}$	$\bar{v}_{s7}$
−1	1	$\bar{v}_{s6}$	$\bar{v}_{s1}$	$\bar{v}_{s2}$	$\bar{v}_{s3}$	$\bar{v}_{s4}$	$\bar{v}_{s5}$
1	−1	$\bar{v}_{s3}$	$\bar{v}_{s4}$	$\bar{v}_{s5}$	$\bar{v}_{s6}$	$\bar{v}_{s1}$	$\bar{v}_{s2}$
0	−1	$\bar{v}_{s7}$	$\bar{v}_{s0}$	$\bar{v}_{s7}$	$\bar{v}_{s0}$	$\bar{v}_{s7}$	$\bar{v}_{s0}$
−1	−1	$\bar{v}_{s5}$	$\bar{v}_{s6}$	$\bar{v}_{s1}$	$\bar{v}_{s2}$	$\bar{v}_{s3}$	$\bar{v}_{s4}$

The control block diagram of DTC for the induction motor drive is depicted in Figure 3.26. The torque command T_e^* and flux command λ_s^* are compared with the respective estimated values, and the errors (ΔT_e and $\Delta\lambda_s$) are processed through the respective hysteresis controllers, which are defined as

$$\delta T_e = \begin{cases} 1 & \text{for } \Delta T_e > H_T \\ 0 & \text{for } -H_T < \Delta T_e < H_T \\ -1 & \text{for } \Delta T_e < -H_T \end{cases} \tag{3.55}$$

$$\delta\lambda_s = \begin{cases} 1 & \text{for } \Delta\lambda_s > H_\lambda \\ -1 & \text{for } \Delta\lambda_s < -H_\lambda \end{cases} \tag{3.56}$$

where H_T is the hysteresis of the three-level torque controller and H_λ is the hysteresis of the two-level flux controller. The estimator serves to calculate the torque and stator flux linkage as well as the angle for sector selection based on the measured voltages and currents. This DTC induction motor drive can easily operate in all four quadrants, and provide the torque response comparable with that using FOC.

As compared with the FOC, the DTC for the induction motor drive possesses some definite advantages:

- There is no need to use coordinate transformations that are computationally intensive.
- There is no need to perform feedback current control.
- The motor torque can be directly controlled so that the torque response is much faster.

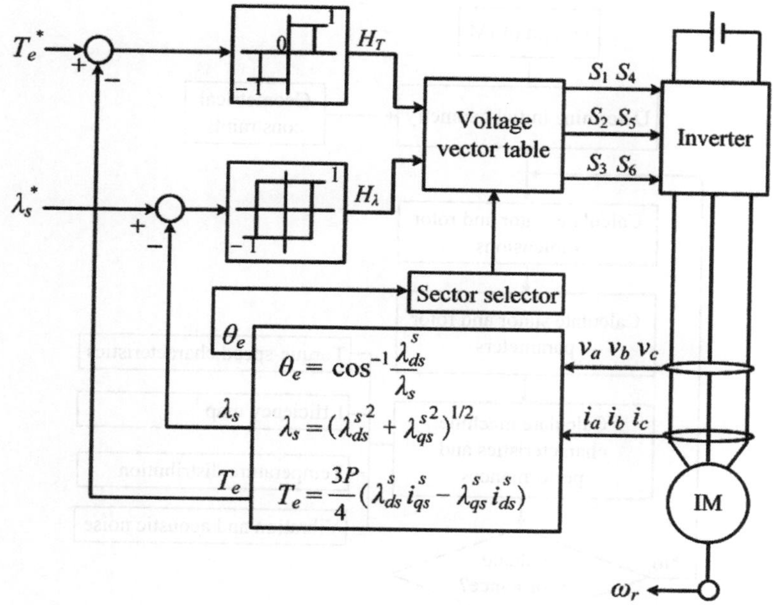

Figure 3.26 DTC block diagram for induction motor drive

However, the DTC suffers from the drawbacks of sluggish response during start-up and high torque ripple. There are several possible solutions such as the incorporation of fuzzy logic into the selection strategy of stator voltage vectors and the use of predictive algorithm to select the inverter switching states.

3.5 Design Criteria of Induction Motor Drives for EVs

The induction motors for electric propulsion is principally similar to that for industrial application. Nevertheless, these induction motors need to be specially designed. Laminated thin silicon cores should be used for both the stator and rotor to reduce the iron loss, while copper bars should be adopted for the rotor cage to reduce the copper loss. Stator coils should adopt the Class C insulation, and be directly cooled by oil with low viscosity. All housings should be made of cast aluminum to reduce the total motor weight. Reasonable high-voltage low-current machine design should be employed to reduce the cost and size of the power inverter, although the voltage level of the motor is usually limited by the number, weight, and type of batteries. Low stray reactance is also necessary to favor flux-weakening operation.

Figure 3.27 shows the flowchart to depict the design process of induction motors for EV propulsion, which involves an iterative loop to optimize the machine dimensions and parameters. Firstly, the machine geometry, dimensions, and parameters are initialized by using analytical and experiential equations. Secondly, the machine performances are determined by using finite element analysis, including the electromagnetic, thermal, and force field analyses. Finally, performance evaluations will be assessed and used to iteratively tune the machine dimensions.

The initialization of IM design is mainly based on analytical and experiential equations (Say, 1976; Chen, 1982). Firstly, main dimensions of the motor are determined. The stator inside diameter D_{si} can be expressed as

$$D_{si} = \left(\frac{2p}{\pi\rho} \frac{6.1 \times 10^3}{\alpha_p K_f K_w} \frac{1}{AB} \frac{P'}{n} \right)^{\frac{1}{3}} \quad (3.57)$$

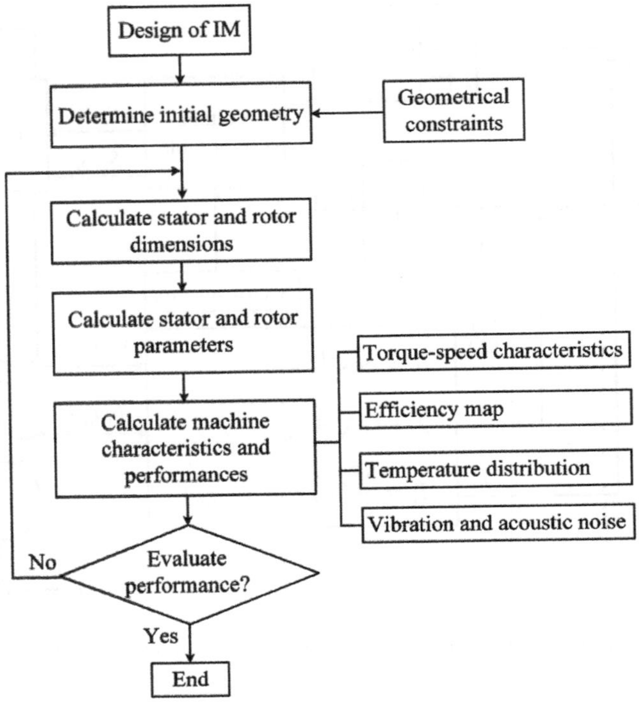

Figure 3.27 Design process of induction motors

where p is the number of pole-pairs, ρ is the aspect ratio (length-to-pole pitch ratio), which is in the range of 0.82–1.2, α_p is the pole-arc factor which can be chosen from 0.66 to 0.71, K_f is the air-gap flux waveform factor which is normally chosen as 1.11, K_w is the distributed winding factor which is about 0.96 for the single-layer winding and 0.92 for the double-layer winding, A is the electric loading which is in the range of 15–50 kA/m, B is the magnetic loading which is in the range of 0.5–0.8 T, P' is air-gap apparent power, and n is the rotor speed. How to select the value in the range is essentially based on experience. In addition, P' is given by

$$P' = \varepsilon \frac{P}{\eta \cos \phi} \tag{3.58}$$

where ε is the ratio of back EMF to rated voltage, which is in the range of 0.85–0.95, η is the efficiency, ϕ is the power factor angle, and P is the rated output power. Once D_{si} is determined, the stator outside diameter D_{so} can be estimated based on the relationship between D_{si}/D_{so} and the number of poles: for instance, 0.5–0.56 for two poles, 0.63–0.64 for four poles, and 0.68–0.78 for eight poles. Then, the effective core length L can be estimated by

$$L = \frac{1}{D_{si}^2} \frac{6.1 \times 10^3}{\alpha_p K_f K_w} \frac{1}{AB} \frac{P\prime}{n} \tag{3.59}$$

From experience, the air-gap length δ can be numerically expressed as

$$\delta = D_{si} \left(1 + \frac{9}{2p}\right) \times 10^{-3} \tag{3.60}$$

Hence, the rotor diameter D_r can be deduced as

$$D_r = D_{si} - 2\delta \tag{3.61}$$

Secondly, the dimensions of teeth and slots in the stator and rotor are determined. The stator slot width b_{ss} and slot height h_{ss} can be estimated by

$$b_{ss} = (0.45 \sim 0.62)t_s \tag{3.62}$$

$$h_{ss} = (3.5 \sim 5.5)b_{ss} \tag{3.63}$$

where t_s is the stator tooth pitch which is given by

$$t_s = \frac{\pi D_{si}}{Z_s} \tag{3.64}$$

where Z_s is the number of stator slots. Similarly, the rotor slot width b_{sr} and slot height h_{sr} can be estimated by

$$b_{sr} = (0.45 \sim 0.62)t_r \tag{3.65}$$

$$h_{sr} = (3.5 \sim 5.5)b_{sr} \tag{3.66}$$

where t_r is the rotor tooth pitch which is given by

$$t_r = \frac{\pi D_r}{Z_r} \tag{3.67}$$

where Z_r is the number of rotor slots. Meanwhile, the rotor tooth width b_{tr} can be estimated by

$$b_{tr} = \frac{t_r B}{K_i B_{tr}} \tag{3.68}$$

where K_i is the stacking factor of iron core which is in the range of 0.96–0.98, and B_{tr} is the nominal rotor tooth flux density in the range of 1.25–1.6 T. Then, the rotor yoke height, h_{yr} can be estimated by

$$h_{yr} = \frac{\tau\alpha_p}{2K_i}\frac{B}{B_{yr}} \tag{3.69}$$

where B_{yr} is the nominal rotor yoke flux density which is normally equal or less than 1 T, and τ is the pole pitch which can be expressed as

$$\tau = \frac{\pi D_{si}}{2p} \tag{3.70}$$

Thirdly, the number of turns of stator winding per phase N_s can be expressed as

$$N_s = \frac{\pi D_{si} A}{6 I_s} \tag{3.71}$$

where I_s is the stator current per phase. It can be found that when the number of turns is reduced, the electric loading decreases, whereas the magnetic loading increases. Consequently, the power factor, starting torque, and starting current will be affected. Thus, N_s is a useful parameter that can be adjusted to facilitate achieving the desired performance.

It should be noted that the above design equations are derived based on the conventional IM structure, namely the inner-rotor topology. For the design of outer-rotor induction motors, those design equations need to be modified accordingly.

3.6 Design Example of Induction Motor Drives for EVs

When the EV adopts the single-motor configuration, the induction motor is generally designed for high-speed operation while using a planetary gear to step down the motor speed so as to match with the wheel speed. On the basis of the requirements of a typical passenger EV, the specifications of the corresponding induction motor are listed in Table 3.3.

Aiming to achieve high-speed operation while limiting the switching frequency, the three-phase four-pole IM topology is adopted. On the basis of the specifications, the geometric dimensions and parameters of this induction motor can be initialized by using the aforementioned design equations, and are listed in Table 3.4. As mentioned earlier, this initialization can greatly facilitate the iterative process to optimize the machine dimensions and parameters.

In order to take into account the saturation and fringing effects, the electromagnetic finite element analysis is inevitable. The electromagnetic field distribution of the induction motor at the rated condition is shown in Figure 3.28 in which there is no significant magnetic saturation and the current density of stator winding is equal to 4.6 A/mm^2.

When the motor drive operates at or below the base speed, it provides constant-torque operation. The phase current waveform of the motor drive operating at the rated speed of 3000 rpm under the rated load is shown in Figure 3.29. The corresponding torque waveform is shown in Figure 3.30. It can be observed that the designed motor drive can produce the desired rated torque of 32 Nm as listed in Table 3.3.

Table 3.3 Specifications of planetary-geared induction motor design

DC voltage	360 V
Rated power	10 kW
Rated speed	3000 rpm
Rated torque	32 Nm
Constant-torque operation	0–3000 rpm
Constant-power operation	3000–7000 rpm
Gear ratio	10 : 1

Table 3.4 Initialization of planetary-geared induction motor design

Number of phases	3
Number of poles	4
Number of stator slots	24
Number of rotor slots	34
Stator outside diameter	248 mm
Rotor diameter	135.3 mm
Air-gap length	0.44 mm
Core length	106.8 mm
Stator slot width	8 mm
Stator slot height	28 mm
Rotor slot width	5.5 mm
Rotor slot height	22 mm
Rotor yoke thickness	34.7 mm
Number of turns per phase	128
Slot-fill factor	60%
Lamination material	50A230

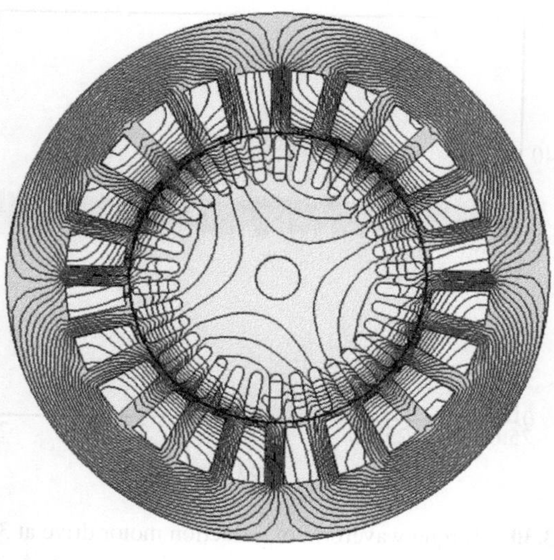

Figure 3.28 Electromagnetic field distribution of induction motor

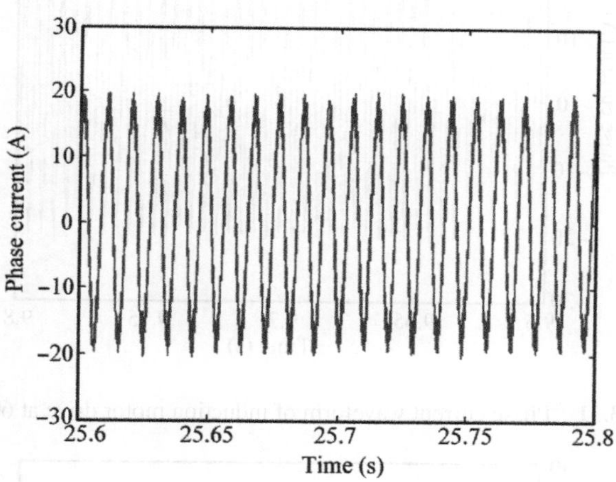

Figure 3.29 Phase current waveform of induction motor drive at 3000 rpm

Subsequently, the designed motor drive performs flux-weakening control to achieve high-speed constant-power operation. The phase current waveform of the motor drive operating at twice the rated speed, that is, 6000 rpm, under the rated load is shown in Figure 3.31. The corresponding torque waveform is shown in Figure 3.32. It can be observed that the designed motor drive can offer the desired constant-power operation, which can achieve 16 Nm at 6000 rpm. The overall torque-speed capability of this motor drive is plotted in Figure 3.33 in which the rotor flux starts to be weakened by using FOC when the motor speed is beyond the rated speed. It illustrates that the designed induction motor drive can operate in both the constant-torque and constant-power regions, and the constant-power range can be significantly extended by using flux-weakening control, which is highly favorable for EV application.

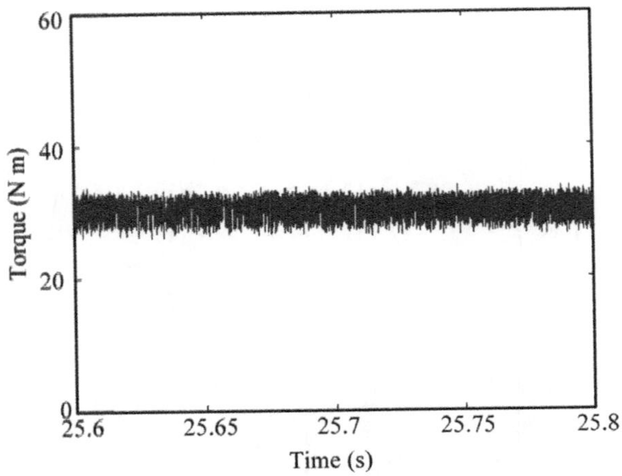

Figure 3.30 Torque waveform of induction motor drive at 3000 rpm

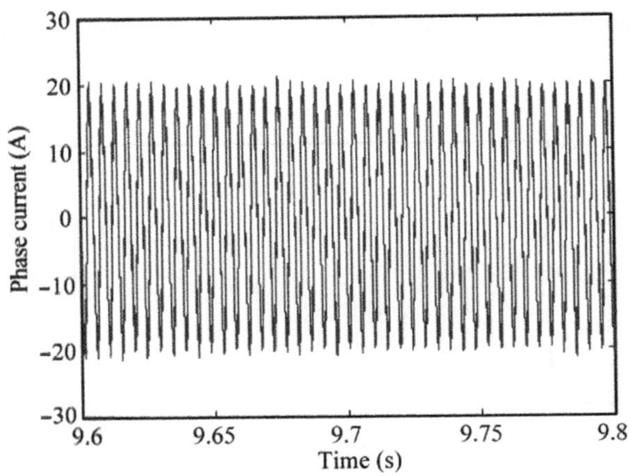

Figure 3.31 Phase current waveform of induction motor drive at 6000 rpm

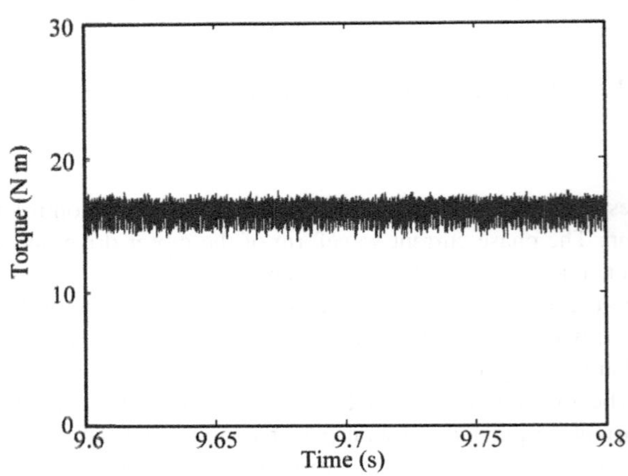

Figure 3.32 Torque waveform of induction motor drive at 6000 rpm

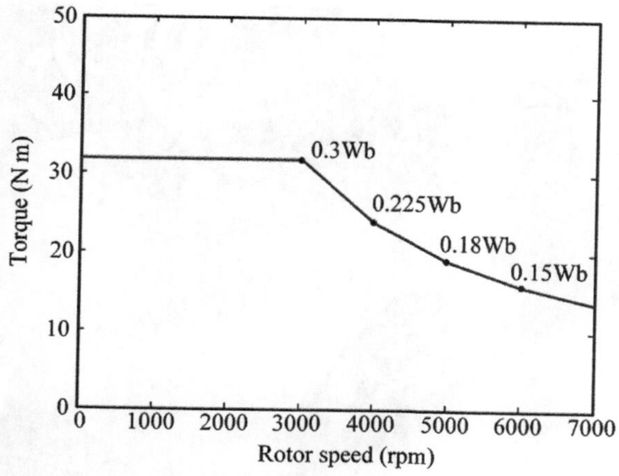

Figure 3.33 Torque-speed capability of induction motor drive

3.7 Application Examples of Induction Motor Drives in EVs

The Ford Siemens EV motor was one of the most well-developed induction motors for EVs. It was a three-phase water-cooled induction motor with the output power of 33 kW (rated) and 67 kW (peak), and weighed about 65 kg (Zuglet, 2014). This motor was targeted for rear-wheel drive in which the gear reduction and the differential are integrated with the motor. This motor and its derivatives were successfully applied to different models of EVs such as the flagship EV of Ford Motor–Ford Ranger EV. The Ford Ranger EV was a pickup truck which was powered by a six-pole induction motor drive that could achieve a maximum speed of 13 000 rpm while equipped with a 3 : 1 single-speed gear reduction.

Showcasing the most advanced propulsion system in 1997, the EV1 was the flagship EV of General Motors (GM). It was a front-wheel-drive electric sports car which adopted a three-phase induction motor drive and a single-speed transaxle with dual-reduction of 10.95 : 1. It could offer an axle torque of 149 Nm from 0 to 7000 rpm and a propulsion power of 102 kW from 7000 to 14 000 rpm as well as a top speed of 128 km/h and an acceleration from 0 to 96 km/h in less than 9 seconds. While customer reaction to the EV1 was positive, the EV1 was horrifically expensive so that GM decided to terminate its production in 2002. The resurrection of high-performance electric sports cars happened at the San Francisco International Auto Show in 2006, where the Tesla Roadster exhibited its streamlined body and outstanding performance. As shown in Figure 3.34, the Tesla Roadster was produced by Tesla Motors between 2008 and 2012. It adopted a three-phase four-pole induction motor drive that could produce a maximum power of 215 kW, a maximum torque of 400 Nm, and a maximum speed of 14 000 rpm (Tesla, 2010). It weighed less than 32 kg and delivered an average efficiency of 90%. The whole EV weighed about 1200 kg and was a rear-wheel drive. Equipped with a single-speed fixed-gear transmission of 8.28 : 1, it could offer an acceleration from 0 to 96 km/h in 3.7 seconds, and the top speed of 200 km/h. Installed with the lithium-ion battery of 53 kWh, it could run 393 km per charge based on the US Environmental Protection Agency (EPA) combined cycle.

3.8 Matured Technology for EVs?

The induction motor was probably invented by Galileo Ferraris in 1885 and by Nikola Tesla in 1887 independently. Over the last 130 years, it was continually improved and had already been dominant for industrial applications. Having been fueled by computer-aided design tools, power electronics technology, and sophisticated control strategies, the induction motor drive becomes a matured technology for EVs.

Figure 3.34 Tesla Roadster with induction motor drive

Nevertheless, with ever-increasing demand on high power density and high efficiency of motor drives for EV propulsion, the induction motor drive is being superseded by the permanent magnet (PM) brushless motor drive. Luckily, because of high PM cost, shortage of supply of rare-earth PM material, and thermal instability of PMs, there is a rekindling interest in the induction motor drive for EVs.

Because of its technological maturity, further advancement of the induction motor drive for EV propulsion is incremental only. Nevertheless, there are some interesting research activities on the induction motor drive for EVs. For instance, the development of double-rotor induction motor drive can potentially offer a compact and robust EV propulsion system with the inherent feature of electronic differential.

With the definite advantages of low cost, robust structure, and mature technology, it is anticipated that the induction motor drive can maintain a reasonable market share of EV motor drives in the foreseeable future.

References

Bose, B.K. (1992) *Modern Power Electronics: Evolution, Technology, and Application*, IEEE Press, New York.

Bose, B.K. (2001) *Modern Power Electronics and AC Drives*, Prentice-Hall, Upper Saddle River, NJ.

Bose, B.K. (2006) *Power Electronics and Motor Drives: Advances and Trends*, Academic Press, Burlington, MA.

Chen, S. (1982) *Electric Machine Design (Chinese)*, China Machine Press, Beijing.

Divan, D.M. (1986) The resonant DC link converter – a new concept in static power conversion. Proceedings of IEEE Industry Application Society Annual Meeting, pp. 648–656.

Dubey, G.K. (1989) *Power Semiconductor Controlled Drives*, Prentice-Hall, Englewood Cliffs, NJ.

Ehsani, M., Gao, Y. and Emadi, A. (2009) *Modern Electric, Hybrid Electric, and Fuel Cell Vehicles: Fundamentals, Theory, and Design*, 2nd edn, CRC Press.

Lai, J.S. (1997) Resonant snubber-based soft-switching inverters for electric propulsion drives. *IEEE Transactions on Industrial Electronics*, **44**, 71–80.

Novotny, D.W. and Lipo, T.A. (1996) *Vector Control and Dynamics of AC Drives*, Oxford University Press, Oxford.

Say, M.G. (1976) *Alternating Current Machines*, Pitman, London.

Sul, S.-K. (2011) *Control of Electric Machine Drive Systems*, Wiley-IEEE Press, Hoboken, NJ.

Tesla (2010) 2010 Tesla Roadster Performance Specs, Tesla Motors, http://web.archive.org/web/20100313074144/http://www.teslamotors.com/performance/perf_specs.php (accessed September 2014).

Vas, P. (1998) *Sensorless Vector and Direct Torque Control*, Oxford University Press, Oxford.

Zuglet (2014) Ford Siemens Motor, Zuglet, http://www.zuglet.com/ev/fordsiemens/fordsiemens.html (accessed September 2014).

4

Permanent Magnet Brushless Motor Drives

Among various types of motor drives, permanent magnet (PM) brushless motor drives, especially the PM synchronous motor drive, are currently the most attractive motor drives for electric vehicle (EV) propulsion. Their key features, namely high power density and high efficiency, are attributed to the use of high-energy PM material. Actually, they are becoming dominant in the market share of EV motor drives. However, these PM brushless motor drives still have some shortcomings such as high cost and thermal instability of the PM material.

In this chapter, two major PM brushless motor drives, namely the PM synchronous and PM brushless DC (BLDC) types, are presented. Their PM materials, system configurations, machine topologies, inverter topologies, and control strategies are described. The corresponding design criteria, design examples, and application examples are also discussed for EV propulsion.

4.1 PM Materials

Definitely, PM materials are the key of PM brushless motor drives that provide the machines with lifelong excitation. The development of PM materials dates back over many centuries. As depicted in Figure 4.1, the progress of PM materials proceeded in a stepwise manner, and each PM material was developed and improved before being superseded by a new one (Gutfleisch, 2000). Particularly, the corresponding energy product has had a drastic improvement starting from the 1980s.

Currently, there are four major types of PM materials that are widely used for motor drives:

- *Ferrite*: The ferrite magnet was invented in the 1930s. It has been widely used as commercial magnets for the last few decades because of the abundance of raw materials and low production cost. It also offers the advantage of high electrical resistance, which can suppress the corresponding eddy-current loss. However, it suffers from drawbacks such as high temperature coefficient and low energy density, leading to be sensitive to temperature variations and bulky in size, respectively, for application to PM machines.
- *Alnico*: It is a nickname of iron-based aluminum-nickel-cobalt (Al-Ni-Co) alloy, which was invented in the 1940s. It was the first modern PM material offering high remanence. Because of its high Curie temperature, it can be used at high operating temperature. Unfortunately, its coercivity is very low so that it is very easy to be demagnetized, which limits the corresponding application to PM machines.

Electric Vehicle Machines and Drives: Design, Analysis and Application, First Edition. K. T. Chau.

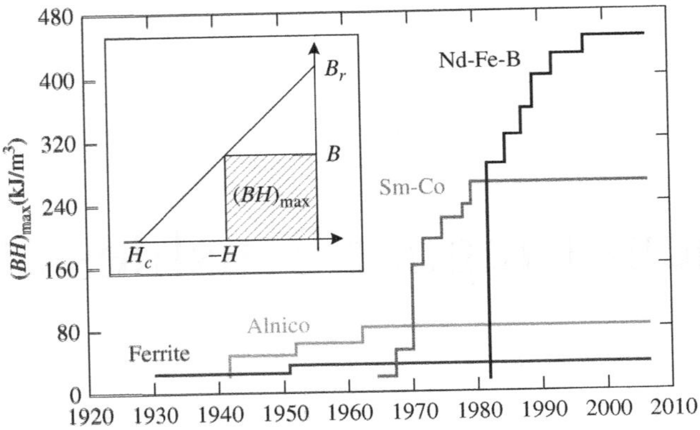

Figure 4.1 Development of PM materials

Nevertheless, some special machines, such as the memory machine, positively utilize this property so that the magnetization level can be online tuned.

- *Samarium-cobalt (Sm-Co)*: Invented in the 1960s, this rare-earth PM material has merits such as high remanence, high coercivity, high energy density, high Curie temperature, and low temperature coefficient. It is well suited for application to PM machines, which desire high power density, high efficiency, and high stability. However, its cost is the key drawback. Particularly, the rare-earth element, samarium, is very expensive.
- *Neodymium-iron-boron (Nd-Fe-B)*: First produced in 1984, this rare-earth PM material has better magnetic properties than samarium-cobalt. Since neodymium is a relatively less expensive rare-earth element, the corresponding cost becomes reasonable for application to PM machines. The relatively low Curie temperature (345 °C) is the major concern, which limits its use for high-temperature applications. Currently, this PM material is almost exclusively used for EV motor drives. Because of such an exponential growth in demand, the price is highly volatile, sometimes unreasonably expensive.

Figure 4.2 shows typical demagnetization characteristics of the above PM materials in which the remanence, B_r denotes the strength of the magnetic field produced, and the coercivity, H_c denotes the resistance to becoming demagnetized. The maximum product of an operating point on the characteristic is dubbed as the energy product $(BH)_{max}$, which actually represents the corresponding energy density. Additionally, the PM properties are usually temperature-dependent, and vary according to the temperature coefficient. Exposure to a temperature called the Curie temperature, T_c, the PM material completely loses its magnetization. Therefore, the operating temperature range must be taken into account during the design of PM machines. A brief summary of typical PM properties is given in Table 4.1.

4.2 System Configurations

The system configuration of PM brushless motor drives for electric propulsion is similar to that of induction motor drives. Major alternatives such as single-motor and multiple-motor configurations as well as geared and gearless transmissions can be found. Basically, the single-motor system configuration consists of a PM brushless machine, a voltage-fed inverter, an electronic controller, and some sensors as shown in Figure 4.3. Nevertheless, there are some variations:

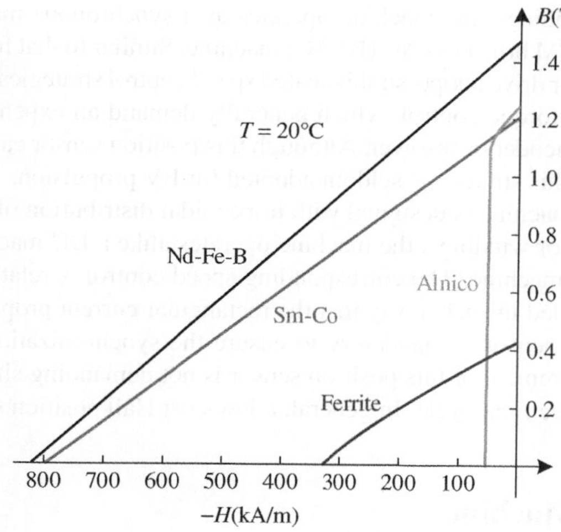

Figure 4.2 Demagnetization curves of PM materials

Table 4.1 Properties of PM materials

	Ferrite	Alnico	Sm-Co	Nd-Fe-B
Remanence, B_r (T)	0.43	1.25	1.21	1.47
Coercivity, H_c (kA/m)	330	51	796	820
Energy product, $(BH)_{max}$ (kJ/m^3)	35	44	271	422
Temperature coefficient of B_r (%/°C)	−0.18	−0.02	−0.03	−0.11
Temperature coefficient of H_c (%/°C)	0.2	0.01	−0.22	−0.65
Curie temperature, T_c (°C)	450	860	825	345

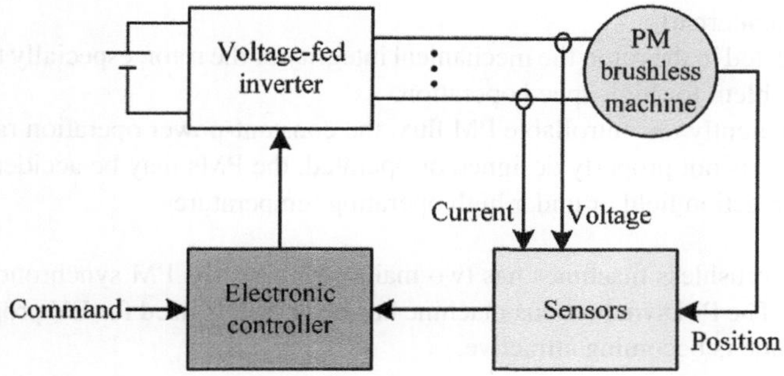

Figure 4.3 Basic configuration of PM brushless motor drive

- When the PM brushless machine is designed with sinusoidal distribution of air-gap flux and distributed arrangement of stator windings, the machine operates as a synchronous machine, the so-called PM synchronous machine or PM brushless AC (BLAC) machine. Similar to that for induction motor drives, the PM synchronous motor drive adopts sophisticated speed control strategies such as the field-oriented control (FOC) and direct torque control, which generally demand an expensive position sensor such as the absolute position encoder or resolver. Although this position sensor can be ignored theoretically for open-loop operation, this strategy is seldom adopted for EV propulsion.
- When the PM brushless machine is designed with trapezoidal distribution of air-gap flux and concentrated arrangement of stator windings, the machine operates alike a DC machine but without brushes, the so-called PM BLDC machine. The corresponding speed control is relatively simple in which the stator currents are controlled in such a way that the rectangular current properly aligns with the trapezoidal flux. The position sensor is mandatory to ensure the synchronization of the current with the flux. The resolution requirement of this position sensor is not demanding since only six commutation instants are needed per electrical cycle. In general, a low-cost Hall position sensor is adopted.

4.3 PM Brushless Machines

The PM brushless machines have superseded induction machine in recent years, becoming the preferred machines for EV propulsion. Their major advantages are as follows:

- Since the magnetic field is excited by high-energy PMs, the overall weight and volume can be significantly reduced for a given output power, leading to higher power density.
- Because of the absence of rotor copper loss, they inherently offer higher efficiency.
- Since the heat mainly arises in the stator, it can be more efficiently dissipated to the surroundings, leading to easier cooling.
- As the PM excitation has no risk of manufacturing defects, overheating, or mechanical damage, they inherently offer higher reliability.
- Because of absence of copper windings in the rotor, they possess lower electromechanical time constant and hence better dynamic response.

However, these machines still suffer from some drawbacks as follows:

- Since the high-energy PMs are based on rare-earth elements, the machine cost is much higher than that of the induction counterpart.
- As the PMs are located in the rotor, the mechanical integrity of the rotor, especially the surface-mounted PM type, has a problem for high-speed operation.
- Because of the inherently uncontrollable PM flux, the constant-power operation range is limited.
- In case the machine is not properly designed or operated, the PMs may be accidentally demagnetized by high armature reaction field or under high operating temperatures.

The family of PM brushless machines has two main members: the PM synchronous machine and the PM BLDC machine. The PM synchronous machine has been widely used for EV propulsion; meanwhile, the PM BLDC machine is becoming attractive.

4.3.1 Structure of PM Brushless Machines

The basic structure of the PM brushless machine, including both the PM synchronous and PM BLDC types, is depicted in Figure 4.4. It mainly consists of a stator incorporated with the three-phase armature winding and a rotor incorporated with PM poles. Compared with the induction machine, it is relatively

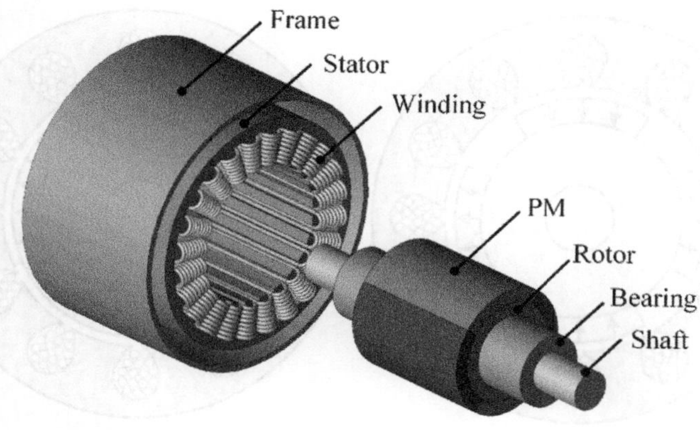

Figure 4.4 PM brushless machine exploded diagram

simpler because of the absence of cage bars and end-rings. In addition, since the associated heat loss in the rotor is not significant, it generally does not require to mount fan blades on the rotor or shaft for forced-air cooling.

4.3.1.1 PM Synchronous Machine Structure

By replacing the field winding of the wound-rotor synchronous machine with PM poles, the PM synchronous machine can readily be generated. Similar to the conventional synchronous machine, the stator has the three-phase distributed armature winding in the stator. When the armature winding is fed with three-phase sinusoidal currents, a synchronously rotating sinusoidal air-gap flux is created. Hence, the PM rotor with the same number of poles as the rotating air-gap flux always rotates in synchronism, which depends on the applied frequency.

On the basis of the placement of PMs in the rotor, the PM synchronous machine can be further classified as the surface-mounted, surface-inset, interior-radial, and interior-circumferential topologies while the stator is the same as shown in Figure 4.5:

- For the surface-mounted topology as shown in Figure 4.5a, the PMs are simply glued on the rotor surface using epoxy adhesive, thus offering the advantage of simplicity of manufacture. Since the permeability of PMs is near to that of air, the effective air-gap is the sum of the actual air-gap length and the radial thickness of the PMs. Hence, the corresponding armature reaction field is small and the stator winding inductance is low. However, since the d-axis and q-axis stator winding inductances are nearly the same, its reluctance torque is almost zero. In addition, there is a possibility that the PMs may fly apart during high-speed operation.
- For the surface-inset topology as shown in Figure 4.5b, the PMs are inset on the rotor surface. Thus, the q-axis inductance becomes higher than the d-axis inductance, hence producing an additional reluctance torque. In addition, since the PMs are inset in the rotor, it can offer better mechanical integrity than the surface-mounted one to withstand the centrifugal force at high-speed operation.
- For the interior-radial topology as shown in Figure 4.5c, the PMs are radially magnetized and buried inside the rotor. Compared with the surface-inset one, this topology enables the PMs well protected from flying apart, thus further improving the mechanical integrity for high-speed operation. Also, because of its d–q saliency, an additional reluctance torque is generated. Differing from the surface-inset one, this interior-radial topology adopts linear PMs that are easier for insertion and are easily machinable.

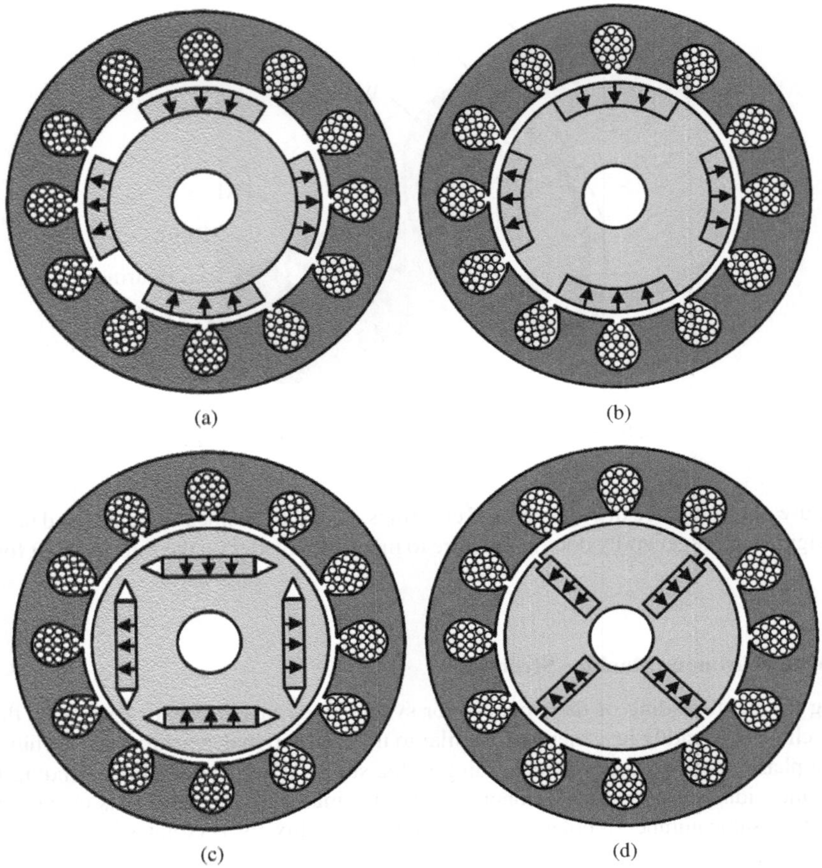

Figure 4.5 PM synchronous machine topologies: (a) surface-mounted, (b) surface-inset, (c) interior-radial, and (d) interior-circumferential

- For the interior-circumferential topology as shown in Figure 4.5d, the PMs are circumferentially magnetized and buried inside the rotor. It takes the definite advantage that the air-gap flux density can be higher than the PM remanence, the so-called flux-focusing or flux-concentration. In addition, it holds the merits of good mechanical integrity and additional reluctance torque. However, because of significant flux leakage at the inner ends of PMs, a nonmagnetic shaft or collar is generally required, which may degrade the torsional stiffness of the rotor shaft.

The above-mentioned inner-rotor PM synchronous machine topologies can readily be extended to the outer-rotor ones. The principle of operation of the outer-rotor PM synchronous machine is the same as the inner-rotor counterpart. The outer-rotor machine topology is particularly preferred to in-wheel drive because the corresponding outer rotor has a large radial diameter which can accommodate a large number of PM poles, hence offering low-speed high-torque direct-drive capability.

4.3.1.2 PM Brushless DC Machine Structure

By topologically inverting the stator and rotor of the PM DC machine, the PM BLDC machine can be generated. The key is to use electronic commutation to replace the mechanical commutation. Hence, the most obvious advantage of the PM BLDC machine is the removal of commutator and brushes, leading to eliminate many problems associated with them.

The PM BLDC machine has a similar structure as the PM synchronous machine, namely the three-phase stator windings and PM rotor. However, the PM BLDC machine generally adopts the concentrated winding instead of the distributed winding, and has the trapezoidal air-gap flux density distribution instead of the sinusoidal flux distribution. Because of the use of concentrated windings, the end-windings can be significantly shortened, hence saving the copper material and associated copper loss. In general, the PM BLDC machine adopts the surface-mounted PM rotor as depicted in Figure 4.6, which takes the definite advantages of simplicity in both structure and control. Nevertheless, other types of PM rotors such as the surface-inset, interior-radial, and interior-circumferential topologies can also be adopted, provided the air-gap flux density distribution is close to trapezoidal; otherwise, the torque production will be deteriorated.

4.3.2 Principle of PM Brushless Machines

The principle of PM brushless machines, including the PM synchronous machine and the PM brushless machine, needs to be separately described because they are very different. That is, the principle of PM synchronous machine is based on the interaction of sinusoidal back electromotive force (EMF) waveforms and sinusoidal armature current waveforms, whereas the principle of PM BLDC machine is based on the interaction of trapezoidal back EMF waveforms and rectangular armature current waveforms.

4.3.2.1 PM Synchronous Machine Operation

The PM synchronous machine has the balanced three-phase sinusoidal distribution of air-gap flux and sinusoidal distribution of armature winding. So, the three-phase induced back EMF waveforms can be expressed as

$$e_a = E_m \sin(\omega t) \tag{4.1}$$

$$e_b = E_m \sin(\omega t - 120°) \tag{4.2}$$

$$e_c = E_m \sin(\omega t - 240°) \tag{4.3}$$

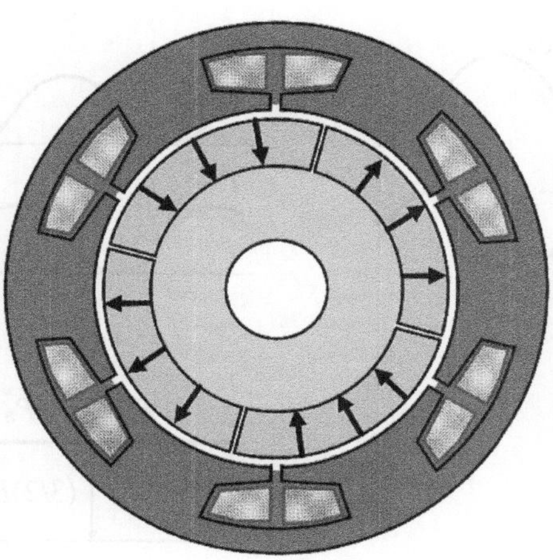

Figure 4.6 PM brushless DC machine topology

where E_m is the amplitude of back EMF waveforms and ω is the angular frequency. For proper operation, the machine is fed by balanced three-phase sinusoidal currents as given by

$$i_a = I_m \sin(\omega t - \phi) \tag{4.4}$$

$$i_b = I_m \sin(\omega t - 120^\circ - \phi) \tag{4.5}$$

$$i_c = I_m \sin(\omega t - 240^\circ - \phi) \tag{4.6}$$

where I_m is the amplitude of current waveforms and ϕ is the phase difference between the current and back EMF waveforms. Hence, as depicted in Figure 4.7, the converted electrical power can be calculated by Nam (2010) as

$$P_e = e_a i_a + e_b i_b + e_c i_c = \frac{3E_m I_m}{2} \cos\phi \tag{4.7}$$

Thus, the developed torque of this PM synchronous machine is given by

$$T_e = \frac{P_e}{\omega_r} = \frac{3E_m I_m}{2\omega_r} \cos\phi \tag{4.8}$$

which is constant at a given speed ω_r. Obviously, the developed torque can be maximized by controlling the phase difference between the armature current and back EMF waveforms to be zero.

4.3.2.2 PM Brushless DC Machine Operation

The PM BLDC machine adopts the concentrated armature winding in the stator with 120° coil span. As the PMs in the rotor cover 180° pole arc, the three-phase-induced back EMF waveforms are trapezoidal with 120° flat top regions. For proper operation, the machine is generally fed by balanced three-phase rectangular currents with 120° conduction periods, which coincide with the flat top regions of the back EMF. As depicted in Figure 4.8, each phase power is simply equal to $E_m I_m$ over the conduction period of 120°, and zero over the nonconduction period of 60°. At each instant, one phase power is equal to zero,

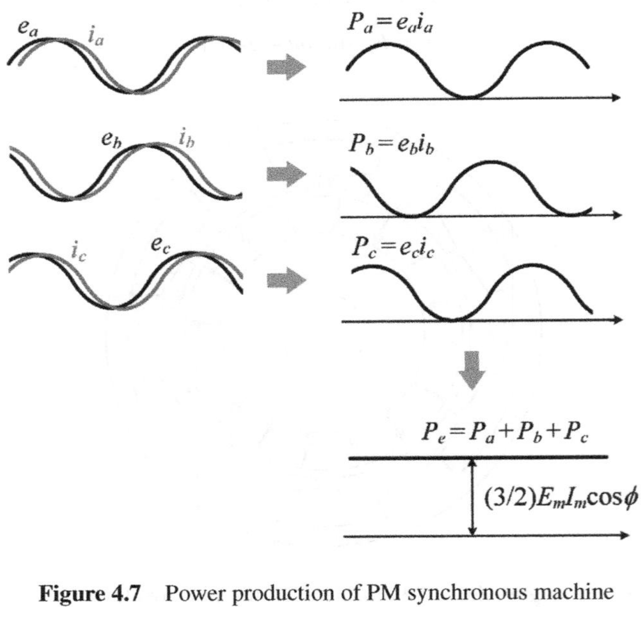

Figure 4.7 Power production of PM synchronous machine

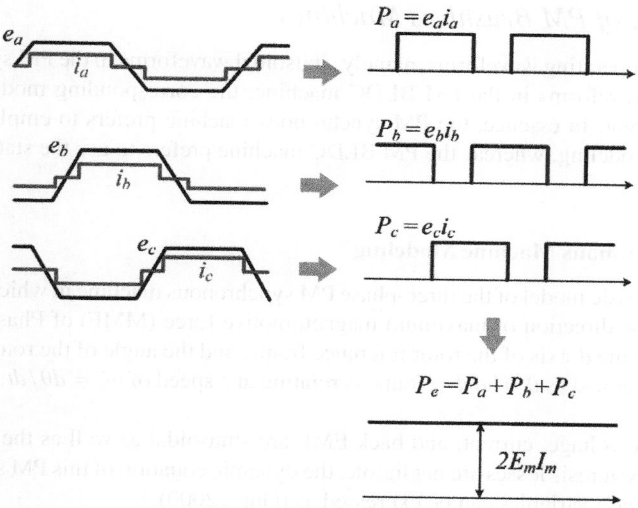

Figure 4.8 Power production of PM brushless DC machine

while the other two phases are $E_m I_m$. Hence, the converted electrical power is simply the summation of all phase powers as given by Nam (2010) as

$$P_e = e_a i_a + e_b i_b + e_c i_c = 2E_m I_m \tag{4.9}$$

Thus, the developed torque of this PM BLDC machine can be expressed as

$$T_e = \frac{P_e}{\omega_r} = \frac{2E_m I_m}{\omega_r} \tag{4.10}$$

which is constant at a given speed, ω_r. Differing from the PM synchronous machine, the PM BLDC machine produces the maximum torque naturally.

On the basis of the same current rating of the power devices for the two types of PM brushless machines, their current amplitudes are both equal to the current rating, I_p. Hence, the ratio of their power outputs (PM BLDC power to PM synchronous power) can be easily deduced from Eqs. (4.7) and (4.9) as given by

$$\text{Power ratio} = \frac{2E_m I_p}{(3/2)E_m I_p \cos\phi} = \frac{1.33}{\cos\phi} \tag{4.11}$$

which indicates that the PM BLDC machine can offer at least 33% higher power capability than the PM synchronous machine.

On the basis of the same copper loss of the armature windings for the two types of PM brushless machines (Krishnan, 2010), the current amplitude of the PM synchronous machine is I_p, whereas the current amplitude of the PM BLDC machine becomes $(\sqrt{3}/2)I_p$. Hence, the corresponding ratio of their power outputs is given by

$$\text{Power ratio} = \frac{2E_m(\sqrt{3}/2)I_p}{(3/2)E_m I_p \cos\phi} = \frac{1.15}{\cos\phi} \tag{4.12}$$

which indicates that the PM BLDC machine can offer at least 15% higher power density than the PM synchronous machine.

Therefore, the PM BLDC machine can offer higher power density and hence higher torque density than the PM synchronous machine by at least 15% or 33%.

4.3.3 Modeling of PM Brushless Machines

Because of different operating waveforms, namely sinusoidal waveforms in the PM synchronous machine and nonsinusoidal waveforms in the PM BLDC machine, the corresponding modeling techniques are fundamentally different. In essence, the PM synchronous machine prefers to employ *d–q* coordination transformation for modeling, whereas the PM BLDC machine prefers to use the state-space equation for modeling.

4.3.3.1 PM Synchronous Machine Modeling

Figure 4.9 depicts a basic model of the three-phase PM synchronous machine in which the stator reference axis is selected as the direction of maximum magnetomotive force (MMF) of Phase A, the direction of PM flux is chosen as the *d*-axis of the rotor reference frame, and the angle of the rotor *q*-axis with respect to the stator axis is defined as θ. The *d–q* frame is rotating at a speed of $\omega_e = d\theta/dt$, while the stator axes are fixed in space.

Assuming that the voltage, current, and back EMF are sinusoidal as well as the magnetic saturation, eddy currents, and hysteresis losses are negligible, the dynamic equation of this PM synchronous machine model in terms of phase variables can be expressed as (Ohm, 2000)

$$\begin{cases} v_a = Ri_a + p\lambda_a \\ v_b = Ri_b + p\lambda_b \\ v_c = Ri_c + p\lambda_c \end{cases} \tag{4.13}$$

where v_a, v_b, and v_c are the instantaneous three-phase stator voltages; i_a, i_b, and i_c are the instantaneous three-phase stator currents; R is the armature resistance; and λ_a, λ_b, and λ_c are the instantaneous three-phase flux linkages which can be written as

$$\begin{cases} \lambda_a = L_{aa}i_a + L_{ab}i_b + L_{ac}i_c + \lambda_{ma} \\ \lambda_b = L_{ab}i_a + L_{bb}i_b + L_{bc}i_c + \lambda_{mb} \\ \lambda_c = L_{ac}i_a + L_{bc}i_b + L_{cc}i_c + \lambda_{mc} \end{cases} \tag{4.14}$$

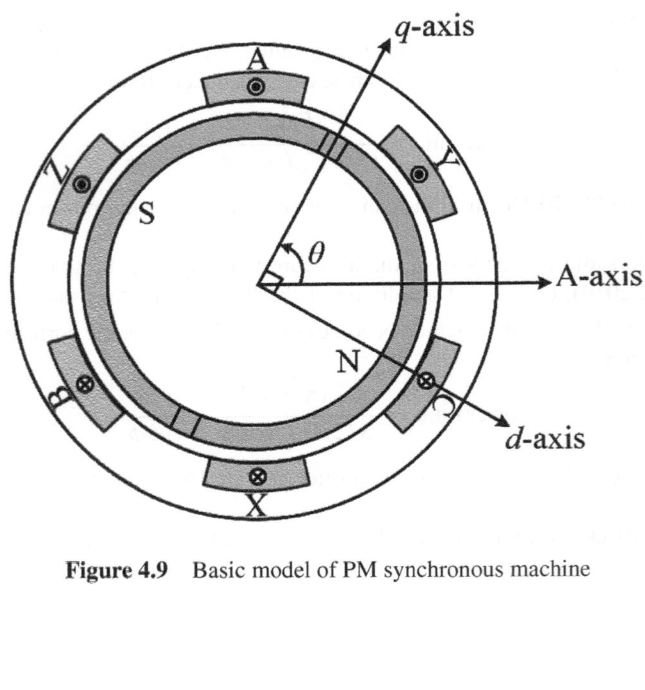

Figure 4.9 Basic model of PM synchronous machine

where L_{ij} ($i = a, b, c$ and $j = a, b, c$) are the mutual inductances which are symmetrical functions of angle θ, and λ_{ma}, λ_{mb}, and λ_{mc} are the instantaneous three-phase PM flux linkages.

By applying the Park's transformation to voltages, currents, and flux linkages in Eqs. (4.13) and (4.14), the dynamic equation can be written as

$$\begin{cases} v_d = Ri_d + p\lambda_d - \omega_e\lambda_q \\ v_q = Ri_q + p\lambda_q + \omega_e\lambda_d \end{cases} \tag{4.15}$$

where v_d and v_q are, respectively, the d-axis and q-axis components of stator voltages, i_d and i_q are, respectively, the d-axis and q-axis components of stator currents, and λ_d and λ_q are the d-axis and q-axis components of flux linkages, which can be expressed as

$$\begin{cases} \lambda_d = L_d i_d + \lambda_m \\ \lambda_q = L_q i_q \end{cases} \tag{4.16}$$

where L_d and L_q are called the d-axis and q-axis synchronous inductances, respectively and λ_m is the PM flux linkage. Substituting Eq. (4.16) into Eq. (4.15) yields

$$\begin{cases} v_d = (R + pL_d)\, i_d - \omega_e L_q i_q \\ v_q = (R + pL_q) i_q + \omega_e L_d i_d + \omega_e \lambda_m \end{cases} \tag{4.17}$$

which constitutes to the dynamic equivalent circuit of this PM synchronous machine as shown in Figure 4.10.

On the basis of the three-phase variables, the instantaneous input power P_i of this machine is given by

$$P_i = v_a i_a + v_b i_b + v_c i_c \tag{4.18}$$

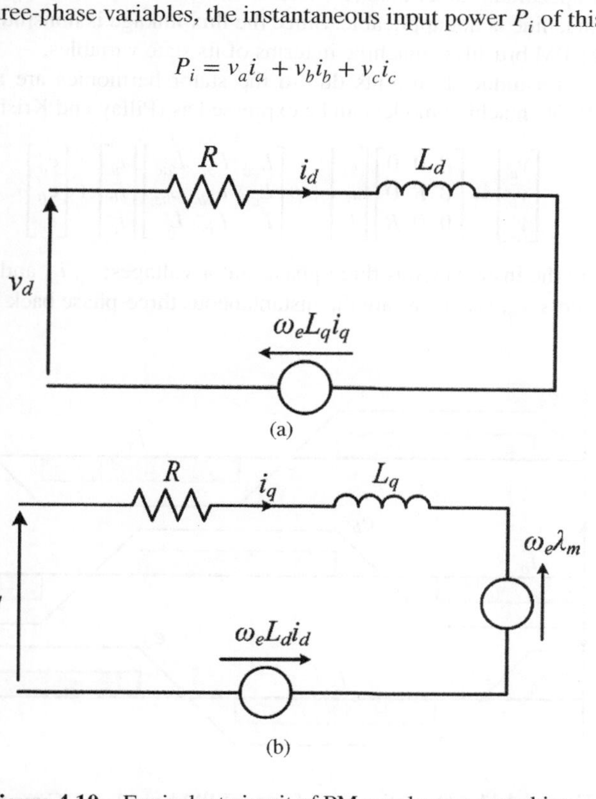

Figure 4.10 Equivalent circuit of PM synchronous machine

which can be written in terms of d-axis and q-axis variables as

$$P_i = \frac{3}{2}(v_d i_d + v_q i_q) \tag{4.19}$$

The output power P_o can be obtained by replacing v_d and v_q in Eq. (4.19) with the relevant voltages in Eq. (4.17) and flux linkages in Eq. (4.16), which yields

$$P_o = \frac{3}{2}(-\omega_e \lambda_q i_d + \omega_e \lambda_d i_q) \tag{4.20}$$

Hence, the developed torque T_e is the output power divided by the mechanical speed of $\omega_r = \omega_e / (P/2)$ as given by

$$T_e = \frac{3}{2}\frac{P}{2}(\lambda_m i_q + (L_d - L_q) i_d i_q) \tag{4.21}$$

where P is the number of poles. It can be observed that the developed torque is composed of two torque components. The first component corresponds to the reaction torque between the PM flux and q-axis armature current, while the second component corresponds to the reluctance torque due to the difference between d-axis and q-axis inductances.

4.3.3.2 PM Brushless DC Machine Modeling

Figure 4.11 shows the operating waveforms of the three-phase PM BLDC machine in which the back EMF waveforms are trapezoidal, rather than sinusoidal. Thus, the d–q modeling technique developed for the PM synchronous machine is not applicable. Since the flux linkage is nonsinusoidal, it is necessary to derive the model of the PM brushless machine in terms of its state variables.

Assuming that the rotor-induced currents due to the stator harmonics are negligible, the dynamic equation of this PM BLDC machine model can be expressed as (Pillay and Krishnan, 1988)

$$\begin{bmatrix} v_a \\ v_b \\ v_c \end{bmatrix} = \begin{bmatrix} R & 0 & 0 \\ 0 & R & 0 \\ 0 & 0 & R \end{bmatrix} \begin{bmatrix} i_a \\ i_b \\ i_c \end{bmatrix} + p \begin{bmatrix} L_{aa} & L_{ab} & L_{ac} \\ L_{ab} & L_{bb} & L_{bc} \\ L_{ac} & L_{bc} & L_{cc} \end{bmatrix} \begin{bmatrix} i_a \\ i_b \\ i_c \end{bmatrix} + \begin{bmatrix} e_a \\ e_b \\ e_c \end{bmatrix} \tag{4.22}$$

where v_a, v_b, and v_c are the instantaneous three-phase stator voltages; i_a, i_b, and i_c are the instantaneous three-phase stator currents; e_a, e_b, and e_c are the instantaneous three-phase back EMFs; R is the armature

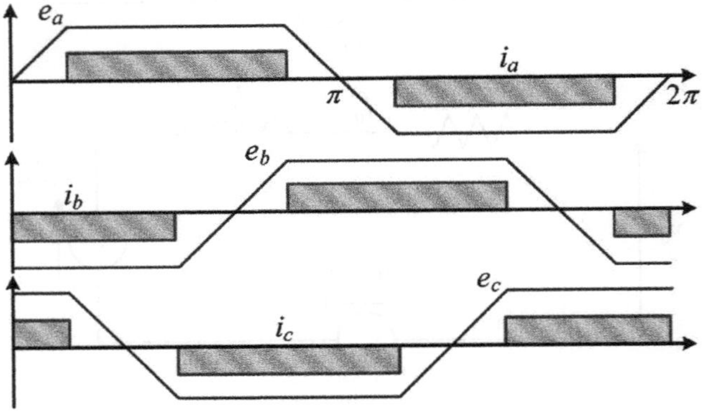

Figure 4.11 Operating waveforms of PM brushless DC machine

resistance; and L_{ij} ($i = a, b, c$ and $j = a, b, c$) are the mutual inductances which are symmetrical. As the PM BLDC machine generally adopts the surface-mounted PMs, there is no variation in the rotor reluctance with respect to the position. Hence, the self-inductances of all phases are equal to L, and the mutual inductances between phases are the same and equal to M. The dynamic equation can be written as

$$\begin{bmatrix} v_a \\ v_b \\ v_c \end{bmatrix} = \begin{bmatrix} R & 0 & 0 \\ 0 & R & 0 \\ 0 & 0 & R \end{bmatrix} \begin{bmatrix} i_a \\ i_b \\ i_c \end{bmatrix} + \begin{bmatrix} L & M & M \\ M & L & M \\ M & M & L \end{bmatrix} p \begin{bmatrix} i_a \\ i_b \\ i_c \end{bmatrix} + \begin{bmatrix} e_a \\ e_b \\ e_c \end{bmatrix} \quad (4.23)$$

Since $i_a + i_b + i_c = 0$, it yields $M\, i_b + M\, i_c = -M\, i_a$. Thus, the dynamic equation can be rewritten as

$$\begin{bmatrix} v_a \\ v_b \\ v_c \end{bmatrix} = \begin{bmatrix} R & 0 & 0 \\ 0 & R & 0 \\ 0 & 0 & R \end{bmatrix} \begin{bmatrix} i_a \\ i_b \\ i_c \end{bmatrix} + \begin{bmatrix} (L-M) & 0 & 0 \\ 0 & (L-M) & 0 \\ 0 & 0 & (L-M) \end{bmatrix} p \begin{bmatrix} i_a \\ i_b \\ i_c \end{bmatrix} + \begin{bmatrix} e_a \\ e_b \\ e_c \end{bmatrix} \quad (4.23)$$

which indicates that the phase voltage equation of this PM BLDC machine is similar to the armature voltage equation of the PM DC machine. Consequently, the equivalent circuit of this PM BLDC machine is depicted in Figure 4.12. Considering that all three phases are balanced and identical, a simplified equivalent circuit of one phase is shown in Figure 4.13, where $L_l = (L - M)$ is the leakage inductance.

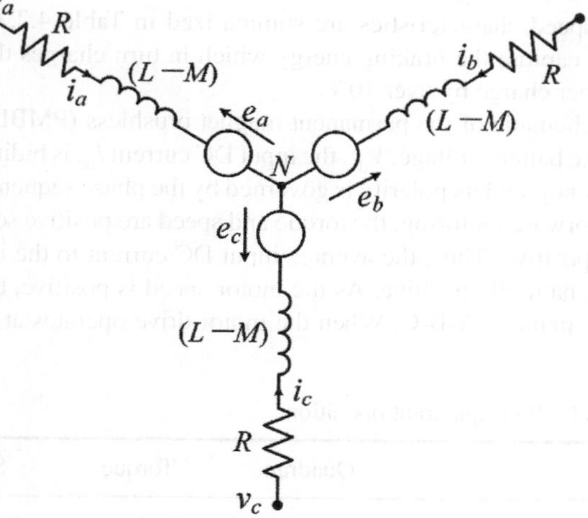

Figure 4.12 Equivalent circuit of PM brushless DC machine

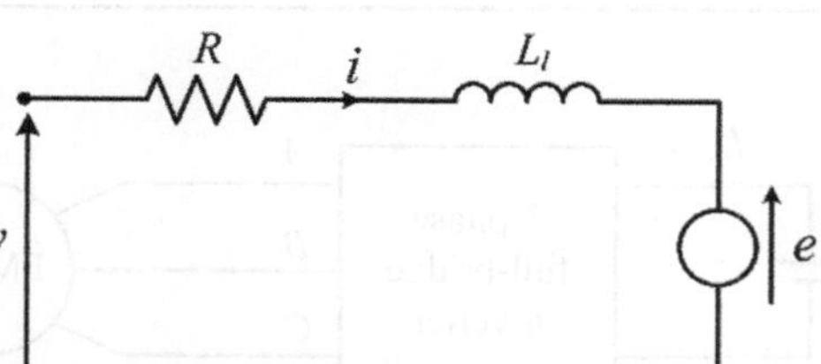

Figure 4.13 Simplified equivalent circuit of PM brushless DC machine

The developed torque T_e is the output power divided by the mechanical speed, which is given by

$$T_e = \frac{P_e}{\omega_r} = \frac{P}{2}\left(\frac{e_a i_a + e_b i_b + e_c i_c}{\omega_e}\right) \tag{4.24}$$

where P is the number of poles. As each phase back EMF is a bipolar trapezoidal waveform displaced by 120° (electrical), each current waveform needs to be 120° on positive and 120° on negative within a period to ensure the synchronism and hence produce the desired steady torque without torque pulsations.

4.4 Inverters for PM Brushless Motors

While the PM synchronous motor and PM BLDC motor are fed by sinusoidal current and rectangular current, respectively, the corresponding inverter topologies are essentially the same – a three-phase full-bridge inverter. Nevertheless, their switching schemes are different, namely the pulse-width modulation (PWM) control for BLAC operation of the PM synchronous motor, and the stepwise control for BLDC operation of the PM BLDC motor.

4.4.1 Inverter Requirements

EV propulsion requires the motor drives offering four-quadrant characteristics for forward motoring, forward regeneration, backward motoring, and backward regeneration. The definition of four quadrants and the corresponding torque-speed characteristics are summarized in Table 4.2. Particularly, the forward regeneration is essential to capture the braking energy which in turn charges the EV battery, leading to increase the driving range per charge by over 10%.

Figure 4.14 shows the schematic of the permanent magnet brushless (PMBL) AC or DC motor drive where the input voltage is the battery voltage, V_{dc}, the input DC current I_{dc} is bidirectional, the motor speed depends on the stator frequency, and its polarity is governed by the phase sequence. When the motor drive operates at Quadrant I for forward motoring, the torque and speed are positive so that the input and output powers of the inverter are positive. Thus, the average input DC current to the inverter has to be positive since the battery voltage is naturally positive. As the motor speed is positive, the phase sequence of the stator terminals is positive, namely, A-B-C. When the motor drive operates at Quadrant IV for forward

Table 4.2 Four-quadrant operation

	Quadrant	Torque	Speed
Forward motoring	I	Positive	Positive
Backward regeneration	II	Positive	Negative
Backward motoring	III	Negative	Negative
Forward regeneration	IV	Negative	Positive

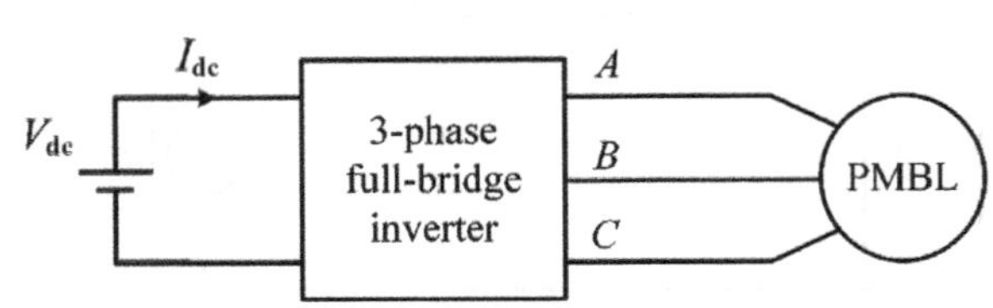

Figure 4.14 Schematic of PM brushless motor drive

Table 4.3 Inverter requirements for four-quadrant operation

	Quadrant	DC current	AC sequence
Forward motoring	I	Positive	Positive
Backward regeneration	II	Negative	Negative
Backward motoring	III	Positive	Negative
Forward regeneration	IV	Negative	Positive

regeneration, the speed is positive but the torque is negative so that the input and output powers of the inverter are negative. Thus, the average input DC current becomes negative so that the braking energy is converted to charge the battery. As the motor speed is still positive, the stator phase sequence is positive. The operations at Quadrants III and II are similar to that at Quadrants I and IV, respectively, except that the motor speed is negative and the corresponding stator phase sequence is negative, namely, A-C-B. The input DC current and stator AC phase sequence for four-quadrant operation of the PM BLAC or BLDC motor drive are summarized in Table 4.3.

4.4.2 *Switching Schemes for Brushless AC Operation*

The power inverter topology used for the PM synchronous motor drive operating at the BLAC mode for EV propulsion is essentially the same as that for the induction motor drive. That is, the preferred power inverter is based on the voltage-fed full-bridge topology. As the inverter has to act on the commands of magnitude, frequency, and phase of the output voltages and currents, there are various switching schemes to achieve such control. Among them, the hysteresis current control and space-vector modulation are widely used for the PM synchronous motor drive. Detailed discussion of these two PWM switching schemes can be found in Section 3.3.1.

In the hysteresis current control scheme, the actual current is forced to track the reference current within a hysteresis band. The switching frequency and peak-to-peak current ripple are governed by the width of the hysteresis band. When the width of the hysteresis band increases, the switching frequency decreases but the current ripple increases. On the other hand, when the hysteresis band decreases, the switching frequency increases whereas the current ripple decreases, leading to a more sinusoidal current waveform but associated with a higher switching loss. As a result, the width of the hysteresis band has to be properly selected to compromise between the current harmonics and switching loss. Since the hysteresis current control has the advantages of simple to implement, fast transient response, direct limitation of device current, and practical insensitivity to machine parameters, it is widely adopted by the PM synchronous motor drive for EV propulsion. However, its major drawbacks are the relatively high switching frequency and hence high switching loss, and the time-varying switching frequency may result in undesirable current harmonics.

In the space-vector modulation scheme, the three-phase voltage quantities (v_a, v_b, v_c) are first combined to derive a voltage space vector which is then resolved into any two adjacent active vectors: namely six active vectors $\overline{V}_1(100)$, $\overline{V}_2(110)$, $\overline{V}_3(010)$, $\overline{V}_4(011)$, $\overline{V}_5(001)$, $\overline{V}_6(101)$ and two zero vectors $\overline{V}_0(000)$ and $\overline{V}_7(111)$. The six nonzero vectors shape the axes of a hexagonal and the two zero vectors are at the origin of the hexagon. Consequently, the voltage vector can be formulated in terms of the eight space vectors. Therefore, rather than using a separate modulator for each of the three phases, the interaction between the three phases is taken into account. The inverter not only controls the magnitude and angular velocity but also the angular position of the voltage vector. This position variable is actually the key to offer the advantages of lowering harmonics in both the output voltage and current as well as the switching loss. Compared with the hysteresis current control, the space-vector modulation is more complicated to be implemented.

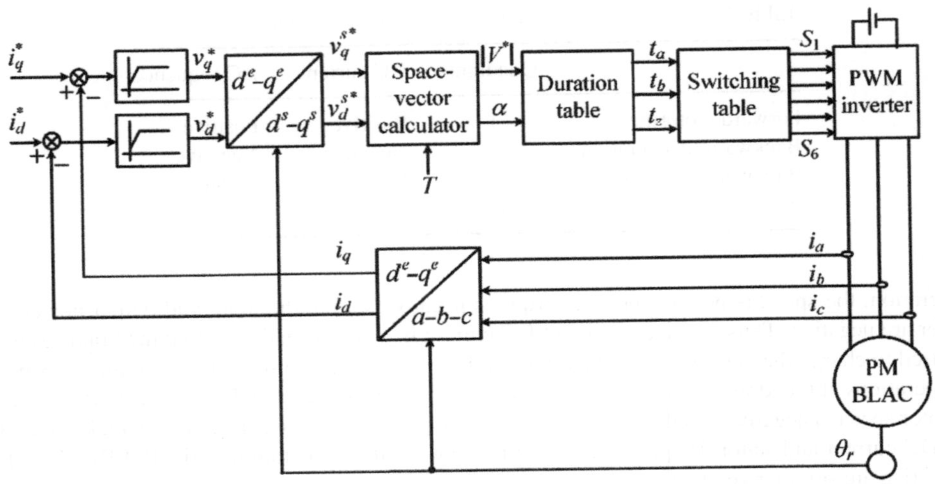

Figure 4.15 Implementation of space-vector modulation in PM synchronous motor drive

A generic implementation of space-vector modulation in the PM synchronous motor drive is shown in Figure 4.15 (Krishnan, 2010). On the basis of the errors between the d-axis and q-axis current commands and three-phase feedback currents, the d-axis and q-axis voltage commands in the rotor reference frame are resulted. After coordinate transformation, the resulting voltage commands in the stator reference frame are fed into the space-vector calculator which in turn calculates the desired magnitude and angle of the reference voltage vector at a given instant. Then, the sector is selected and the corresponding time durations of the required two switching vectors and zero vector are retrieved from a predefined table of durations. Consequently, the driving signals of power devices are extracted from another predefined table of switching devices. Notice that the sampling time T is an optional input to the space-vector calculator to enable varying the switching frequency when it is desirable.

4.4.3 Switching Schemes for Brushless DC Operation

The PM BLDC machine is driven by strokes in accordance with the rotor position. These strokes must be properly applied to the active phases of the three-phase armature winding so that the angle between the stator flux and the rotor flux is kept close to 90° to develop the maximum torque. There are two kinds of switching schemes to drive this PM BLDC machine (Safi, Acarnley, and Jack, 1995; Hu, Sathiakumar, and Shrivastava, 2009):

- Two-phase 120° conduction scheme
- Three-phase 180° conduction scheme.

In the two-phase 120° conduction scheme, at any instant, only two phases are conducted with the conduction interval of 120° while the remaining phase is nonconducted as depicted in Figure 4.16. The corresponding switching sequence is summarized in Table 4.4. For normal operation, the phase current waveform is near rectangular in shape and easily reaches the current demand as shown in Figure 4.17.

In the three-phase 180° conduction scheme, at any instant, all three phases are conducted with the conduction interval of 180° as depicted in Figure 4.18. The corresponding switching sequence is summarized in Table 4.5. For normal operation, the phase current waveform is near quasi-square in shape and easily reaches the demanded current as shown in Figure 4.19.

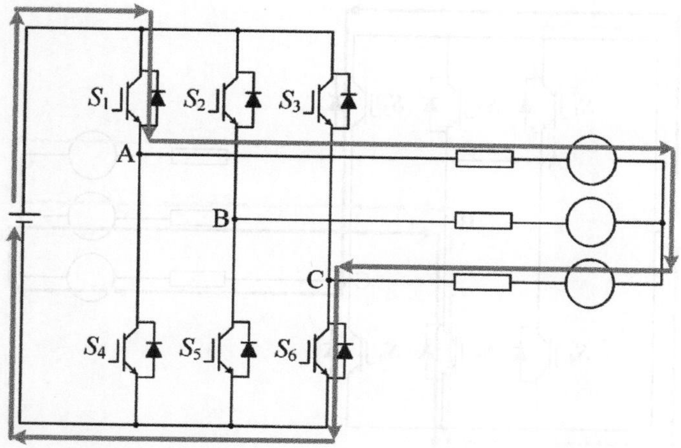

Figure 4.16 Current flow under two-phase 120° conduction scheme

Table 4.4 Switching sequence under two-phase 120° conduction scheme

Intervals (°)	Devices on	Phases A, B, C
0–60	S_1, S_6	+, 0, −
60–120	S_2, S_6	0, +, −
120–180	S_2, S_4	−, +, 0
180–240	S_3, S_4	−, 0, +
240–300	S_3, S_5	0, −, +
300–360	S_1, S_5	+, −, 0

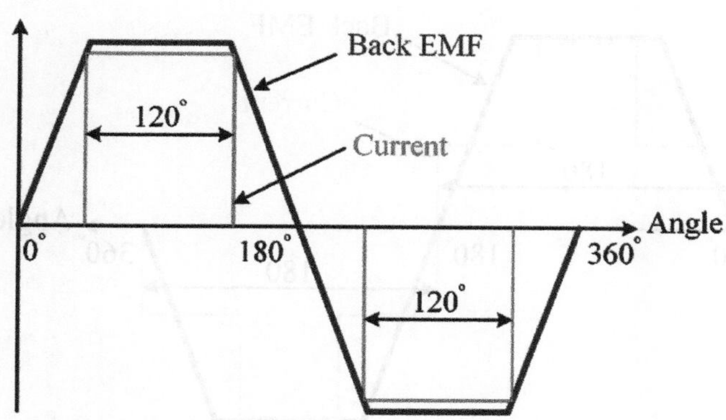

Figure 4.17 Phase back EMF and current waveforms under two-phase 120° conduction scheme

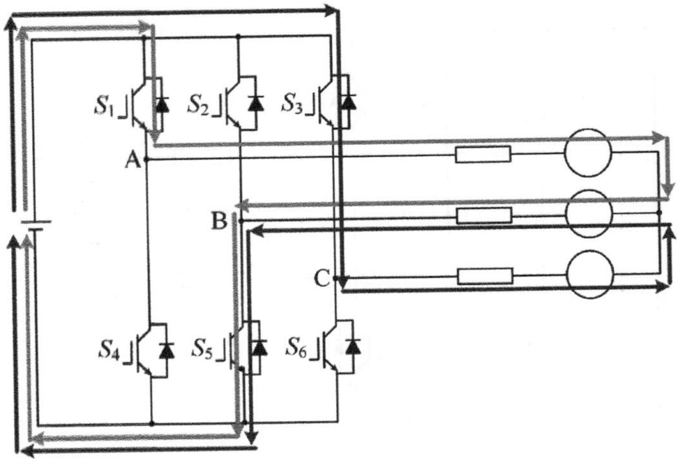

Figure 4.18 Current flow under three-phase 180° conduction scheme

Table 4.5 Switching sequence under three-phase 180° conduction scheme

Intervals (°)	Devices on	Phases A, B, C
0–60	S_1, S_5, S_3	+, −, +
60–120	S_1, S_5, S_6	+, −, −
120–180	S_1, S_2, S_6	+, +, −
180–240	S_4, S_2, S_6	−, +, −
240–300	S_4, S_2, S_3	−, +, +
300–360	S_4, S_5, S_3	−, −, +

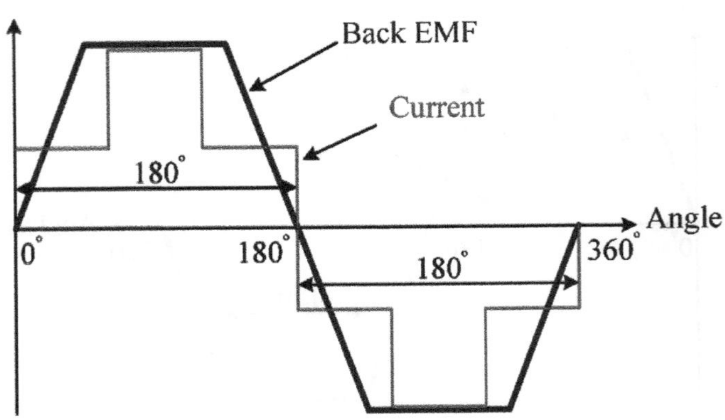

Figure 4.19 Phase back EMF and current waveforms under three-phase 180° conduction scheme

Between the 120° conduction and 180° conduction schemes, the former takes the advantage that it yields the highest torque capability for the same peak phase current which can maximize the developed torque and efficiency, whereas the latter requires only switching one device on or off in each switching sequence which can minimize the switching loss and chance of short circuit. In the 180° conduction scheme, as there are some intervals that the phase current is fed where the back EMF is not constant, the corresponding torque ripple is larger than that in the 120° conduction scheme. Therefore, the 120° conduction scheme is generally more preferable.

4.5 PM Brushless Motor Control

Since the operating waveforms of the PM synchronous and PM BLDC machines are different, the corresponding motor control strategies are naturally different. Essentially, the PM synchronous motor can adopt those control strategies that have been developed for the induction motor because both types of motors are based on sinusoidal waveforms, whereas the PM brushless motor needs to adopt dedicated control strategies because of its nonsinusoidal operating waveforms.

4.5.1 PM Synchronous Motor Control

As mentioned earlier, the PM synchronous motor can adopt those sophisticated control strategies that have been developed for the induction motor, such as the FOC and direct torque control. Between them, the FOC has been widely applied to the PM synchronous motor for EV propulsion. Meanwhile, differing from the induction motor, the PM synchronous motor adopts PM field excitation which is inherently uncontrollable. In order to provide the constant-power operation for EV cruising, the flux-weakening control of the PM synchronous motor is highly desirable. On the other hand, in order to get rid of the costly position encoder, the position sensorless control has been actively developed for the PM synchronous motor. Therefore, these three prominent control strategies will be discussed.

4.5.1.1 Field-Oriented Control of PM Synchronous Motor

The FOC of the PM synchronous motor is similar with that of the induction motor, except that their torque-production mechanisms are different. Taking into account the inertia J and viscous friction B of the entire system, the motion equation can be expressed as

$$T_e = T_l + J\frac{d\omega_r}{dt} + B\omega_r \tag{4.25}$$

where T_l is the mechanical load torque. Substituting Eqs. (4.17) and (4.21) into Eq. (4.25), the system dynamics of the PM synchronous motor in the *d–q* frame can be deduced as illustrated in Figure 4.20.

In order to implement the FOC, the sensed currents are first transformed from the *a–b–c* frame into the *d–q* frame. For such transformation, the flux angle θ_e is required which is deduced from the position sensor such as a position encoder or resolver. In general, the deviation between the reference speed and feedback speed is used to deduce the *q*-axis current reference via the speed proportional-integral (PI) regulator. Then, two PI controllers are used to regulate the *d*-axis current and *q*-axis current, hence the flux and torque, respectively (Nam, 2010). Mathematically, the corresponding v_d and v_q are given by

$$\begin{cases} v_d = G_d\left(i_d^* - i_d\right) - \omega_e L_q i_q \\ v_q = G_q(i_q^* - i_q) + \omega_e L_d i_d + \omega_e \lambda_m \end{cases} \tag{4.26}$$

where G_d and G_q are the transfer functions of relevant PI controllers. Consequently, the resulting v_d and v_q are fed into the space-vector PWM modulator to produce the desired switching signals for the power inverter. The whole block diagram of this PM synchronous machine using FOC is shown in Figure 4.21.

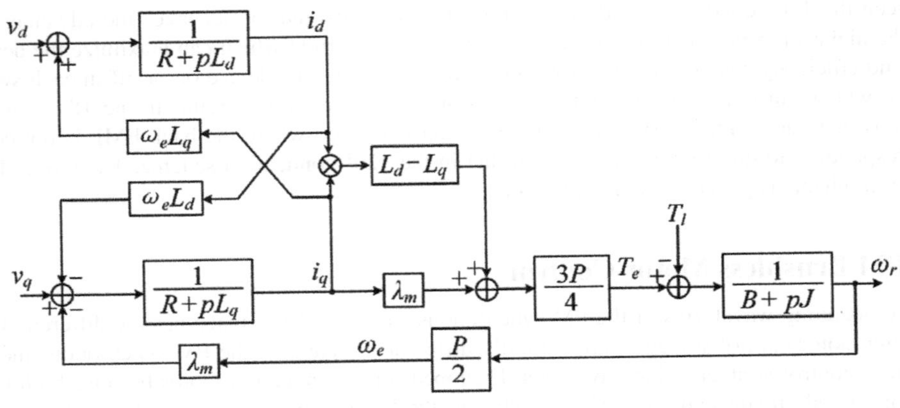

Figure 4.20 System dynamics of PM synchronous motor

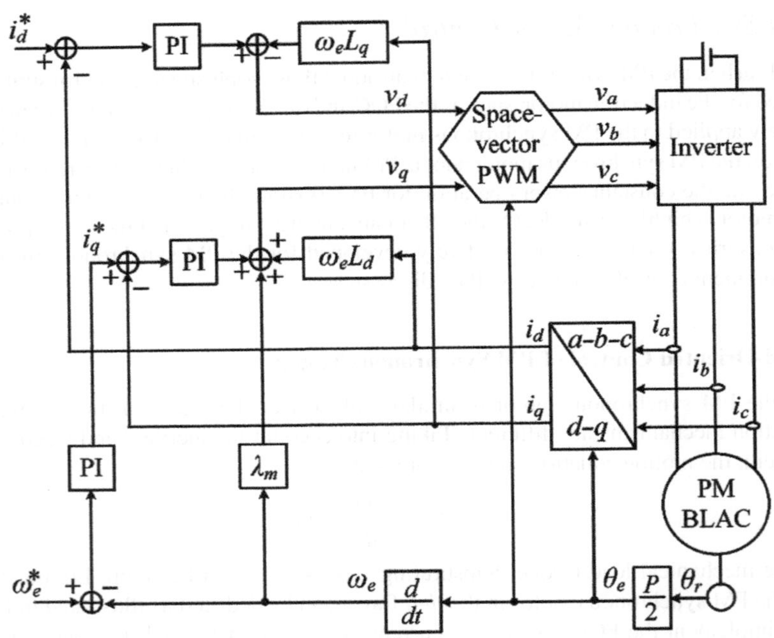

Figure 4.21 Field-oriented control of PM synchronous motor

4.5.1.2 Flux-Weakening Control of PM Synchronous Motor

At the base speed of the PM synchronous machine, the terminal voltage reaches the rated voltage. Since the back EMF grows with the speed, the speed range can only be extended by reducing the air-gap flux, the so-called flux-weakening operation. Thus, the torque decreases while the speed increases, leading to keep the power constant, the so-called constant-power operation.

In order to realize the flux-weakening control of the PM synchronous motor, the current and voltage vectors, as given by Eq. (4.17), should be controlled in such a way that the d-axis armature current is negative, while the q-axis armature current is positive. Figure 4.22 depicts the corresponding vectors when

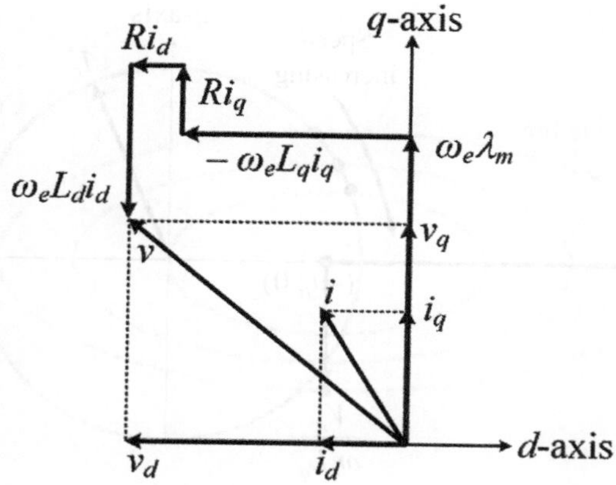

Figure 4.22 Current and voltage vectors of flux-weakening control of PM synchronous motor

$i_d < 0$ and $i_q > 0$. It can be observed that the total flux linkage and hence the back EMF can be compensated by the induced voltage with negative *d*-axis armature current (Schiferl and Lipo, 1990; Soong and Ertugrul, 2002). Thus, by increasing i_d in the negative direction, the back EMF can be significantly weakened.

For high-speed operation, the voltage drop across the stator resistance is negligible. Thus, the steady-state relationship of Eq. (4.17) can be written as

$$\begin{cases} v_d = -\omega_e L_q i_q \\ v_q = \omega_e L_d i_d + \omega_e \lambda_m \end{cases} \tag{4.27}$$

Taking V_r as the rated voltage limit, the *d*- and *q*-axis voltages are governed by

$$v_d^2 + v_q^2 \leq V_r^2 \tag{4.28}$$

Substituting Eq. (4.27) in Eq. (4.28) and expressing the PM flux linkage λ_m as a product of *d*-axis inductance L_d and a virtual field current source i_f, it yields

$$\frac{(i_d + i_f)^2}{V_r^2/(\omega_e L_d)^2} + \frac{i_q^2}{V_r^2/(\omega_e L_q)^2} \leq 1 \tag{4.29}$$

which represents a set of ellipses with different speeds. The higher the speed ω_e, the smaller the ellipse is resulted (Nam, 2010), leading to shrink to the center $(-i_f, 0)$ as depicted in Figure 4.23 in which the *d*-axis and *q*-axis currents are also governed by the rated current I_r as given by

$$i_d^2 + i_q^2 \leq I_r^2 \tag{4.30}$$

The intersection points of the ellipses and the circle represent the solution pairs of (i_d, i_q) to achieve the desired flux-weakening operation. It can be observed that when the speed becomes very high, the *d*-axis current approaches $-I_r$ while the *q*-axis current approaches zero. Hence, using Eq. (4.16), the criterion of offering flux-weakening operation at infinite speed can be deduced as

$$\frac{L_d I_r}{\lambda_m} = 1 \tag{4.31}$$

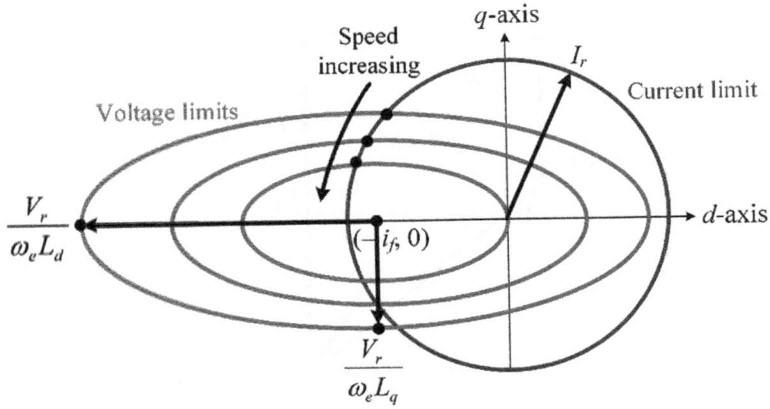

Figure 4.23 Voltage and current limits of flux-weakening control of PM synchronous motor

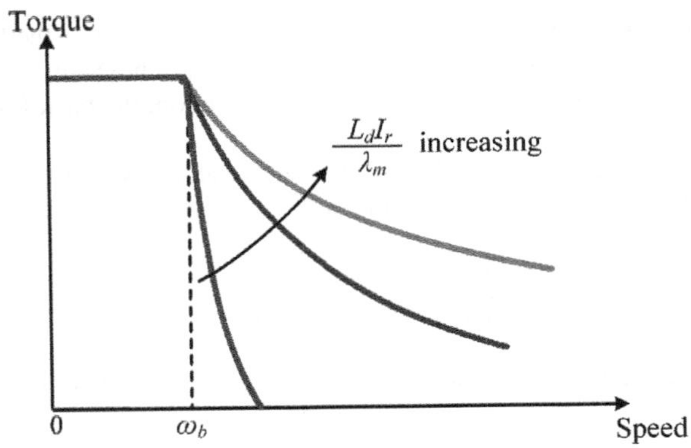

Figure 4.24 Torque-speed capabilities of flux-weakening control of PM synchronous motor

Although it is possible to design the PM synchronous motor satisfying the above criterion, it is generally with $L_d I_r / \lambda_m < 1$ because the d-axis inductance is relatively low as a consequence of the PM permeability approximately equal to unity. Figure 4.24 depicts the torque-speed capabilities of the PM synchronous motor beyond the base speed ω_b at different values of the ratio $L_d I_r / \lambda_m$. It can be found that the higher the ratio is adopted, the better the flux-weakening capability can be provided. In other words, if the PM flux linkage is set to a high value in order to achieve a high torque capability at low speeds, the speed range for flux-weakening operation will be sacrificed. On the other hand, if the rated current is very high, that is the motor is liquid cooled, the criterion Eq. (4.31) will also be satisfied even for the surface-mounted PM synchronous motor that has a low d-axis inductance and a high PM flux linkage. However, in general, it is easier for the interior PM synchronous motor to satisfy the criterion because it has a higher d-axis inductance.

4.5.1.3 Position Sensorless Control of PM Synchronous Motor

In general, the PM synchronous motor requires an accurate position sensor for the implementation of sophisticated control strategies. This position sensor is usually based on an optical encoder or a resolver, which sometimes costs as much as a low-power machine. Hence, in recent years, the position sensorless control of the PM synchronous motor has been actively developed, especially for those cost-sensitive applications. However, the cost of position sensor for EV motor drives, especially the PM synchronous motor drive, is relatively insignificant. The precision, reliability, and maturity of position sensorless technology are of concern. Therefore, the position sensorless technology is seldom adopted by the PM synchronous motor for EV propulsion. Although the position sensorless control of the PM synchronous motor is not highly desirable for EV propulsion, the corresponding technology can be considered as a fault-tolerant control during sensor failure.

There are many position sensorless techniques that have been developed for the PM synchronous motor. These techniques can be classified by various ways (Li and Zhu, 2008). Typically, they can be classified into four main methods (Krishnan, 2010):

- Current model adaptive scheme
- External signal injection scheme
- Current model-based injection scheme
- PWM carrier component scheme

4.5.2 PM Brushless DC Motor Control

As the PM BLDC motor is driven when the stator flux and the rotor flux is kept close to 90°, it can readily produce the maximum torque per ampere in the region of constant-torque operation. This enables the PM BLDC motor taking a definite advantage over the PM synchronous motor which relies on using the *d–q* coordinate transformation. However, in the absence of *d–q* coordinate transformation or FOC, the control scheme for constant-power operation of the PM BLDC motor is much more complicated than that of the PM synchronous motor (Chan *et al.*, 1998). Thus, the phase-advance angle control will be discussed which can enable the PM BLDC motor offering constant-power operation for EV cruising. Meanwhile, as the position sensorless control has been actively developed for the PM BLDC motor, it will also be briefed for possible application to EVs.

4.5.2.1 Phase-Advance Angle Control of PM Brushless DC Motor

When the PM BLDC motor operates at speeds higher than the base speed, it runs out of time to enable the phase current reaching the desired level because of the small difference between the applied voltage and back EMF (Safi, Acarnley, and Jack, 1995; Chan *et al.*, 1996). At this speed, the phase current may just reach the desired level and then be turned off at the end of the conduction period. By purposely advancing the turn-on angle of the phase current, dubbed as the phase-advance angle control, the phase current can have sufficient time to rise up and keep the current in phase with the back EMF under high-speed operation. Figure 4.25 illustrates the effect of phase-advance angle control. It can be found that the constant-power operation region can be significantly extended by gradually increasing the phase-advance angle.

This phase-advance angle control can be explained by using the voltage equation of the *j*th phase as given by

$$\begin{aligned} v_j\left(\omega t+\theta_o-\frac{(j-1)\pi}{3}\right) &= Ri_j\left(\omega t+\theta_o-\frac{(j-1)\pi}{3}\right) \\ &\quad + L\frac{di_j}{dt}\left(\omega t+\theta_o-\frac{(j-1)\pi}{3}\right)+e_j\left(\omega t-\frac{(j-1)\pi}{3}\right) \end{aligned} \tag{4.32}$$

where $j = 1–3$, v_j is the applied voltage, i_j is the phase current, e_j is the back EMF, R is the phase resistance, L is the phase self-inductance, and θ_o is the phase-advance angle or called the advanced conduction angle of the applied voltage leading ahead the back EMF. The control mechanism is to purposely employ

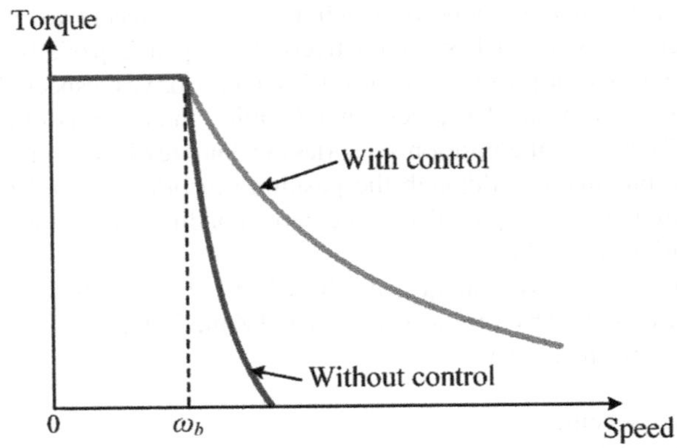

Figure 4.25 Torque-speed capabilities of phase-advance angle control of PM brushless DC motor

the inductance EMF across the phase self-inductance ($L\, ^{di_j}/_{dt}$) to counteract the back EMF, which may be even larger than the applied voltage at high speeds. This inductance EMF, proportional to the derivative of phase current, is controlled by the phase-advance angle. When this angle has a suitable positive value, at the beginning of conduction with a lower back EMF, the phase current quickly rises up and the inductance EMF is positive, which indicates that electromagnetic energy is stored in the phase winding. The phase current reaches the maximum value when the back EMF equals the applied voltage. After that, the back EMF is larger than the applied voltage. The phase winding begins to release the electromagnetic energy and the phase current decreases gradually. Since the corresponding inductance EMF becomes negative, it assists the applied voltage to counteract the back EMF. By varying the phase-advance angle, the phase current waveform is shaped in such a way that the phase current is properly regulated. Therefore, this phase-advance angle control can produce an equivalent flux-weakening effect to achieve the constant-power operation of the PM BLDC motor.

Between the 120° conduction and 180° conduction schemes for the PM BLDC motor, the former takes the advantage of higher torque capability at speeds below the base speed, whereas the latter has better constant-power operation performance at speeds higher than the base speed since it virtually extends the interval for which the phase winding is excited. Figure 4.26 compares the torque-speed capabilities of the

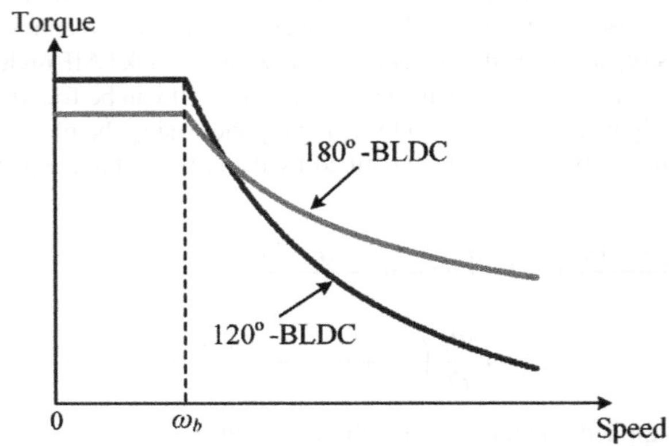

Figure 4.26 Comparison of torque-speed capabilities of two conduction schemes of PM brushless DC motor

PM BLDC motor using the 120° conduction and 180° conduction schemes with the use of phase-advance angle control for operating speeds above the base speed. It confirms that the 180° conduction scheme is more suitable for constant-power operation (Zhu and Howe, 2007).

4.5.2.2 Position Sensorless Control of PM Brushless DC Motor

Similar to the PM synchronous motor, the PM BLDC motor requires a position sensor to obtain the rotor position information in order to perform electronic commutation properly. Differing from the PM synchronous motor, which desires accurate position sensing using an encoder or a resolver, the PM BLDC motor usually requires a less demanding one, either a three-element Hall-effect or optical sensor. Thus, the position sensorless control of the PM BLDC motor is less desirable than that of the PM synchronous motor. Taking into account the fact that the cost of such position sensor is insignificant as compared with the total cost of the EV motor drive, the use of position sensorless control for the PM BLDC motor in EVs is not attractive. Nevertheless, the position sensorless technology can be considered useful for fault-tolerant control of the PM BLDC motor during sensor failure.

There are many position sensorless techniques that have been developed to estimate the commutation signals for the PM BLDC motor (Kim, Lee, and Ehsani, 2007). Typically, they can roughly be classified into five main methods (Krishnan, 2010):

- Estimation using machine model
- Induced EMF from sensing coils
- Inactive phase EMF sensing
- Third harmonic-induced EMF detection
- Artificial intelligent control.

4.6 Design Criteria of PM Brushless Motor Drives for EVs

PM brushless motors have become the preferred choice for EV propulsion because of their high efficiency and high power density. The key drawback of these motors is the expensive PM material. Thus, it is essential to use as little PM material as possible in order to reduce the cost without sacrificing the motor performance. In recent years, finite element analysis has been extensively used for PM motor designs (Rasmussen *et al.*, 2000). However, the preliminary motor dimensions should first be determined before one can proceed to finite element analysis; otherwise, the design process will be ad-hoc and tedious.

The sizing of PMs is one of the critical tasks of PM brushless motor design. Some analytical approaches have been developed to calculate the volume and size of PMs (Gieras and Wing, 2002), which can serve as a preliminary design for finite element analysis. As detailed in Mi (2006), the total PM volume V_m required for the PM synchronous machine can be estimated by

$$V_m = C_V \frac{P_1}{f B_r H_c} \tag{4.33}$$

where C_V is a coefficient which is in the range of 0.54–3.1 with a typical value of 2, P_1 is the input power, f is the supply frequency, B_r is the PM remanence, and H_c is the PM coercivity. When the motor adopts a series-magnet structure such as the surface-mounted, surface-inset, or interior-radial PM structures, the thickness (or height) h_m of each PM pole is given by

$$h_m = K_s \frac{K_a F_{ad}}{H'_c} \tag{4.34}$$

where K_s is the safety ratio, which can be chosen as 1.1, K_a is the maximum possible per-unit armature current, which is in the range of 4–8, F_{ad} is the d-axis armature reaction MMF, and H'_c is the

PM coercivity at the maximum operating temperature. On the other hand, when the motor adopts a parallel-magnet structure such as the interior-circumferential PM structure, the corresponding thickness is halved. Consequently, when the PMs are rectangular in shape, no matter for the series-magnet or parallel-magnet structure, the width (or breadth) b_m of each PM pole is given by

$$b_m = \frac{V_m}{2ph_m l_e} \tag{4.35}$$

where p is the number of pole-pairs and l_e is the effective PM length along the shaft direction, which is usually the same as the rotor stack length. If the motor adopts a series-magnet structure with arc-shaped PMs rather than rectangular PMs, the PM width will be accordingly represented by a PM radius or roughly labeled by the average width. Furthermore, for the PM BLDC machine, the volume and dimensions of PMs can be derived similarly.

The split ratio, the ratio of the stator inside diameter to the stator outside diameter, is another important design parameter for PM brushless machines, because it can significantly influence the corresponding torque capability and efficiency. The optimal split ratio for both the PM synchronous and BLDC machines has been analytically derived, which takes into account the influence of the winding disposition, air-gap flux density distribution, back EMF waveform, stator tooth-tip and end-winding (Pang, Zhu, and Howe, 2006).

When the PM brushless motor is designed to operate at the BLAC mode, it is the PM synchronous motor where the PM flux linkage and back EMF are sinusoidal. Assuming that the winding resistance is negligible, the output power P can be expressed as

$$P = \eta \frac{m}{T} \int_o^T e(t)i(t)dt = \eta \frac{m}{2} E_m I_m \tag{4.36}$$

where η is the efficiency, m is the number of phases, T is the period of back EMF, $e(t)$ and $i(t)$ are respectively the instantaneous value of back EMF and phase current, and E_m and I_m are respectively the amplitude of back EMF and phase current of the armature winding. The flux linkage of PMs on the armature winding ψ_{pm} can be expressed as

$$\psi_{\mathrm{pm}} = \psi_m \cos(\pi\theta_r/\theta_{\mathrm{cr}}) = \psi_m \cos(p\theta_r) \tag{4.37}$$

where ψ_m is the amplitude of PM flux linkage, θ_r is the angle of rotor position, p is the number of PM pole-pairs, and θ_{cr} is the PM pole-pitch. And, ψ_m can be expressed as

$$\psi_m = \sqrt{2}\frac{1}{p} k_w N_{\mathrm{ph}} B D_{\mathrm{si}} l_e \tag{4.38}$$

where k_w and N_{ph} are respectively the winding factor and number of turns of the armature winding, B is the magnetic loading, D_{si} is the stator inside diameter, and l_e is the effective axial length. Hence, e can be deduced as

$$e(t) = -d\psi_{\mathrm{pm}}/dt = \sqrt{2} N_{\mathrm{ph}} \omega_r k_w B l_e D_{\mathrm{si}} \sin(p\theta_r) \tag{4.39}$$

where ω_r is the rotor speed. The corresponding amplitude E_m is given by

$$E_m = \sqrt{2} N_{\mathrm{ph}} \omega_r k_w B l_e D_{\mathrm{si}} \tag{4.40}$$

On the other hand, the current amplitude I_m can be deduced as

$$I_m = \frac{\sqrt{2} A \pi D_{\mathrm{si}}}{2mN_{\mathrm{ph}}} \tag{4.41}$$

where A is the electric loading. Therefore, based on Eqs. (4.36)–(4.41), P can be rewritten as

$$P = \frac{\pi}{2}\eta k_w ABD_{si}^2 l_e \omega_r \tag{4.42}$$

Therefore, the output torque T can be obtained as

$$T = \frac{\pi}{2}\eta k_w ABD_{si}^2 l_e \tag{4.43}$$

In order to investigate the influence of split ratio on the output torque, the magnetic and electric loadings should be expressed in terms of size parameters. Thus, the amplitude of magnetic loading B_m is expressed as

$$B_m = \frac{4}{\pi}B_{pm}\sin\left(\frac{\pi}{2}\alpha_p\right) \tag{4.44}$$

where α_p is the pole-arc factor of PMs and B_{pm} is the operating point of PMs at no load. In addition, B_{pm} is constrained by

$$h_m = \mu_r \frac{l_g}{\left(\frac{B_r}{B_{pm}} - 1\right)} \tag{4.45}$$

where μ_r is the relative permeability of PMs, B_r is the remanence of PMs, l_g is the air-gap length, and h_m is the thickness of PMs. On the other hand, the electric loading can be expressed as

$$A = \frac{JA_s k_p N_s}{\pi D_{si}} \tag{4.46}$$

where J is the stator slot current density, k_p is the slot packing factor, N_s is the number of stator slots, and A_s is the stator slot area which is given by

$$A_s = \left\{\frac{\pi D_{so}^2}{4N_s}\left[\left(1 - \frac{2h_c}{D_{so}}\right)^2 - \left(\frac{D_{si}}{D_{so}}\right)^2\right] - w_t\left(\frac{D_{so} - D_{si}}{2} - h_c\right)\right\} \tag{4.47}$$

where h_c is the stator yoke thickness and w_t is the stator tooth width. The tooth width and yoke thickness of the stator can be expressed as

$$w_t = h_c = \frac{1}{2}\frac{B_m}{B_{max}}\frac{D_{si}}{p} \tag{4.48}$$

where B_{max} is the maximum magnetic flux density allowable in the stator core. Substituting Eq. (4.48) into Eq. (4.47), the stator slot area can be obtained as

$$A_s = \frac{\pi D_{so}^2}{4N_s}\left\{\left[\frac{1}{p}\left(\frac{1}{p} + \frac{N_s}{p\pi}\right)\left(\frac{B_m}{B_{max}}\right)^2 + \frac{N_s}{p\pi}\left(\frac{B_m}{B_{max}}\right) - 1\right]\left(\frac{D_{si}}{D_{so}}\right)^2 - \left(\frac{2}{p} + \frac{N_s}{p\pi}\right)\left(\frac{B_m}{B_{max}}\right)\left(\frac{D_{si}}{D_{so}}\right) + 1\right\} \tag{4.49}$$

And, D_{si} is constrained by

$$D_{si} = D_{sh} + 2h_r + 2h_m + 2l_g \tag{4.50}$$

where D_{sh} is the shaft diameter and h_r is the rotor yoke thickness.

Substituting Eqs. (4.44), (4.46), and (4.50) into Eq. (4.43), the variation of output torque with respect to the split ratio D_{si}/D_{so} can be obtained. In general, there is an optimal split ratio for the maximum output torque, which corresponds to a particular magnetic loading. When the PM brushless motor is designed to operate at the BLDC mode, the optimal split ratio can also be derived in a similar manner.

Recently, the double-stator PM brushless motor has been actively developed for EV application because it can offer exceptionally high torque density (Niu, Chau, and Yu, 2009). From the structural point of view, the double-stator PM brushless motor is composed of two coaxial PM brushless motors: its inner motor is an outer-rotor PM brushless motor, while its outer motor is a conventional PM brushless motor. Its total output torque is the resultant of the torque components produced by the interactions of the inner and outer armature windings with the inner and outer air-gap flux densities, respectively. Thus, the aforementioned design criteria for the optimal split ratio of a conventional PM brushless motor can readily be extended to the double-stator PM brushless motor to achieve the maximum torque density (Wang *et al.*, 2011).

4.7 Design Examples of PM Brushless Motor Drives for EVs

Because of inherently high power density and high torque density, PM brushless motor drives are particularly attractive for in-wheel drive where the usable space to accommodate the motor drive is highly limited. There are two major types of in-wheel drives: a high-speed motor coupled with a built-in reduction gear and a low-speed outer-rotor direct-drive motor. Therefore, two PM brushless motor drives, the planetary-geared PM synchronous motor drive and the outer-rotor PM BLDC motor drive, will be designed and analyzed for exemplification.

4.7.1 Planetary-Geared PM Synchronous Motor Drive

When the PM synchronous motor is designed for high-speed operation, the required size and weight can be significantly reduced, leading to achieve high power density and high torque density. Meanwhile, in order to match the motor speed and the wheel speed, a single-step planetary gear is integrated with the motor in such a way that the sun gear is coupled with the PM rotor and the ring gear is mounted with the wheel rim. Of course, the overall power density and torque density of this PM synchronous motor drive, including the high-speed motor and planetary gear, should be higher than those of a low-speed direct-drive PM synchronous motor drive.

On the basis of the requirements of a typical passenger EV, the specifications of the planetary-geared PM synchronous motor drive are listed in Table 4.6.

Aiming to have a simple and robust structure to withstand high-speed operation, the three-phase four-pole interior-radial PM synchronous machine topology is adopted. Figure 4.27 shows its exploded diagram. On the basis of the specifications, the geometric dimensions and parameters of this motor can be initialized as listed in Table 4.7.

In order to take into account the saturation and fringing effects, the finite element analysis is extensively used for PM brushless motor designs. The electromagnetic field distribution of the PM synchronous motor at the rated condition is shown in Figure 4.28. It can be observed that when the armature current density is equal to 4.2 A/mm^2, there is no significant magnetic saturation.

Table 4.6 Specifications of planetary-geared PM synchronous motor drive

DC voltage	360 V
Rated power	6.3 kW
Rated speed	3000 rpm
Rated torque	20 N m
Constant-torque operation	0–3000 rpm
Constant-power operation	3000–9000 rpm
Gear ratio	10 : 1

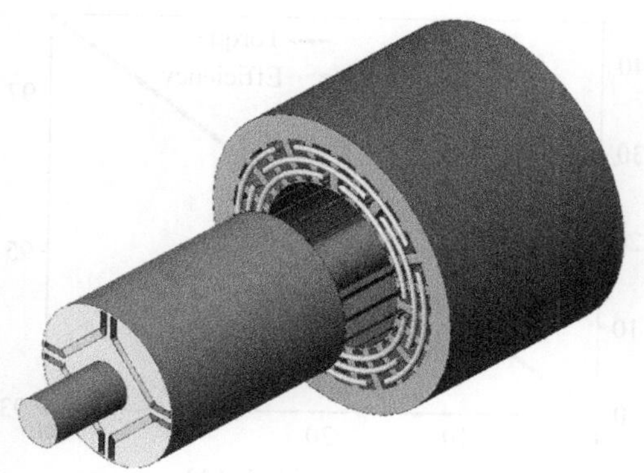

Figure 4.27 Three-phase four-pole interior-radial PM synchronous motor

Table 4.7 Initialization of PM synchronous motor design

Phase number	3
Pole number	4
Slot number	36
Stator outside diameter	140 mm
Rotor diameter	76.4 mm
Shaft diameter	22.9 mm
Air-gap length	0.3 mm
Core length	100 mm
Stator yoke thickness	12 mm
Rotor yoke thickness	7.5 mm
PM dimensions	$3.82 \times 30.6 \times 100\ \text{mm}^3$
Turns per slot	28
Slot-fill factor	60%
Lamination material	35JN210

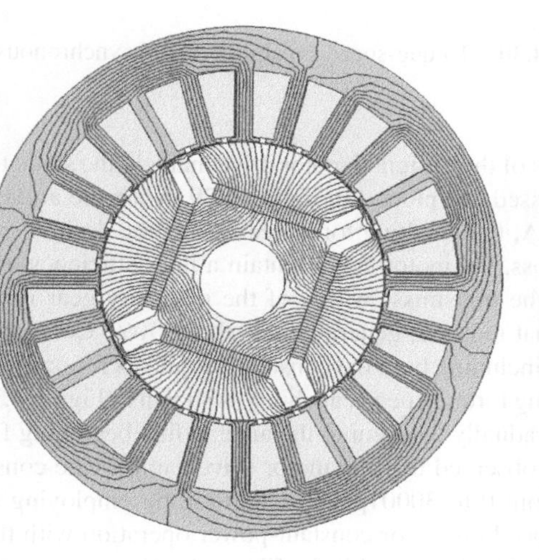

Figure 4.28 Electromagnetic field distribution of PM synchronous motor

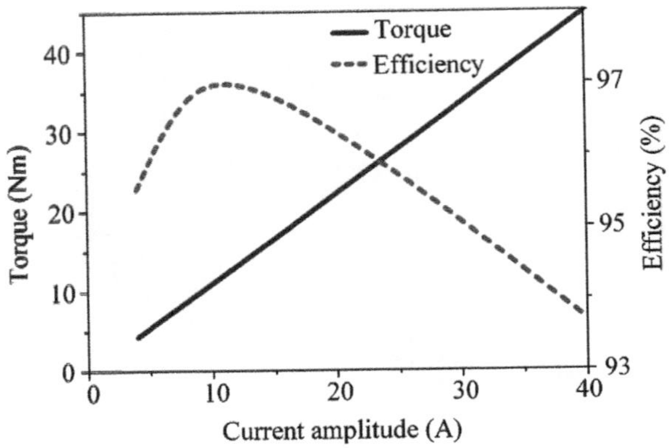

Figure 4.29 Static characteristics of torque and efficiency versus current of PM synchronous motor drive

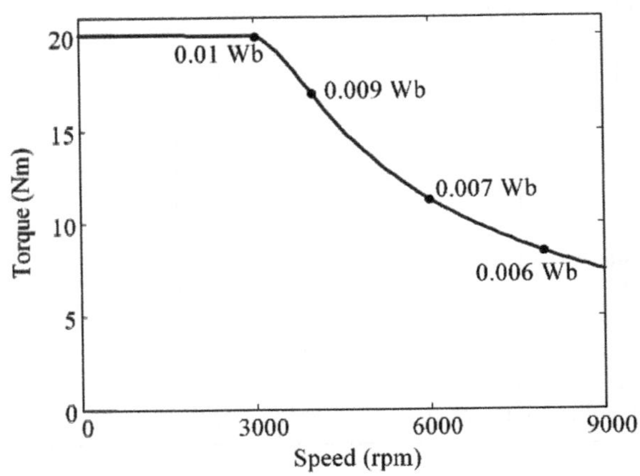

Figure 4.30 Torque-speed capability of PM synchronous motor drive

The static characteristics of the designed motor drive, namely the static torque and efficiency versus the armature current, are assessed. As plotted in Figure 4.29, when the armature current amplitude is equal to the rated value of 18.3 A, the resulting torque can reach the desired value of 20 N m. Because of the absence of rotor copper loss, the motor can maintain a high efficiency, mostly beyond 95% for normal operation. Notice that if the transmission loss of the planetary gear is taken into account, the overall efficiency will be somewhat reduced, generally still above 90%. Consequently, the operating regions of the designed motor drive, including both the constant-torque operation and constant-power operation, are assessed. The corresponding torque-speed capability is shown in Figure 4.30 in which the constant-power operation is realized by gradually weakening the air-gap flux beginning from 0.01 Wb with the increase of motor speed. It can be observed that the motor drive can provide constant-torque operation with the rated torque of 20 N m from 0 to 3000 rpm. In addition, by employing flux-weakening control, it can significantly extend the speed range for constant-power operation with the rated power of 6.3 kW from 3000 to 9000 rpm, which is highly favorable for EV application.

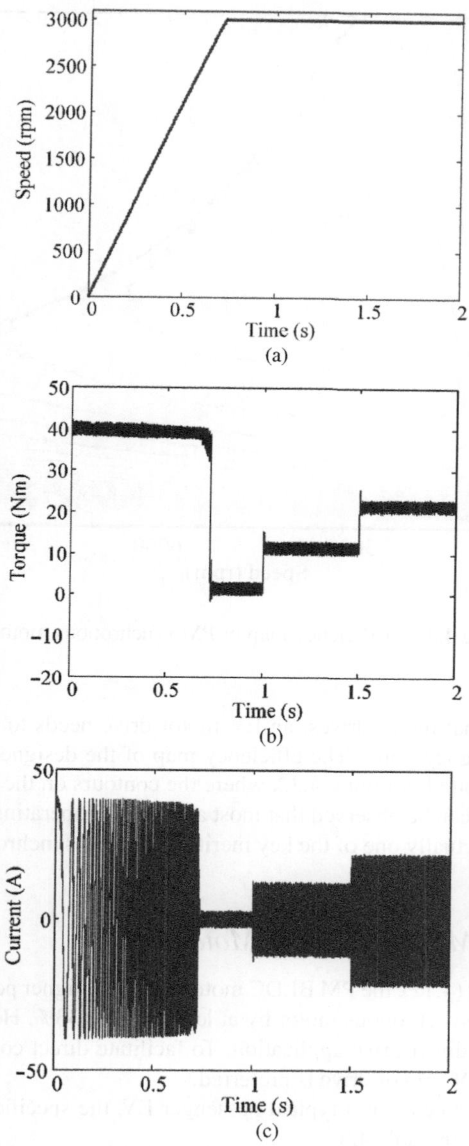

Figure 4.31 Transient responses of PM synchronous motor drive (a) speed, (b) torque, and (c) current

In order to evaluate the transient responses of the designed motor drive, it is first started up at no-load to the speed of 3000 rpm. After reaching the steady state, a half-load torque of 10 N m is suddenly applied. Then, after reaching the new steady state, a full-load torque of 20 N m is suddenly applied. As illustrated in Figure 4.31, it can be observed that the required start-up time is about 0.73 seconds, and the response time to sudden load changes is very short. In addition, the torque ripple at the rated torque is within 20%, which is very acceptable. Such low-percentage torque ripple is mainly due to the inherent feature of sinusoidal current and flux waveforms.

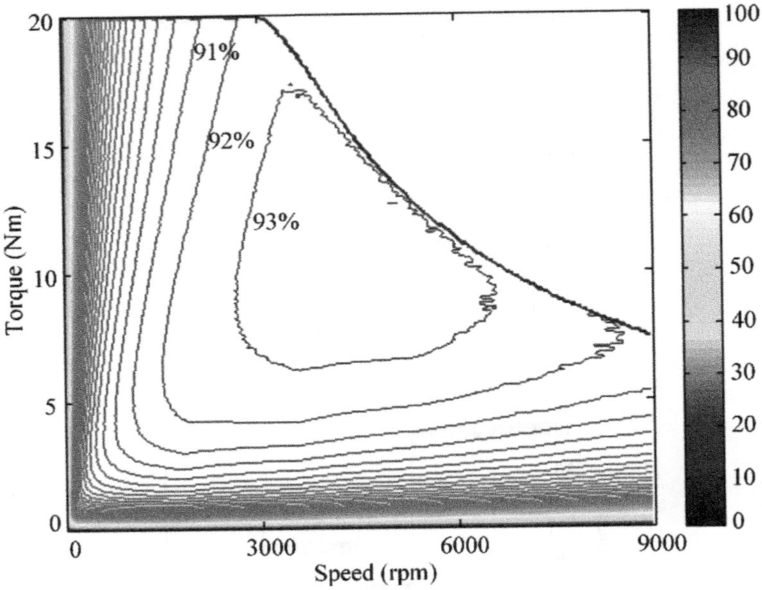

Figure 4.32 Efficiency map of PM synchronous motor drive

Differing from conventional motor drives, an EV motor drive needs to offer a good efficiency map, rather than an optimal efficiency point. The efficiency map of the designed motor drive embracing the entire operating region is plotted in Figure 4.32, where the contours on the torque-speed plane represent the operating efficiencies. It can be observed that most areas of the operating region can offer efficiencies higher than 90%, which is actually one of the key merits of the PM synchronous motor drive.

4.7.2 Outer-Rotor PM Brushless DC Motor Drive

As derived in Eqs. (4.11) and (4.12), the PM BLDC motor can offer higher power density and hence higher torque density than the PM synchronous motor by at least 15% or 33%. Hence, the PM BLDC motor is very attractive for in-wheel direct-drive application. To facilitate direct coupling between the rotor and rim, the outer-rotor PM BLDC motor drive is preferred.

On the basis of the requirements of a typical passenger EV, the specifications of the outer-rotor PM BLDC motor drive are listed in Table 4.8.

Table 4.8 Specifications of outer-rotor PM brushless DC motor drive

DC voltage	360 V
Rated power	6.3 kW
Rated speed	300 rpm
Rated torque	200 N m
Constant-torque operation	0–300 rpm
Constant-power operation	300–900 rpm

Figure 4.33 Three-phase 38-pole surface-mounted PM brushless DC motor

Table 4.9 Initialization of PM brushless DC motor design

Phase number	3
Pole number	38
Slot number	36
Rotor outside diameter	250 mm
Stator diameter	207.4 mm
Shaft diameter	80 mm
Air-gap length	0.3 mm
Core length	100 mm
Stator yoke thickness	20 mm
Rotor yoke thickness	15 mm
PM dimensions	$17.7 \times 6 \times 100\ \text{mm}^3$
Turns per slot	48
Slot-fill factor	60%
Lamination material	35JN210

In order to provide the desired low-speed direct-drive operation, a large number of PM poles are required. Thus, the three-phase 38-pole surface-mounted PM BLDC machine topology is adopted. Figure 4.33 shows its exploded diagram. On the basis of the specifications, the geometric dimensions and parameters of this motor can be initialized as listed in Table 4.9. On the inner rotor surface, there are 38 PM pieces mounted alternately to form 38 poles, and two adjacent poles make up a pair of poles so that the flux paths of different pole pairs are independent. This multipole magnetic circuit arrangement enables to reduce the magnetic iron yoke, resulting in the reduction of volume and weight. The coil span of the armature winding in the stator is designed to be equal to the slot pitch, thus the end-winding and hence copper material can be significantly reduced, leading to further reduction of volume and weight. By using the fractional number of slots per pole per phase, the magnetic force between the stator and rotor at any rotating position is uniform, thus minimizing the cogging torque that usually occurred in PM BLDC motors (Wang *et al.*, 2002).

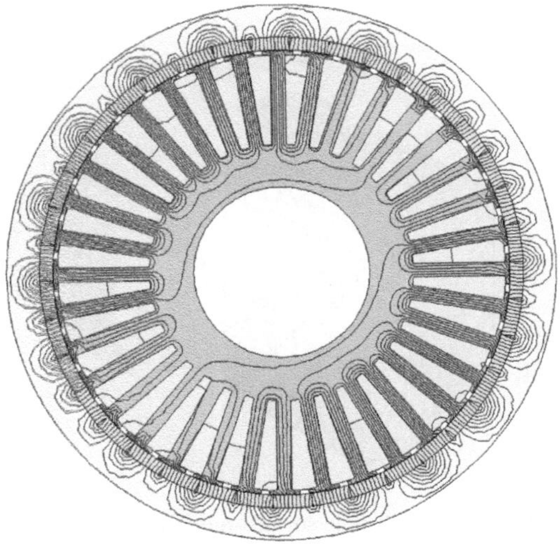

Figure 4.34 Electromagnetic field distribution of PM brushless DC motor

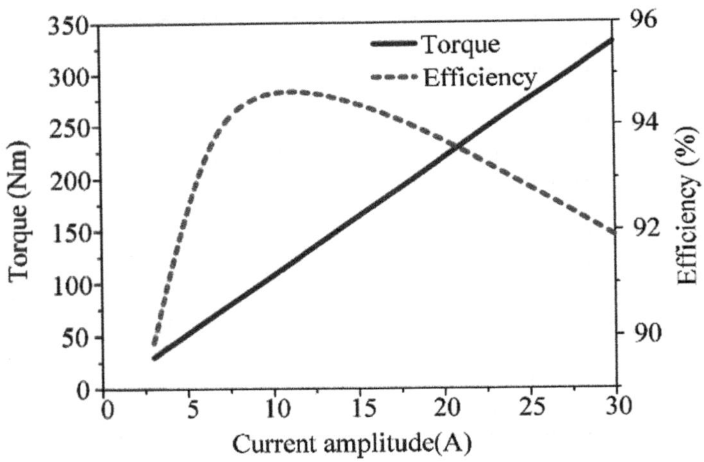

Figure 4.35 Static characteristics of torque and efficiency versus current of PM brushless DC motor drive

Firstly, the electromagnetic field distribution of the PM BLDC motor at the rated condition is analyzed as shown in Figure 4.34. It can be observed that when the armature current density is equal to 5.7 A/mm^2, there is no significant magnetic saturation. Then, the static torque and efficiency versus the armature current amplitude are plotted in Figure 4.35. It can be found that when the armature current amplitude is equal to the rated value of 18.2 A, the resulting torque can reach the desired value of 200 N m. Meanwhile, the motor can maintain a high efficiency, mostly beyond 90%, for normal operation. It should be noted that the efficiency of this PM BLDC motor drive is slightly lower than the previous PM synchronous motor drive because of the higher copper loss under the low-speed design. Nevertheless, because of the absence of gear loss, the overall efficiency of this PM BLDC motor drive is similar to that of the previous PM synchronous motor drive.

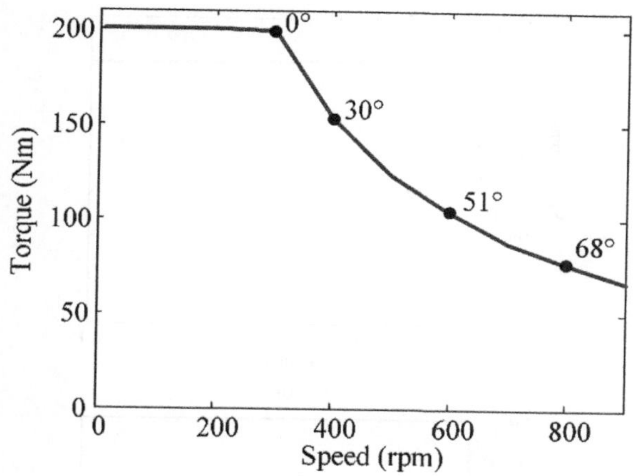

Figure 4.36 Torque-speed capability of PM brushless DC motor drive

Consequently, the torque-speed capability of the designed motor drive, including both the constant-torque operation and constant-power operation, is shown in Figure 4.36. It can be found that the motor drive can provide constant-torque operation with the rated torque of 200 N m from 0 to 300 rpm, and constant-power operation with the rated power of 6.3 kW from 300 to 900 rpm. Meanwhile, by using the phase-advance angle control, the motor drive can significantly extend the speed range for constant-power operation, which is highly favorable for EV application.

The transient responses of the designed PM BLDC motor drive is shown in Figure 4.37, where the motor is first started up at no-load to the speed of 300 rpm, then to a sudden half-load torque of 100 N m, and finally to a sudden full-load torque of 200 N m. It can be observed that the required start-up time and the response time to sudden load changes are very short. It should be noted that the required start-up time of this PM BLDC motor drive is much shorter than that of the previous PM synchronous motor drive, which is simply due to its much larger starting torque under the same no-load torque. When taking into account the gearing effect, their response times should be similar. On the other hand, it can be found that the torque ripple of the PM BLDC motor is higher than that of the previous PM synchronous motor drive, which is actually due to the finite time for nonideal rectangular current commutation.

Finally, the efficiency map of the designed PM BLDC motor drive embracing the entire operating region is plotted in Figure 4.38. As expected, most areas of the operating region can offer efficiencies higher than 90%, which is highly desirable for EV propulsion.

4.8 Application Examples of PM Brushless Motor Drives in EVs

The Nissan Leaf, which stands for a Leading, Environmentally friendly, Affordable, Family car, is a five-door hatchback EV introduced by Nissan in 2010 as shown in Figure 4.39. Currently, it has the largest market share of worldwide EVs sold since 2010. Using its on-board 24 kW h lithium-ion battery, it can offer a driving range of 135 km per charge based on the US Environmental Protection Agency (EPA) profile. The corresponding propulsion system adopts an 80 kW and 254 N m front-axle PM synchronous motor drive (Nissan, 2014). With the single-speed transmission, it can offer the top speed of over 150 km/h.

The Smart Fortwo Electric Drive (ED) was launched in 2007, which is an electric version of the conventional Smart Fortwo. Daimler AG plans to mass-produce it with availability in major markets worldwide. As shown in Figure 4.40, it features a 55 kW PM BLDC motor drive, which can offer the top speed of 120 km/h with the single-speed transmission (Millikin, 2011). On the basis of a lithium-ion battery pack of

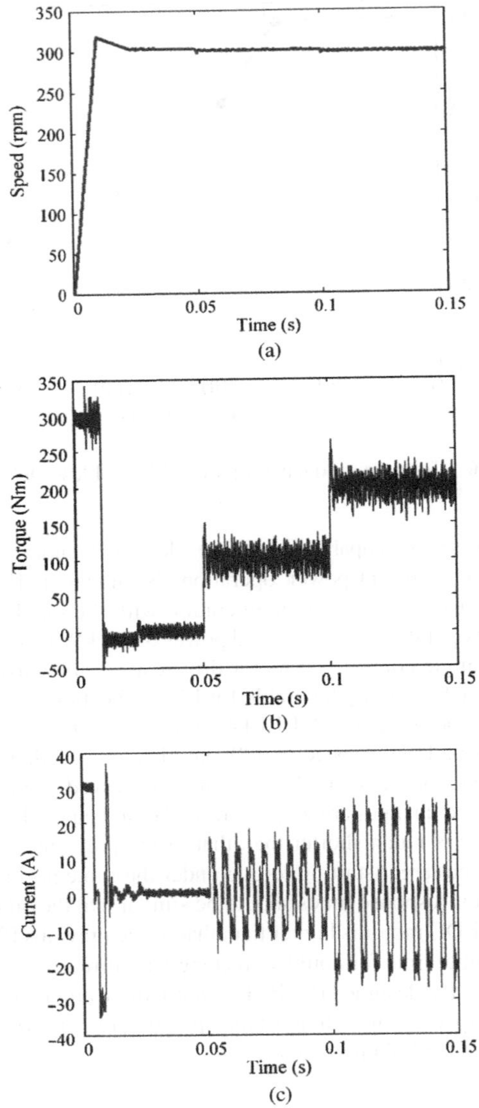

Figure 4.37 Transient responses of PM brushless DC motor drive: (a) speed, (b) torque, and (c) current

17.6 kW h, it can run 140 km per charge. With the torque of 130 N m, it can accelerate from 0 to 100 km/h in 11.5 seconds, which is essentially on par with that of the conventional Smart Fortwo.

4.9 Preferred Technology for EVs?

Among the three main types of commercially available EV motor drives, namely the DC, induction, and PM brushless, the PM brushless motor drives take the definite advantages of higher efficiency, higher power density, and higher torque density, which are highly desirable for EV propulsion. Currently, there is no doubt that the PM brushless motor drives are the preferred technology for EVs.

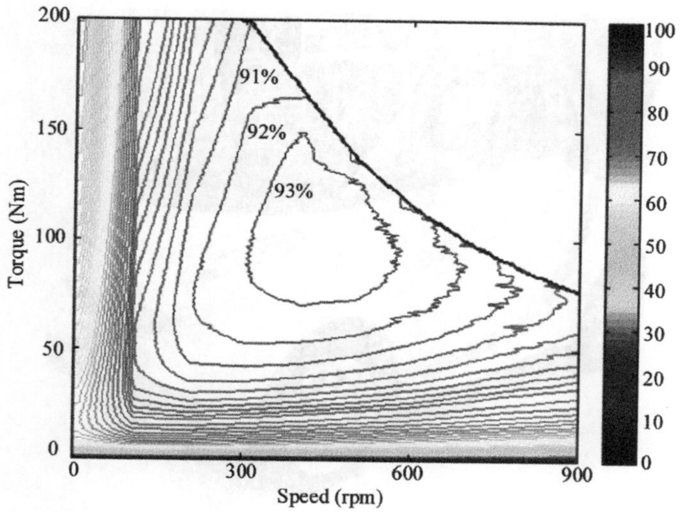

Figure 4.38 Efficiency map of PM brushless DC motor drive

Figure 4.39 Nissan Leaf

Between the two members of PM brushless motor drives, the PM synchronous motor drive is more mature than the PM BLDC motor drive while they have their individual merits. That is, the PM synchronous motor drive has relatively higher efficiency and lower torque ripple, whereas the PM BLDC motor drive has relatively higher power density and higher torque density. It is anticipated that both members will equally share the market of PM brushless motor drives.

The development of PM brushless motor drives heavily depends on the PM materials. In recent years, because of the high initial cost, shortage of supply, and thermal stability problem of PM materials, the rapid development of PM brushless motor drives has been quenched, leading to open up two new research and development directions for EV motor drives: the stator-PM motor drives and advanced magnetless motor drives. Nevertheless, the development of PM brushless motor drives is being further extended to the double-stator topologies to achieve exceptionally high torque density for EVs or the double-rotor topologies to achieve power splitting for hybrid EVs.

Figure 4.40 Smart Fortwo ED (*Source*: Courtesy Wikimedia Commons, http://en.wikipedia.org/wiki/Smart_Fortwo_ed#mediaviewer/File:2013_Smart_Fortwo_Electric_Drive_--_2012_NYIAS.JPG)

References

Chan, C.C., Chau, K.T., Jiang, J.Z. *et al.* (1996) Novel permanent magnet motor drives for electric vehicles. *IEEE Transactions on Industrial Electronics*, **43**, 331–339.

Chan, C.C., Xia, W., Jiang, J.Z. *et al.* (1998) Permanent-magnet brushless drives. *IEEE Industry Applications Magazine*, **4**, 16–22.

Gieras, J.F. and Wing, M. (2002) *Permanent Magnet Motor Technology: Design and Applications*, 2nd, Revised and Expanded edn, Marcel Dekker, Inc., New York.

Gutfleisch, O. (2000) Controlling the properties of high energy density permanent magnetic materials by different processing routes. *Journal of Physics D: Applied Physics*, **33**, R157–R172.

Hu, B., Sathiakumar, S., and Shrivastava, Y. (2009) 180-degree commutation system of permanent magnet brushless DC motor drive based on speed and current control. Proceedings of International Conference on Intelligent Computation Technology and Automation, pp. 723–726.

Kim, T., Lee, H.W. and Ehsani, M. (2007) Position sensorless brushless DC motor/generator drives: review and future trends. *IET Electric Power Applications*, **1**, 557–564.

Krishnan, R. (2010) *Permanent Magnet Synchronous and Brushless DC Motor Drives*, CRC Press, Boca Raton, FL.

Li, Y. and Zhu, H. (2008) Sensorless control of permanent magnet synchronous motor – A survey. Proceedings of IEEE Vehicle Power and Propulsion Conference, pp. 1–8.

Mi, C.C. (2006) Analytical design of permanent-magnet traction-drive motors. *IEEE Transactions on Magnetics*, **42**, 1861–1866.

Millikin, M. (2011) Third-generation smart fortwo electric drive to launch worldwide in spring 2012; first use of battery packs from Daimler JV Deutsche Accumotive and motor from JV EM-motive. Green Car Congress, http://www.greencarcongress.com/2011/08/smart-20110816.html (accessed September 2014).

Nam, K.H. (2010) *AC Motor Control and Electric Vehicle Applications*, CRC Press, Boca Raton, FL.

Nissan (2014) 2015 Nissan LEAF, Nissan, http://www.nissanusa.com/electric-cars/leaf/ (accessed September 2014).

Niu, S., Chau, K.T. and Yu, C. (2009) Quantitative comparison of double-stator and traditional permanent magnet brushless machines. *Journal of Applied Physics*, **105**, 07F105-1–07F105-3.

Ohm, D.Y. (2000) *Dynamic Model of PM Synchronous Motors*, Drivetech, pp. 1–10.

Pang, Y., Zhu, Z.Q. and Howe, D. (2006) Analytical determination of optimal split ratio for permanent magnet brushless motors. *IEE Electric Power Applications*, **153**, 7–13.

Pillay, P. and Krishnan, R. (1988) Modeling of permanent magnet motor drives. *IEEE Transactions on Industrial Electronics*, **35**, 537–541.

Rasmussen, K.F., Davies, J.H., Miller, T.J.E. *et al.* (2000) Analytical and numerical computation of air-gap magnetic fields in brushless motors with surface permanent magnets. *IEEE Transactions on Industry Applications*, **36**, 1547–1554.

Safi, S.K., Acarnley, P.P. and Jack, A.G. (1995) Analysis and simulation of the high-speed torque performance of brushless DC motor drives. *IEE Proceedings – Electric Power Applications*, **142**, 191–200.

Schiferl, R.F. and Lipo, T.A. (1990) Power capability of salient pole permanent magnet synchronous motors in variable speed drive application. *IEEE Transactions on Industry Applications*, **26**, 115–123.

Soong, W.L. and Ertugrul, N. (2002) Field-weakening performance of interior permanent-magnet motors. *IEEE Transactions on Industry Applications*, **38**, 1251–1258.

Wang, Y., Chau, K.T., Chan, C.C. and Jiang, J.Z. (2002) Transient analysis of a new outer-rotor permanent-magnet brushless DC drive using circuit-field-torque time-stepping finite element method. *IEEE Transactions on Magnetics*, **38**, 1297–1300.

Wang, Y., Cheng, M., Chen, M. *et al.* (2011) Design of high-torque-density double-stator permanent magnet brushless motors. *IET Electric Power Applications*, **5**, 317–323.

Zhu, Z.Q. and Howe, D. (2007) Electrical machines and drives for electric, hybrid, and fuel cell vehicles. *Proceedings of the IEEE*, **95**, 746–765.

5

Switched Reluctance Motor Drives

The switched reluctance (SR) machine is a specific kind of doubly salient machine. The earliest recorded one was built by Davidson in Scotland in 1838 to propel a locomotive. The SR motor drive could not provide its full potential until the advent of power electronics, digital control, and sensory circuitry. The modern SR motor drive is becoming attractive for electric vehicle (EV) propulsion because of its low system cost, robust machine structure, reliable converter topology, control simplicity, and high efficiency. On the other hand, it suffers from the drawbacks of large torque ripple, high machine nonlinearities, and annoying acoustic noise.

In this chapter, the SR motor drives, including their system configurations, SR machines, SR converters, and control strategies, are discussed. The corresponding design criteria, design examples, and application examples for EV propulsion are also provided.

5.1 System Configurations

In general, the SR motor drive consists of four main components: the SR machine, SR converter, sensor, and controller as shown in Figure 5.1. Among them, the SR machine plays a key role, that is, conversion of electrical energy into mechanical motion. Meanwhile, the SR converter is powered by a DC source, a DC link, or a battery pack, and the SR machine is fed by the SR converter, which is generally different from the inverter adopted by AC machines. The controller feeds driving signals to the power devices of the converter in accordance with the command signal, control algorithm, and sensory feedback. Consequently, with the use of proper control strategies, high performance can be achieved in the SR motor drive.

For EV propulsion, the system configuration of the SR motor drive can be single-motor or multiple-motor. The former adopts a single motor to propel the driving wheels, whereas the latter uses multiple motors to independently drive individual wheels. As shown in Figure 5.2, the single-motor configuration has the merits of using only one SR machine and one SR converter, which can minimize the corresponding size, weight, and cost. Of course, the SR machine and converter have to be designed to provide the desired capabilities for vehicular operation. Since a differential is needed to adjust the relative speeds of the driving wheels for rounding a curved road, the use of mechanical gearing is inevitable. The SR machine is normally designed for high-speed operation, typically over 10 000 rpm, so as to reduce the size and weight. In order to match the wheel speed, typically less than 1000 rpm, a fixed gear (FG) is usually employed to step down the SR motor speed by about 10 times.

Compared with the induction and permanent magnet (PM) brushless machines, the SR machine takes the definite advantages of robust structure and heat tolerance, especially the absence of copper windings or PMs in the rotor. Thus, it is becoming attractive to adopt the multiple-motor configuration of

Electric Vehicle Machines and Drives: Design, Analysis and Application, First Edition. K. T. Chau.

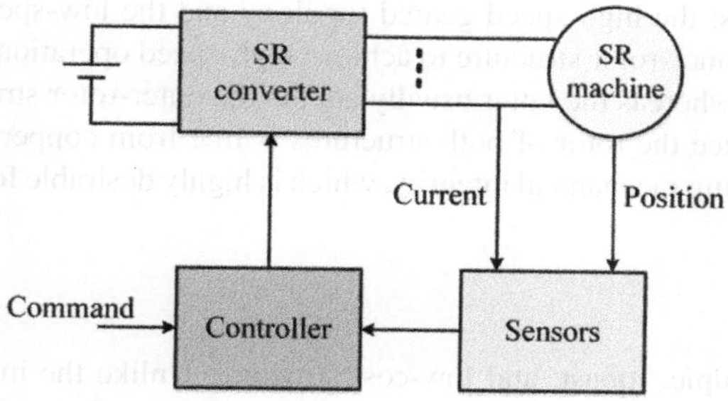

Figure 5.1 Basic configuration of SR motor drive

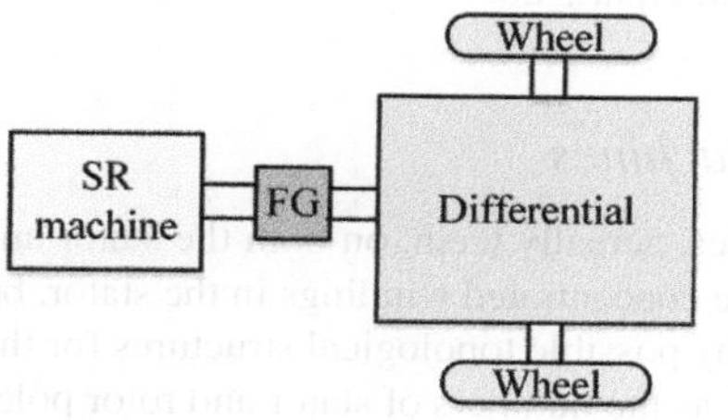

Figure 5.2 Single-motor SR motor drive configuration for EV propulsion

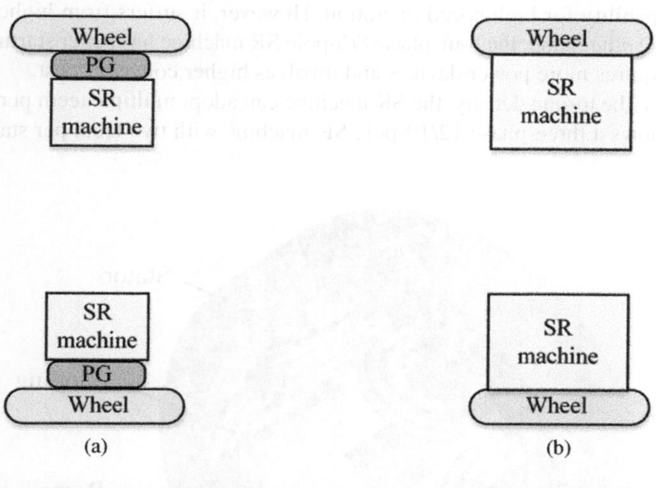

Figure 5.3 Dual-motor SR motor drive configurations for EV propulsion: (a) high-speed geared structure and (b) low-speed gearless structure

SR motor drives for in-wheel EV propulsion. In general, either quad-motor configuration or dual-motor configuration is adopted for passenger cars. The dual-motor one is preferred when taking into account the system complexity and cost. As shown in Figure 5.3, the dual-motor configuration employs two SR machines to separately power two driving wheels. Since the two motors are independently controlled, the differential action can be electronically achieved, hence eliminating the bulky and heavy differential. While each SR machine equally shares the desired capabilities for vehicular operation, the machine can adopt two different designs: the high-speed geared topology and the low-speed gearless topology. The former usually adopts the inner-rotor structure to achieve high-speed operation and then drives the wheel via a planetary gear (PG), whereas the latter usually adopts the outer-rotor structure to directly drive the wheel without gearing. Since the rotor of both structures is free from copper windings or PMs, the SR motor drive offers outstanding mechanical integrity, which is highly desirable for in-wheel EV propulsion.

5.2 SR Machines

The SR machine has a simple, robust, and low-cost structure. Unlike the induction and PM brushless machines, it is particularly suitable for high-speed operation without suffering from mechanical failure due to high centrifugal force. Meanwhile, it is also very suitable for in-wheel drive due to its high mechanical integrity. However, the SR machine generally suffers from lower torque density, higher torque ripple, and larger acoustic noise than its counterparts do.

5.2.1 Structure of SR Machines

The SR machine has salient poles, actually teeth, on both the stator and rotor as shown in Figure 5.4 in which it has installed multiphase concentrated windings in the stator, but with no copper winding or PM piece in the rotor. There are many possible topological structures for the SR machine, mainly depending on the number of phases as well as the numbers of stator and rotor poles.

Figure 5.5 shows two basic SR machine topologies: the three-phase 6/4-pole topology, which has six stator poles and four rotor poles and the four-phase 8/6-pole topology, which has eight stator poles and six rotor poles. The three-phase 6/4-pole SR machine has the advantages of lower cost and better phase-advancing capability for high-speed operation. However, it suffers from higher torque ripple and acoustic noise. On the other hand, the four-phase 8/6-pole SR machine has better starting torque and lower torque ripple, but requires more power devices and involves higher converter cost.

In order to improve the torque density, the SR machine can adopt multiple teeth per stator pole (Miller, 1993). Figure 5.6 shows a three-phase 12/10-pole SR machine with two teeth per stator pole. It is based

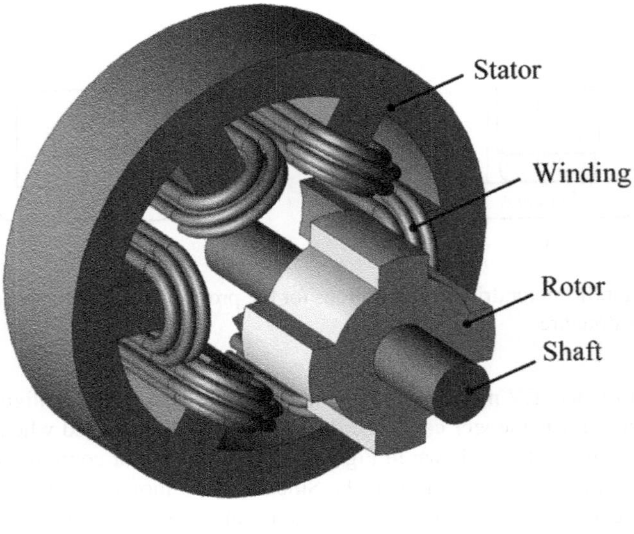

Figure 5.4 SR machine exploded diagram

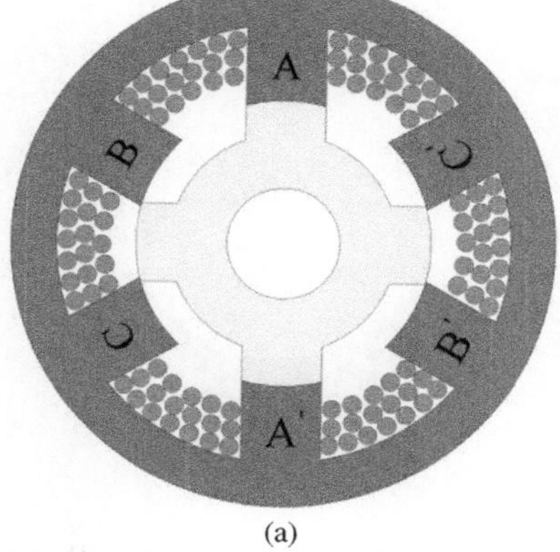

(a)

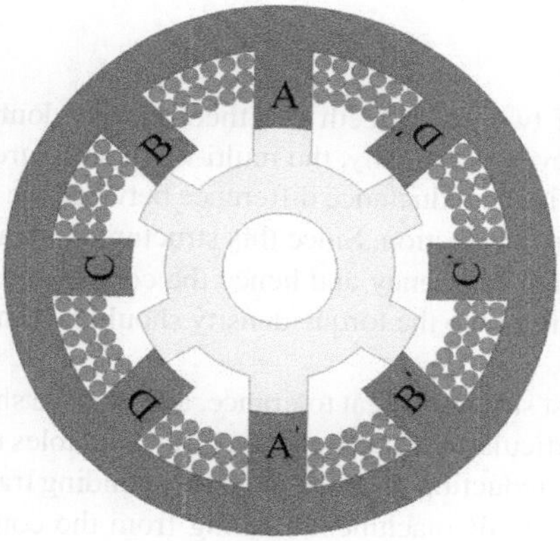

(b)

Figure 5.5 Basic SR machine topologies: (a) three-phase 6/4-pole and (b) four-phase 8/6-pole

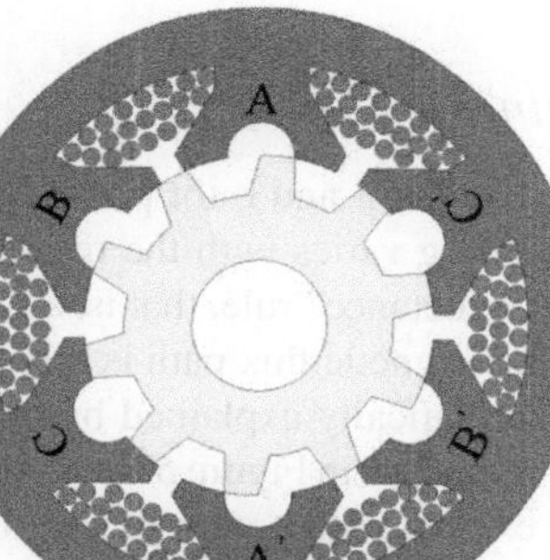

Figure 5.6 Three-phase 12/10-pole multi-tooth SR machine topology

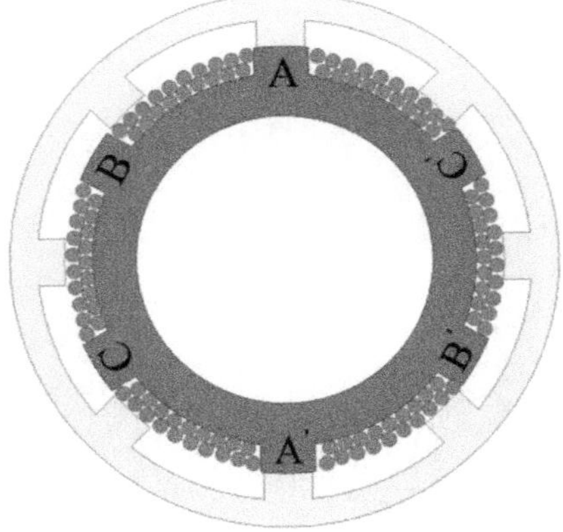

Figure 5.7 Three-phase 6/8-pole outer-rotor SR machine topology

on the principle that the use of two stator teeth can theoretically double the torque production under the same electric loading. However, practically, the multi-tooth structure leads to a reduction of area for copper windings and a decrease in the inductance difference between the aligned and unaligned positions, which in turn decrease the torque production. Since this structure generally offers a smaller step angle, it essentially increases the switching frequency and hence the core losses. Therefore, the benefit on using multiple teeth per stator pole to improve the torque density should be limited to low-speed design such as for direct-drive EV propulsion.

Because of the merits of robust structure, heat tolerance, and flexible shape, the SR machine is attractive for in-wheel drive in EVs. In particular, the outer-rotor topology enables the rotor mounted with the rim of the wheel, hence eliminating the reduction gear and the corresponding transmission loss. Figure 5.7 shows a three-phase 6/8-pole outer-rotor SR machine. Differing from the conventional SR machine topology that the rotor has a smaller diameter than the stator, this outer-rotor topology has a larger rotor diameter. Hence, the number of rotor poles is generally selected to be larger than the number of stator poles in order to reduce the torque ripple directly imposed on the wheel. Meanwhile, this outer-rotor SR machine preferably adopts the pancake-shaped design, namely large radial diameter and short axial length, so that it can readily be implanted into the wheel.

5.2.2 *Principle of SR Machines*

Because of the salient nature of both the stator and rotor poles of the SR machine, the reluctance of the magnetic flux path for each phase winding varies with the rotor position. Thus, the principle of torque production is based on the "minimum reluctance" rule, that is, a rotor pole tends to align with the excited stator pole so that the reluctance of the magnetic flux path is minimum.

The torque production can be mathematically explained by using the principle of electromechanical energy conversion (Krishnan, 2001). As shown in Figure 5.8, the stator poles are wound with concentrated windings of N turns. When they are excited with a current i, a magnetic flux ψ is established. The flux versus magnetomotive force (MMF) characteristics are plotted for the unaligned position between the stator and rotor poles at $\theta = \theta_1$ and the aligned position between the stator and rotor poles at $\theta = \theta_2$. It can be seen that the flux versus MMF characteristic for the unaligned position is almost linear because the reluctance of the air-gap is large and dominant, whereas the characteristic for the aligned position is nonlinear because the reluctance of the magnetic flux path suffers from magnetic saturation.

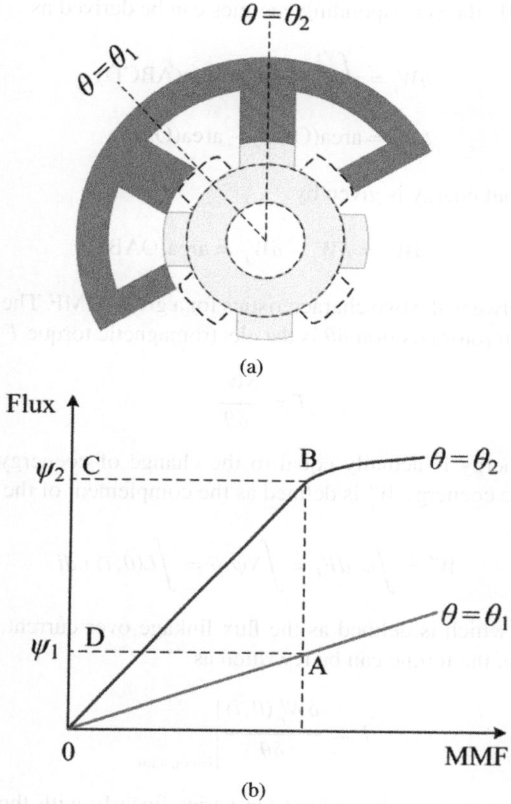

Figure 5.8 Principle of SR machine torque production: (a) rotor positions and (b) flux characteristics

Upon exciting the phase winding, the input energy W_i is expressed as

$$W_i = \int e\, i\, dt \tag{5.1}$$

where e is the induced electromotive force (EMF) which is given by

$$e = \frac{d(N\psi)}{dt} \tag{5.2}$$

Hence, the input energy can be rewritten as

$$W_i = \int F_i\, d\psi \tag{5.3}$$

where $F_i = N\, i$ is the MMF excited in the phase winding. This input energy is equal to the sum of the magnetic field energy stored in the phase winding W_f and the output mechanical energy converted to motion W_o:

$$W_i = W_f + W_o \tag{5.4}$$

For incremental changes, Eq. (5.4) can be rewritten as

$$\delta W_i = \delta W_f + \delta W_o \tag{5.5}$$

For a given current and MMF, the corresponding energies can be derived as

$$\delta W_i = \int_{\psi_1}^{\psi_2} F_i d\psi = \text{area(ABCD)} \tag{5.6}$$

$$\delta W_f = \text{area(OBC)} - \text{area(OAD)} \tag{5.7}$$

Hence, the incremental output energy is given by

$$\delta W_o = \delta W_i - \delta W_f = \text{area(OAB)} \tag{5.8}$$

which is actually the area between the two characteristics for a given MMF. The incremental output energy with respect to the change in rotor position $\delta\theta$ is the electromagnetic torque T which can be expressed as

$$T = \frac{\delta W_o}{\delta \theta} \tag{5.9}$$

This incremental output energy is actually equal to the change of coenergy between the aligned and unaligned positions, and the coenergy W_f' is defined as the complement of the magnetic field energy:

$$W_f' = \int \psi \, dF_i = \int N\psi di = \int L(\theta, i) \, i \, di \tag{5.10}$$

where L is the inductance which is defined as the flux linkage over current, and is a function of rotor position and current. Hence, the torque can be rewritten as

$$T = \left. \frac{\delta W_f'(\theta, i)}{\delta \theta} \right|_{i=\text{constant}} \tag{5.11}$$

In case of no magnetic saturation, the inductance varies linearly with the rotor position for a given current so that the torque is given by

$$T = \frac{1}{2} i^2 \frac{dL(\theta)}{d\theta} \tag{5.12}$$

In case of magnetic saturation, the torque can no longer be expressed as a simple algebraic equation. Actually, it needs to be expressed as an integral equation:

$$T = \int_0^i \frac{\partial L(\theta, i)}{\partial \theta} i di \tag{5.13}$$

On the basis of the aforementioned principle of torque production, the principle of rotation can readily be explained by using the three-phase 6/4-pole SR machine as shown in Figure 5.9. When the phase A winding is excited, the rotor tends to rotate clockwise to decrease the reluctance of the flux path until the rotor poles I and I′ align with the stator poles A and A′ where the reluctance of the flux path has a minimum value. Then, the phase A is switched off and the phase B is switched on so that the reluctance torque tends to make the rotor poles II and II′ align with the stator poles B and B′. Similarly, the excitation of phase C winding results in the alignment of rotor poles I′ and I with stator poles C and C′, respectively. The torque direction is always toward the nearest aligned position. Hence, by conducting the phase windings in the sequence of A-B-C according to the rotor position feedback from the position sensor, the rotor can continuously rotate clockwise. On the contrary, by conducting the phase windings in the sequence of A-C-B, the rotor rotates anticlockwise. It can also be observed that the rotor rotates one revolution by switching 12 times, that is, the number of phases times the number of rotor poles. Therefore, the larger the number of rotor poles, the higher the switching frequency per phase is required.

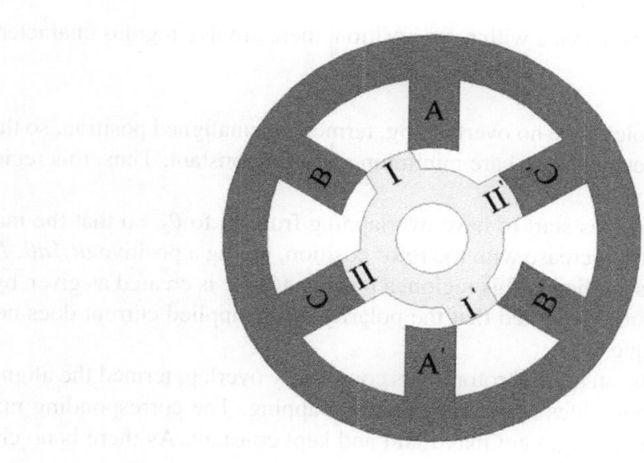

Figure 5.9 Principle of SR machine rotation

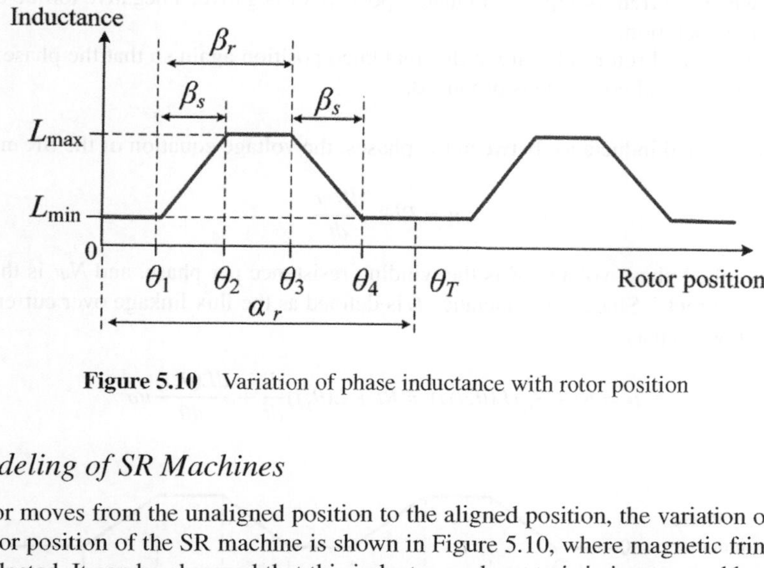

Figure 5.10 Variation of phase inductance with rotor position

5.2.3 Modeling of SR Machines

When the rotor moves from the unaligned position to the aligned position, the variation of phase inductance with rotor position of the SR machine is shown in Figure 5.10, where magnetic fringing and saturation are neglected. It can be observed that this inductance characteristic is governed by the number of rotor poles as well as the stator and rotor pole arcs. The relevant rotor positions θ_1, θ_2, θ_3, θ_4, and θ_T are governed by Krishnan (2001)

$$\theta_1 = (\alpha_r - (\beta_s + \beta_r))/2 \tag{5.14}$$

$$\theta_2 = \theta_1 + \beta_s \tag{5.15}$$

$$\theta_3 = \theta_2 + (\beta_r - \beta_s) \tag{5.16}$$

$$\theta_4 = \theta_3 + \beta_s \tag{5.17}$$

$$\theta_T = \theta_4 + \theta_1 = \alpha_r \tag{5.18}$$

where $\alpha_r = 2\pi/N_r$ is the rotor pole pitch, N_r is the number of rotor poles, β_s is the stator pole arc, and β_r is the rotor pole arc.

Along the variation of phase inductance with rotor position, there are five regions characterized by the following features:

- $0 - \theta_1$: The stator and rotor poles have no overlapping, termed the unaligned position, so that the magnetic flux and hence the inductance (L_{min}) are minimum and kept constant. Thus, this region does not contribute to torque production.
- $\theta_1 - \theta_2$: The stator and rotor poles start to have overlapping from θ_1 to θ_2, so that the magnetic flux and hence the phase inductance increase with the rotor position, giving a positive $dL/d\theta$. As shown in Figure 5.11, when a current is applied in this region, a positive torque is created as given by Eq. (5.12) for motoring operation. It should be noted that the polarity of the applied current does not affect the polarity of the developed torque.
- $\theta_2 - \theta_3$: Within this region, the stator and rotor poles completely overlap, termed the aligned position, and the movement of rotor pole does not change the overlapping. The corresponding magnetic flux and hence the phase inductance (L_{max}) are maximum and kept constant. As there is no change in the inductance, there is no torque production.
- $\theta_3 - \theta_4$: The stator and rotor poles start to reduce overlapping from θ_3 to θ_4, so that the magnetic flux and hence the phase inductance decrease with the rotor position, giving a negative $dL/d\theta$. As shown in Figure 5.11, when a current is applied, no matter positive or negative, a negative torque can be created for regeneration operation.
- $\theta_4 - \theta_T$: The stator and rotor poles are at the unaligned position again so that the phase inductance is kept constant at L_{min} and no torque is produced.

Neglecting the mutual inductance between the phases, the voltage equation of the SR machine can be expressed as

$$u = Ri + \frac{dN\psi}{dt} \tag{5.19}$$

where u is the applied phase voltage, R is the winding resistance per phase, and $N\psi$ is the flux linkage due to the phase current i. Since the inductance L is defined as the flux linkage over current, the voltage equation can be rewritten as

$$u = Ri + \frac{d}{dt}(L(\theta, i)\, i) = Ri + L(\theta, i)\frac{di}{dt} + \frac{dL(\theta, i)}{d\theta} i\omega \tag{5.20}$$

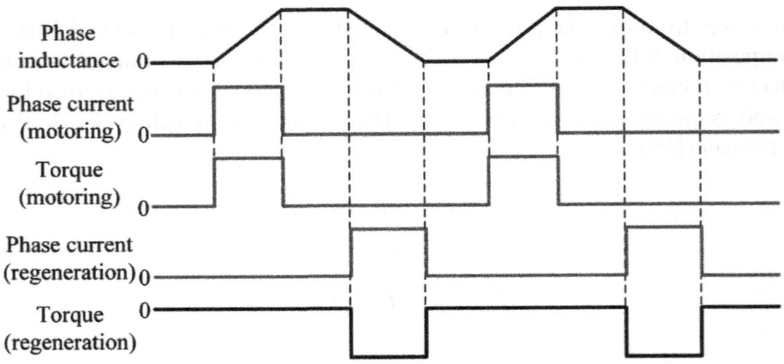

Figure 5.11 Relationship between inductance, current, and torque profiles

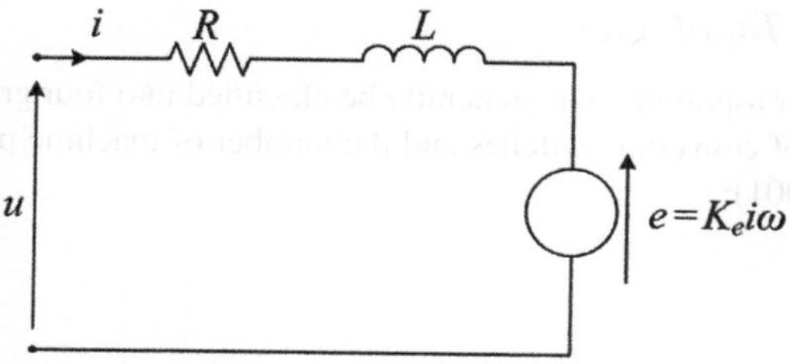

Figure 5.12 Per-phase model of SR machine

where the three terms on the right-hand side represent the resistive voltage drop, inductive voltage drop, and back EMF of the SR machine, respectively. The corresponding back EMF, e can be written as

$$e = \frac{dL(\theta, i)}{d\theta} i\omega = K_e i\omega \tag{5.21}$$

where K_e is defined as the back EMF constant of the SR machine. Hence, the per-phase model of the SR machine is resulted as shown in Figure 5.12. It can be observed that this model is similar to that of the series DC machine.

As mentioned earlier, the developed torque per phase depends on the current and rate of change of inductance with the rotor position. Therefore, the resultant output torque, T_e is the summation of torques developed at various phases:

$$T_e = \sum_{i=1}^{m} T(i, \theta) \tag{5.22}$$

where m is the number of phases of the SR machine. It should be noted that it is hardly possible to achieve the ideal inductance profile due to the presence of magnetic saturation. This saturation causes the inductance profile to curl near the top and thus causes a diminishing return on torque production. In addition, each phase can produce a positive torque only in half a rotor pole pitch, resulting in a large torque ripple. This creates the well-known problems of speed oscillation and acoustic noise. Nevertheless, this torque ripple can be alleviated by increasing the number of phases or adopting sophisticated control means. Consequently, the equation of motion is given by

$$T_e = J\frac{d\omega}{dt} + B\omega + T_L \tag{5.23}$$

where J is the moment of inertia, B is the coefficient of viscous friction, T_L is the load torque, and ω is the rotor speed, which is actually the rate of change of rotor position $d\theta/dt$.

5.3 SR Converters

Differing from other AC machines, the SR machine produces torque, which is independent of the current polarity. Thus, conventional bridge inverters adopted by AC machines are not used for SR machines. Actually, many converter topologies have been specifically developed for SR machines, the so-called SR converters. The key feature is that the stator windings are connected in series with the upper and lower switches of the converter, which can prevent the shoot-through fault that exists in the inverter of AC machines.

5.3.1 SR Converter Topologies

In general, the SR converter topologies can generally be classified into four groups based on the relationship between the number of converter switches and the number of machine phases m (Ehsani, Gao, and Emadi, 2009; Krishnan, 2001):

- $2m$-switch SR converter.
- m-switch SR converter.
- $(m + 1)$-switch SR converter.
- $1.5m$-switch SR converter.

Figure 5.13a shows the asymmetric bridge converter topology, which is the most common $2m$-switch SR converter. It uses two switches and two diodes per phase. The main advantage of this converter is the control flexibility which allows that all machine phases can be controlled independently. This particularly facilitates high-speed operation where there may be considerable overlap between the adjacent phase

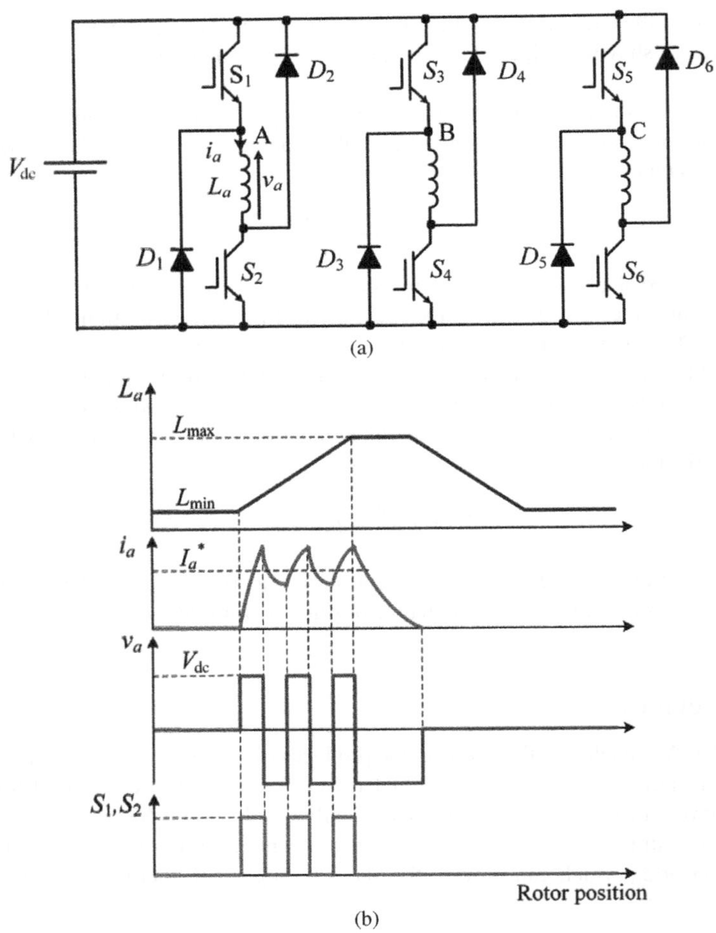

Figure 5.13 Asymmetric bridge SR converter: (a) topology and (b) operating waveforms

currents. The main drawback is the need of more power devices than its counterparts, which increases the converter cost and packaging size. In order to control the developed torque, the phase current needs to be effectively controlled. Although this current can be varied by using either voltage pulse width modulation (PWM) control or current PWM control, the latter is preferred because it can directly control the current magnitude. The operating waveforms of this asymmetric bridge converter are depicted in Figure 5.13b. When the switches S_1 and S_2 are turned on, the current of phase A starts to increase. If the current rises above the upper limit of the hysteresis current controller, the S_1 and S_2 will be turned off. Then, the current decreases by freewheeling via the diodes D_1 and D_2, leading to return the magnetic energy stored in the phase winding to the DC source V_{dc}. If the current drops below the lower limit of the hysteresis current controller, both S_1 and S_2 will be turned on again. Consequently, the current is forced to swing between the hysteresis current window to maintain the average current at the desired level I_a^* until the end of conduction interval. It should be noted that if the turn-off process involves only turning off S_1 while holding on S_2, the current will decrease slowly by freewheeling via S_2 and D_1. The voltage across the winding becomes zero. This can reduce the switching frequency and hence the switching loss.

Figure 5.14a shows a typical m-switch SR converter topology, which is dubbed as the R-dump SR converter. It uses only one switch and one diode per phase, but involving additional passive components R and C. The main advantage is the minimum component count and hence the converter cost. However, it suffers from the drawbacks of inefficient operation and inflexible control. As shown in Figure 5.14b, the operating waveforms of this R-dump converter are similar to those of the asymmetric converter except during the turn-off process. That is, if the current rises above the upper limit of the hysteresis current controller, the switch S_1 will be turned off. Therefore, the current decreases by freewheeling via the diode D_1 in such a way that the magnetic energy stored in the phase winding is charging the capacitor C to the DC supply voltage and dissipating in the resistor R. This reduces the overall converter efficiency. In addition, the zero voltage across the phase winding does not exist in this topology, thus reducing its control flexibility. Moreover, the selection of R is critical. If R is too small, the fall time of the current will be lengthened, which may produce negative torque when the current is still nonzero under the negative slope region of the phase inductance. On the contrary, if R is too large, the turn-off transient voltage needs to be included in the rating of the switch.

Figure 5.15a shows a typical $(m + 1)$-switch SR converter topology, which is dubbed as the C-dump SR converter. It has the advantage of requiring minimum switches for achieving independent phase current control. The main disadvantage of this topology is that the negative voltage across the phase winding is limited by the difference between the voltage across the capacitor and the DC supply voltage, which slows down the current commutation. The operating waveforms of this C-dump converter are depicted in Figure 5.15b. When the current rises above the upper limit of the hysteresis current controller, the switch S_1 is turned off. This enables the diode D_1 to be forward biased, and the current decreases in such a way that the magnetic energy stored in the phase winding is charging the capacitor C, and then transferring to the DC source via the switch S.

Figure 5.16a shows the 1.5m-switch SR converter topology that shares switches of each couple of nonadjacent phases, hence requiring only three switches and three diodes for two phases. It has the advantage of using less than two switches per phase while offering independent phase current control. However, this topology is limited to an even number of phases of SR machines. The operating waveforms of this converter are shown in Figure 5.16b. The switches S_x and S_y carry the currents of phases A and C and phases B and D, respectively. Accordingly, their ratings are greater than those of the phase switches S_1, S_2, S_3, and S_4. This grouping of phase windings can guarantee the independent control of their currents.

5.3.2 Soft-Switching SR Converter Topologies

In order to improve the performances of SR motor drives such as reducing the torque ripple or improving the dynamic response, the SR converter needs to operate at higher switching frequencies. However,

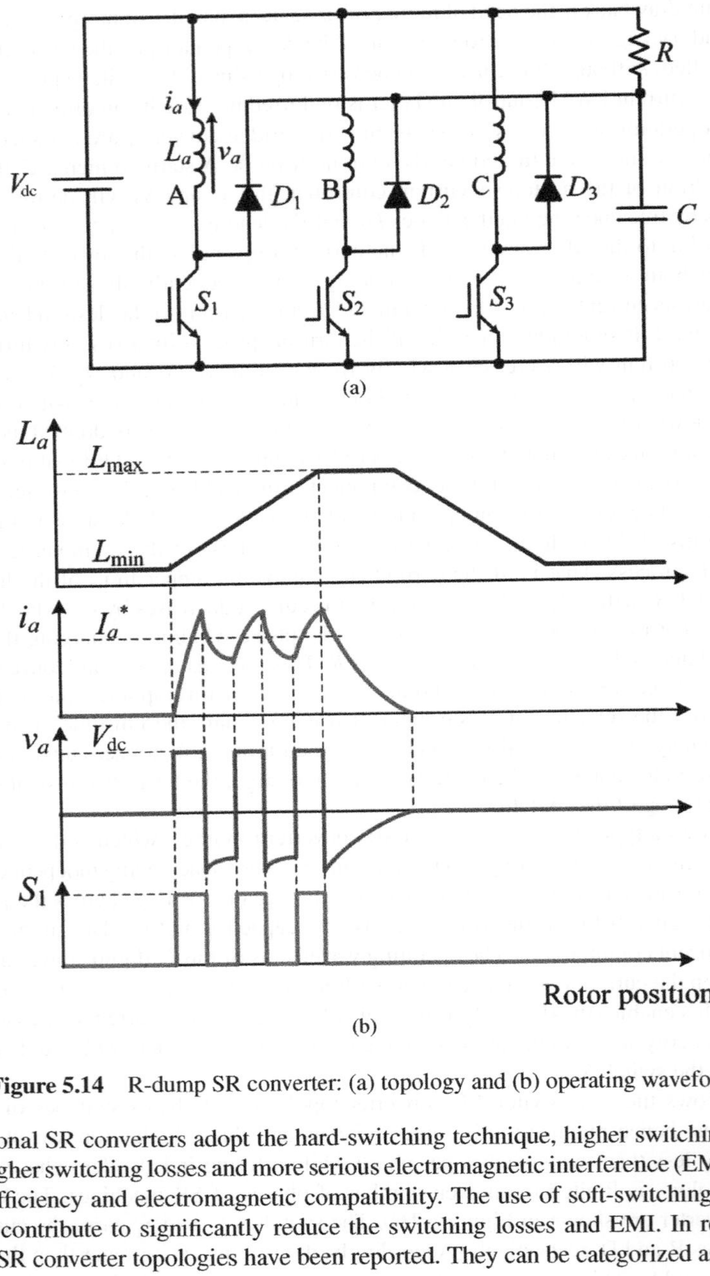

Figure 5.14 R-dump SR converter: (a) topology and (b) operating waveforms

as the conventional SR converters adopt the hard-switching technique, higher switching frequency definitely causes higher switching losses and more serious electromagnetic interference (EMI), thus degrading the converter efficiency and electromagnetic compatibility. The use of soft-switching techniques in SR converters can contribute to significantly reduce the switching losses and EMI. In recent years, many soft-switching SR converter topologies have been reported. They can be categorized as

- resonant switch SR converter;
- resonant DC-link SR converter;
- zero-transition SR converter.

The resonant switch SR converter (Murai, Cheng, and Yoshida, 1999) employs an inductor-capacitor (LC) resonant circuit and an auxiliary switch to enable the main switches turning on or off at zero-voltage

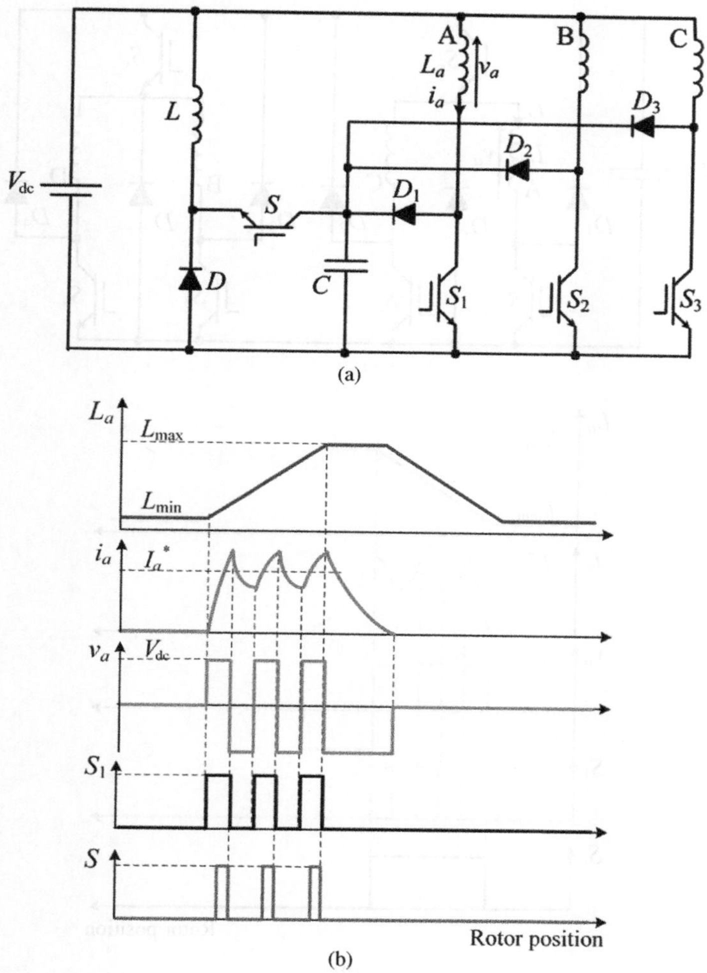

Figure 5.15 C-dump SR converter: (a) topology and (b) operating waveforms

switching (ZVS) condition or zero-current switching (ZCS) condition. Hence, it can achieve theoretically zero switching loss. However, this kind of converter suffers from the limitations that the PWM switching control cannot be performed, and the switching frequency needs to be synchronized with the resonant frequency of the auxiliary circuit. Consequently, it can only provide pulse-packet current to the SR machine, leading to aggravate the torque ripple and audible noise. The resonant capacitor also increases the voltage stress on the main switch (Luo and Zhan, 2005).

The resonant DC-link SR converter (Gallegos-Lopez *et al.*, 1997) incorporates an LC resonant circuit and an auxiliary switch at the DC link. By turning the DC-link voltage into pulsating voltage with zero voltage notches, the main switches can be turned on or off at those voltage notches, achieving ZVS and hence theoretically zero switching losses. However, this converter also lacks the PWM switching control capability. Inevitably, the resonant DC-link operation is accomplished at the expense of voltage stress and DC-link component losses. The drawback of voltage stress can be alleviated through the use of the actively clamped resonant DC-link SR converter topology (Rolim *et al.*, 1999).

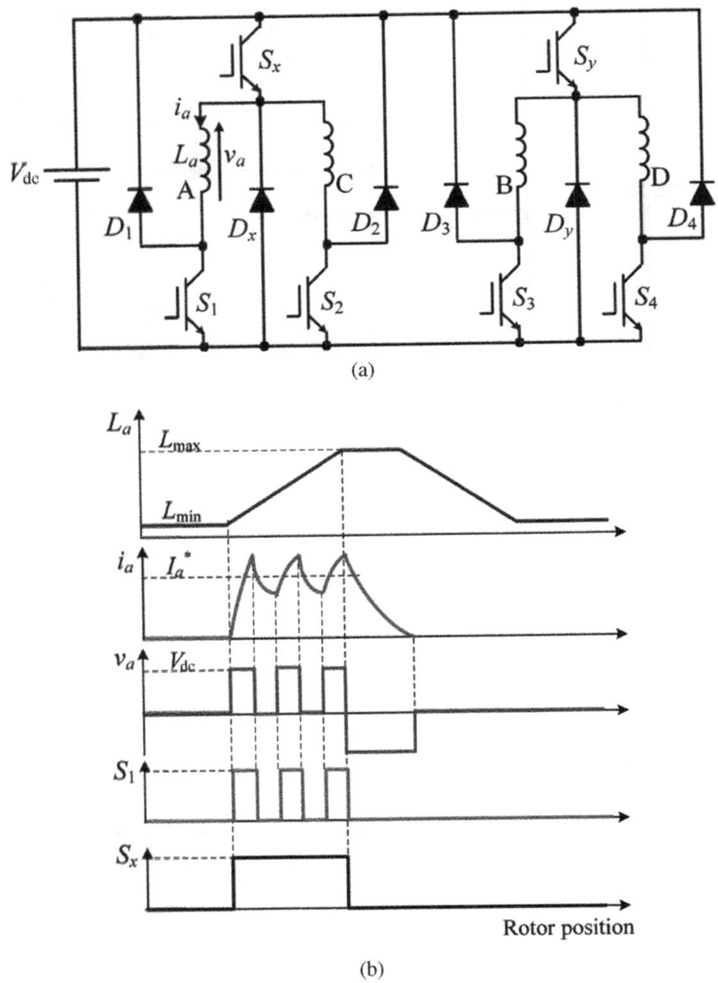

Figure 5.16 1.5m-switch SR converter: (a) topology and (b) operating waveforms

By incorporating both the auxiliary resonant circuit and PWM switching control, the SR converter can achieve ZVS or ZCS while offering PWM control, dubbed as the zero-voltage transition (ZVT) or zero-current transition (ZCT), respectively. Owing to the ability of making PWM switching control, the zero-transition SR converters possess high controllability and theoretically zero switching loss (Chao, 2009). Two zero-transition SR converters are shown in Figure 5.17. The ZVT SR converter possesses the advantage that all main switches and diodes can achieve ZVS while the corresponding device voltage and current stresses are kept at unity. It is a very desirable feature for power conversion where power metal-oxide-semiconductor field-effect transistors (MOSFETs) are used because of their severe capacitive turn-on losses. On the other hand, the ZCT SR converter possesses the advantage that both the main and auxiliary switches can always maintain ZCS with minimum current and voltage stresses. It is particularly suitable for power conversion using insulated gate bipolar transistors (IGBTs) as power devices since they generally suffer from severe inductive turn-off losses (Ching, 2009).

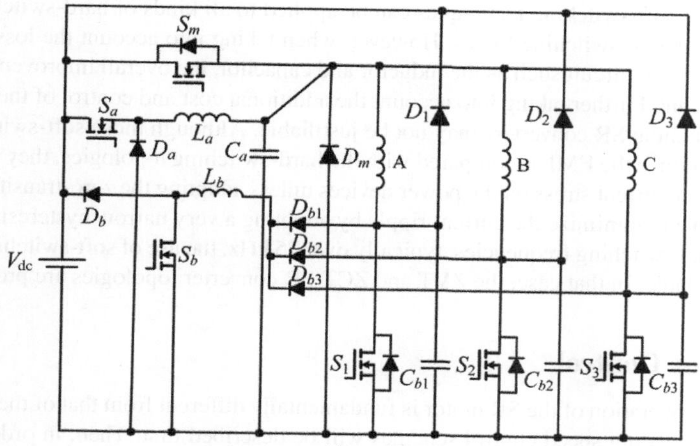

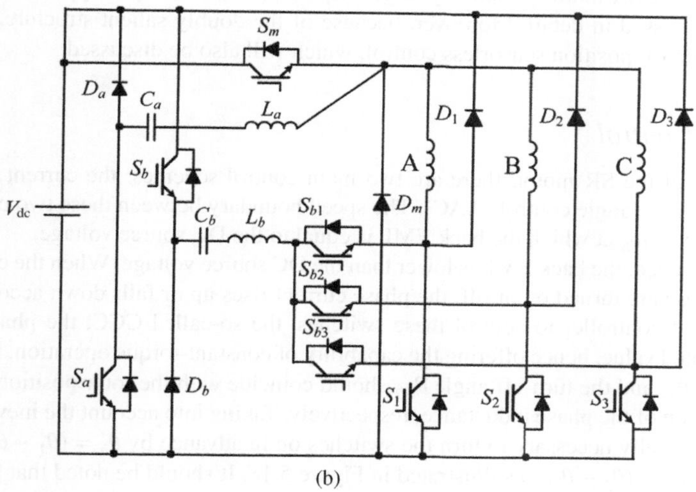

Figure 5.17 Zero-transition SR converters: (a) ZVT topology and (b) ZCT topology

5.3.3 *Comparison of SR Converters for EVs*

In order to assess the preferable type of SR converters for EVs, there are five major factors: cost, power density, efficiency, controllability, and reliability. Among those hard-switching SR converters, the asymmetric SR converter and the C-dump SR converter are most acceptable when considering these five factors holistically. Specifically, the asymmetric converter takes the definite advantages of high controllability and high reliability but suffers from higher cost of power devices, whereas the C-dump SR converter takes the advantages of requiring less switches and independent phase current control capability but has the drawbacks of using an additional capacitor and slower current commutation. In recent years, the increase of power density of power devices with simultaneously decreasing cost has been a constant trend in the market. As the cost of SR motor drives has been dominated by the machine cost rather than the converter cost, the pursuit of smaller number of power devices with the sacrifice of controllability and reliability is no longer necessary for selecting the SR converter. Therefore, the asymmetric SR converter is the preferred topology, actually the most commonly adopted one for EV propulsion.

Theoretically, the soft-switching techniques can be applied to all kinds of hard-switching SR converters, hence achieving zero switching losses. However, when taking into account the losses in the passive elements of the resonant circuit such as the inductor and capacitor, the overall improvement of efficiency may not be significant. Further taking into account the additional cost and control of the resonant circuit, the use of soft-switching SR converters may not be justifiable. Although those soft-switching topologies can inherently minimize the EMI as compared with the hard-switching topologies, they generally impose additional voltage or current stress on the power devices unless adopting the zero-transition topologies. If it is highly desirable to minimize the current ripple by adopting a very narrow hysteresis current window and hence very high switching frequencies, typically over 25 kHz, the use of soft-switching SR converters may become justifiable. In that case, the ZVT and ZCT SR converter topologies are preferred.

5.4 SR Motor Control

As the principle of operation of the SR motor is fundamentally different from that of the conventional AC motors, the corresponding speed control schemes will be described first. Then, in order to alleviate the major drawback of the SR motor – relatively large torque ripples, the torque-ripple minimization (TRM) control will be discussed in detail. Moreover, because of the doubly salient structure, the SR motor is particularly suitable for position sensorless control, which will also be discussed.

5.4.1 Speed Control

For speed control of the SR motor, there are two main control schemes: the current chopping control (CCC) and the advance angle control (AAC). The speed boundary between these two control schemes is called the base speed, ω_b at which the back EMF is equal to the DC source voltage.

Below the base speed, the back EMF is lower than the DC source voltage. When the converter switches of a particular phase are turned on or off, the phase current rises up or falls down accordingly. By using a hysteresis current controller to control these switches, the so-called CCC, the phase current can be regulated at the rated value, hence offering the capability of constant-torque operation. In the CCC mode, the turn-on angle θ_{on} and the turn-off angle θ_{off} should coincide with the rotor positions θ_1 and θ_2 of the positive slope region of the phase inductance, respectively. Taking into account the inevitable rise time or fall time, it is practically necessary to turn the switches on in advance by $\theta_a = (\theta_1 - \theta_{on})$ and turn them off in advance by $\theta_b = (\theta_2 - \theta_{off})$ as illustrated in Figure 5.18. It should be noted that both θ_a and θ_b are dependent on the phase current and rotor speed.

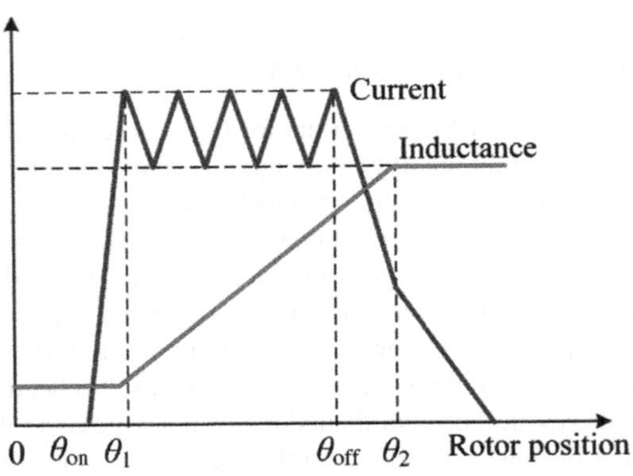

Figure 5.18 Current chopping control for constant-torque operation

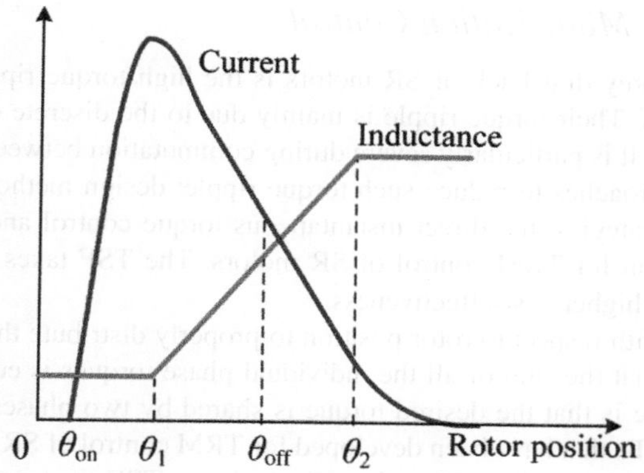

Figure 5.19 Advance angle control for constant-power operation

Above the base speed, the back EMF is higher than the DC source voltage so that the phase current is limited by the back EMF. In order to feed the current into the phase winding, the switches of a particular phase need to be turned on well ahead of the unaligned position, the so-called AAC as illustrated in Figure 5.19. Since the back EMF increases with the rotor speed, the phase current inevitably decreases and hence the torque drops. The higher the rotor speed, the larger the advancing of turn-on angle is required. In general, the turn-on angle is advanced in such a way that the torque drops inversely with respect to the rotor speed, leading to offer the capability of constant-power operation. This advancing of turn-on angle is limited by the position of the preceding negative slope region of the phase inductance. Meanwhile, the turn-off angle is also advanced to ensure that the current quenches before the succeeding negative slope region of the phase inductance.

Beyond the critical speed of the motor ω_c, at which both θ_{on} and θ_{off} reach their limit values, no phase advancing is allowable and the torque drops significantly. The SR motor can no longer keep constant-power operation and offers the natural mode of operation.

Figure 5.20 depicts the torque-speed capability of the SR motor covering all operating regions, namely the constant-torque operation, constant-power operation, and natural operation, which are desirable for EV low-speed urban driving, medium-speed suburban driving, and high-speed highway cruising, respectively.

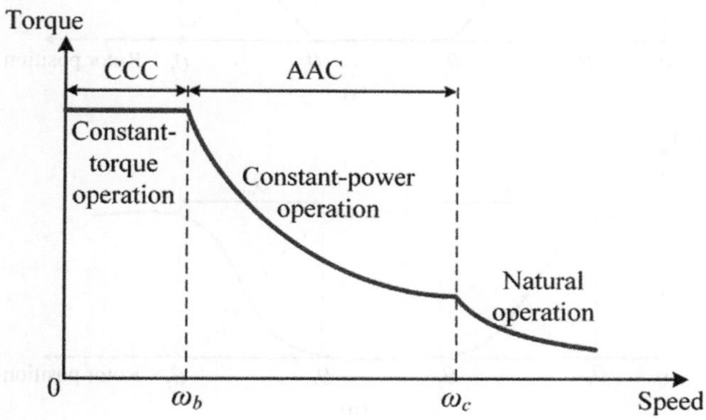

Figure 5.20 Torque-speed capability of SR motor

5.4.2 Torque-Ripple Minimization Control

As mentioned earlier, the key drawback of SR motors is the high torque ripple as compared to other conventional AC machines. Their torque ripple is mainly due to the discrete excitation of phase windings for torque production. It is particularly severe during commutation between two adjacent phases. In general, there are two approaches to reduce such torque ripple: design methods and control strategies. Among various control strategies, the direct instantaneous torque control and torque-sharing function (TSF) have gained attention for TRM control of SR motors. The TSF takes the definite advantage of easier implementation and higher cost-effectiveness.

The TSF is a function with respect to rotor position to properly distribute the reference torque among all phases in such a way that the sum of all the individual phase torques is equal to the desired torque. The basis of this technique is that the desired torque is shared by two phases during the commutation period. There are many TSFs that have been developed for TRM control of SR motors. They can be basically classified as the linear and nonlinear functions. In the linear TSF, the torque produced by the phases during commutation changes linearly with the rotor position as depicted in Figure 5.21a, whereas in the nonlinear TSF, the torque during commutation changes nonlinearly with the rotor position as depicted in Figure 5.21b, where θ_i, θ_f, and θ_{ov} denote the initial angle, final angle, and overlap angle of the commutation region, respectively. These angles are chosen based on the inductance profile of the SR motor. There are three major nonlinear functions that have been adopted as the nonlinear TSF: the cubic function, cosine function, and exponential function. Other kinds of nonlinear TSFs are essentially derived from them.

During the commutation interval, the torque can be expressed as the sum of the torque due to the incoming phase I and the torque due to the outgoing phase II:

$$T = T_I + T_{II} \tag{5.24}$$

where

$$T_I = f_I(\theta)\,T \tag{5.25}$$

$$T_{II} = f_{II}(\theta)\,T \tag{5.26}$$

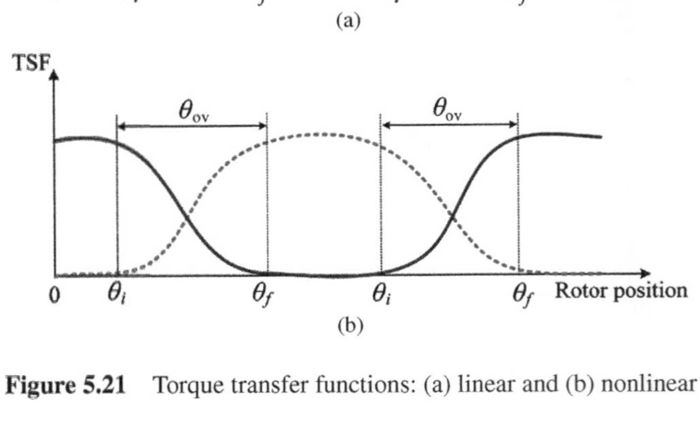

Figure 5.21 Torque transfer functions: (a) linear and (b) nonlinear

where $f_I(\theta)$ and $f_{II}(\theta)$ are the TSFs to the phases I and II, respectively, and $0 \le \theta < \theta_T/2$, where θ_T is the conduction period per phase.

In the linear TSF (Schramm, Williams, and Green, 1992), the incoming and outgoing functions during phase commutation have constant slopes with respect to the rotor position, which are defined as

$$f_I(\theta) = \begin{cases} 0 & \text{for } 0 \le \theta < \theta_i \\ (\theta - \theta_i)/\theta_{\text{ov}} & \text{for } \theta_i \le \theta < \theta_f \\ 1 & \text{for } \theta_f \le \theta < \theta_T/2 \end{cases} \tag{5.27}$$

$$f_{II}(\theta) = \begin{cases} 1 & \text{for } 0 \le \theta < \theta_i \\ 1 - (\theta - \theta_i)/\theta_{\text{ov}} & \text{for } \theta_i \le \theta < \theta_f \\ 0 & \text{for } \theta_f \le \theta < \theta_T/2 \end{cases} \tag{5.28}$$

This linear TSF takes the advantage of simple implementation, but is unable to reflect the nonlinear features of torque production and magnetic saturation of the SR motor. Thus, the variation in the rate of inductance change, particularly at the initial and final stages of phase commutation, causes significant errors between the torque reference input and its actual output.

In the cubic TSF (Sahoo, Panda, and Xu, 2005), the incoming and outgoing functions during commutation are in the form of cubic polynomials with respect to the rotor position, which are defined as

$$f_I(\theta) = \begin{cases} 0 & \text{for } 0 \le \theta < \theta_i \\ 3\left[(\theta - \theta_i)/\theta_{\text{ov}}\right]^2 - 2[(\theta - \theta_i)/\theta_{\text{ov}}]^3 & \text{for } \theta_i \le \theta < \theta_f \\ 1 & \text{for } \theta_f \le \theta < \theta_T/2 \end{cases} \tag{5.29}$$

$$f_{II}(\theta) = \begin{cases} 1 & \text{for } 0 \le \theta < \theta_i \\ 1 - 3\left[(\theta - \theta_i)/\theta_{\text{ov}}\right]^2 + 2[(\theta - \theta_i)/\theta_{\text{ov}}]^3 & \text{for } \theta_i \le \theta < \theta_f \\ 0 & \text{for } \theta_f \le \theta < \theta_T/2 \end{cases} \tag{5.30}$$

In the cosine TSF (Husain and Ehsani, 1996), the incoming and outgoing functions during commutation are denoted as

$$f_I(\theta) = \begin{cases} 0 & \text{for } 0 \le \theta < \theta_i \\ \left[1 - \cos \pi (\theta - \theta_i)/\theta_{\text{ov}}\right]/2 & \text{for } \theta_i \le \theta < \theta_f \\ 1 & \text{for } \theta_f \le \theta < \theta_T/2 \end{cases} \tag{5.31}$$

$$f_{II}(\theta) = \begin{cases} 1 & \text{for } 0 \le \theta < \theta_i \\ \left[1 + \cos \pi (\theta - \theta_i)/\theta_{\text{ov}}\right]/2 & \text{for } \theta_i \le \theta < \theta_f \\ 0 & \text{for } \theta_f \le \theta < \theta_T/2 \end{cases} \tag{5.32}$$

In the exponential TSF (Illic-Spong, Miller, and MacMinn, 1987), the incoming and outgoing functions during commutation are denoted as

$$f_I(\theta) = \begin{cases} 0 & \text{for } 0 \le \theta < \theta_i \\ 1 - \exp\left[-\left((\theta - \theta_i)/\theta_{\text{ov}}\right)^2\right] & \text{for } \theta_i \le \theta < \theta_f \\ 1 & \text{for } \theta_f \le \theta < \theta_T/2 \end{cases} \tag{5.33}$$

$$f_{II}(\theta) = \begin{cases} 1 & \text{for } 0 \le \theta < \theta_i \\ \exp\left[-\left((\theta - \theta_i)/\theta_{\text{ov}}\right)^2\right] & \text{for } \theta_i \le \theta < \theta_f \\ 0 & \text{for } \theta_f \le \theta < \theta_T/2 \end{cases} \tag{5.34}$$

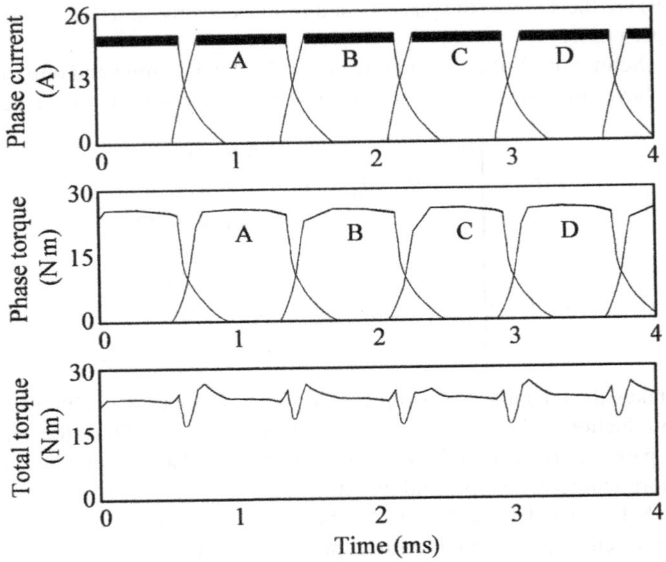

Figure 5.22 Operating waveforms without using torque-ripple minimization control

Compared with the linear TSF, the nonlinear TSFs including the cubic, cosine, and exponential functions have higher degree of freedom to shape the nonlinear slopes in the commutation region, leading to achieve much better matching between the torque reference input and its actual output, and hence minimizing the torque ripple. Furthermore, the nonlinear TSFs can be further extended to maximize the speed range and minimize the copper loss (Xue, Cheng, and Ho, 2009), and to optimize the efficiency (Lee *et al.*, 2009). A comparison of different TSFs for SR machines in EV propulsion has also been studied, confirming the usefulness of TSFs for TRM control (Pop *et al.*, 2012).

Taking a four-phase 8/6-pole SR machine for exemplification, the waveforms of phase currents, phase torques, and total torque without using the TRM control are simulated as shown in Figure 5.22. Then, after applying a typical TSF, the cosine TSF, for TRM control, the corresponding waveforms are simulated as shown in Figure 5.23. It can be observed that the torque ripple can be significantly suppressed by using the TRM control.

5.4.3 Position Sensorless Control

In order to achieve effective operation and control, the SR motor needs proper synchronization between the excitation of phase windings and the rotor position. Usually, a position sensor or encoder is mounted on the shaft to feed back the information of rotor position to the controller. This position sensor is relatively costly as compared with the controller, hence triggering the development of position sensorless technology for the SR motor. For low-cost application, it is particularly desirable to make use of position sensorless technology to get rid of the costly position sensor. However, for EV application, the cost of position sensor in the whole motor drive is not a major concern. Therefore, EV propulsion systems seldom adopt the position sensorless technology. Nevertheless, taking into account fault-tolerant control, the use of position sensorless control for the SR motor drive can effectively continue the operation in case the position sensor loses its function. This can improve the fault-tolerant capability of EVs.

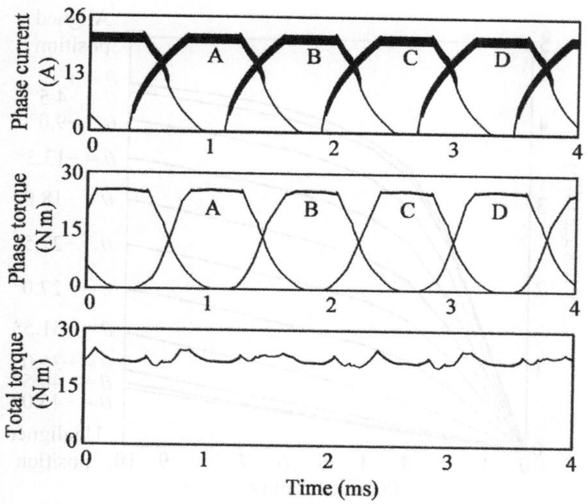

Figure 5.23 Operating waveforms with torque-ripple minimization control using cosine torque-sharing function

Many position sensorless control methods were developed over the past two decades. The commonly adopted position sensorless control methods can be categorized as

- Flux linkage-based method.
- Inductance-based method.
- Signal injection-based method.
- Current-based method.
- Observer-based method.

The flux linkage-based method (Lyons, MacMinn, and Preston, 1991) is to make use of the three-dimensional relationship between the flux linkage, phase current, and rotor position as depicted in Figure 5.24. That is, based on the measured voltage and current, the flux linkage can be calculated. Then, according to the relationship, the rotor position can be deduced. This method takes the advantage of simplicity. It is more suitable for estimating the rotor position during the high-speed AAC mode. At the low-speed CCC mode, the error in winding resistance or measured current may cause significant errors in the flux linkage and hence rotor position estimations.

The inductance-based method (Suresh *et al.*, 1999) utilizes the relationship between the phase inductance, phase current, and rotor position as depicted in Figure 5.25. Similar to the flux linkage-based method, this method relies on the accuracy of winding resistance and measured current, and is suitable for high-speed operation only. On the other hand, the rotor position at low-speed operation can be estimated by the incremental inductance (Gao, Salmasi, and Ehsani, 2004). On the basis of the relationship between the incremental inductance, phase current, and rotor position, the rotor position can be estimated. This method takes the merit that it does not require any additional sensor. However, it is applicable only at low speeds where the back EMF can be neglected.

The signal injection-based method (Ehsani *et al.*, 1994) is to apply a low-amplitude sinusoidal voltage to the inactive phase winding, and then measure the resulting amplitude and phase angle of the phase current to deduce the phase inductance, hence providing the information of rotor position. These amplitude modulation and phase modulation approaches desire an additional voltage source to produce the low-amplitude

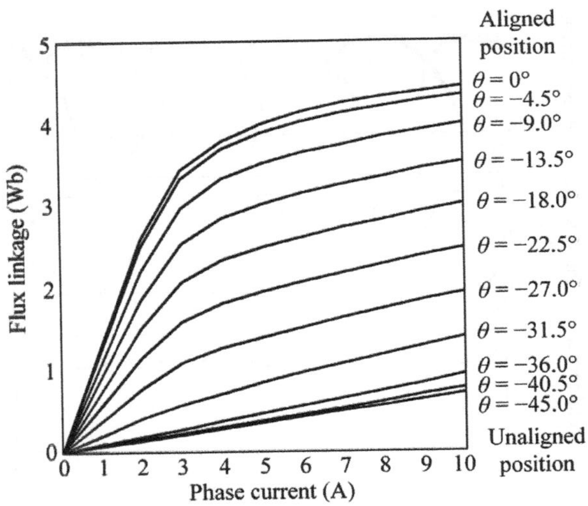

Figure 5.24 Relationship between flux linkage, phase current, and rotor position of SR motor

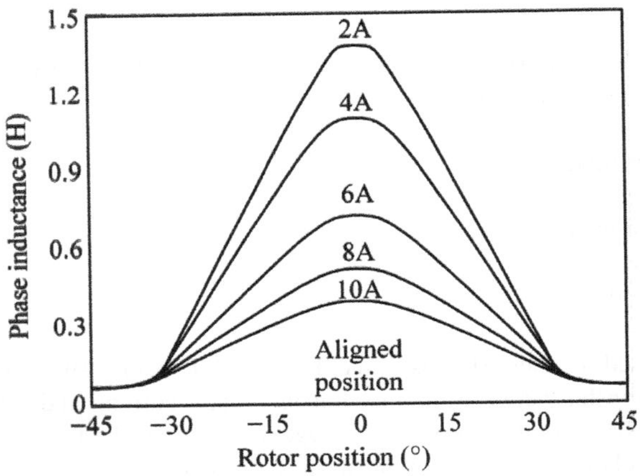

Figure 5.25 Relationship between phase inductance, phase current, and rotor position of SR motor

sinusoidal voltage. On the other hand, the SR converter can be used to inject a short-period voltage pulse. The rate of change of the resulting phase current can also be used to deduce the phase inductance and hence the rotor position. Although this diagnostic pulse approach does not desire an additional voltage, it suffers from the drawbacks of the generation of negative torque due to the injected current and the limited resolution at high speeds due to the time delay caused by the diagnostic phase.

The current-based method (Panda and Amaratunga, 1991) makes use of the idea that the rate of change of current depends on the incremental inductance, which in turn depends on the rotor position. This incremental inductance can be expressed as the rise or fall time of the phase current, leading to facilitate real-time implementation with no additional cost. However, the accuracy of the calculation of the incremental inductance may be deteriorated by the variations of winding resistance and back EMF during high-speed operation. On the other hand, the rate of change of current also depends on the phase inductance, which in turn depends on the rotor position (Yang, Kim, and Krishnan, 2012). Instead of online calculating the inductance, by comparing two consecutive turn-on times of the PWM phase current during the CCC mode, the aligned and unaligned rotor positions can be detected. This method is simple to implement and insensitive to system parameter variations.

The observer-based method (Lumsdaine and Lang, 1990) makes use of an observer that is derived based on the state-space equations of the SR motor drive for estimating the rotor position. In general, the input and output of this observer are the phase voltage and phase current, respectively, while the state variables include the stator flux linkage, rotor position, and rotor speed. The phase current estimated by this observer is compared with the measured phase current in such a way that the resulting current error is employed to adjust the parameters of the observer. The current error is zero when the observer model can accurately describe the dynamic behavior of the SR motor drive. Hence, the rotor position estimated by the observer well represents the actual rotor position. This method needs real-time implementation of complex algorithms, which requires considerable computational power. In addition, it generally suffers from a significant error at the time of starting because the observer takes time to converge. By further incorporating a sliding-mode scheme (Zhan, Chan, and Chau, 1999), the convergence and robustness against disturbances and variations can be improved.

5.5 Design Criteria of SR Motor Drives for EVs

Although the SR motor possesses simplicity in construction, it is not simple in design and analysis. Because of the heavy saturation of pole tips and the fringing effect of poles and slots, the design of the SR motor suffers from a great difficulty in using the magnetic circuit approach. In most cases, the electromagnetic finite element method (FEM) is employed to determine the machine parameters and performances.

The design criteria of the SR motor for EV propulsion is much more stringent than those for industrial applications. These criteria are as follows:

- The SR motor needs to offer high torque density, hence minimizing the machine weight and size for the desired rated and peak torque capabilities of EVs. The corresponding torque density should be comparable with that of the PM brushless motor and at least higher than that of the induction motor.
- It is well known that the SR motor suffers from relatively high torque ripple and hence acoustic noise, which affect its application to EVs desiring smooth and silent motion. The machine design should minimize the torque ripple and acoustic noise, while maintaining the torque density.
- It needs to offer a good efficiency map, namely high efficiency over wide torque and speed ranges, thus maximizing the driving range of EVs. The corresponding efficiency map should be comparable with that of the PM brushless motor and at least higher than that of the induction motor.
- The SR motor drive needs to offer good dynamic performance. Particularly, when the multiple-motor configuration of SR motor drives is adopted for in-wheel EV propulsion, it is mandatory to have good dynamic performance.
- Since the SR motor does not involve copper windings or PMs in the rotor, the effect of temperature distribution in the rotor is not so critical as compared with that of the induction motor or PM brushless motor. Nevertheless, if the SR motor adopts the outer-rotor structure for in-wheel EV propulsion, the corresponding design needs to consider the temperature effect in the stator.

Figure 5.26 shows the flowchart to depict the design process of the SR motor, which involves iterative loops to optimize the machine dimensions and parameters. Firstly, the machine geometry and dimensions are initialized by some analytical or experiential equations. Secondly, the machine static performances are determined by using the electromagnetic FEM, which will be assessed and used to adjust the machine dimensions. Thirdly, the machine dynamic performances are obtained by using the electromagnetic FEM, which will be evaluated and fed back to tune the machine dimensions. Finally, if necessary, the acoustic noise and temperature of the SR motor are analyzed by using the structural and thermal FEMs, which will be appraised to fine-tune the machine dimensions.

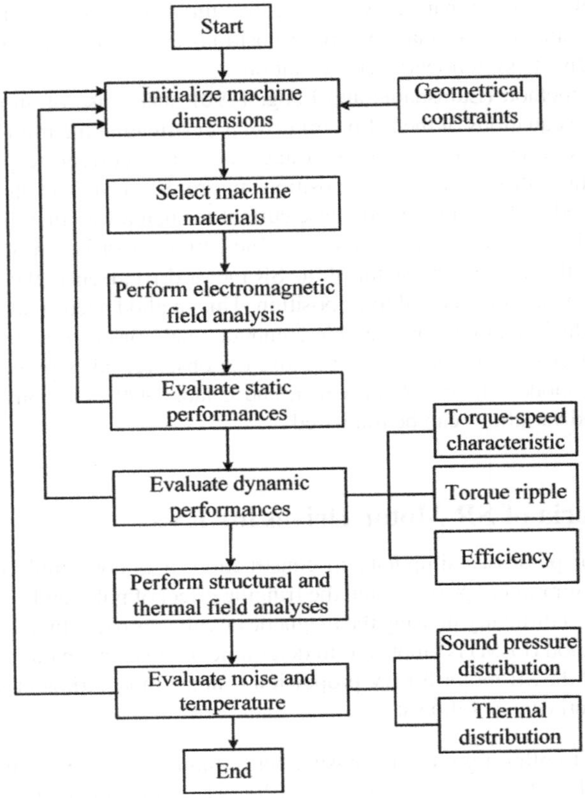

Figure 5.26 Design process of SR motor

5.5.1 *Machine Initialization*

The initialization of machine geometry and dimensions is the key to kick off the design process. A good initialization can significantly reduce the iterative process and design time. A detailed guideline on the initialization of SR motors can be found in Miller (1993).

5.5.1.1 Sizing Rotor Diameter and Core Length

The sizing of the SR motor starts from the output torque equation. The relationship between the output torque, T_e, rotor diameter, D_r, and core length, L_c is generally governed by

$$T_e = K\, D_r^2\, L_c \tag{5.35}$$

where K is the output coefficient proportional to the product of the electric and magnetic loadings. A typical range of K for SR motors for EV application is 10.3–34.5 kN m/m^3. Thus, once the torque requirement is specified, the required $D_r^2\, L_c$ can be deduced. In order to determine the rotor diameter and length separately, it is necessary to select the rotor diameter/length ratio, which is typically close to unity. For EV propulsion, this ratio is more flexible and mainly governed by the available space under the hood or within the wheel.

5.5.1.2 Setting Air-Gap Length

There is no doubt that a small or short air-gap of SR motors creates a high inductance ratio, leading to achieve high torque density. However, the smaller the air-gap, the higher the difficulty in manufacturing is resulted, hence increasing the production cost. As a rule of thumb, the air-gap of SR motors should be similar to, but not greater than that of induction motors with a comparable rotor diameter.

5.5.1.3 Sizing Stator Diameters

When the air-gap length g is selected, the stator inside diameter D_{si} of the SR motor can readily be deduced by:

$$D_{si} = D_r + 2g \tag{5.36}$$

Then, the stator outside diameter D_{so} can be estimated based on the following experiential relationship:

$$D_{so} = (1.5\text{–}1.8)\, D_{si} \tag{5.37}$$

This relationship varies over quite a wide range, depending on the number of stator and rotor poles. In general, the larger the number of poles, the smaller the stator outside diameter is required. Inevitably, this outside diameter is governed by the available space in the EV so that a compromise has to be made between the outside diameter and the length of the SR motor.

5.5.1.4 Selecting Phase and Pole Numbers

An appropriate selection of the number of phases and poles of the SR motor is important to satisfy the desired performance. The number of stator poles, N_s and the number of rotor poles, N_r are generally governed by

$$N_s = 2\,k\,m \tag{5.38}$$

$$N_r = 2\,k\,(m + q) \tag{5.39}$$

where m is the number of phases, k is a positive integer, and q is a positive or negative integer. When the rotor speed is ω, the switching frequency f_s of a particular phase is given by

$$f_s = \frac{\omega}{2\pi} N_r \tag{5.40}$$

To minimize the switching frequency and iron losses, the number of rotor poles should be selected as small as possible. On the contrary, the higher the number of rotor poles, the lower the torque ripple is resulted. Meanwhile, by increasing the number of phases, the torque ripple can be reduced, but the SR converter cost will be increased. Thus, many combinations have been developed for SR motors as listed in Table 5.1.

The three-phase 6/4-pole and four-phase 8/6-pole SR machine topologies are most commonly used. The three-phase 6/4-pole SR machine offers the advantages of simplicity and good AAC capability. However, the corresponding large torque ripple reduces the starting torque for EV launching. On the other hand, the four-phase 8/6-pole SR machine has a relatively higher cost while possessing a smaller torque ripple to favor the launching. Moreover, by increasing the number of poles per phase such as the three-phase 12/8-pole topology, the torque ripple can also be reduced while maintaining the converter cost. In case of using an outer-rotor SR machine for in-wheel EV propulsion, the rotor pole number is generally higher than the stator pole number because the torque ripple is the major concern for direct-drive application while the switching frequency is inherently limited by the low rotor speed.

Table 5.1 Common combinations of phase and pole numbers of SR motors

m	N_s	N_r	k	q
3	6	4	1	−1
3	6	8	1	+1
3	6	10	1	+2
3	12	8	2	−1
3	18	12	3	−1
3	24	16	4	−1
3	24	32	4	+1
4	8	6	1	−1
4	16	12	2	−1
4	16	20	2	+1
5	10	4	1	−3
5	10	6	1	−2
5	10	8	1	−1

5.5.1.5 Sizing Pole Arcs

Since the output torque depends on the slot area to accommodate the stator winding, the stator pole arc, β_s is generally made smaller than the rotor pole arc, β_r so that a slightly larger slot area can be provided without sacrificing the aligned inductance. It yields

$$\beta_s < \beta_r \tag{5.41}$$

In order to enable torque production at all rotor positions, the stator or rotor pole arc, whichever is less, should be larger than the step angle of each phase as given by

$$\min(\beta_s, \beta_r) > \frac{2\pi}{mN_r} \tag{5.42}$$

In order to minimize the inductance at the unaligned rotor position, the total arc length of the rotor and stator poles should be shorter than the rotor pole pitch as governed by

$$\beta_s + \beta_r < \frac{2\pi}{N_r} \tag{5.43}$$

On the basis of the pole arcs, the stator pole width, t_s and rotor pole width, t_r can readily be obtained as

$$t_s = (D_r + 2g)\sin\frac{\beta_s}{2} \tag{5.44}$$

$$t_r = D_r \sin\frac{\beta_r}{2} \tag{5.45}$$

5.5.1.6 Sizing Pole Heights and Yoke Thicknesses

In order to provide a low inductance at the unaligned rotor position, the rotor pole height h_r should be much greater than the air-gap length, but very high value of h_r causes significant magnetic fringing. Hence, the rule of thumb for its initialization is given by

$$h_r = \frac{t_s}{2} \tag{5.46}$$

The rotor yoke thickness, y_r is governed by the condition of peak rotor flux without magnetic saturation. The rule of thumb for its initialization is given by

$$y_r > \frac{t_r}{2} \tag{5.47}$$

The stator yoke thickness, y_s is subject to the same constraint as the rotor yoke thickness. Hence, the rule of thumb for its initialization is given by

$$y_s > \frac{t_s}{2} \tag{5.48}$$

Having initialized the rotor pole height as well as the rotor and stator yoke thicknesses, the stator pole height, h_s can be deduced by

$$h_s = \frac{1}{2}(D_{\text{so}} - D_{\text{si}} - 2y_s) \tag{5.49}$$

5.5.1.7 Sizing Shaft Diameter

A large shaft diameter is necessary to provide the stiffness to withstand the lateral vibration and torsional rigidity. It also helps reduce the lateral vibration and hence the resulting acoustic noise. After sizing the rotor diameter, pole height, and yoke thickness, the shaft diameter D_{sh} can readily be expressed as

$$D_{\text{sh}} = D_r - 2(h_r + y_r) \tag{5.50}$$

5.5.1.8 Estimating Number of Turns

At speed ω, the maximum flux linkage per pole $N\psi_{\max}$ that can be produced by the stator winding can be estimated as

$$N\psi_{\max} = \frac{V_{\text{dc}}}{\omega}\frac{2\pi}{mN_r} \tag{5.51}$$

where V_{dc} is the applied DC source voltage and N is number of turns per pole. If this maximum flux linkage is sufficient to cause saturation in the iron core, it can be written as

$$N\psi_{\max} = \frac{N\,N_s}{m}\,t_s L_c B_s \tag{5.52}$$

where B_s is the saturation flux density in the iron core (typically 1.7 T for standard lamination material). Hence, the number of turns per pole can be estimated as

$$N = \frac{2\pi\,V_{\text{dc}}}{\omega\,N_r\,N_s\,t_s\,L_c\,B_s} \tag{5.53}$$

5.5.1.9 Estimating Current Density

In order to have the desired average torque T_{av}, the required ampere-turns per pole of the stator winding can be expressed as

$$N\,I_{\text{rms}} = \frac{T_{\text{av}}}{B_s\,D_r\,L_c} \tag{5.54}$$

where I_{rms} is the root-mean-square (RMS) current. With two coil sides per slot, the current density J_{rms} can be obtained as

$$J_{\text{rms}} = \frac{N\,I_{\text{rms}}}{k_s\,(A_s/2)} \tag{5.55}$$

where A_s is the area of each slot and k_s is the slot-fill factor, which is the ratio of copper area to slot area. The allowable current density is governed by using various cooling methods, namely totally enclosed (<5 A/mm^2), fan cooled (<10 A/mm^2), and liquid cooled (<30 A/mm^2) unless there are special measures to optimize the thermal conductivity between the coils and the stator. Meanwhile, the slot-fill factor depends on the wire shape, wiring technique, and slot geometry. This slot-fill factor ranges from 40% when using loosely bundled coils, to 60% when using preformed coils with rectangular shape, and up to 80% when using preformed coils with shape adapted to the slot (Carstensen, 2007).

It should be noted that the aforementioned design equations are derived based on the conventional SR machine structure, namely those inner-rotor topologies. For the design of outer-rotor SR machine topologies, those equations need to be modified accordingly.

5.5.2 *Suppression of Acoustic Noise*

The vibration force and hence the acoustic noise are well-known problems in SR motors (Ehsani, Gao, and Emadi, 2009). As reflected by the torque ripple, the vibration force is mainly caused by the varying magnetic forces between the stator and rotor poles. The formation of radial force, F_r and tangential force, F_θ components of this vibration force is illustrated in Figure 5.27 – the radial force component increases gradually from the minimum value at the unaligned position to the maximum value at the aligned position, whereas the tangential force component increases from zero at the unaligned position to the maximum value and returns to zero at the aligned position.

The vibration force and acoustic noise in SR motors can readily be calculated by using the FEM. To determine the relationship between this vibration force and the machine dimensions, various analytical formulas have been derived. The results show that the radial force component is the most dominant source to cause the deformation of the stator and, therefore, the stator vibration and acoustic noise.

Instead of using the TRM control such as the torque-sharing functions to reduce the torque ripple and hence the stator vibration, some design-oriented methods can be employed to suppress the stator vibration and hence the acoustic noise:

- Since the stator yoke thickness plays an important role in the dynamic behavior of the stator deformation, an increase in this yoke thickness will increase the natural frequency and decrease the deformation of the stator. This will also reduce the chance of mechanical resonance of the whole machine structure even at high speeds. However, there is a penalty that a thick stator yoke will reduce the slot area, which in turn increases the current density. Therefore, the selection of stator yoke thickness needs a compromise between the current density and acoustic noise.

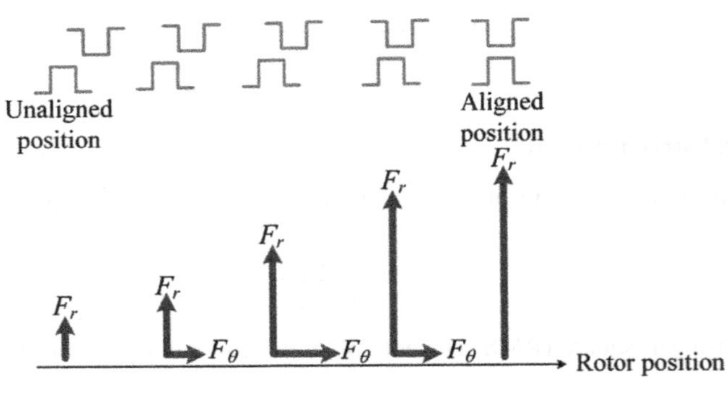

Figure 5.27 Vibration force components in SR motor

- As the radial force component depends on the radial air-gap flux density, an increase in air-gap length will directly reduce the radial air-gap flux density and hence the radial vibration. However, this accompanies with a reduction of torque density. Hence, there is a trade-off between torque density and acoustic noise in selecting the air-gap length.
- Skewing the rotor pole can reduce the radial force, which produces unwanted vibration and noise. However, utilizing this technique inevitably affects the machine performances, especially the torque density. Nevertheless, through careful selection of the skew angle, the reduction in radial force can be much lower than the sacrifice of the average torque. Therefore, if it is highly desirable to reduce vibration and noise while admitting to have a lower average torque, the rotor skewing is an effective way.

5.6 Examples of SR Motor Drives for EVs

Two major types of SR motor drives are selected for exemplification. The first one is based on a high-speed SR motor internally coupled with a planetary gear for speed reduction. The second one is based on a low-speed SR motor for in-wheel direct-drive.

5.6.1 Planetary-Geared SR Motor Drive

When the EV adopts the single-motor configuration, the SR motor drive is generally designed for high-speed operation while using a planetary gear to step down the motor speed so as to match the wheel speed. On the basis of the requirements of a typical passenger EV, the specifications of the corresponding SR motor drive are listed in Table 5.2.

To suppress the torque ripple, the four-phase 16/12-pole SR machine topology is selected. Figure 5.28 shows its exploded diagram. On the basis of the specifications, the geometric dimensions and parameters

Table 5.2 Specifications of planetary-geared SR motor drive

DC voltage	360 V
Rated power	12.6 kW
Rated speed	3000 rpm
Rated torque	40 N m
Constant-torque operation	0–3000 rpm
Constant-power operation	3000–9000 rpm
Gear ratio	10 : 1

Figure 5.28 Four-phase 16/12-pole SR motor

of this SR motor can be initialized by using the aforementioned design equations, and are listed in Table 5.3.

In order to take into account the saturation and fringing effects, the electromagnetic finite element analysis is inevitable. The electromagnetic field distribution of this SR motor design example is shown in Figure 5.29. Hence, both static and dynamic performances of the SR motor can be assessed. Consequently, a complete set of magnetization curves of the flux linkage versus the applied current with the rotor at various positions is evaluated. On the basis of the flux linkage characteristics, the inductance

Table 5.3 Initialization of four-phase 16/12-pole SR motor

Phase number	4
Stator pole number	16
Rotor pole number	12
Stator outside diameter	220 mm
Rotor outside diameter	125 mm
Rotor inside diameter	66.4 mm
Air-gap length	0.3 mm
Core length	100 mm
Stator pole arc	12°
Rotor pole arc	15°
Stator yoke thickness	20 mm
Rotor yoke thickness	20 mm
Turns per pole	50
Slot-fill factor	60%
Rated current density	4.82 A/mm^2
Peak current density	7.23 A/mm^2
Lamination material	50JN700

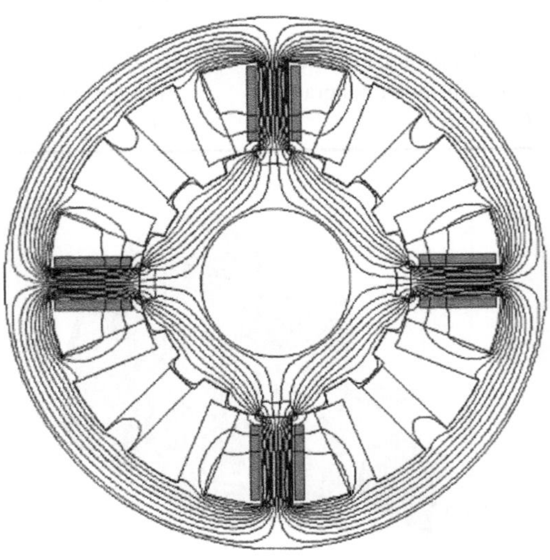

Figure 5.29 Electromagnetic field distribution of four-phase 16/12-pole SR motor

characteristics can readily be deduced by using Eq. (5.10). The inductance varies widely as a function of both rotor position and phase current. Particularly, the inductance at the aligned position decreases with the increase of current due to the effect of saturation. Knowing the inductance characteristics, the torque characteristics can be deduced by using Eq. (5.13). As shown in Figure 5.30, the torque-angle curves are nonlinear with the torque capability increasing with the phase current.

A typical set of phase torque and total torque waveforms during the CCC mode of operation is shown in Figure 5.31. It can be observed that the phase torque is rectangular within the first half period and completely vanishes in the second half period. Meanwhile, a typical set of phase torque and total torque waveforms during the AAC mode of operation is shown in Figure 5.32. It can be observed that the phase torque is single-pulse in nature within the first half period and vanishes in the second half period. Both of these sets of waveforms agree well with the theoretical waveforms.

The designed SR motor drive can perform the constant-torque operation using the CCC mode and the constant-power operation using the AAC mode. During the constant-torque operation, the turn-on and turn-off angles are essentially kept constant, whereas during the constant-power operation, they are optimally tuned. The corresponding torque-speed capability is shown in Figure 5.33. It can be observed

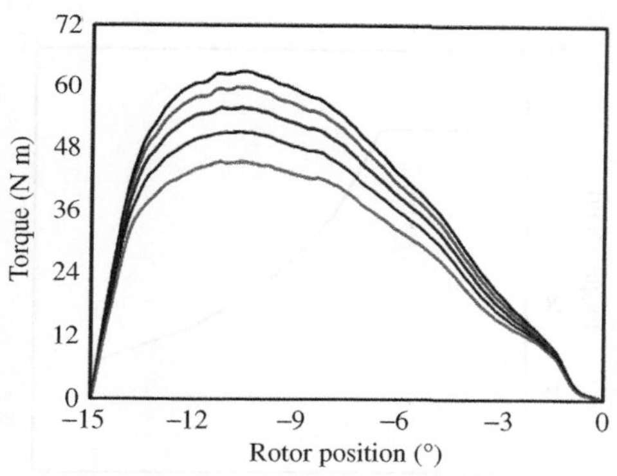

Figure 5.30 Torque-angle curves of four-phase 16/12-pole SR motor

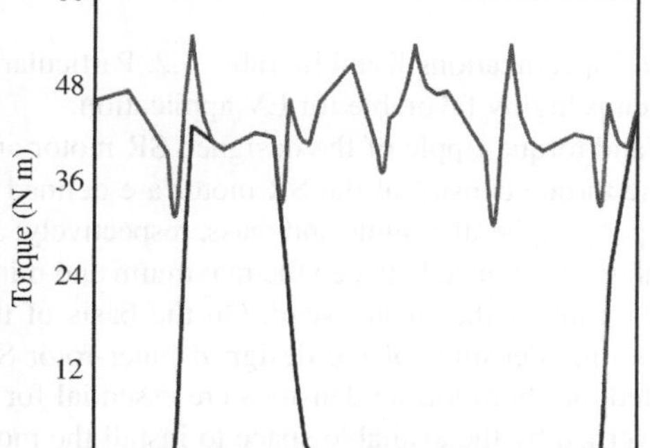

Figure 5.31 Phase torque and total torque waveforms during CCC mode of four-phase 16/12-pole SR motor

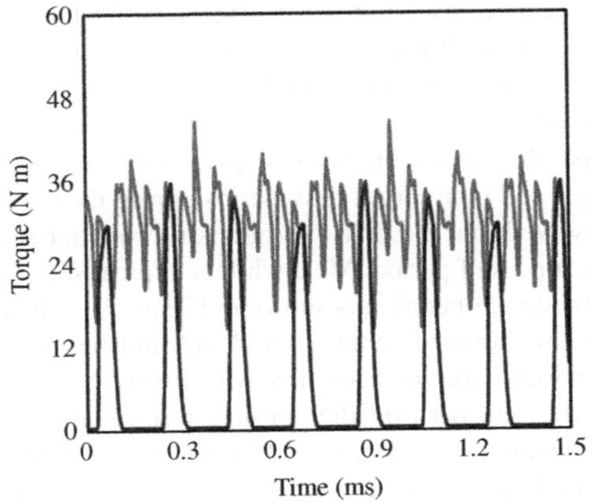

Figure 5.32 Phase torque and total torque waveforms during AAC mode of four-phase 16/12-pole SR motor

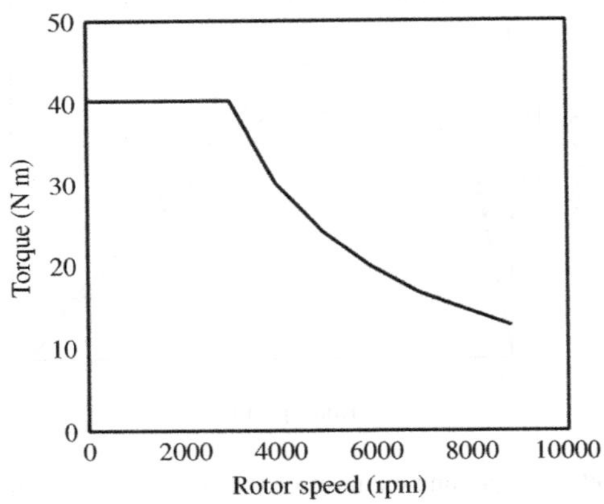

Figure 5.33 Torque-speed capability of four-phase 16/12-pole SR motor

that it fully satisfies the desired specifications listed in Table 5.2. Particularly, the constant-power range is significantly extended, which is highly favorable for EV application.

Finally, the torque density and torque ripple of the designed SR motor are evaluated. The volumetric torque density and gravimetric torque density of the SR motor are defined as the averaged steady-state torque at the rated conditions per peripheral volume and mass, respectively; and the corresponding torque ripple is defined as the percentage difference between the maximum and minimum instantaneous torques with respect to the averaged torque at the steady state. On the basis of the machine dimensions and the torque performance, the torque densities of the designed inner-rotor SR motor are 10.5 N m/l and 1.35 N m/kg. It should be noted that both torque densities are essential for EV application. That is, the volumetric one is directly governed by the available space to install the motor, whereas the gravimetric one significantly affects the driving range and hill climbing capability. On the other hand, the corresponding torque ripple is calculated as 44%. Although this level of torque ripple is normal and acceptable for SR motor drives, it may cause annoying during EV launching or low-speed operation where its presence can be physically felt by the driver.

5.6.2 *Outer-Rotor In-Wheel SR Motor Drive*

If the EV prefers to get rid of the transmission gear and differential gear, the dual-motor configuration is adopted. Thus, each SR motor is generally designed with an outer-rotor topology for low-speed operation so as to directly drive each wheel. On the basis of the requirements of a typical passenger EV, the specifications of the corresponding SR motor drive are listed in Table 5.4 in which the space to accommodate each SR motor is limited by the size of the wheel. For instance, the 225/40R18 wheel denotes that the axial length is 225 mm and the rim diameter is 457 mm (18 in.).

In order to suppress the torque ripple, the four-phase 16/20-pole outer-rotor SR machine topology is adopted. Figure 5.34 shows its exploded diagram. On the basis of the specifications, the geometric dimensions and parameters of this SR motor can be initialized as listed in Table 5.5.

Similar to the previous example, the electromagnetic finite element analysis is inevitable. The electromagnetic field distribution of this SR motor design example is depicted in Figure 5.35. Consequently, the static performances such as the torque-angle curves of this motor are deduced as shown in Figure 5.36. The dynamic performances such as the torque waveforms during the CCC mode of operation and the AAC mode of operation are simulated as shown in Figures 5.37 and 5.38, respectively. It can be observed that all these curves and waveforms agree well with the theoretical ones. Subsequently, the torque-speed capability of this SR motor drive is resulted as shown in Figure 5.39. It is obvious that it fully satisfies the desired specifications as listed in Table 5.4.

Table 5.4 Specifications of outer-rotor in-wheel SR motor drive

DC voltage	360 V
Rated power	6.3 kW
Rated speed	300 rpm
Rated torque	200 N m
Constant-torque operation	0–300 rpm
Constant-power operation	300–900 rpm
Wheel	225/40R18

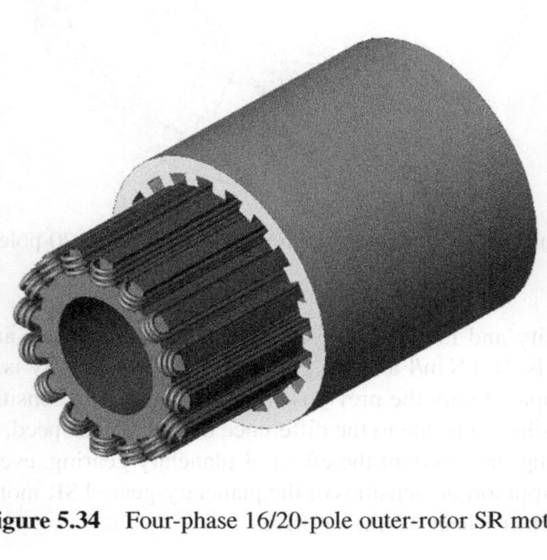

Figure 5.34 Four-phase 16/20-pole outer-rotor SR motor

Table 5.5 Initialization of four-phase 16/20-pole outer-rotor SR motor

Phase number	4
Stator pole number	16
Rotor pole number	20
Stator outside diameter	180 mm
Rotor outside diameter	230 mm
Rotor inside diameter	100 mm
Air-gap length	0.3 mm
Core length	225 mm
Stator pole arc	10°
Rotor pole arc	8°
Stator yoke thickness	20
Rotor yoke thickness	20
Turns per pole	50
Slot-fill factor	60%
Rated current density	5.23 A/mm
Peak current density	8.36 A/mm
Lamination material	50JN700

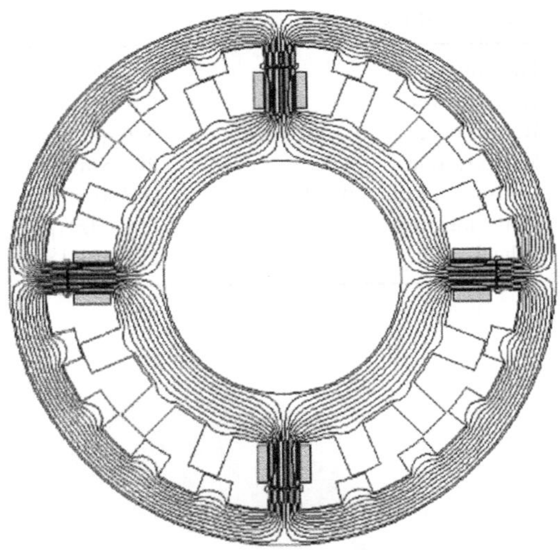

Figure 5.35 Electromagnetic field distribution of four-phase 16/20-pole outer-rotor motor

Finally, the torque density and torque ripple of the designed SR motor are calculated. That is, the volumetric torque density is 21.4 N m/l and the gravimetric torque density is 2.74 N m/kg, whereas the torque ripple is 35%. Compared with the previous example, the torque densities of this SR motor seem to be better. Actually, this illusion is due to the difference of their rated speed, one rated at 3000 rpm and one rated at 300 rpm. Taking into account the effect of planetary gearing, even with the additional gear volume and weight, the output torque densities of the planetary-geared SR motor drive are actually better than those of the outer-rotor in-wheel SR motor drive.

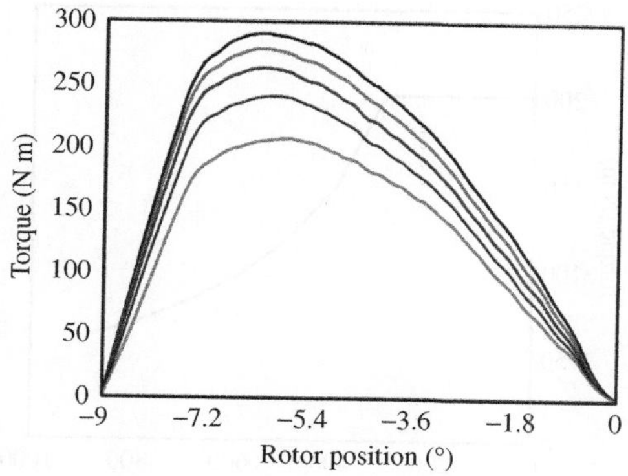

Figure 5.36 Torque-angle curves of four-phase 16/20-pole outer-rotor motor

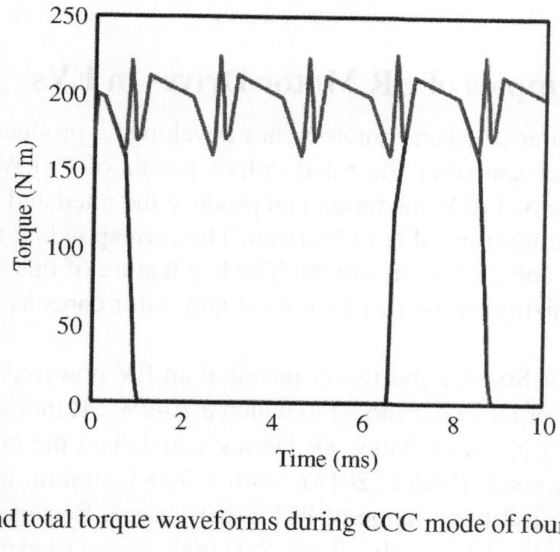

Figure 5.37 Phase torque and total torque waveforms during CCC mode of four-phase 16/20-pole outer-rotor motor

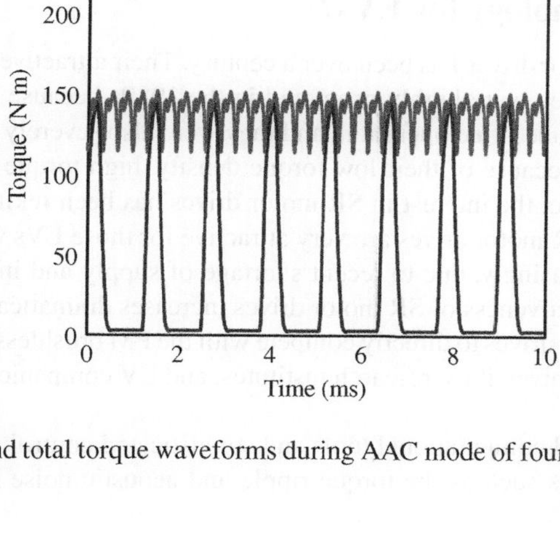

Figure 5.38 Phase torque and total torque waveforms during AAC mode of four-phase 16/20-pole outer-rotor motor

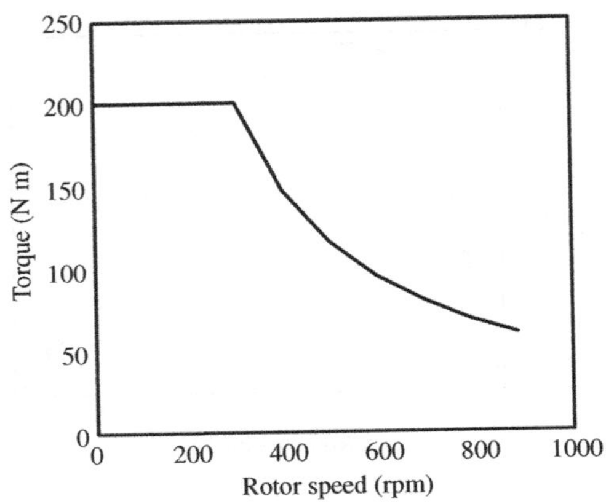

Figure 5.39 Torque-speed capability of four-phase 16/20-pole outer-rotor motor

5.7 Application Examples of SR Motor Drives in EVs

Nidec, a Japanese manufacturer of electric motors, has developed a product list of SR motors for EVs. For instance, its Z019 model can offer the rated output power of 19 kW, and achieve up to 29 kW (Nidec, 2014). With a voltage of 170 V, the motor can produce the rated and maximum torques of 40 and 86 N m, respectively. Its maximum speed is 11 000 rpm. The corresponding specifications are 177 mm in diameter, 234 mm in length, and 26.5 kg in weight. The key feature of this SR motor is the reduction of vibration and noise by optimizing the shape of the rotor and stator cores as well as the timing of current control.

At the 2013 Geneva Motor Show, Land Rover unveiled an EV powered by an SR motor drive. This EV is based on the Land Rover Defender model in which a 70-kW SR motor is installed. This SR motor, which is developed by Harrogate-based Nidec SR Drives, can deliver the rated power over a wide speed range for all-terrain traction needs (Nidec, 2013). With a 300-V lithium-ion battery having a capacity of 27 kWh, the EV can offer a driving range of 80 km per charge. Because of the SR motor delivering a starting torque of 330 N m, the EV simply adopts the single-speed transmission with the gear ratio of 2.7 : 1 for four-wheel drive. The corresponding top speed can achieve 112 km/h.

5.8 Maturing Technology for EVs?

The development of SR motor drives has been over a century. Their attractiveness has exercised many ups and downs. For instance, they were highly promoted in the 1980s because of their robust construction, simple control, and maintenance-free operation, and they were also severely challenged by PM brushless motor drives in the 1990s because of their low torque density, high torque ripple, and noisy operation. Starting from the last decade, the interest in SR motor drives has been rekindled. Particularly, there has been a consensus that the SR motor drives are very attractive for those EVs where cost and reliability are of particular concern. Increasingly, due to recent shortage of supply and increase of price of rare-earth materials for PMs, the attractiveness of SR motor drives increases dramatically, leading to be considered as the most promising motor drives to directly compete with the PM brushless motor drives for EV propulsion. Consequently, many universities, research institutes, and EV companies are actively developing SR motor drives for EVs.

Being fueled by advanced computational tools and sophisticated control strategies, most of the shortcomings of SR motor drives such as the torque ripple and acoustic noise have been solved or at least alleviated. The corresponding torque and power densities have also been improved to overtake those of induction motor drives. In short, the SR motor drive technology is becoming mature. Therefore, in recent years, some SR motor drives that are specially developed for EV propulsion have been commercially available. Meanwhile, some practical applications of SR motor drives to EVs have also been reported. It is anticipated that more and more EV companies will have their commercial EV models adopting SR motor drives in the near future.

References

Carstensen, C. (2007) Eddy currents in windings of switched reluctance machines. Dr-Ing Dissertation. RWTH Aachen University, Germany, Shaker.

Chao, K.-H. (2009) A novel soft-switching converter for switched reluctance motor drives. *WSEAS Transactions on Circuits and Systems*, **8**, 411–421.

Ching, T.W. (2009) Analysis of soft-switching converters for switched reluctance motor drives for electric vehicles. *Journal of Asian Electric Vehicles*, **7**, 893–898.

Ehsani, M., Gao, Y. and Emadi, A. (2009) *Modern Electric, Hybrid Electric, and Fuel Cell Vehicles: Fundamentals, Theory, and Design*, 2nd edn, CRC Press, Boca Raton, FL.

Ehsani, M., Husain, I., Mahajan, S. and Ramani, K.R. (1994) New modulation encoding techniques for indirect rotor position sensing in switched reluctance motors. *IEEE Transactions on Industry Applications*, **30**, 85–91.

Gallegos-Lopez, G., Kjaer, P.C., Miller, T.J.E., and White, G.W. (1997) Simulation study of resonant dc link inverter for current-controlled switched reluctance motors. Proceedings of International Conference on Power Electronics and Drive Systems, pp. 757–761.

Gao, H., Salmasi, F.R. and Ehsani, M. (2004) Inductance model-based sensorless control of the switched reluctance motor drive at low speed. *IEEE Transactions on Power Electronics*, **19**, 1568–1573.

Husain, I. and Ehsani, M. (1996) Torque ripple minimization in switched reluctance drives by PWM current control. *IEEE Transactions on Power Electronics*, **11**, 83–88.

Illic-Spong, M., Miller, T.J.E. and MacMinn, S.R. (1987) Instantaneous torque control of electric motor drives. *IEEE Transactions on Power Electronics*, **2**, 55–61.

Krishnan, R. (2001) *Switched Reluctance Motor Drives: Modeling, Simulation, Analysis, Design, and Applications*, CRC Press, Boca Raton, FL.

Lee, D.-H., Liang, J., Lee, Z.-G. and Ahn, J.-W. (2009) A simple nonlinear logical torque sharing function for low-torque ripple SR drive. *IEEE Transactions on Industrial Electronics*, **56**, 3021–3028.

Lumsdaine, A. and Lang, J.H. (1990) State observer for variable reluctance motors. *IEEE Transactions on Industrial Electronics*, **37**, 133–142.

Luo, J. and Zhan, Q. (2005) A novel soft-switching converter for switched reluctance motor: analysis, design and experimental results. Proceedings of IEEE International Conference on Electric Machines and Drives, pp. 1955–1961.

Lyons, J.P., MacMinn, S.R., and Preston, M.A. (1991) Flux/current methods for SRM rotor position estimation. Proceedings of IEEE Industry Applications Society Annual Meeting, pp. 482–487.

Miller, T.J.E. (1993) *Switched Reluctance Motors and Their Control*, Magna Physics Publishing, Hillsboro, OH.

Murai, Y., Cheng, J. and Yoshida, M. (1999) New soft-switched/switched-reluctance motor drive circuit. *IEEE Transactions on Industry Applications*, **35**, 78–85.

Nidec (2013) Nidec SR Drive Motors Power New Land Rover, Nidec Motor, http://www.srdrives.com/land-rover.shtml (accessed September 2014).

Nidec (2014) SR Motor – Automotive SRMZ019, Nidec, http://www.nidec.com/en-NA/product/motor/category/A040/B070/P0000140/ (accessed September 2014).

Panda, S.K. and Amaratunga, G.A.J. (1991) Analysis of the waveform detection technique for indirect rotor position sensing of switched reluctance motor drives. *IEEE Transactions on Energy Conversion*, **6**, 476–483.

Pop, A.-C., Petrus, V., Martis, C.S. *et al.* (2012) Comparative study of different torque sharing functions for losses minimization in switched reluctance motors used in electric vehicles propulsion. Proceedings of International Conference on Optimization of Electrical and Electronic Equipment, pp. 356–365.

Rolim, L.G.B., Suemitsu, W.I., Watanabe, E.H. and Hanitsch, R. (1999) Development of an improved switched reluctance motor drive using a soft-switching converter. *IEE Proceedings-Electric Power Applications*, **146**, 488–494.

Sahoo, S.K., Panda, S.K. and Xu, J.X. (2005) Indirect torque control of switched reluctance motors using iterative learning control. *IEEE Transactions on Power Electronics*, **20**, 200–208.

Schramm, D.S., Williams, B.W., and Green, T.C. (1992) Torque ripple reduction of switched reluctance motors by PWM phase current optimal profiling. Proceedings of IEEE Power Electronics Specialists Conference, pp. 857–860.

Suresh, G., Fahimi, B., Rahman, K.M., and Ehsani, M. (1999) Inductance based position encoding for sensorless SRM drives. Proceedings of IEEE Power Electronics Specialists Conference, pp. 832–837.

Xue, X.D., Cheng, K.W.E. and Ho, S.L. (2009) Optimization and evaluation of torque-sharing functions for torque ripple minimization in switched reluctance motor drives. *IEEE Transactions on Power Electronics*, **24**, 2076–2090.

Yang, H.-Y., Kim, J.-H. and Krishnan, R. (2012) Low-cost position sensorless switched relutance motor drive using a single-controllable switch converter. *Journal of Power Electronics*, **12**, 75–82.

Zhan, Y.J., Chan, C.C. and Chau, K.T. (1999) A novel sliding mode observer for indirect position sensing of switched reluctance motor drives. *IEEE Transactions on Industrial Electronics*, **46**, 390–397.

6

Stator-Permanent Magnet Motor Drives

Differing from traditional permanent magnet (PM) brushless motor drives, stator-PM motor drives takes the definite advantages that all PM materials are located in the stator while the rotor is simply iron core with salient poles, leading to offer higher robustness for high-speed operation and better thermal stability for PM materials. These two features are highly desirable for electric vehicle (EV) motor drives that need to operate at harsh working environment. Currently, there is a wide variety of stator-PM machines, and some of them are considered viable for EV propulsion.

In this chapter, various stator-PM motor drives, including the doubly-salient permanent magnet (DSPM), flux-reversal permanent magnet (FRPM), flux-switching permanent magnet (FSPM), hybrid-excited permanent magnet (HEPM), and flux-mnemonic permanent magnet (FMPM) topologies, are presented. The corresponding design criteria, design examples, and potential applications are also given.

6.1 Stator-PM versus Rotor-PM

The concept of stator-PM machines can be traced back to 1955 when Rauch and Johnson proposed probably the first stator-PM machine in which both the stator and rotor have salient poles, and the PMs are located in the stator (Rauch and Johnson, 1955). Conceptually, they employ the polarized reluctance principle in such a way that the torque or electromotive force (EMF) is resulted from the variable reluctance action of stator and rotor saliencies on the unipolar flux produced by PMs in the stator, hence offering a very simple and robust rotor structure. However, because of the limitation of high-energy PM materials at that time, the stator-PM machines were unattractive (Cheng *et al.*, 2011).

Although PM brushless motor drives, including both the PM synchronous and PM brushless DC (BLDC) types, have been widely accepted for EV propulsion, the corresponding rotor-PM nature is still their key drawback. That is, it is very demanding to mount the PM pieces in the rotor which desires to withstand high-speed rotation and severe mechanical vibration. In addition, the rotor temperature under the trapped environment is so high that may violate the thermal stability requirement of PM materials. Therefore, the development trend of PM machines is focused on moving the PMs from the rotor to the stator, leading to stimulate the development of stator-PM motor drives.

Electric Vehicle Machines and Drives: Design, Analysis and Application, First Edition. K. T. Chau.

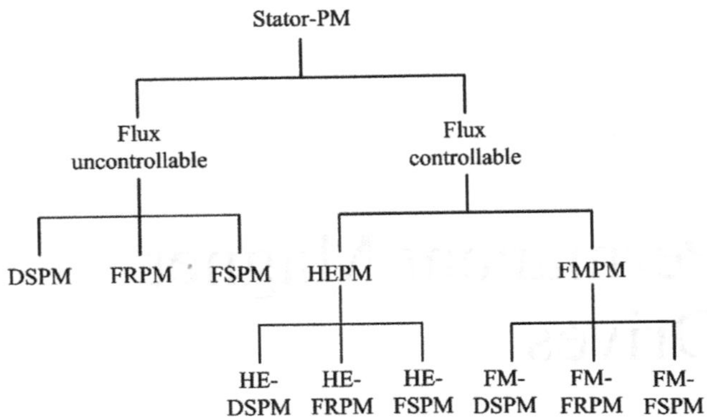

Figure 6.1 Viable members of stator-PM machines

Focusing on those stator-PM machines that are viable for EV propulsion, there are three major types:

- DSPM machine
- FRPM machine
- FSPM machine.

As these three types of stator-PM machines are based on PM excitation, they are classified into the same group – flux uncontrollable. Additionally, with the inclusion of independent field winding or magnetizing winding in the stator for flux control, the stator-PM machines become flux controllable, which can be further classified as

- HEPM
- FMPM.

These flux-controllable techniques can be applied to the three major types of stator-PM machines to form various topologies, such as the hybrid-excited flux-switching permanent magnet (HE-FSPM) machine or flux-mnemonic doubly-salient permanent magnet (FM-DSPM) machine. A tree diagram of viable stator-PM machine topologies is depicted in Figure 6.1.

6.2 System Configurations

The stator-PM motor drive consists of four main components: the stator-PM machine, converter, sensor, and controller. There are various system configurations of the stator-PM motor drives, depending on the types of stator-PM machines and the associated power converters.

Among them, the flux-uncontrollable group, including the DSPM, FRPM, and FSPM machines, needs only one external supply, hence called the singly-fed stator-PM machines; whereas the flux-controllable group, including the HE-DSPM (hybrid-excited doubly-salient permanent magnet), HE-FRPM (hybrid-excited flux-reversal permanent magnet), and HE-FSPM as well as the FM-DSPM, FM-FRPM (flux-mnemonic flux-reversal permanent magnet), and FM-FSPM (flux-mnemonic flux-switching permanent magnet) machines desire two external supplies, hence called the doubly-fed stator-PM machines. The system configuration of those singly-fed stator-PM motor drives is depicted in Figure 6.2 in which the armature winding in the stator is solely fed by a power inverter. Meanwhile, the system configuration of those doubly-fed stator-PM motor drives is depicted in Figure 6.3 in which the armature winding is

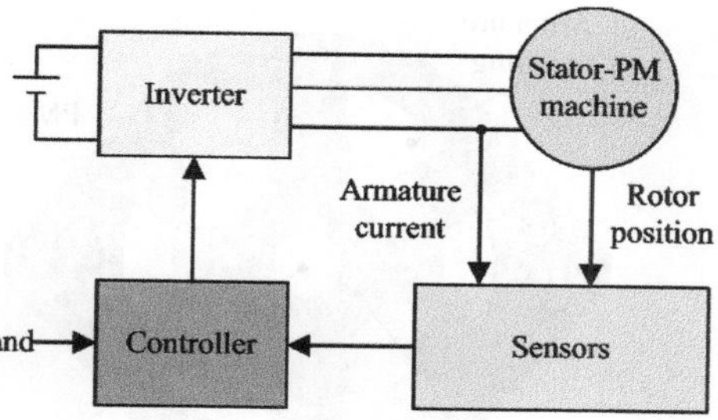

Figure 6.2 Configuration of singly-fed stator-PM motor drives

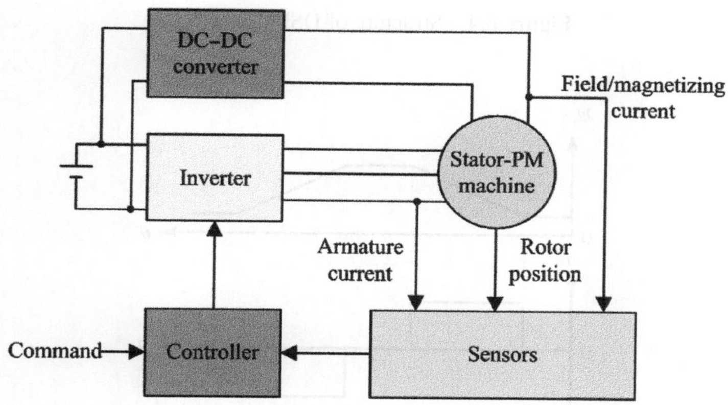

Figure 6.3 Configuration of doubly-fed stator-PM motor drives

fed by an inverter while the field winding is fed by a DC–DC converter. Between the singly-fed and doubly-fed stator-PM motor drives, the former has the definite merit of simplicity, while the latter has additional controllability to improve the operating performances. Of course, both the machine topology and converter circuitry of the doubly-fed stator-PM motor drives are more complicated and costly.

It should be noted that the DC–DC converter for the doubly-fed stator-PM motor drives should allow for bidirectional current flow. That is, for flux strengthening, the field or magnetizing current should be positive, whereas for flux weakening, the field or magnetizing current should be negative. Between the two doubly-fed families, the HEPM machines desire continual DC field current for field excitation whereas the FMPM machines need only temporary current pulses for PM magnetization or demagnetization. Thus, their DC–DC converter designs and ratings are different.

6.3 Doubly-Salient PM Motor Drives

The DSPM motor drive is relatively the most mature type of stator-PM motor drives. As depicted in Figure 6.4, the DSPM machine can be considered as a combination of switched reluctance (SR) machine and PM brushless machine in such a way that the PMs are inserted in the stator back-iron and the rotor is simply iron core with salient poles. Meanwhile, the stator has salient poles wound with the concentrated winding, making the PM flux linkage unipolar. Since the stator can be easily cooled by natural or forced-air cooling, the thermal instability of PMs can be avoided. Hence, this DSPM machine is able to work at high temperature, facilitating its application to EVs.

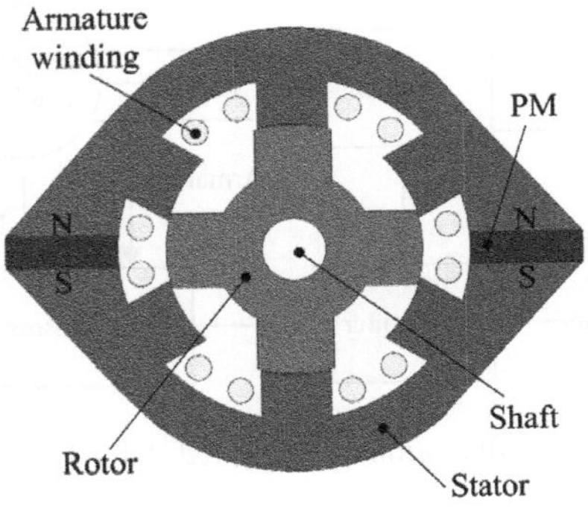

Figure 6.4 Structure of DSPM machine

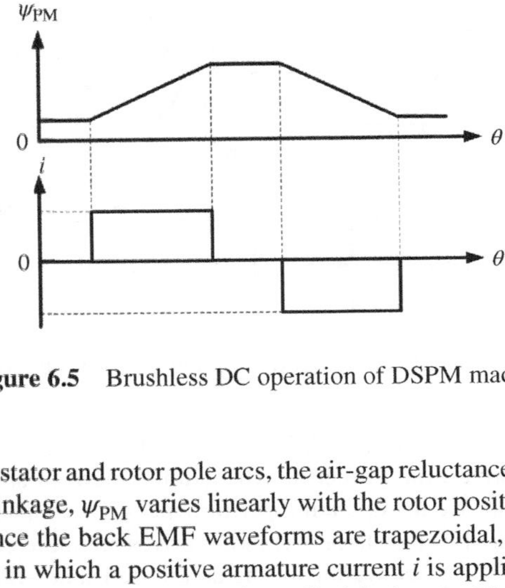

Figure 6.5 Brushless DC operation of DSPM machine

By properly designing the stator and rotor pole arcs, the air-gap reluctance seen by the PMs are invariant. Consequently, the PM flux linkage, ψ_{PM} varies linearly with the rotor position as shown in Figure 6.5. As the PM flux linkage and hence the back EMF waveforms are trapezoidal, the DSPM machine generally operates at the BLDC mode in which a positive armature current i is applied during the PM flux linkage increasing zone while a negative armature current is applied during the PM flux linkage decreasing zone. Notice that the zero-current interval between the positive and negative currents is purposely provided to ensure successful current reversal.

There is a wide range of possible combinations of the phase number, stator pole number, and rotor pole number that can be chosen for DSPM machine design. In accordance with the basic operating principle of DSPM machines, the general relationships between the stator pole number N_s, rotor pole number N_r, and phase number m are governed by

$$\begin{cases} N_s = 2mk \\ N_r = N_s \pm 2k \end{cases} \tag{6.1}$$

where k is a positive integer. In general, k can be considered as the number of PM pole-pairs, N_{PM}. For instance, when $m = 3$ and $k = 1$, it ends up with $N_s = 6$, $N_r = 4$, and $N_{PM} = 1$, which is a three-phase 6/4-pole DSPM machine as shown in Figure 6.4; when $m = 3$ and $k = 2$, it results in $N_s = 12$, $N_r = 8$, and $N_{PM} = 2$, which is a three-phase 12/8-pole DSPM topology as shown in Figure 6.6; and when $m = 4$ and $k = 1$, it comes up with $N_s = 8, N_r = 6$, and $N_{PM} = 1$, which is a four-phase 8/6-pole DSPM topology as shown in Figure 6.7.

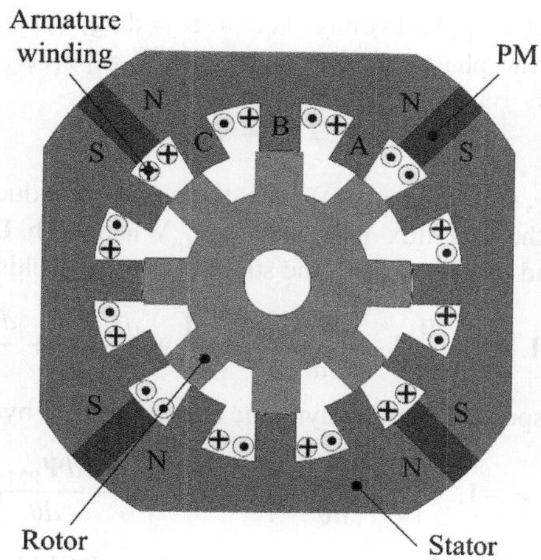

Figure 6.6 Structure of three-phase 12/8-pole DSPM machine

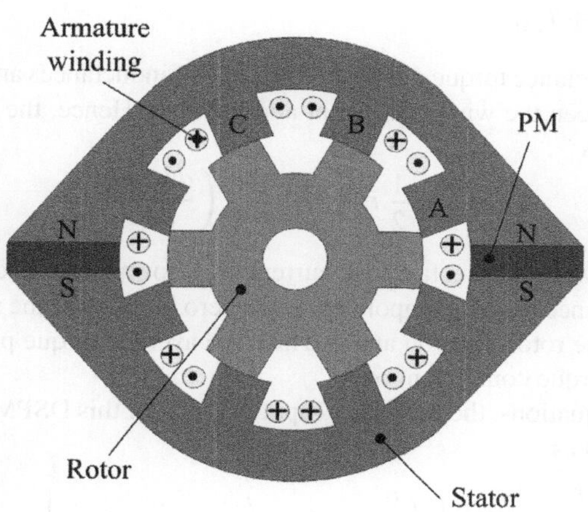

Figure 6.7 Structure of four-phase 8/6-pole DSPM machine

When the machine runs at the speed of n, the commutating frequency of any phase is governed by

$$f_{ph} = N_r n/60 \tag{6.2}$$

To minimize the switching frequency and hence the iron losses in poles and yokes as well as the switching loss in power devices, the number of rotor poles should be selected as small as possible. Thus, the number of rotor poles is usually less than that of stator poles. To make the machine capable of starting itself in either forward or reverse direction, the number of phases should be equal to or more than three.

The system equation describing an m-phase DSPM machine can be expressed as (Cheng, Chau, and Chan, 2001):

$$\mathbf{V} = \mathbf{RI} + \frac{d\boldsymbol{\Psi}}{dt} \tag{6.3}$$

where $\mathbf{V} = [v_1, v_2, \dots, v_m]^T$ is the applied voltage vector, $\mathbf{R} = \text{diag}[r_1, r_2, \dots, r_m]$ is the phase resistance vector, $\mathbf{I} = [i_1, i_2, \dots, i_m]^T$ is the phase current vector, and $\boldsymbol{\Psi} = [\psi_1, \psi_2, \dots, \psi_m]^T$ is the flux linkage vector which can be written as

$$\boldsymbol{\Psi} = \mathbf{LI} + \boldsymbol{\Psi}_{\text{PM}} \tag{6.4}$$

where $\mathbf{L} = [L_{ij}]$ $(i = 1, \dots, m,\ j = 1, \dots, m)$ is the phase inductance matrix and $\boldsymbol{\Psi}_{\mathbf{PM}} = [\psi_{\text{PM1}}, \psi_{\text{PM2}}, \dots, \psi_{\text{PM}m}]^T$ is the PM flux linkage vector. When both $\mathbf{L}$ and $\boldsymbol{\Psi}_{\mathbf{PM}}$ are considered to be spatially dependent only and independent of the stator current, it yields

$$\frac{d\boldsymbol{\Psi}}{dt} = \mathbf{L}\frac{d\mathbf{I}}{dt} + \frac{d\mathbf{L}}{dt}\mathbf{I} + \frac{d\boldsymbol{\Psi}_{\mathbf{PM}}}{dt} = \mathbf{L}\frac{d\mathbf{I}}{dt} + \frac{d\mathbf{L}}{d\theta}\mathbf{I}\omega_r + \frac{d\boldsymbol{\Psi}_{\mathbf{PM}}}{d\theta}\omega_r \tag{6.5}$$

where ω_r is the rotor angular speed. Thus, the dynamic equation given by Eq. (6.3) can be rewritten as

$$\frac{d\mathbf{I}}{dt} = -\mathbf{L}^{-1}\left[\mathbf{R} + \frac{d\mathbf{L}}{d\theta}\omega_r\right]\mathbf{I} + \mathbf{L}^{-1}\left[\mathbf{V} - \frac{d\boldsymbol{\Psi}_{\mathbf{PM}}}{d\theta}\omega_r\right] \tag{6.6}$$

By employing the coenergy method, the developed torque of this DSPM machine can be obtained as

$$\begin{aligned} T_e &= \frac{\partial W'}{\partial \theta} = \frac{\partial}{\partial \theta}\left[\frac{1}{2}\mathbf{I}^T\mathbf{LI} + \boldsymbol{\Psi}_{\mathbf{PM}}^T\mathbf{I}\right] = \frac{1}{2}\mathbf{I}^T\left(\frac{\partial \mathbf{L}}{\partial \theta}\right)\mathbf{I} + \left(\frac{\partial \boldsymbol{\Psi}_{\mathbf{PM}}}{\partial \theta}\right)^T\mathbf{I} \\ &= T_r + T_{\text{PM}} \end{aligned} \tag{6.7}$$

where T_r represents the reluctance torque due to the variation of inductances and T_{PM} is the reaction torque due to the interaction between the winding current and PM flux. Hence, the electromagnetic torque per phase, T_{ph} can be written as

$$T_{\text{ph}} = \frac{1}{2}i^2\left(\frac{dL}{d\theta}\right) + i\left(\frac{d\psi_{\text{PM}}}{d\theta}\right) \tag{6.8}$$

It should be noted that when the rectangular current waveform is applied to the phase winding, the average value of the reluctance torque component equals zero because of the symmetric inductance characteristic with respect to the rotor position angle. Thus, the average torque per phase is governed by the average value of the PM torque component only.

Concerning the power equations, the per-phase input power P of this DSPM machine under the applied voltage U can be expressed as

$$P = \frac{1}{T}\int_0^T vi\,dt = \frac{1}{T}\left[\int_{t_1}^{t_2} UI_m dt + \int_{t_3}^{t_4} (-U)(-I_m)dt\right] = \frac{1}{T}2UI_m\Delta T \tag{6.9}$$

where $T = \theta_{\text{cr}}/\omega_r$, $\Delta T = \theta_w/\omega_r$, $\theta_{\text{cr}} = 2\pi/N_r$ is the rotor pole pitch, $\theta_w = \theta_2 - \theta_1 = \theta_4 - \theta_3$ is the angular displacement of a stroke, N_r is the rotor pole number, ω_r is the rotor angular speed, and $t_1 \sim t_4$ are the time instants corresponding to the rotor positions $\theta_1 \sim \theta_4$. Thus, Eq. (6.9) can also be expressed as

$$P = 2UI_m\frac{\theta_w}{\theta_{\text{cr}}} \tag{6.10}$$

As there are m phases, the total input power P_1 is given by

$$P_1 = mP = 2mUI_m\frac{\theta_w}{\theta_{\text{cr}}} \tag{6.11}$$

Denoting the efficiency as η, the total output power P_2 is written as

$$P_2 = \eta P_1 = 2mUI_m\frac{\theta_w}{\theta_{\text{cr}}}\eta \tag{6.12}$$

Substituting $\theta_{cr} = 2\pi/N_r$ in Eq. (6.12), it yields

$$P_2 = \frac{N_r}{\pi} m k_e E I_m \theta_w \eta \tag{6.13}$$

where $k_e = U/E$ and E is the phase back EMF due to the variation of PM flux linkage. This back EMF can be expressed as

$$E = w\frac{d\psi_{\text{PM}}}{d\theta}\omega_r \approx w\frac{\psi_{\max} - \psi_{\min}}{\theta_w}\omega_r = w\frac{\Delta\psi_{\text{PM}}}{\theta_w}\omega_r \tag{6.14}$$

where w is the number of winding turns in series per phase and $\psi_{\max}$ and $\psi_{\min}$ are the PM flux linked by one coil when the stator pole aligns and unaligns with the rotor pole, respectively. In general, $\Delta\psi_{\text{PM}}$ can further be expressed as

$$\Delta\psi_{\text{PM}} = \psi_{\max} - \psi_{\min} \approx 0.87\psi_{\max} = 0.87k_d\alpha_s\tau_s l_e B_\delta = 0.87k_d\alpha_s\frac{\pi D_{\text{si}}}{N_s} l_e B_\delta \tag{6.15}$$

where k_d is the flux leakage factor, l_e is the stack length, B_δ is the air-gap flux density, $\tau_s = \pi D_{\text{si}}/N_s$ is the stator pole pitch, α_s is the stator pole arc factor, N_s is the stator pole number, and D_{si} is the stator inside diameter. Substituting Eq. (6.15) into Eq. (6.14), the back EMF can be written as

$$E = \frac{0.87\pi k_d w\alpha_s D_{\text{si}} l_e B_\delta}{N_s\theta_w}\omega_r \tag{6.16}$$

On the other hand, the magnitude of the rectangular current waveform can be expressed as

$$I_m = k_i I_{\text{rms}} = k_i\frac{\pi D_{\text{si}} A_s}{2mw} \tag{6.17}$$

where A_s is the electric loading of the stator, I_{rms} is the root-mean-square (RMS) phase current, and $k_i = I_m/I_{\text{rms}}$. Substituting Eqs. (6.16), (6.17), and $\omega_r = 2\pi n_s/60$ into Eq. (6.13) while adopting the general case that $\alpha_s \approx 0.5$, the output power equation of this DSPM machine can be derived as

$$P_2 = \frac{0.87\pi^2}{120}\frac{N_r}{N_s}k_d k_e k_i A_s B_\delta n_s \eta D_{\text{si}}^2 l_e \tag{6.18}$$

where n_s is the rated speed of the machine. This output power equation reveals the relationships between the output power and various design parameters. For instance, it can be found that the output power is directly proportional to the ratio of rotor to stator poles, N_r/N_s. Given A_s and B_δ, the larger the value of N_r/N_s the higher the power density. Hence, the 8/6-pole DSPM machine can offer higher power density than the 6/4-pole one by 12.5%.

When the rotor is purposely skewed to minimize the cogging torque, the characteristic of PM flux linkage and hence the back EMF are more sinusoidal. Typically, the skewing angle δ is selected to be about one-half the stator pole pitch. In order to take into account the effect of rotor skewing, a skewing factor is defined as

$$k_s = \cos\left(\frac{\pi}{2\theta_{\text{cs}}}\delta\right) \tag{6.19}$$

where $\theta_{\text{cs}} = 2\pi/N_s$ is the stator pole pitch. Thus, Eq. (6.18) can be modified as

$$P_2 = \frac{0.87\pi^2}{120}\frac{N_r}{N_s}k_s k_d k_e k_i A_s B_\delta n_s \eta D_{\text{si}}^2 l_e \tag{6.20}$$

Consequently, the DSPM machine can operate at the brushless AC (BLAC) mode as shown in Figure 6.8, where the phase current is fed at 90° phase shift with the PM flux linkage or actually in phase with the back EMF.

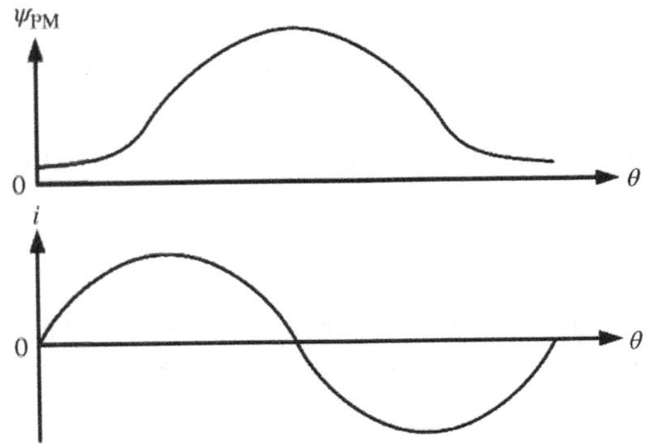

Figure 6.8 Brushless AC operation of DSPM machine

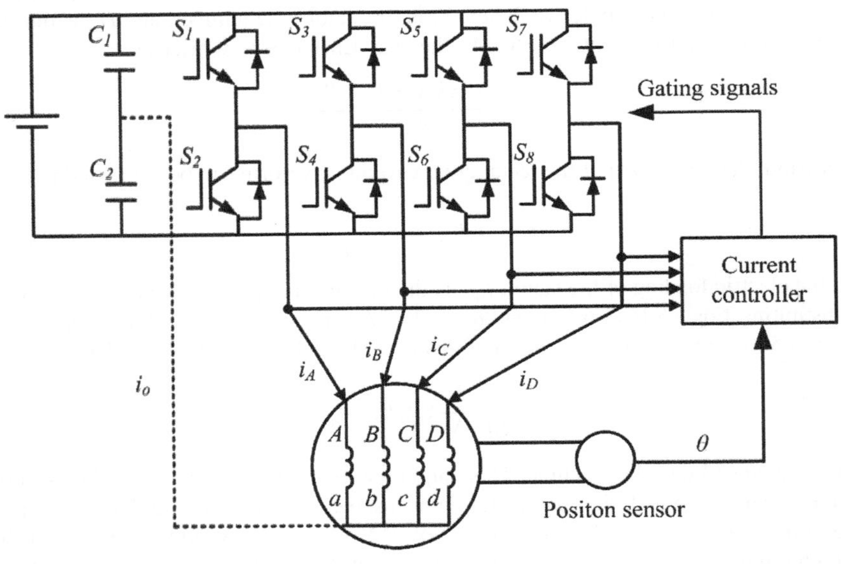

Figure 6.9 Half-bridge inverter with split capacitors for DSPM motor drive

To supply the DSPM motor drive, a bipolar converter is preferred so as to bring the merit of the DSPM machine into full play. To control the phase currents individually for bidirectional operation of the DSPM machine, there are two possible inverter topologies: the full-bridge inverter and the half-bridge inverter with split capacitors. The latter topology is usually selected for the DSPM motor drive because it minimizes the number of power devices. Figure 6.9 shows a four-phase 8/6-pole DSPM motor drive in which the connection between the central point of the split capacitors and the neutral of machine windings is indicated by a dashed line. This connection is mainly used to accommodate the unbalanced current during the commutation period.

In accordance with the operation principle of the DSPM machine, the phase winding should be turned on or off at specific rotor positions. Hence, the rotor position information is indispensable for proper operation of the DSPM machine. Such positions can readily be measured by a simple position sensor as shown in Figure 6.10. It consists of a slotted disc connected to the rotor shaft and two optocouplers

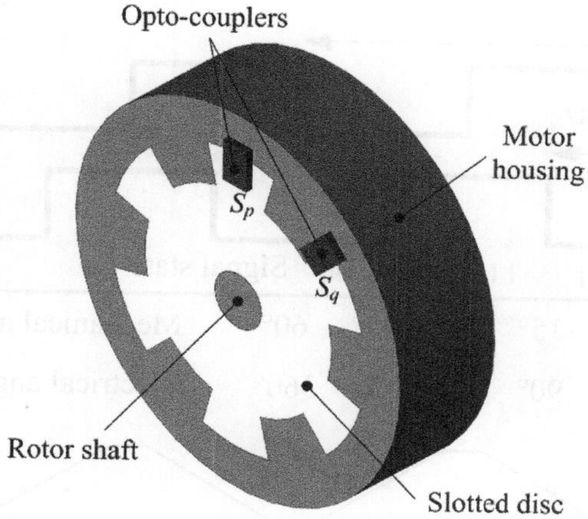

Figure 6.10 Position sensor for DSPM motor drive

Table 6.1 Control logic of 8/6-pole DSPM motor drive

S_qS_p		01	11	10	00
Phase A	S_1	1	1	0	0
	S_2	0	0	1	1
Phase B	S_3	0	1	1	0
	S_4	1	0	0	1
Phase C	S_5	0	0	1	1
	S_6	1	1	0	0
Phase D	S_7	1	0	0	1
	S_8	0	1	1	0

mounted onto the stator housing. The two optocouplers are nominally located 45° apart from each other along the circumference of the disc. The output waveforms of the sensor are shown in Figure 6.11. The sensor generates a signal edge for every 15° of mechanical rotation. The transitions of these outputs determine the specific angles. According to the relationship between the position signals and PM flux linkages, the control logic of the four-phase 8/6-pole DSPM motor drive can be obtained as given in Table 6.1.

For the four-phase 8/6-pole DSPM motor drive, it can be found from the control logic that at any time instant, the power devices in the upper and lower legs of the inverter are conducting in pairs, namely phase A pairs up with phase C, and phase B pairs up with phase D. Hence, it arises the possibility of removing the connection between the central point of the split capacitors and the neutral of the machine windings without significant influence on the behavior of the motor drive provided $\theta_{on}^- = \theta_{on}^+ + \theta_{cr}/2$ and $\theta_{off}^- = \theta_{off}^+ + \theta_{cr}/2$. As a consequence, the problem of voltage unbalance in the split capacitors is eliminated, and the hardware and software of the control system are thus simplified. Moreover, because the positive half cycle and negative half cycle of the phase current are forced to be almost symmetrical, the average value of the reluctance torque approaches zero, minimizing the torque ripple.

The speed control of the DSPM motor drive can readily be implemented by a closed-loop digital proportional-integral (PI) controller, whose output is the torque reference T^* (Cheng *et al.*, 2003).

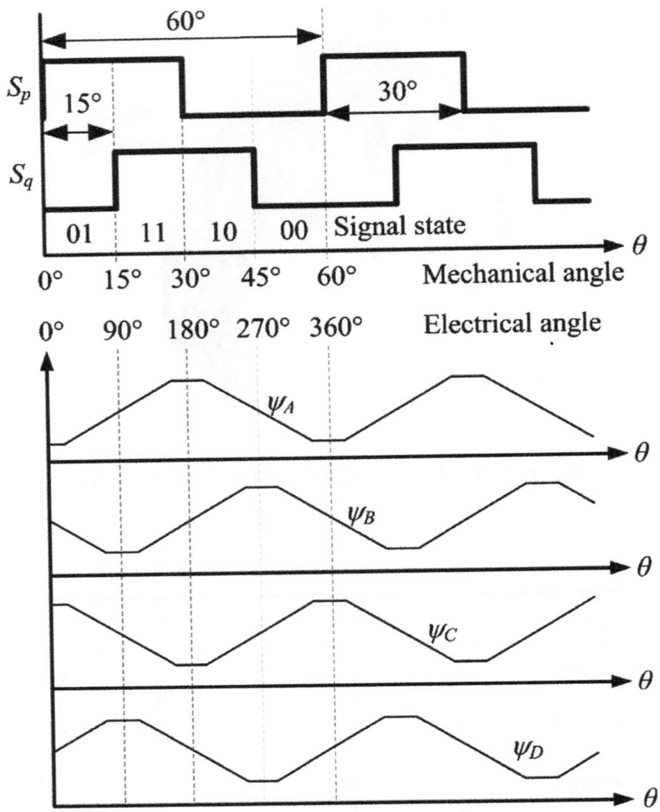

Figure 6.11 Position signals and PM flux linkages of DSPM motor drive

Figure 6.12 shows the block diagram of this speed control system. The corresponding PI controller can be expressed as

$$T^*(k) = K_p e(k) + K_i \sum_{j=0}^{k} e(j) \tag{6.21}$$

where $e(k)$ is the speed error and K_p and K_i are the proportional and integral gains, respectively. While the selection of these gains is a compromise between the requirement for fast response and the need for stable control, it is difficult to have one set of PI parameters satisfying the whole speed and load ranges because the machine parameters vary with both the load current and rotor position. Thus, the adaptive proportional and integral gains are preferred, which can be written as

$$K_p = a_p + b_p(e(t))^2 \tag{6.22}$$

$$K_i = \frac{a_i}{1 + b_i(e(t))^2} \tag{6.23}$$

where a_p, b_p, a_i, and b_i are nonnegative real numbers. Notice that when b_p and b_i are set to zero, it becomes a conventional PI controller.

In order to reduce unnecessary speed regulation, a small dead zone of speed can be introduced into the controller. That is, when the speed error is less than ε_1, the output of the controller $T^*(k)$ adopts the

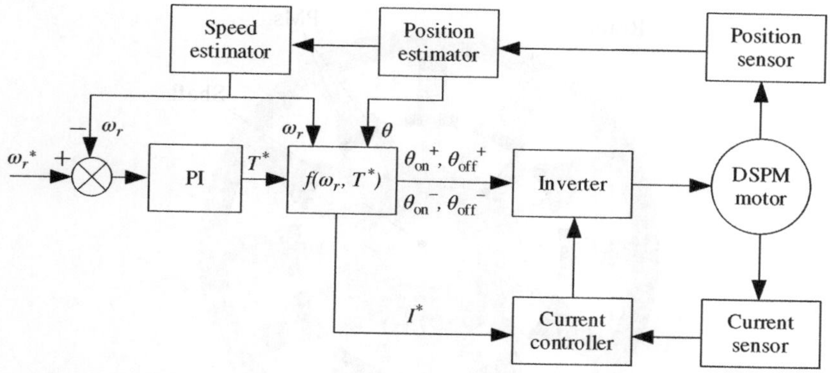

Figure 6.12 Speed control block diagram of DSPM motor drive

previous value $T^*(k-1)$ without corrective action. Meanwhile, in order to improve the dynamic response, the bang-bang control can be combined with the PI control. When the absolute value of the speed error is larger than a predefined value ε_2, the bang-bang control is adopted; otherwise, PI control is performed. That is, in the bang-bang control, when the speed error is positive and the speed is increasing, the output of the controller is set to the maximum value; otherwise, it is set to zero.

For the speed control of DSPM motor drive, the four switching angles and current reference are possible control variables to shape the phase current. When the speed is below the base speed, the current chopping control (CCC) is adopted for the constant-torque operation, in which the four angles are fixed and the torque is controlled by the current reference. Since the current waveforms in the positive and negative strokes are symmetrical in the CCC mode, the average value of the reluctance torque component is zero. Thus, the average torque of the machine is governed by the PM torque component only, which is given by

$$T_{av} = \frac{2m}{\theta_{cr}} \int_{\theta_{on}^+}^{\theta_{off}^+} \left(I \frac{d\psi_{PM}}{d\theta} \right) d\theta \approx \frac{2m}{\theta_{cr}} I(\psi_{PM2} - \psi_{PM1}) \tag{6.24}$$

where I is the magnitude of the rectangular current, m is the phase number, θ_{cr} is the rotor pole pitch angle, ψ_{PM1} and ψ_{PM2} are the PM flux linkages corresponding to θ_{on}^+ and θ_{off}^+, respectively. It can be seen that the average torque is proportional to the current I. Hence, once the torque reference T^* is obtained by using PI regulation, the current reference I^* can readily be specified by using Eq. (6.24).

When the speed is above the base speed, the angle position control (APC) is performed for the constant-power operation. In this high-speed APC mode, the torque is controlled by the conduction angle θ_w, which is given by

$$\theta_w = \theta_{off}^+ - \theta_{on}^+ = \theta_{off}^- - \theta_{on}^- \tag{6.25}$$

The relationship between the torque and conduction angle is nonlinear and complex. It is difficult to get an expression between the torque and conduction angle, because the current waveform at the APC mode is not as regular as that at the CCC mode. Nevertheless, one can get their relationships at some operation points by using steady-state simulation and then fit them into a polynomial as given by

$$\theta_w = f(\omega_r(k-1), T^*(k)) \tag{6.26}$$

6.4 Flux-Reversal PM Motor Drives

The FRPM machine is a kind of stator-PM machine in which the PMs with different polarities are located on the surface of each stator pole, the armature winding generally adopts the concentrated winding

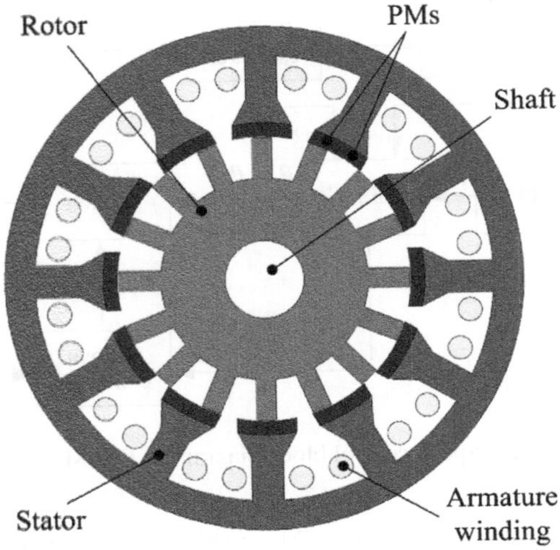

Figure 6.13 Structure of FRPM machine

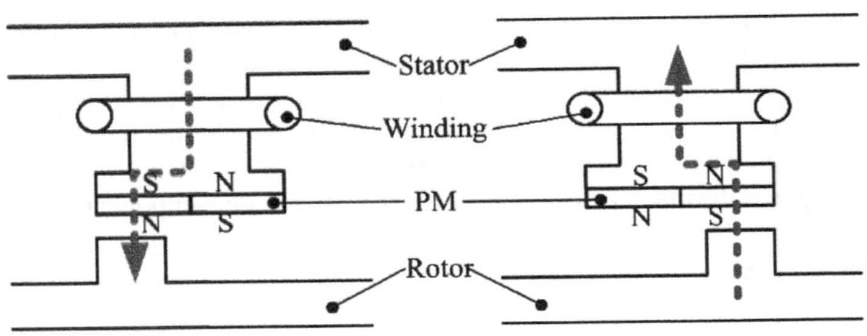

Figure 6.14 PM and winding arrangements of FRPM machine

arrangement, and the rotor is simply iron core with salient poles (Deodhar *et al.*, 1997). Figure 6.13 shows the structure of a three-phase 12/16-pole FRPM machine in which there are 16 stator poles, 12 rotor poles, and 2 PM poles per stator pole. Differing from the DSPM machine, the FRPM machine exhibits bipolar PM flux linkage because the flux linkage with each armature coil reverses polarity as the rotor rotates as depicted in Figure 6.14. Since the bipolar flux linkage variation can have better utilization of iron core than the unipolar counterpart, the FRPM machine inherently offers higher power density than the DSPM machine. Similar to the SR machine, the rotor skewing can also be used to make the air-gap waveform be more sinusoidal, and to reduce the cogging torque.

Since the PMs of this FRPM machine are attached on the surface of stator poles, the induced eddy-current losses in the PMs is a major concern. In addition, it suffers from a relatively low power factor. As the PM thickness equivalently increases the effective air-gap length between the stator poles and rotor poles, the PM thickness, rotor pole arc, and air-gap length have significant effects on the electromagnetic performance of the FRPM machine. Thus, optimal designs on various machine dimensions were investigated, such as the use of concave stator poles and additional flux barriers to reduce the flux leakage (Kim and Lee, 2004), and the technique of rotor teeth pairing to reduce the cogging torque (Kim *et al.*, 2005).

In the three-phase FRPM machine, there are N_s stator poles, N_r rotor poles, and N_{PM} PM pole-pairs in each stator pole (Boldea, Zhang, and Nasar, 2002). As the two-pole pitch angle corresponds to two PM poles of alternate polarities on the stator, it yields

$$2\tau_{PM} = \frac{\pi D_r}{N_r} \tag{6.27}$$

where τ_{PM} is the PM pole pitch and D_r is the rotor diameter. Since the space between two neighboring stator pole PMs is 120° electrical or $2\tau_{PM}/3$, the inner circumference of the stator can be expressed as

$$\left(2N_{PM}\tau_{PM} + \frac{2}{3}\tau_{PM}\right) N_s = \pi(D_r + 2\delta) \tag{6.28}$$

where δ is the air-gap length. Neglecting the air-gap length and substituting Eq. (6.27) in Eq. (6.28), the relationship between N_s, N_r, and N_{PM} can be deduced as

$$\left(N_{PM} + \frac{1}{3}\right) N_s = N_r \tag{6.29}$$

On the basis of Eq. (6.29), there are many possible combinations of N_s, N_r, and N_{PM} that can be selected for the three-phase FRPM machine. For instance, when $N_{PM} = 1$ and $N_s = 12$, it yields $N_r = 16$; when $N_{PM} = 2$ and $N_s = 12$, it results in $N_r = 28$; and when $N_{PM} = 3$ and $N_s = 12$, it ends up with $N_r = 40$. Meanwhile, the rotor speed n_r and the supply frequency f are related by

$$f = n_r N_r \tag{6.30}$$

Therefore, the higher the number of PMs used per stator pole, the higher the number of rotor poles and hence the lower the rotor speed are resulted. For instance, when $N_r = 40$ and $f = 50$ Hz, it yields $n_r = 1.25$ rev/s or 75 rpm.

Similar to the DSPM motor drive, the principle of operation of the FRPM motor drive is illustrated in Figure 6.15, where ψ_{PM} is the instantaneous PM flux linkage and i is the instantaneous applied current. Since the applied current is a rectangular waveform, this operation is called the BLDC mode. That is, a positive current is applied in the PM flux linkage rising zone and a negative current is applied in the PM flux linkage falling zone, hence producing the PM torque at both zones. In general, the variation of inductances with respect to rotor position is so small that the reluctance torque is negligible.

On the other hand, when adopting skewed rotor, the PM flux linkage waveform is more sinusoidal so that a sinusoidal current is fed into the armature winding to create the PM torque as illustrated in Figure 6.16, where the phase current is fed at 90° phase shift with the PM flux linkage. This BLAC mode of operation enables the FRPM motor drive work as the PM synchronous motor drive and can inherit the corresponding control strategies such as vector control and direct torque control.

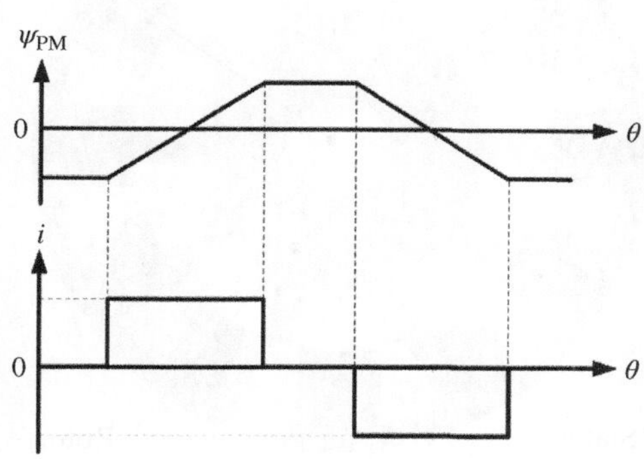

Figure 6.15 Brushless DC operation of FRPM machine

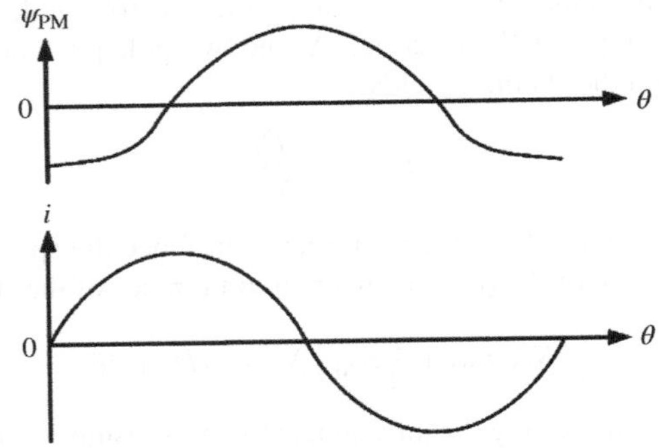

Figure 6.16 Brushless AC operation of FRPM machine

6.5 Flux-Switching PM Motor Drives

The FSPM machine has attracted wide attention in recent years. Among the three major stator-PM machines, the DSPM, FRPM, and FSPM, the stator configuration of the FSPM machine is relatively complex as shown in Figure 6.17. In this topology, each stator pole consists of two adjacent laminated segments and a PM, and each of these segments is sandwiched by two circumferentially magnetized PMs which enables flux focusing (Zhu *et al.*, 2005). In general, the concentrated winding arrangement is adopted, and each coil is wound around the adjacent stator poles. Hence, the polarity of the PM flux linkage in the coil reverses when the rotor pole aligns the successive stator tooth that belongs to the same phase, the so-called flux-switching action, as illustrated in Figure 6.18.

The configuration of this FSPM machine mainly depends on the numbers of stator and rotor poles, which are governed by

$$\begin{cases} N_s = 2mi \\ N_r = N_s \pm 2j \end{cases} \tag{6.31}$$

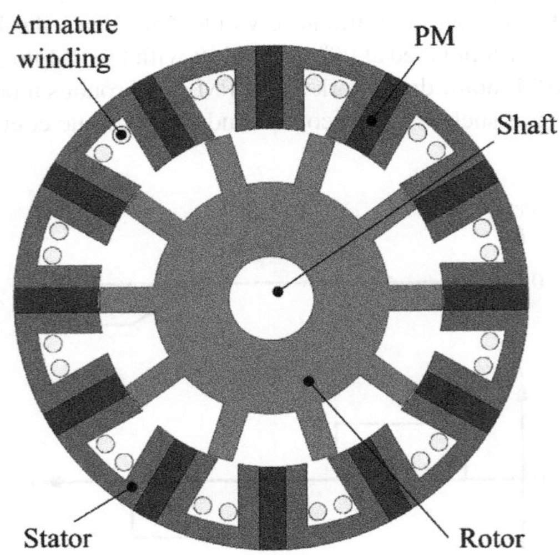

Figure 6.17 Structure of FSPM machine

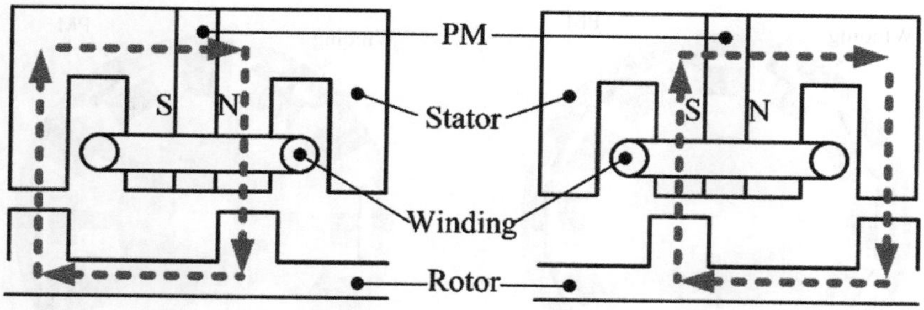

Figure 6.18 PM and winding arrangements of FSPM machine

where N_s is the number of stator poles, N_r is the number of rotor poles, m is the number of phases of the armature winding, and i and j are positive integers. For instance, when $m = 3$, $i = 2$, and $j = 1$ are selected, it yields $N_s = 12$ and $N_r = 10$ or 14, leading to come up with the 12/10-pole and 12/14-pole machine topologies; when $m = 5$, $i = 2$, and $j = 1$ are selected, it ends up with $N_s = 20$ and $N_r = 18$ or 22, leading to have the 20/18-pole and 20/22-pole topologies.

In addition, various winding and pole arrangements of the FSPM machine are developed (Owen *et al.*, 2010) as shown in Figure 6.19. In the original topology, a phase coil is wound around each stator pole and four coils constitute one phase so that each stator slot has coils from two phases, leading to adopt the double-layer winding arrangement. In the second topology, dubbed as alternate poles wound, each slot has coils of only one phase so that only alternate poles are wound. This offers the advantage of using the single-layer winding arrangement, hence reducing the coupling between phases. This facilitates the machine for fault-tolerant operation. In the third topology, dubbed as the alternate poles wound with reduced magnets, the PMs are removed from the unwound stator poles, hence resulting in significant reduction of PM material and cost. The analysis indicates that the use of alternate poles wound for the FSPM machine can maintain the original average torque level, though suffering from relatively high torque ripple, which may need to adopt the rotor skewing to alleviate this drawback. Meanwhile, the alternate poles wound topology with reduced magnets can save one-half of PM material, but incurring a substantial performance penalty.

The use of multi-tooth stator structures has been employed for improving the torque capability of SR machines. This multi-tooth structure can also be extended to the FSPM machine to further improve the torque density (Zhu *et al.*, 2008). The structure of this multi-tooth FSPM machine is similar to that of the original one except that there are effectively more stator teeth on each side of the stator pole mounted with PMs, and also more rotor poles. Consequently, it exhibits higher torque density and lower torque ripple than the original FSPM machine. However, because of the higher armature reaction, it saturates more quickly as the current is increased and thus offers lower torque capability than the original counterpart.

The principle of operation of the FSPM motor drive is similar to that of the FRPM motor drive. Basically, a positive current is applied in the PM flux linkage rising zone and a negative current is applied in the PM flux linkage falling zone, hence producing the PM torque at both zones. Theoretically, the FSPM motor drive can operate in the BLDC mode or BLAC mode. Because of the magnetic reluctance difference between the two pairs of coils comprising a phase, the resultant phase EMF waveforms are relatively more sinusoidal than those of other stator-PM motor drives. Hence, it preferably operates in the BLAC mode, which can utilize the well-known control strategies such as vector control and direct torque control.

6.6 Hybrid-Excited PM Motor Drives

Similar to the conventional rotor-PM machines such as the PM synchronous machine and PM BLDC machine, the stator-PM machines have difficulty in air-gap flux control because of an uncontrollable PM

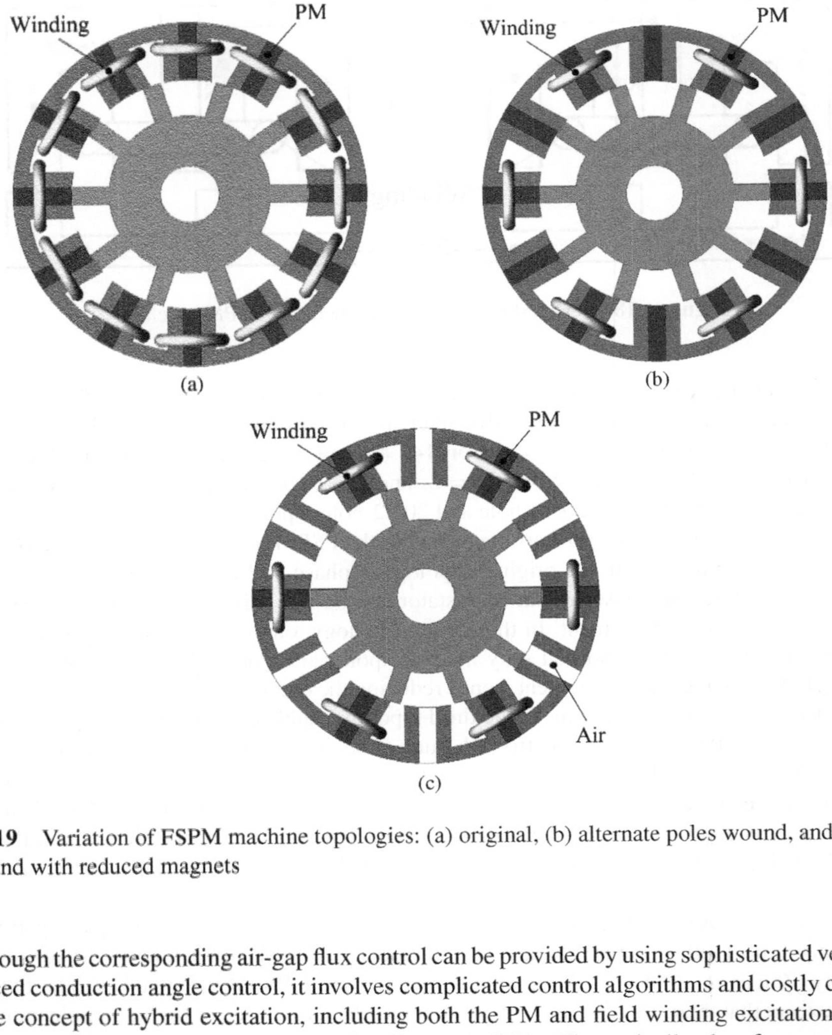

Figure 6.19 Variation of FSPM machine topologies: (a) original, (b) alternate poles wound, and (c) alternate poles wound with reduced magnets

flux. Although the corresponding air-gap flux control can be provided by using sophisticated vector control or advanced conduction angle control, it involves complicated control algorithms and costly control hardware. The concept of hybrid excitation, including both the PM and field winding excitation, can enable the air-gap flux of the stator-PM machines directly controllable. Theoretically, the aforementioned three main types of stator-PM machines, the DSPM, FRPM, and FSPM, can incorporate the concept of hybrid excitation.

By incorporating the hybrid excitation into the DSPM machine, the HE-DSPM machine is resulted as shown in Figure 6.20. The stator incorporates two types of windings: three-phase armature winding and DC field winding, and the PM poles. The rotor has neither PM nor winding, hence offering high mechanical integrity. The three-phase armature winding operates like that for the DSPM machine, whereas the DC field winding not only works as an electromagnet but also as a tool for flux control. In addition, there is an extra air-gap in shunt with each PM. If the DC field winding MMF (magnetomotive force) reinforces the PM MMF, this extra flux path will assist the effect of flux strengthening. On the other hand, if the DC field winding MMF opposes the PM MMF, this extra flux path will favor the PM flux leakage, hence amplifying the effect of flux weakening. As a result, with a proper design of the width of this extra air-gap, a wide flux-regulation range can be obtained by using a small DC field excitation.

Figure 6.21 shows the power circuit for DC field current control, which is actually a typical H-bridge converter. Hence, the magnitude of the DC field current can be easily regulated by adjusting the duty

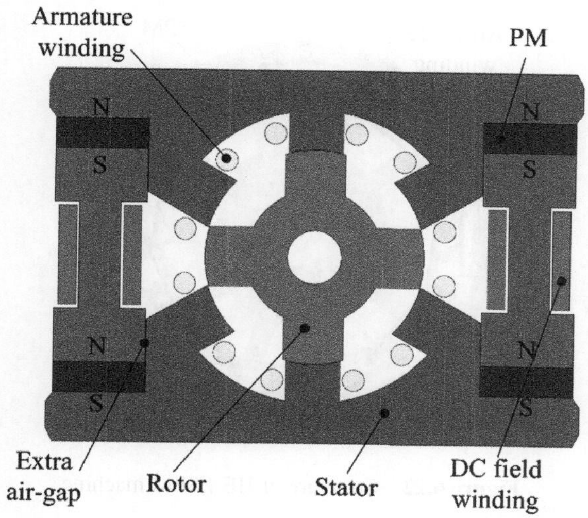

Figure 6.20 Structure of HE-DSPM machine

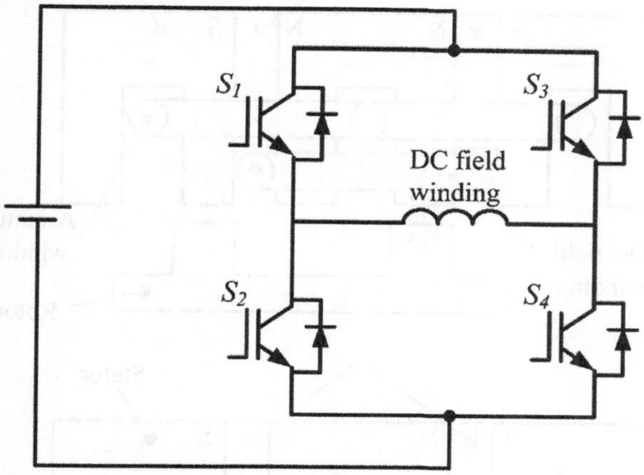

Figure 6.21 Power circuit for DC field winding in HE-DSPM machine

cycle of the conducting pair of power devices. Also, the direction of the DC field current can be simply controlled by selecting different pairs of power devices.

Similarly, by incorporating the hybrid excitation into the FSPM machine, the HE-FSPM machine (Hua, Cheng, and Zhang, 2009) is created as shown in Figure 6.22. The configuration is very similar to that of the FSPM machine except that it is equipped with the DC field winding. In order to accommodate the additional DC field winding, the PM length of the HE-FSPM machine is shorter than that of the original FSPM machine.

The operation principle of this HE-FSPM machine is similar to that of the original FSPM machine except that it can offer the feature of direct air-gap flux control. As depicted in Figure 6.23, by controlling the polarity and value of the DC field current, the air-gap flux density can be easily strengthened or weakened. Meanwhile, this arrangement of hybrid excitation does not need to increase the overall machine size.

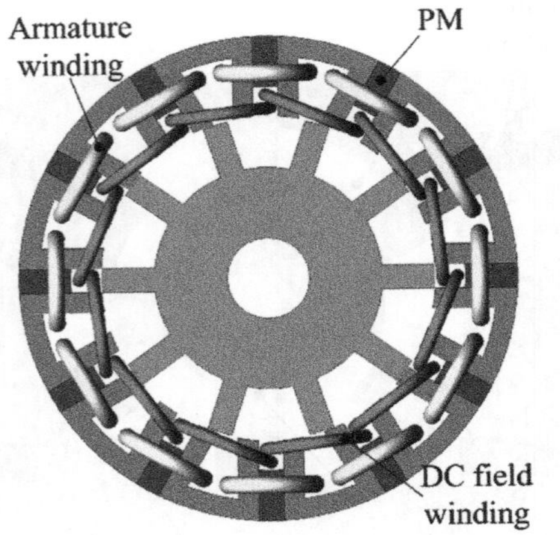

Figure 6.22 Structure of HE-FSPM machine

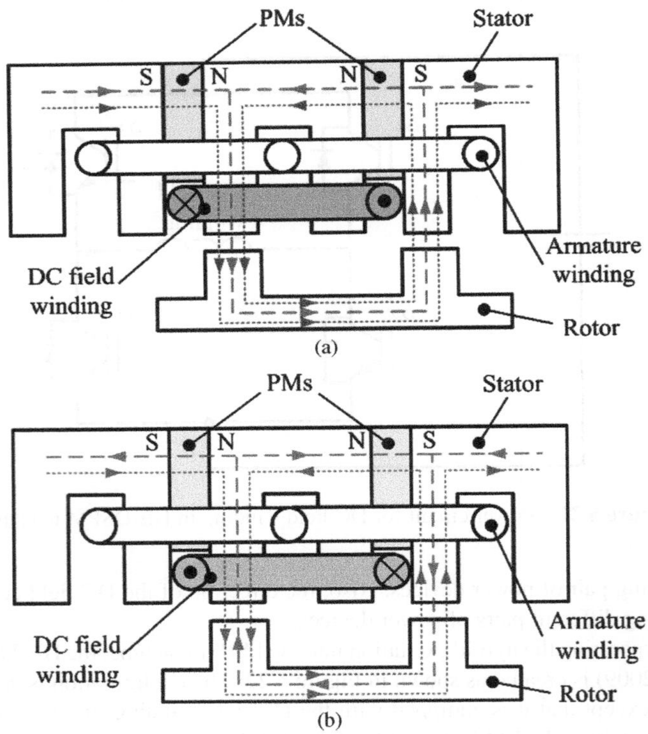

Figure 6.23 Flux control of HE-FSPM machine: (a) flux strengthening and (b) flux weakening

Although the concept of hybrid excitation can theoretically be extended to the FRPM, there is a practical limitation that restricts the development of HE-FRPM machine. Since PMs with different polarities are located on the surface of each stator pole, it is difficult to further insert the DC field winding within each stator pole; otherwise, additional interpoles with the DC field winding may be required.

In summary, the hybrid-excited stator-PM motor drives can offer some attractive features for EV operation:

- By varying the polarity and magnitude of the DC field winding current, the air-gap flux density becomes easily controllable.
- By realizing flux strengthening, the machine can offer the exceptionally high-torque feature, which is very essential for cold cranking HEVs (hybrid electric vehicles) or providing temporary power for vehicular overtaking and hill climbing.
- By realizing flux weakening, the machine can offer the exceptionally wide-speed constant-power feature, which is very essential for EV cruising.
- By online tuning the air-gap flux density, the machine can maintain a constant voltage output under generation or regeneration over a very wide speed range, which is very essential for battery charging in various EVs.
- By online tuning the air-gap flux density, the machine can also offer efficiency-optimizing control, which is highly desirable for EVs.

6.7 Flux-Mnemonic PM Motor Drives

With the advent of stator-PM machines, in which both the PMs and armature windings are located in the stator, the mechanical integrity problem of PM machines is totally solved. The development of these PM machines becomes focusing on how to control the air-gap flux, hence extending the speed range of constant-power operation. By incorporating the DC field winding in the stator, the flux-controllable PM machine, typically the HE-DSPM machine, can enable both flux weakening and flux strengthening. Nevertheless, the corresponding DC field winding needs to be sized for continual excitation, thus degrading the machine power density and efficiency.

The conventional memory machine adopts a traditional rotor-PM machine structure, and utilizes the *d*-axis armature current to perform online magnetization or demagnetization of PMs (Ostovic, 2003). This so-called AC-excited memory machine suffers from the use of complicated vector control, lack of mechanical integrity in the PM rotor, and possibility of accidental demagnetization due to armature reaction. The idea of using DC field winding for the HE-DSPM machine is extended to the memory machine, hence forming the FMPM machine, also dubbed as the DC-excited memory machine (Yu and Chau, 2011a). The concept of flux mnemonic or memory is due to the nature of the aluminum–nickel–cobalt (Al-Ni-Co) PMs in the machine that can be online magnetized or demagnetized to various magnetization levels and then be memorized automatically. It incorporates a small magnetizing winding to temporarily carry the magnetizing current for direct PM magnetization, hence solving the problem of continual field excitation and avoiding the complicated current control.

When the Al-Ni-Co PM material was invented in the 1930s, it was widely employed for PM machines due to its excellent remanence, high thermal stability, and high chemical stability. However, in a conventional PM machine, PMs were expected to have a linear demagnetization curve with a superposed recoil line so as to provide stable performance under normal operating conditions, and a high coercivity so as to avoid irreversible demagnetization under extreme operating conditions. Unfortunately, these two expectations were the shortcomings of the Al-Ni-Co PM. Consequently, the Al-Ni-Co PM was superseded by the samarium-cobalt (Sm-Co) and neodymium-iron-boron (Nd-Fe-B) PMs for application to PM machines. Dramatically, these shortcomings of the Al-Ni-Co PM are positively utilized in the FMPM machine:

- The nonlinearity of their demagnetization characteristics makes the recoil line never superpose on the demagnetization curve. Thus, once the demagnetization current is applied and then removed, the operating point will move along the recoil line and settle at a lower magnetization level, which means that the magnetization level is memorized.
- The low coercivity is the desired feature to enable online demagnetization. This uniqueness cannot be provided by other PM materials, such as the Sm-Co or Nd-Fe-B. Additionally, with both high thermal stability and high chemical stability, the Al-Ni-Co PM is the natural choice for the FMPM machine.

The flux-mnemonic concept can readily be incorporated into all three main types of stator-PM machines, the DSPM, FRPM, and FSPM. Figure 6.24 shows the structure of an FM-DSPM machine that adopts an outer-rotor double-layer-stator topology. The use of five phases rather than three is to enhance the torque smoothness and to provide the capability of fault tolerance. These two features are particularly desirable for EVs. The use of outer rotor rather than the inner rotor is to provide the direct-drive capability for in-wheel propulsion. The inner space of the stator can be fully utilized to accommodate the PMs, armature winding, and magnetizing winding, hence achieving a compact structure. Since the armature winding and PMs are located in different layers of the stator, the PMs can be immune from accidental demagnetization by armature reaction. In addition, the armature adopts fractional-slot windings with the coil span equal to the slot pitch; it not only reduces the cogging torque but also shortens the end-windings. Similar to that of the DSPM machine, the rotor is simply composed of salient poles with no PM or winding. Thus, it is very robust. Differing from the DC field winding of the HE-DSPM machine, the magnetizing winding does not need to handle continual current flow so that the corresponding size is much smaller. Most importantly, it requires only a temporary current pulse for PM magnetization or demagnetization so that the energy consumption for PM magnetization or demagnetization is temporary and insignificant compared with that of the HE-DSPM machine.

Because of the nature of doubly-salient structure, the relationship between the stator pole number N_s, PM pole number N_{PM}, rotor pole number N_r, and phase number m of the FM-DSPM machine is similar to that of the DSPM machine:

$$\begin{cases} N_s = 2mk \\ N_{PM} = \dfrac{N_s}{m} \\ N_r = N_s \pm 2k \end{cases} \tag{6.32}$$

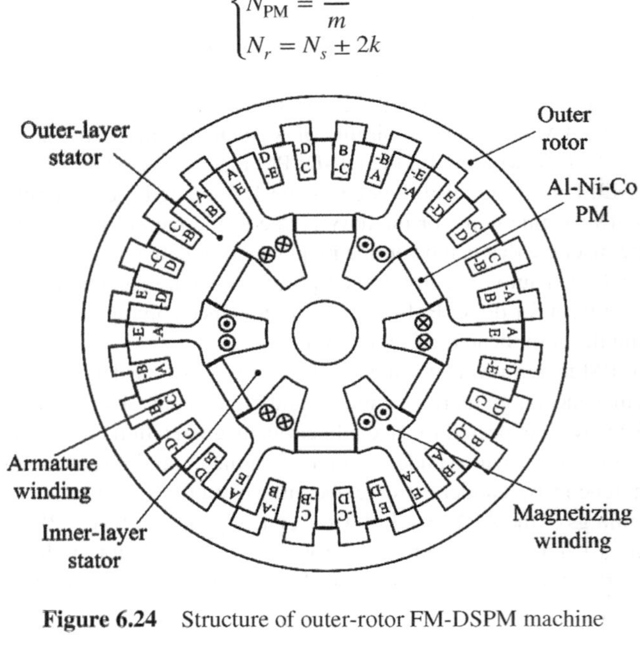

Figure 6.24 Structure of outer-rotor FM-DSPM machine

where k is a positive integer. For instance, when $m = 5$ is selected, $N_s = 10k$, $N_r = 8k$, and $N_{PM} = 2k$ can be deduced. The use of more phases can enhance the torque smoothness and provide the capability of fault tolerance, but it desires more power devices. The use of more rotor poles can generate more torque pulses per revolution, leading to have a smoother resultant torque. On the other hand, since the operating frequency is proportional to the number of rotor poles, it should be limited to avoid causing high core losses. As a compromise, $k = 3$ is chosen, hence resulting in $N_s = 30, N_r = 24$, and $N_{PM} = 6$.

The principle of operation of this FM-DSPM machine is similar to that of the DSPM machine, except that the PM magnetization is tunable. The theoretical waveforms of PM flux linkage ψ_{PM} and self-inductance L are shown in Figure 6.25. In order to make a unidirectional torque all the time, there are two operation modes, depending on the waveform of armature current i. For the first mode, a bipolar current is fed into the armature winding in which a positive current is applied when the flux linkage increases whereas a negative current is applied when the flux linkage decreases. As a result, the PM torque becomes the dominant torque component, while the reluctance one is a parasitic pulsating torque with zero average value. This mode of operation is originally used by the DSPM machine, the so-called DSPM mode of this FM-DSPM machine. For the second mode, a unipolar current is fed into the armature winding in which only the positive current is applied in the period of increasing flux linkage and inductance. Since the period with decreasing flux linkage and inductance is not utilized, the torque density is sacrificed. This mode of operation is originally used by the SR machine, the so-called SR mode of this machine.

The analysis of this FM-DSPM machine is more difficult than that of other stator-PM machines, because it adopts the Al-Ni-Co PM material which exhibits a nonlinear demagnetization characteristic and has a rather narrow B–H loop with a knee in the second quadrant. Thus, a hysteresis model of this Al-Ni-Co PM needs to be adopted. Rather than using a complicated hysteresis model using Preisach theory (Gong *et al.*, 2009), a piecewise-linear hysteresis model is preferred since it can readily be incorporated into the finite element method for both steady-state and transient analysis. Figure 6.26 depicts this hysteresis model in which the major hysteresis loop and all the minor hysteresis loops have the same value of coercivity, H_c but with different values of remanence, B_{rk}, and the initial magnetizing curve partly superposes on the hysteresis loop. The relative permeability, μ_r and recoil permeability, μ_{rec} are the same.

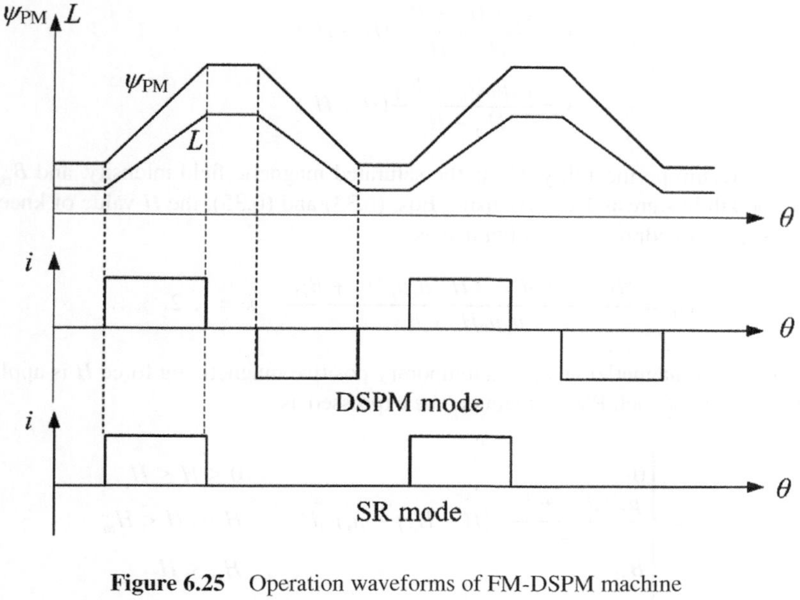

Figure 6.25 Operation waveforms of FM-DSPM machine

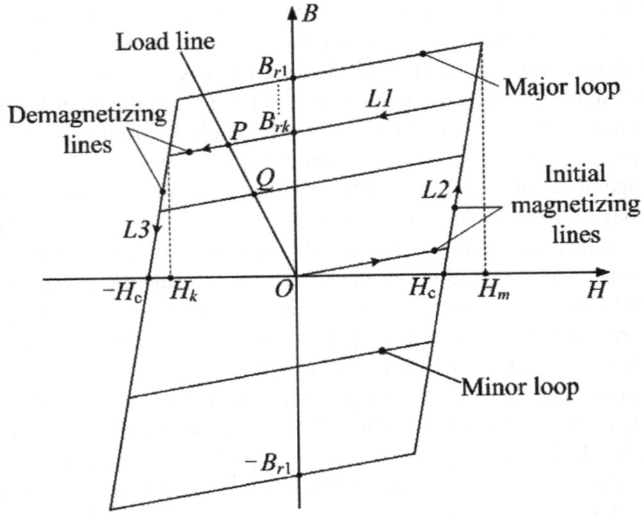

Figure 6.26 Hysteresis model of Al-Ni-Co PM

The main beelines that represent the magnetizing and demagnetizing processes are labeled *L1*, *L2*, and *L3*. During the magnetizing process, the operating point of the Al-Ni-Co PM moves upward along *L2*, then leftward along *L1*, and settles at the operating point *P*. Consequently, in the demagnetizing process, the operating point moves leftward along *L1*, then downward along *L3*, then rightward along the recoil line until settling at the operating point *Q* with a lower flux level. These three beelines, *L1*, *L2*, and *L3*, can be respectively expressed as

$$B = \mu_r\mu_0 H + B_{rk} \quad k = 1, 2, 3, \ldots \tag{6.33}$$

$$B = \frac{\mu_r\mu_0 H_m + B_{r1}}{H_m - H_c}(H - H_c) \tag{6.34}$$

$$B = \frac{\mu_r\mu_0 H_m + B_{r1}}{H_m - H_c}(H + H_c) \tag{6.35}$$

where μ_0 is the vacuum permeability, H_m is the saturated magnetic field intensity, and B_{rk} denotes the remanence of the *k*th hysteresis loop. By using Eqs. (6.33) and (6.35), the *H* value of knee point of the *k*th loop in the second quadrant can be obtained as

$$H_k = \frac{(H_m - H_c)B_{rk} - H_c(\mu_r\mu_0 H_m + B_{r1})}{\mu_r\mu_0 H_c + B_{r1}} \quad k = 1, 2, 3, \ldots \tag{6.36}$$

Firstly, in the initial magnetizing state, a temporary positive magnetizing force *H* is applied. The corresponding remanence of each PM element can be expressed as

$$B_r = \begin{cases} 0 & 0 \le H \le H_c \\ \dfrac{\mu_r\mu_0 H_m + B_{r1}}{H_m - H_c}(H - H_c) - \mu_r\mu_0 H & H_c < H < H_m \\ B_{r1} & H_m \le H \end{cases} \tag{6.37}$$

Secondly, in the working state, the B value of each PM element is calculated from Eq. (6.33) according to its associated B_{rk} and the currently applied negative magnetizing force H. At each time step, the value of B_{rk} is modified until it converges using the under-relax iteration method, which is given by

$$B_r = \begin{cases} B_{rk} & H_k \le H \le 0 \\ \dfrac{\mu_r\mu_0 H_m + B_{r1}}{H_m - H_c}(H + H_c) - \mu_r\mu_0 H & -H_c < H < H_k \\ \mu_r\mu_0\left[H + H_c - \dfrac{(\mu_r\mu_0 H + B_{rk})(H_m - H_c)}{\mu_r\mu_0 H + B_{r1}}\right] + B_{rk} & H \le -H_c \end{cases} \quad (6.38)$$

As mentioned earlier, the FM-DSPM motor drive can operate at either the DSPM mode or SR mode, where bipolar or unipolar armature current is applied, respectively. For normal operation, the PMs are fully or partially magnetized so that the DSPM mode of operation is adopted. Figure 6.27 shows the corresponding speed control block diagram, in which there are two feedback loops, namely the inner current loop with a hysteresis band regulator and the outer speed loop with a PI regulator. Then, with the aid of position sensor, the conduction signal of each phase is combined with the associated hysteresis regulator signal, hence forming the gating signal of each power device.

The flux controller of this FM-DSPM machine is different from that of the HE-DSPM machine, because it just needs to produce a temporary current pulse that is controllable in both magnitude and direction. As illustrated in Figure 6.28, the power stage consists of a buck converter and a bridge converter, where the former functions to control the magnitude I_m of the magnetizing current and the latter serves to control the direction D and duration T of the magnetizing current. For instance, under flux-weakening operation, the required air-gap flux, ψ^* deduced from the motor speed, ω is compared with the estimated ψ from the flux observer, hence the direction of the magnetizing current can be easily determined. The corresponding duration is simply the minimum time required to accomplish the demagnetization or magnetization process, which does not need to be controlled. Consequently, the required magnitude of demagnetizing or magnetizing current, I_m^* is deduced from the relationship between the air-gap flux density and the magnetizing MMF. Finally, the duty cycle, δ of the buck converter is adjusted to tune its output voltage and hence the value of I_m.

The relationships among the back EMFs, position signals, and conduction intervals of this five-phase machine are depicted in Figure 6.29. Since the back EMF waveforms are trapezoidal, the corresponding position signals are adjusted in such a way that they align with the maximum values of the back EMF waveforms. Hence, the power devices in the upper or lower leg of each phase of the inverter are conducted

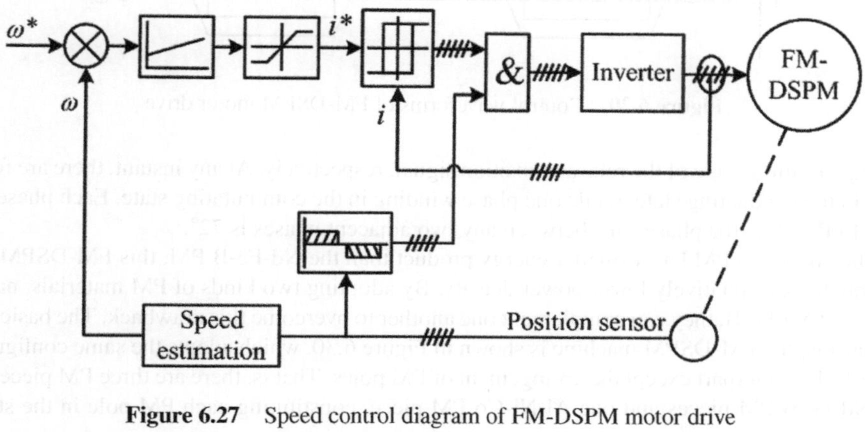

Figure 6.27 Speed control diagram of FM-DSPM motor drive

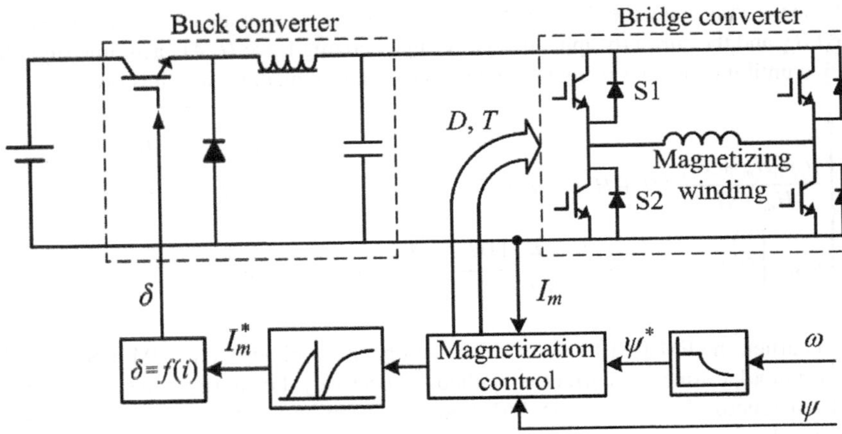

Figure 6.28 Flux control diagram of FM-DSPM motor drive

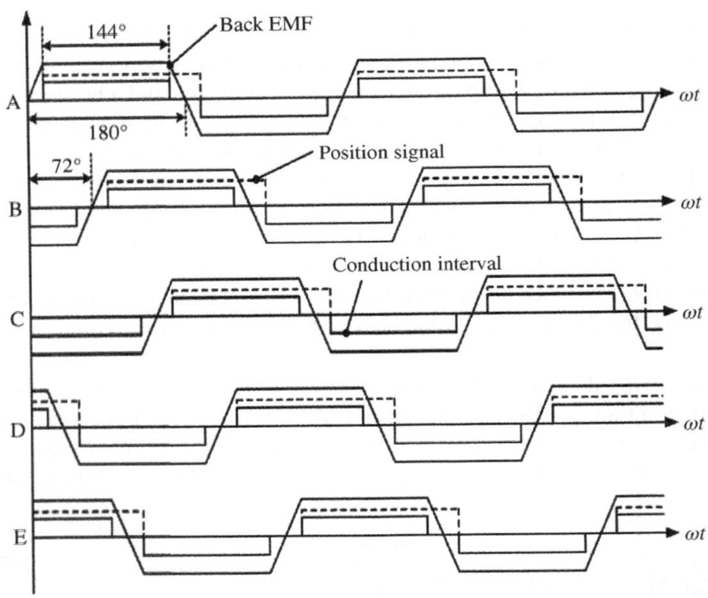

Figure 6.29 Control waveforms of FM-DSPM motor drive

at the rising or falling edge of the relevant position signal, respectively. At any instant, there are four phase windings in the conducting state, while one phase winding in the commutating state. Each phase winding conducts 144°, while the phase shift between any two adjacent phases is 72°.

Since the Al-Ni-Co PM has a smaller energy product than the Nd-Fe-B PM, this FM-DSPM machine has the drawback of relatively lower power density. By adopting two kinds of PM materials, namely the Al-Ni-Co and Nd-Fe-B, they can complement one another to overcome this drawback. The basic structure of this dual-magnet FM-DSPM machine is shown in Figure 6.30, which adopts the same configuration as the single-PM counterpart except the arrangement of PM poles. That is, there are three PM pieces, including two Nd-Fe-B PM pieces and one Al-Ni-Co PM piece, constituting each PM pole in the stator. The

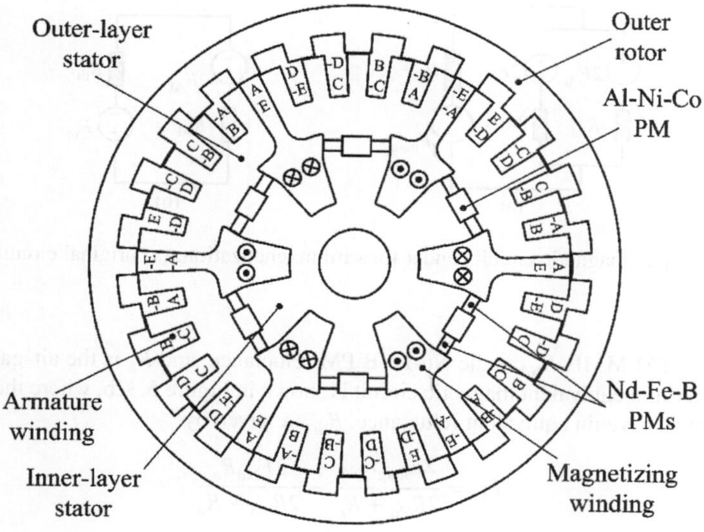

Figure 6.30 Structure of outer-rotor dual-magnet FM-DSPM machine

magnetizing winding is located in the inner space of the stator, which functions to online magnetize the Al-Ni-Co PM in forward or backward direction. The Al-Ni-Co PM provides a tunable PM flux, whereas the Nd-Fe-B PM provides a constant PM flux, hence achieving a strong and adjustable air-gap flux density.

The dual-magnet FM-DSPM machine incorporates two kinds of PMs to provide the MMF. The Al-Ni-Co PM can be forward or backward magnetized by the magnetizing winding. When the machine works under normal operation, the Al-Ni-Co PM is fully forward magnetized in the same magnetization direction as the neighboring Nd-Fe-B PM so that the maximum air-gap flux density can be achieved. Under light-load or high-speed condition, the Al-Ni-Co PM is demagnetized so that the air-gap flux density depends only on the Nd-Fe-B PM field. When necessary, it can also be backward magnetized to cancel out the flux produced by the Nd-Fe-B PM, hence operating as a reluctance machine.

Figure 6.31 depicts the simplified magnetic flux diagram of each PM pole in which the flux linking through two neighboring Nd-Fe-B PM pieces is within one closed flux loop, whereas the flux passing through the Al-Ni-Co PM piece belongs to two closed flux loops (Li *et al.*, 2012). Hence, the equivalent magnetic circuit of the machine describing the forward magnetization is shown in Figure 6.32a, where F_{mag} is the magnetizing winding MMF, F_{Al} is the Al-Ni-Co PM MMF, R_{Al} is the Al-Ni-Co PM reluctance,

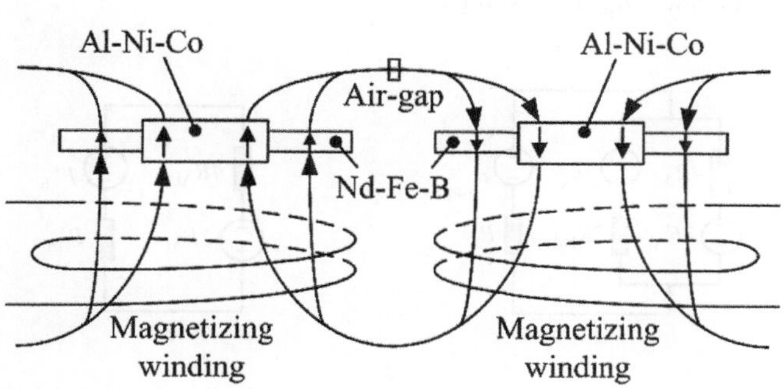

Figure 6.31 Magnetic flux flow of each PM pole-pair

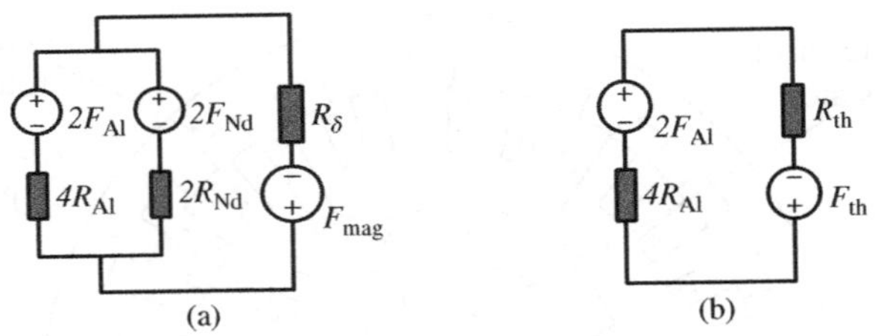

Figure 6.32 Equivalent magnetic circuits under forward magnetization: (a) original circuit and (b) Thevenin circuit

F_{Nd} is the Nd-Fe-B PM MMF, R_{Nd} is the Nd-Fe-B PM reluctance, and R_δ is the air-gap reluctance. The corresponding Thevenin equivalent magnetic circuit is shown in Figure 6.32b, where the Thevenin equivalent MMF, F_{th} and Thevenin equivalent reluctance, R_{th} are given by

$$F_{th} = \frac{2F_{mag}R_{Nd}}{2R_{Nd} + R_\delta} - \frac{2F_{Nd}R_\delta}{2R_{Nd} + R_\delta} \tag{6.39}$$

$$R_{th} = \frac{2R_{Nd}R_\delta}{2R_{Nd} + R_\delta} \tag{6.40}$$

In order to forward magnetize the Al-Ni-Co PM, the magnetizing MMF should be large enough to compensate the Nd-Fe-B PM MMF as governed by Eq. (6.39). Hence, the criterion for forward magnetization of the Al-Ni-Co PM can be derived as

$$\frac{2F_{mag}R_{Nd}}{2R_{Nd} + R_\delta} - \frac{2F_{Nd}R_\delta}{2R_{Nd} + R_\delta} \geq 3H_{Al}2h_{Al} \tag{6.41}$$

where H_{Al} and h_{Al} are the coercivity and thickness of the Al-Ni-Co PM, respectively. Meanwhile, in the absence of magnetizing excitation, it is necessary to ensure that the Al-Ni-Co PM is thick enough to withstand the Nd-Fe-B PM field. Thus, the criterion for the selection of the Al-Ni-Co PM thickness can be deduced as

$$h_{Al} \geq \frac{2F_{Nd}R_\delta}{H_{Al}(2R_{Nd} + R_\delta)} \tag{6.42}$$

The backward magnetization is to reversely magnetize the Al-Ni-Co PM, which serves to weaken or cancel out the magnetic field produced by the Nd-Fe-B PM. The equivalent magnetic circuit of the machine describing the backward magnetization is depicted in Figure 6.33a. The corresponding Thevenin equivalent magnetic circuit is shown in Figure 6.33b, where the Thevenin equivalent MMF, F'_{th} and Thevenin equivalent reluctance, R'_{th} are given by

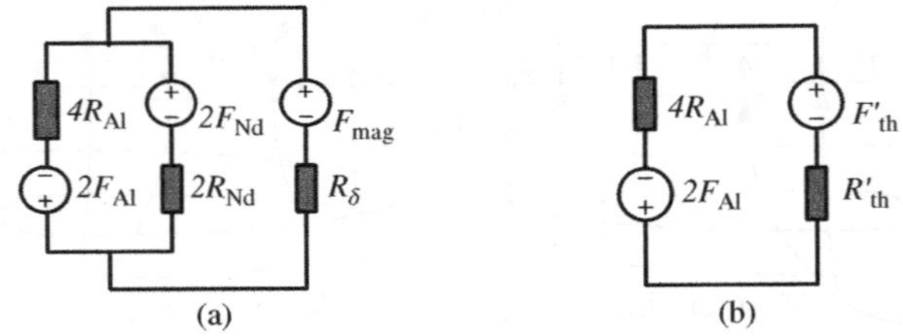

Figure 6.33 Equivalent magnetic circuits under backward magnetization: (a) original circuit and (b) Thevenin circuit

$$F'_{\text{th}} = \frac{2F_{\text{Nd}}R_\delta}{2R_{\text{Nd}} + R_\delta} + \frac{F_{\text{mag}}2R_{\text{Nd}}}{2R_{\text{Nd}} + R_\delta} \tag{6.43}$$

$$R'_{\text{th}} = R_{\text{th}} \tag{6.44}$$

Since the Thevenin equivalent MMF consists of two positive terms, the Nd-Fe-B PM MMF exerts a magnetizing force to assist the magnetizing winding to reversely magnetize the Al-Ni-Co PM. Thus, the required magnetizing MMF for backward magnetization is much lower than that for forward magnetization. From Eq. (6.43), the criterion for backward magnetization of the Al-Ni-Co can be deduced as

$$F_{\text{mag}} \geq 3H_{\text{Al}}h_{\text{Al}}\left(2 + \frac{R_\delta}{R_{\text{Nd}}}\right) - F_{\text{Nd}}\frac{R_\delta}{R_{\text{Nd}}} \tag{6.45}$$

Hence, the thickness of the Al-Ni-Co PMs is governed by

$$h_{\text{Al}} \leq \frac{F_{\text{Nd}}R_\delta + F_{\text{mag}}R_{\text{Nd}}}{3H_{\text{Al}}(2R_{\text{Nd}} + R_\delta)} \tag{6.46}$$

On the basis of the derived equations, the PM dimensions of this outer-rotor dual-magnet FM-DSPM machine can be deduced. Typically, the width and thickness of the Al-Ni-Co PM piece are about twice those of each Nd-Fe-B PM piece. Figure 6.34 depicts typical magnetic field distributions when the Al-Ni-Co PMs are under forward magnetization, no magnetization, and backward magnetization, respectively. It can be observed that the air-gap flux density can be significantly strengthened under forward magnetization, because both the Al-Ni-Co PM piece and two Nd-Fe-B PM pieces simultaneously contribute magnetic flux to the air-gap. Meanwhile, when the Al-Ni-Co PM piece is under no magnetization, the air-gap flux density is solely contributed by the two Nd-Fe-B PM pieces. Under backward magnetization, the Al-Ni-Co PM piece essentially shunts the two Nd-Fe-B PM pieces, thus dramatically suppressing the air-gap flux density (Li *et al.*, 2011).

6.8 Design Criteria of Stator-PM Motor Drives for EVs

The design criteria of various types of stator-PM machines are similar to that of the DSPM machine (Cheng *et al.*, 2001). By rearranging Eq. (6.18), the corresponding sizing equation can be deduced as

$$D_{\text{si}}^2 l_e = \frac{P_2}{\frac{0.87\pi^2}{120}\frac{N_r}{N_s}k_s k_d k_e k_i A_s B_\delta n_s \eta} \tag{6.47}$$

where the general relationship between N_s and N_r of the DSPM machine is given by Eq. (6.1). Since it is generally valid that $\theta_w \approx \theta_{\text{cr}}/3$, the relationship between I_{rms} and I_m can be expressed as

$$I_{\text{rms}} = \sqrt{\frac{1}{T}\int_0^T i^2 dt} = \sqrt{\frac{1}{\theta_{\text{cr}}}2I_m^2\theta_w} = \sqrt{\frac{2}{\theta_{\text{cr}}}\frac{1}{3}\theta_{\text{cr}}I_m} = \sqrt{\frac{2}{3}}I_m \tag{6.48}$$

Thus, k_i can be calculated as

$$k_i = \frac{I_m}{I_{\text{rms}}} = \sqrt{\frac{3}{2}} \tag{6.49}$$

In general, $k_d = 0.90 - 0.93$, $k_e = 1.2 - 2.0$, $k_s = 1$ (without rotor skewing), $A_s = 10 - 30$ kA/m, and $B_\delta = 1.5$ T are selected. Given the rated P_2, n_s, and η, the main dimensions of the machine, namely, D_{si} and l_e can be deduced. Once the main dimensions are determined, other structural dimensions such as the stator outside diameter, pole heights, and pole arcs can be specified similarly to the SR machine.

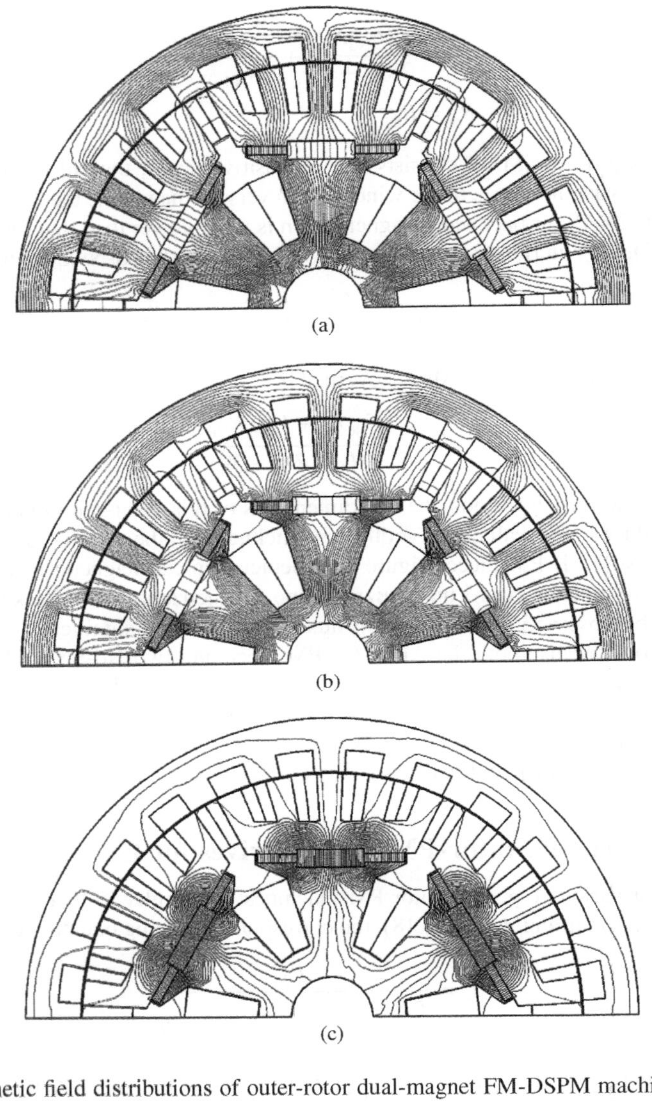

(a) (b) (c)

Figure 6.34 Magnetic field distributions of outer-rotor dual-magnet FM-DSPM machine: (a) forward magnetization, (b) no magnetization, and (c) backward magnetization

The initial sizing of PMs is essential since it directly affects the machine performance and cost. Although the PM size and location of various stator-PM machines are different from one another, the corresponding sizing approach is very similar, mainly based on the magnetic circuit approach. For exemplification, the equivalent magnetic circuit of a four-phase 8/6-pole DSPM machine is shown in Figure 6.35, in which the iron core is assumed to be of infinite permeability. From the magnetic circuit, it yields

$$\psi_a = \psi_\delta \frac{\Lambda_a}{\Lambda_a + \Lambda_b + \Lambda_c + \Lambda_d} = \psi_\delta \frac{\Lambda_a}{\Lambda_\delta} \tag{6.50}$$

$$\psi_\delta = \psi_{\mathrm{PM}} \frac{\Lambda_\delta}{\Lambda_\delta + \Lambda_\sigma} \tag{6.51}$$

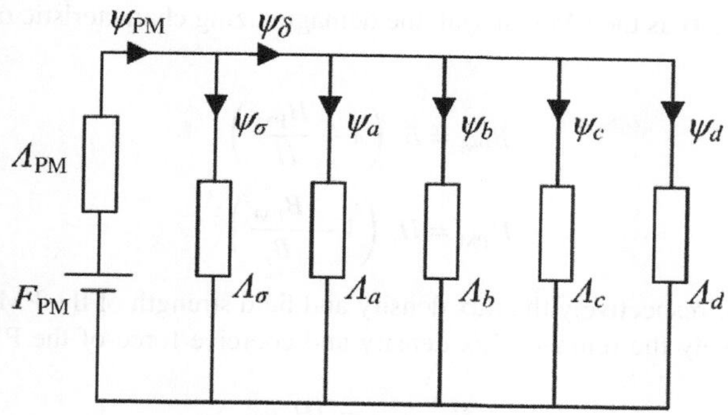

Figure 6.35 Equivalent magnetic circuit of four-phase DSPM machine

where Λ_a, Λ_b, Λ_c, and Λ_d are the permeances of phases A, B, C, and D, respectively, Λ_δ is the sum of the permeances of four phases, Λ_σ is the leakage permeance of PMs, ψ_a is the flux of phase A, ψ_δ is the air-gap flux equal to the sum of four phase fluxes, and ψ_{PM} is the PM flux. Making use of Eqs. (6.50) and (6.51), it deduces

$$\psi_{\text{PM}} = \frac{\Lambda_\delta + \Lambda_\sigma}{\Lambda_\delta}\frac{\Lambda_\delta}{\Lambda_a}\psi_a = \sigma\frac{\Lambda_\delta}{\Lambda_a}\psi_a \tag{6.52}$$

where σ is the PM leakage factor which is defined as the ratio of PM flux to air-gap flux:

$$\sigma = \frac{\psi_{\text{PM}}}{\psi_\delta} = \frac{\Lambda_\delta + \Lambda_\sigma}{\Lambda_\delta} \tag{6.53}$$

which depends on the motor configuration, typically $\sigma = 1.4 - 1.5$.

The permeances Λ_a and Λ_δ can be expressed as

$$\Lambda_a = \mu_0 \frac{D_{\text{si}}\alpha_a l_e}{4g_0} \tag{6.54}$$

$$\Lambda_\delta = \mu_0 \frac{D_{\text{si}}\alpha_\delta l_e}{4g_0} \tag{6.55}$$

where α_a is the overlapping angle between the stator pole of phase A and a rotor pole, α_δ is the sum of overlapping angles between four-phase stator poles and rotor poles, g_0 is the air-gap length, and μ_0 is the permeability of free space. When the rotor pole arc, β_r is given by

$$\beta_r = 2\theta_{\text{cs}} - \theta_{\text{cr}} \tag{6.56}$$

then α_δ becomes constant, hence the operating point of PMs does not change with the rotor position. Even Eq. (6.56) is not satisfied, the variation of this operating point does not introduce a significant error. So the PM flux in Eq. (6.52) can be calculated at a particular rotor position that the stator pole of phase A fully aligns with a rotor pole. That is, the flux of phase A has its maximum value $\psi_{a\max}$. Then, α_δ and α_a can be expressed as

$$\alpha_\delta = 2\beta_s + \beta_r + \theta_{\text{cr}} - 2\theta_{\text{cs}} \tag{6.57}$$

$$\alpha_a = \min(\beta_s, \beta_r) = \beta_s \tag{6.58}$$

Substituting Eqs. (6.54) and (6.55) into Eq. (6.52) and making use of Eqs. (6.57) and (6.58), it deduces

$$\psi_{\text{PM}} = \frac{\sigma B_\delta D_{\text{si}} l_e}{2}(2\beta_s + \beta_r + \theta_{\text{cr}} - 2\theta_{\text{cs}}) \tag{6.59}$$

When adopting Nd-Fe-B as the PM material, the demagnetizing characteristic of PMs is almost linear. Thus, it yields

$$B_{\text{PM}} = B_r \left(1 - \frac{H_{\text{PM}}}{H_c}\right) \tag{6.60}$$

$$H_{\text{PM}} = H_c \left(1 - \frac{B_{\text{PM}}}{B_r}\right) \tag{6.61}$$

where B_{PM} and H_{PM} are respectively the flux density and field strength of the PM operating point, while B_r and H_c are respectively the remnant flux density and coercive force of the PM. By using Ampere's Law, it yields

$$H_{\text{PM}} h_{\text{PM}} = 2H_\delta g_0 \tag{6.62}$$

where H_δ is the field strength in the air-gap and h_{PM} is the PM thickness in the direction of magnetization. From Eqs. (6.61) and (6.62), it deduces

$$h_{\text{PM}} = \frac{2B_\delta g_0}{\mu_0 H_c \left(1 - \frac{B_{\text{PM}}}{B_r}\right)} \tag{6.63}$$

Typically, $B_{\text{PM}}/B_r = 0.7 - 0.95$. By using Eq. (6.59), the surface area S_{PM} of each piece of PMs can be expressed as

$$S_{\text{PM}} = \frac{\psi_{\text{PM}}}{2B_{\text{PM}}} = \frac{\sigma B_\delta D_{\text{si}} l_e}{4B_{\text{PM}}} (2\beta_s + \beta_r + \theta_{\text{cr}} - 2\theta_{\text{cs}}) \tag{6.64}$$

As the magnet length, l_{PM} in the axial direction is usually the same as the stack length, the magnet width, w_{PM} can be expressed as

$$w_{\text{PM}} = \frac{S_{\text{PM}}}{l_{\text{PM}}} = \frac{\sigma B_\delta D_{\text{si}}}{4B_{\text{PM}}} (2\beta_s + \beta_r + \theta_{\text{cr}} - 2\theta_{\text{cs}}) \tag{6.65}$$

The electromagnetic torque per phase of the DSPM machine under the BLDC mode can generally be expressed as

$$T_{\text{ph}} = i \left(\frac{d\psi_{\text{PM}}}{d\theta}\right) + \frac{1}{2} i^2 \left(\frac{dL}{d\theta}\right) = T_{\text{PM}} + T_r \tag{6.66}$$

where the average value of the reluctance torque component T_r is equal to zero, and the average value of the PM torque component T_{PM} is dominant in the average torque per phase, which is given by

$$T_{\text{ph}} = \frac{1}{\theta_{\text{cr}}} \int_0^{\theta_{\text{cr}}} \left(i \frac{d\psi_{\text{PM}}}{d\theta}\right) d\theta = \frac{2}{\theta_{\text{cr}}} I_m \Delta\psi_{\text{PM}} w \tag{6.67}$$

where I_m is the magnitude of the rectangular current waveform. Hence, the total average torque T_{av} of m phases is given by

$$T_{\text{av}} = \frac{2m}{\theta_{\text{cr}}} I_m \Delta\psi_{\text{PM}} w \tag{6.68}$$

When neglecting the friction and windage losses, it yields

$$T_{\text{av}} \frac{2\pi n_s}{60} = P_2 \tag{6.69}$$

Substituting Eqs. (6.12) and (6.68) into Eq. (6.69), it deduces

$$w = \frac{U\theta_w \eta}{\Delta\psi_{\text{PM}} \frac{2\pi n_s}{60}} \tag{6.70}$$

By substituting Eq. (6.15) into Eq. (6.70), the required number of winding turns per phase can be obtained as

$$w = \frac{U\theta_w \eta}{0.87k_d \frac{\alpha_s \pi D_{si}}{N_s} l_e B_\delta \frac{2\pi n_s}{60}} \tag{6.71}$$

Since the realistic current waveform is trapezoidal, rather than rectangular, the corresponding θ_w is usually smaller than the ideal one. Taking into account a correction factor of 0.8, the modified number of winding turns is expressed as

$$w = \frac{0.8U\theta_w \eta}{0.87k_d \frac{\alpha_s \pi D_{si}}{N_s} l_e B_\delta \frac{2\pi n_s}{60}} \tag{6.72}$$

The aforementioned design criteria of the DSPM machine can readily be extended to other stator-PM machines, including the FRPM, FSPM, HE-DSPM, HE-FRPM, HE-FSPM, FM-DSPM, FM-FRPM, and FM-FSPM, but with some modifications, such as the design criteria of additional DC field winding, additional magnetizing winding, and additional PM material.

6.9 Design Examples of Stator-PM Motor Drives for EVs

There are many types of stator-PM motor drives: the DSPM, FRPM, FSPM, HE-DSPM, HE-FRPM, HE-FSPM, FM-DSPM, FM-FRPM, and FM-FSPM. Two recently developed stator-PM motor drives, namely based on the outer-rotor HE-DSPM and FM-DSPM machines, are elaborated.

6.9.1 Outer-Rotor Hybrid-Excited DSPM Motor Drive

Differing from the aforementioned HE-DSPM machine, this machine adopts an outer-rotor topology as shown in Figure 6.36. There are two main reasons to adopt the outer-rotor topology rather than the inner-rotor one. First, if the conventional inner-rotor topology is used, the stator yoke needs to be significantly enlarged to accommodate both the PMs and DC field winding. However, if the outer-rotor topology is adopted, the inner space of the machine, the space beneath the armature winding, can be fully utilized to accommodate both the PMs and DC field winding, hence minimizing the overall size. Second,

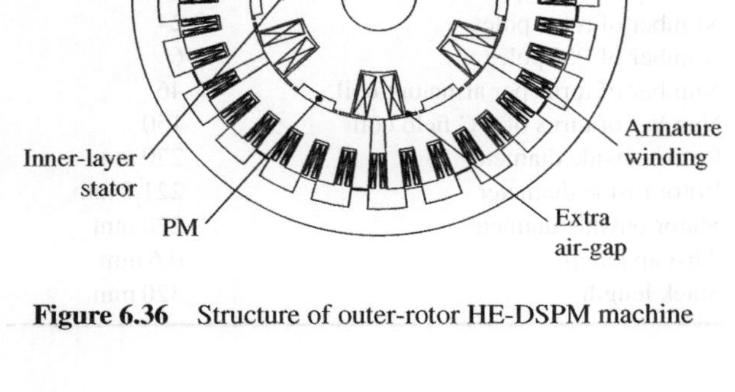

Figure 6.36 Structure of outer-rotor HE-DSPM machine

if adopting the conventional topology, both the PMs and DC field winding in the stator have high flux leakage. However, if adopting the outer-rotor topology, both the PMs and DC field winding are embraced by the rotor, hence solving the problem of high flux leakage. Additionally, there are two layers of stator: the outer layer of the stator is to accommodate the armature winding wound on the stator salient poles and the inner layer of the stator is to accommodate both the PMs and DC field winding. Meanwhile, the outer rotor is simply composed of salient poles without winding or PM (Chau *et al.*, 2006).

Because of the additional DC field winding, this HE-DSPM machine definitely offers a lower power density than the DSPM machine. It is really a trade-off between the ability of flux control and the reduction of power density. Since it is more difficult to effectively cool the inside stator, the outer-rotor topology has the drawback of thermal dissipation.

The choice of pole numbers is governed by Eq. (6.1). For $m = 3$ and $k = 3$, it results in $N_s = 36$ and $N_r = 24$, leading to be a three-phase 36/24-pole HE-DSPM machine. It should be noted that if k is too large, N_r will be very large, which will significantly increase the operating frequency and hence the iron loss. On the contrary, if k is too small, the operating frequency will be insufficient to limit the rectified output voltage ripple. In order to improve the effectiveness of using the DC field winding to perform flux control, an extra air-gap with an iron bridge is added in shunt with the PM. These extra air-gaps function to bypass the PM flux when two field excitations are opposite, hence amplifying the effect of flux weakening. The key design data of this outer-rotor HE-DSPM machine are listed in Table 6.2.

The principle of operation of this HE-DSPM machine is similar to that of the DSPM machine, except that the flux is controllable. Thus, when the DC field winding is excited to produce additional flux in the same direction as the PM flux, the air-gap flux linkage is strengthened, hence temporarily producing high torque for EV launching or overtaking. On the other hand, when the DC field winding is excited to produce flux in the opposite direction to the PM flux, the air-gap flux linkage is weakened to enable high-speed constant-power operation for EV cruising. Figure 6.37 shows its air-gap flux density distributions under flux weakening ($F_{DC} = -350$ A-turns), no flux control ($F_{DC} = 0$), and flux strengthening ($F_{DC} = +1000$ A-turns), which indicates that a range of nine times of air-gap flux regulation can be realized.

While the use of DC field winding for field excitation in the HE-DSPM machine inevitably involves additional copper loss compared with the original DSPM machine, it can make use of this flux control capability to compensate such loss and improve the system efficiency (Liu *et al.*, 2009). The loss analysis of this HE-DSPM machine is described below:

Table 6.2 Key design data of outer-rotor HE-DSPM machine

Rated power	8 kW
Rated speed	1000 rpm
Constant-torque operation	0–1000 rpm
Constant-power operation	1000–4000 rpm
Number of phases	3
Number of stator poles	36
Number of rotor poles	24
Number of PM poles	6
Number of turns per armature coil	46
Number of turns of DC field coil	150
Rotor outside diameter	270 mm
Rotor inside diameter	221.2 mm
Stator outside diameter	220 mm
Air-gap length	0.6 mm
Stack length	320 mm

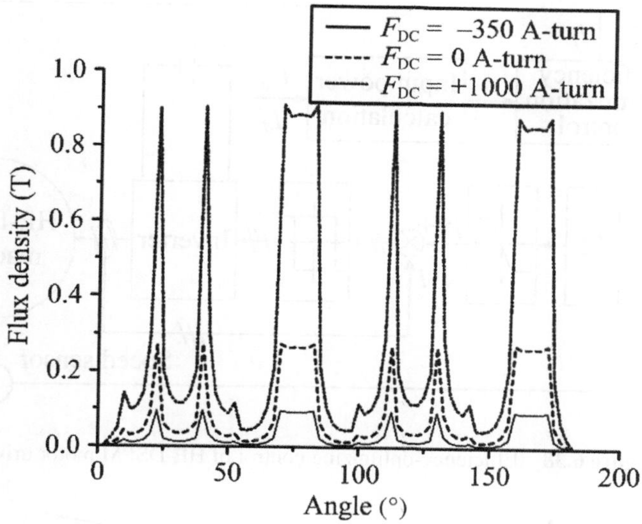

Figure 6.37 Air-gap flux density distributions of outer-rotor HE-DSPM machine with different DC field excitations

- There are copper losses in both the armature winding and DC field winding, which are governed by the armature current and field current, respectively. For a given operating point, the desired armature current is affected by the air-gap flux density, which can be directly regulated by tuning the DC field current. Thus, these copper losses are indirectly or directly related to the DC field current.
- The iron losses are composed of the magnetic hysteresis loss and eddy-current loss. The magnetic hysteresis loss is caused by the time-varying magnetization of ferromagnetic material. Under a given speed, this magnetic hysteresis loss varies with the air-gap flux density, which can certainly be adjusted by the DC field current. Meanwhile, the eddy-current loss is due to the existence of eddy currents in the iron core. Under a given speed, the eddy-current loss is certainly related to the air-gap flux density. Thus, these iron losses are certainly adjustable by the DC field current.
- The stray losses are additional losses caused by the load current. Such losses are generally quite small but should not be neglected. Since the stator current is affected by changing the air-gap flux density, the stray losses can be altered by the DC field current.
- The mechanical losses are due to friction and windage, which mainly depend on the operating speed and environment, and are independent of those electrical variables.

Except the mechanical losses, all other losses can be regulated by tuning the DC field current under the given load torque and speed. Hence, the optimal efficiency of the HE-DSPM motor drive can readily be achieved by online tuning the DC field current.

Figure 6.38 shows an efficiency-optimizing control scheme of the HE-DSPM motor drive. The key is to adjust the DC field winding current to online tune the air-gap flux, hence achieving minimum input power under the given load torque and speed. Rather than using analytical loss equations to minimize the total losses, self-searching control (SSC) is preferred, which can eliminate the uncertainties of loss equations and avoid the parameter variations due to prolonged operation. The corresponding SSC algorithm is illustrated in Figure 6.39. The algorithm initially adopts the DC field current value based on the historical data, which have been pre-stored as a look-up table. Then, under the given load torque and speed, the DC field current is purposely perturbed. If the initial point, P_1 is on the left-hand side of the minimum input power point, the increment of the DC field current will reduce the input power. On the contrary, if the initial point,

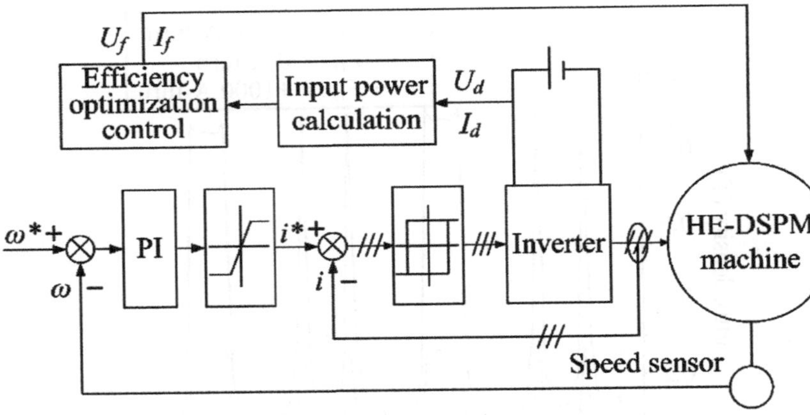

Figure 6.38 Efficiency-optimizing control of HE-DSPM motor drive

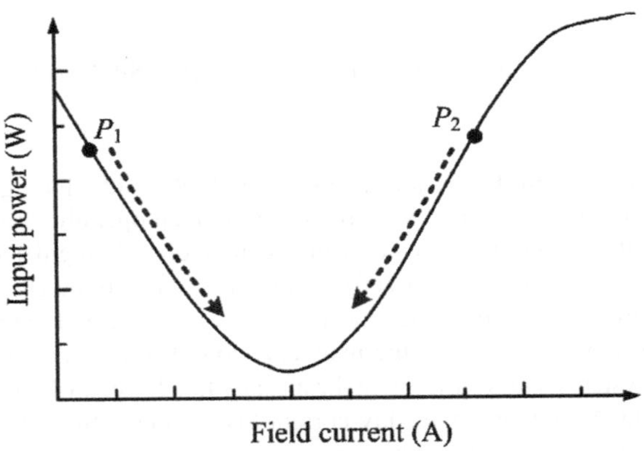

Figure 6.39 Self-searching control algorithm for efficiency-optimizing control of HE-DSPM motor drive

P_2 is on the right-hand side of the minimum input power point, the decrement of the DC field current will reduce the input power. Thus, if there is a decrease in input power, the subsequent perturbation will be kept unchanged until reaching the minimum input power point; otherwise, the perturbation will be reversed. Additionally, once the optimal condition is obtained, the corresponding value will be memorized, and will be used as the initial value for next search. This arrangement can greatly enhance the speed of searching. It should be noted that since the input power is measured at the DC link, the efficiency optimization is for the entire motor drive, including the HE-DSPM machine and power inverter.

The efficiency-optimizing control of the HE-DSPM motor drive is first assessed when the air-gap flux density is below the optimal value. Figure 6.40 shows the input power and field current responses under different load torques and speeds. It can be seen that in Condition I (10 N m at 300 rpm), the DC field current gradually increases until the input power progressively reaches the minimum. When the operation is suddenly changed to Condition II (5 N m at 200 rpm), a new search is activated. Consequently, the input power is also minimized by tuning the DC field current. When the operation is suddenly switched to Condition III, which is actually the same as Condition I, the initial operation point is directly retrieved from the memory. Therefore, it virtually takes no time to achieve the minimum input power. Then, the efficiency optimization is evaluated when the air-gap flux density is above the optimal value as illustrated

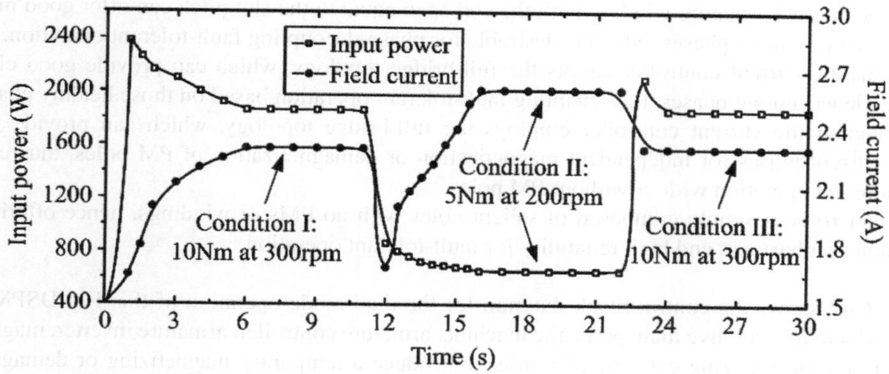

Figure 6.40 Responses of efficiency-optimizing control of HE-DSPM motor drive with increasing air-gap flux density

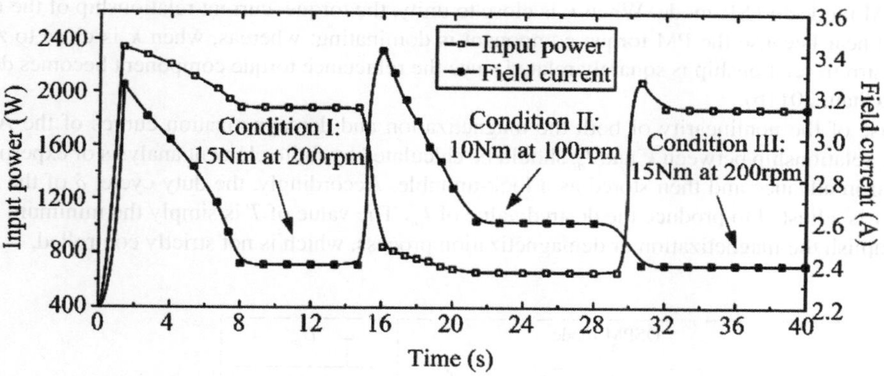

Figure 6.41 Responses of efficiency-optimizing control of HE-DSPM motor drive with decreasing air-gap flux density

in Figure 6.41. It can be found that in Condition I (15 N m at 200 rpm), the DC field current decreases until the input power gradually arrives at the minimum. When the operation is suddenly changed to Condition II (10 N m at 100 rpm), another new search is triggered, and comes up with another minimum. When the operation suddenly returns to the same case as Condition I, the initial operation point is directly retrieved from the memory so that it quickly settles at the minimum input power again.

6.9.2 *Outer-Rotor Flux-Mnemonic DSPM Motor Drive*

The structure of a five-phase outer-rotor FM-DSPM machine has been described as shown in Figure 6.24. The corresponding operation principle has also been discussed. One unique feature of this motor drive is the capability of high fault tolerance as summarized below:

- The use of five phases can allow for fault-tolerant operation under the loss of one phase or even two phases, hence improving the reliability of the whole system.
- The armature winding and PMs are located in different layers of the stator so that the PMs can be immune from accidental demagnetization by armature reaction, even under short-circuit fault.

- The concentrated armature winding with the coil span equal to the slot pitch can offer good magnetic independence among phases, which is desirable for phase-decoupling fault-tolerant operation.
- The armature current controller adopts the full-bridge topology, which can provide good electrical independence among phases, thus enabling fault-tolerant operation based on those healthy phases.
- The magnetizing current controller employs the full-bridge topology, which can provide bidirectional current pulses for independent magnetization or demagnetization of PM poles, thus enabling fault-tolerant operation with or without PM poles.
- The outer rotor is simply composed of salient poles with no PMs or windings, hence offering high mechanical robustness and high reliability for fault-tolerant operation.

Figure 6.42 shows the control block diagram for the dual-mode operation of this FM-DSPM motor drive, in which there are five main parts: the machine, armature controller, armature inverter, magnetizing controller, and magnetizing converter. In order to produce a temporary magnetizing or demagnetizing current pulse with controllable magnitude, the magnetizing converter consists of a DC–DC converter and a bridge converter, where the former functions to control the magnitude I_m of the current pulse and the latter serves to control the direction D and duration T of the current pulse. The coefficient of PM flux linkage κ, where $\kappa \in [0, 1]$, is the key factor determining the operation modes of this motor drive, namely the DSPM mode and SR mode. When κ is close to unity, the torque–current relationship of the machine is more linear because the PM torque component is dominating; whereas, when κ is close to zero, the torque–current relationship is squarely related since the reluctance torque component becomes dominant (Yu and Chau, 2011b).

Because of the nonlinearity of both the magnetization and demagnetization curves of the Al-Ni-Co PMs, the relationship between κ and I_m should be calculated by finite element analysis or experimentally measured in advance and then stored as a look-up table. Accordingly, the duty cycle, δ of the DC–DC converter is adjusted to produce the desired value of I_m. The value of T is simply the minimum duration to accomplish the magnetization or demagnetization process, which is not strictly controlled.

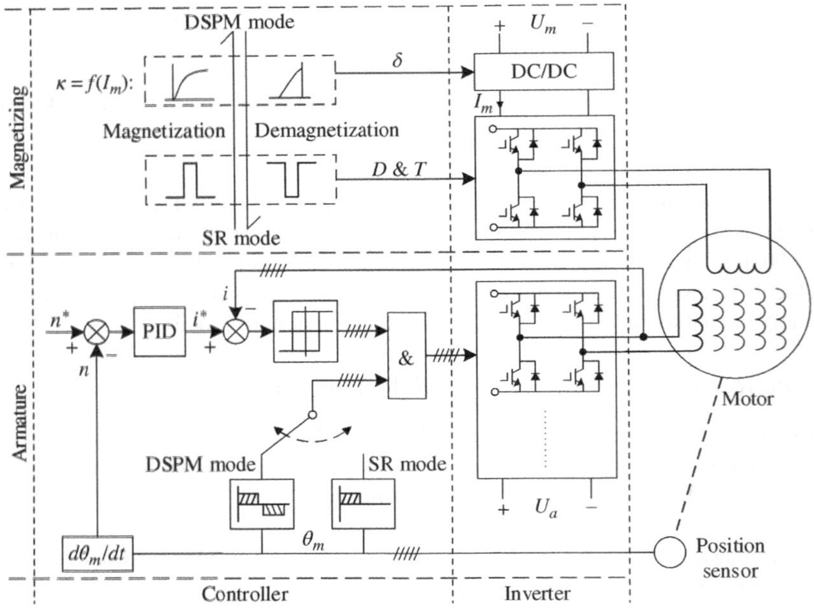

Figure 6.42 Dual-mode operation of outer-rotor FM-DSPM motor drive

For the armature controller, it adopts a conventional dual closed-loop control scheme, where the outer speed loop adopts a proportional-integral-derivative (PID) regulator to perform speed control, and the inner current loop uses a hysteresis regulator to perform CCC. By comparing the speed command with the actual speed, the armature current command is resulted. Then by comparing the current command with the actual current, the hysteresis regulator signal is generated. By logically combining the hysteresis regulator signal with the phase conduction signal, which depends on the mode of operation, the gating signal of each power device in the armature inverter can be obtained. In order to consolidate a conventional half-bridge inverter used for the DSPM motor drive and a conventional asymmetric half-bridge inverter used for the SR motor drive, the full-bridge inverter topology is adopted to power the armature winding.

The key design data of this outer-rotor FM-DSPM machine are listed in Table 6.3. In order to assess the capability of online magnetization and demagnetization, the transient responses of the no-load EMF of the motor drive at the speed of 100 rpm are simulated during two mode transitions, namely from the DSPM mode to the SR mode and vice versa, as shown in Figure 6.43. It can be observed that the corresponding duration of current pulse is only about 0.1 second. And the PMs can be completely demagnetized by using a negative current pulse with a magnitude of 2.4 A so as to activate the SR mode, whereas the PMs can be completely re-magnetized by using a positive current pulse with a magnitude of 10 A so as to reactivate the DSPM mode.

Additionally, the operating performances under the DSPM mode and the SR mode are assessed as shown in Figure 6.44 in which a bipolar armature current is applied under the DSPM mode while a unipolar armature current is applied under the SR mode. It can be found that the armature current of 1.2 A under the DSPM mode and that of 5.5 A under the SR mode produce the same output torque of about 20 N m. It can also be observed that the output torque ripple under the SR mode is larger than that under the DSPM mode, which is due to its feature of half-stroke torque production.

When an open-circuit fault occurs, such as the winding open circuit and power device open circuit, it results in disabling the faulty phase. Under this circumstance, the motor drive works as in the DSPM mode while utilizing the remaining healthy phases for remediation. In general, without purposely adopting the rotor skewing, the FM-DSPM machine offers an essentially trapezoidal back EMF so that the motor drive normally runs in the DSPM mode under the BLDC operation. The remediation strategy for fault-tolerant control under open-circuit faults is based on the use of BLAC operation for the remaining healthy phases so that the resulting average torque can be maintained. The strategy is to first derive the relationship between the BLDC operation and BLAC operation of the FM-DSPM motor drive to produce the same

Table 6.3 Key design data of outer-rotor FM-DSPM machine

Rated power	4 kW
Rated speed	600 rpm
Constant-torque operation	0–600 rpm
Constant-power operation	600–2000 rpm
Number of phases	3
Number of stator poles	36
Number of rotor poles	24
Number of PM poles	6
Number of turns per armature coil	60
Number of turns of magnetizing coil	250
Rotor outside diameter	270 mm
Rotor inside diameter	221.2 mm
Stator outside diameter	220 mm
Air-gap length	0.6 mm
Stack length	320 mm

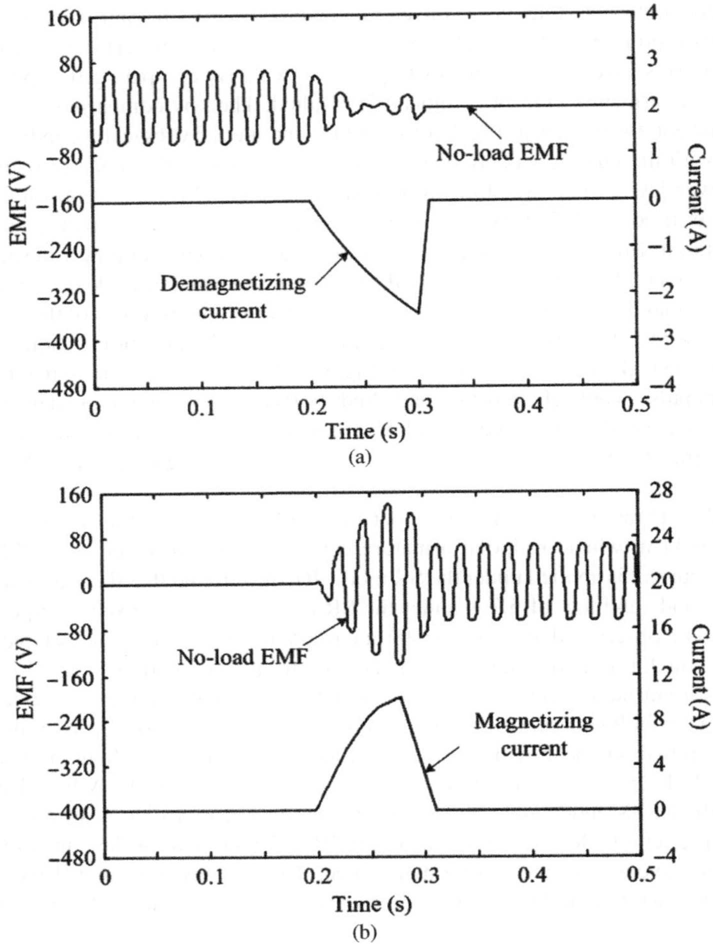

Figure 6.43 Transient responses of outer-rotor FM-DSPM motor drive: (a) DSPM mode to SR mode and (b) SR mode to DSPM mode

torque, and then to derive the relationship of BLAC operations with different numbers of healthy phases (Yu and Chau, 2011c).

Under the BLDC operation in the DSPM mode, the PM torque component is predominant whereas the reluctance torque component is negligible. The developed torque T_1 of this motor drive is governed by the average value of the PM torque component as given by

$$T_1 = \frac{P_1}{\omega} = \frac{5}{2\pi\omega}\left(\int_{\theta_1}^{\theta_2} E_{1m}I_{1m}d\theta + \int_{\theta_3}^{\theta_4} \left(-E_{1m}\right)\left(-I_{1m}\right)d\theta\right) = \frac{4E_{1m}I_{1m}}{\omega} \tag{6.73}$$

where P_1 is the electromagnetic power during BLDC operation, ω is the angular speed, E_{1m} is the amplitude of the trapezoidal back EMF, I_{1m} is the amplitude of the rectangular current, and $\theta_i(i = 1, 2, 3, 4)$ stands for the turn-on or turn-off electrical rotor position in the BLDC operation with $\theta_2 - \theta_1 = \theta_4 - \theta_3 = 144°$ as depicted in Figure 6.45. Similarly, under the BLAC operation, the developed torque T_2 can

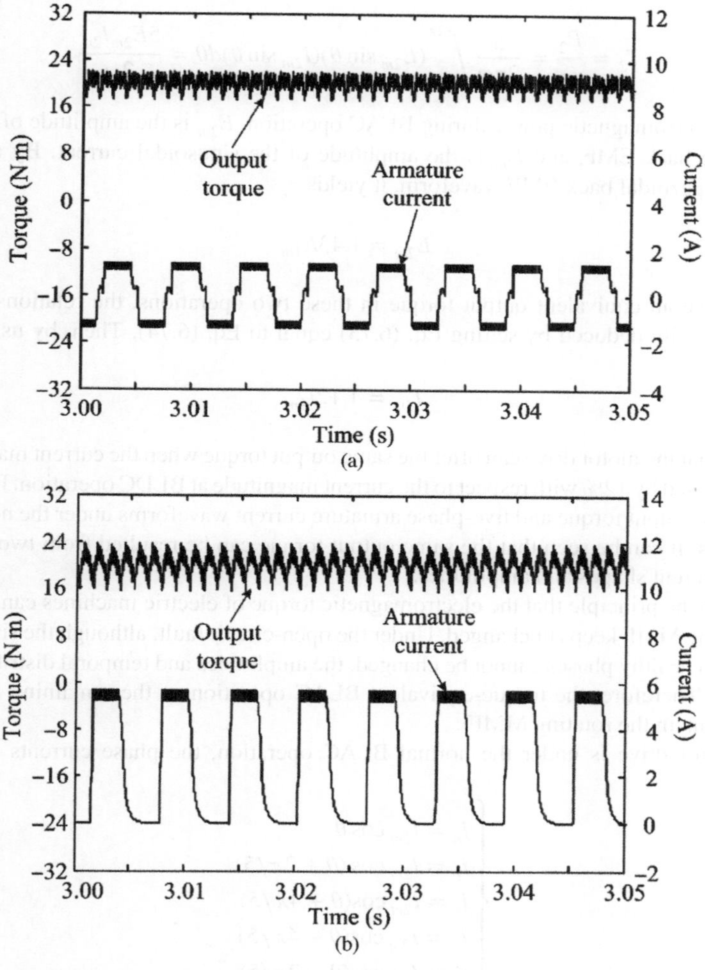

Figure 6.44 Steady-state waveforms of outer-rotor FM-DSPM motor drive: (a) DSPM mode and (b) SR mode

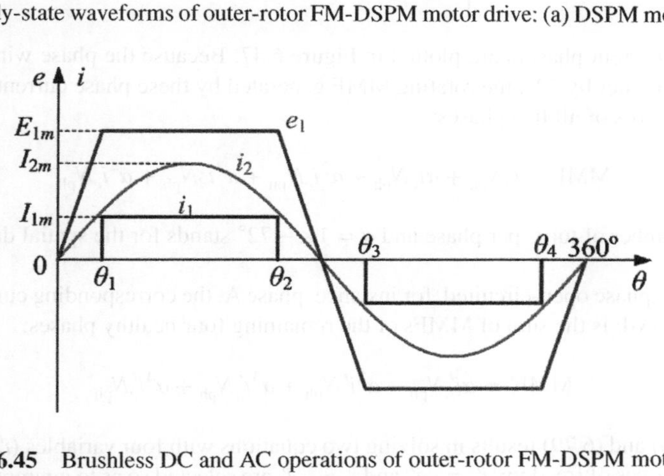

Figure 6.45 Brushless DC and AC operations of outer-rotor FM-DSPM motor drive

be expressed as

$$T_2 = \frac{P_2}{\omega} = \frac{5}{2\pi\omega}\int_0^{2\pi}(E_{2m}\sin\theta)(I_{2m}\sin\theta)d\theta = \frac{5E_{2m}I_{2m}}{2\omega} \tag{6.74}$$

where P_2 is the electromagnetic power during BLAC operation, E_{2m} is the amplitude of the fundamental component of the back EMF, and I_{2m} is the amplitude of the sinusoidal current. By applying Fourier analysis to the trapezoidal back EMF waveform, it yields

$$E_{2m} = 1.43E_{1m} \tag{6.75}$$

In order to achieve an equivalent output torque in these two operations, the relationship between I_{1m} and I_{2m} can be readily deduced by setting Eq. (6.73) equal to Eq. (6.74). Then, by using Eq. (6.75), it results in

$$I_{2m} = 1.12I_{1m} \tag{6.76}$$

which indicates that the motor drive can offer the same output torque when the current magnitude at BLAC operation is increased by 12% with respect to the current magnitude at BLDC operation. Figure 6.46 shows the corresponding output torque and five-phase armature current waveforms under the normal BLDC and BLAC operations. It can be seen that the same output torque can be resulted from two sets of armature currents with different shapes and amplitudes.

On the basis of the principle that the electromagnetic torque of electric machines can be maintained as long as the rotating MMF keeps unchanged. Under the open-circuit fault, although the sinusoidal distribution of currents in healthy phases cannot be changed, the amplitudes and temporal distribution of currents are controllable. Therefore, the torque-equivalent BLAC operation of the remaining healthy phases is employed to maintain the rotating MMF.

When the motor drive is under the normal BLAC operation, the phase currents are sinusoidal as given by

$$\begin{cases} i_a = I_{2m}\cos\theta \\ i_b = I_{2m}\cos(\theta + 2\pi/5) \\ i_c = I_{2m}\cos(\theta + 4\pi/5) \\ i_d = I_{2m}\cos(\theta - 4\pi/5) \\ i_e = I_{2m}\cos(\theta - 2\pi/5) \end{cases} \tag{6.77}$$

The corresponding current phasors are plotted in Figure 6.47. Because the phase windings are spatially displaced from each other by 72°, the rotating MMF generated by these phase currents can be expressed as the sum of the MMFs of all five phases:

$$\text{MMF} = i_aN_{\text{ph}} + \alpha i_bN_{\text{ph}} + \alpha^2 i_cN_{\text{ph}} + \alpha^3 i_dN_{\text{ph}} + \alpha^4 i_eN_{\text{ph}} \tag{6.78}$$

where N_{ph} is the number of turns per phase and $\alpha = 1\angle -72°$ stands for the spatial distribution of phase currents.

When there is one phase open-circuited, for instance, phase A, the corresponding current becomes zero. Thus, the rotating MMF is the sum of MMFs of the remaining four healthy phases:

$$\text{MMF}' = \alpha i_b'N_{\text{ph}} + \alpha^2 i_c'N_{\text{ph}} + \alpha^3 i_d'N_{\text{ph}} + \alpha^4 i_e'N_{\text{ph}} \tag{6.79}$$

Equating Eqs. (6.78) and (6.79) results in solving two equations with four variables (i_b', i_c', i_d', i_e'). In order to eliminate two degrees of freedom, $i_b' = -i_d'$ and $i_c' = -i_e'$ are selected so as to guarantee that the solution

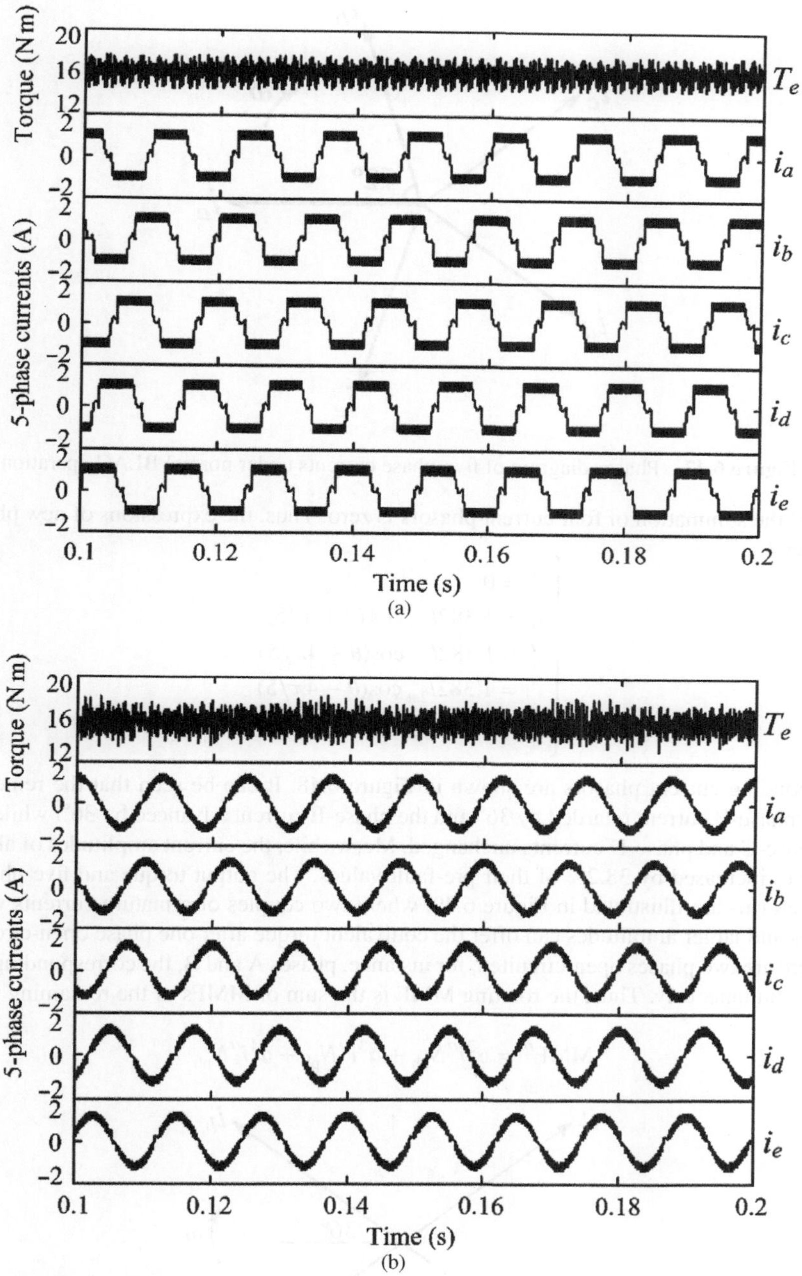

Figure 6.46 Output torque and armature current waveforms of outer-rotor FM-DSPM motor drive: (a) brushless DC operation and (b) brushless AC operation

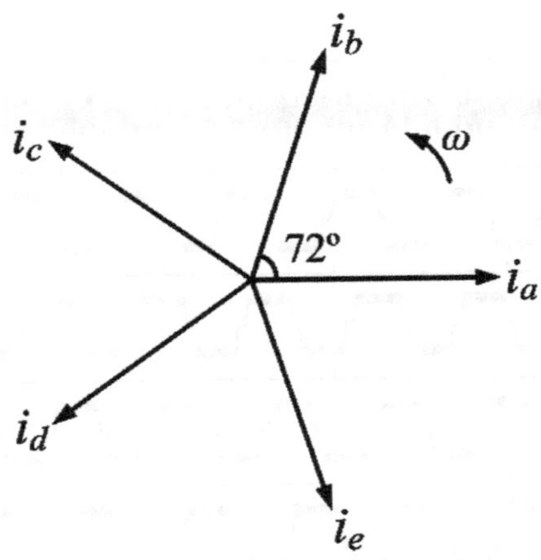

Figure 6.47 Phasor diagram of five-phase currents under normal BLAC operation

is unique and the summation of four current phasors is zero. Thus, the expressions of new phase currents can be deduced as

$$\begin{cases} i'_a = 0 \\ i'_b = 1.382 I_{2m} \cos(\theta + \pi/5) \\ i'_c = 1.382 I_{2m} \cos(\theta + 4\pi/5) \\ i'_d = 1.382 I_{2m} \cos(\theta - 4\pi/5) \\ i'_e = 1.382 I_{2m} \cos(\theta - \pi/5) \end{cases} \quad (6.80)$$

The corresponding current phasors are shown in Figure 6.48. It can be seen that the remediation is to regulate the phase-B current retarded by 36° and the phase-E current advanced by 36°, while keeping the angles of phase-C and phase-D currents unchanged. Meanwhile, the current amplitudes of all the remaining phases are increased by 38.2% of their pre-fault values. The output torque and five-phase armature current waveforms are illustrated in Figure 6.49, where two couples of armature currents with reversed phase angles and larger amplitudes can offer the equivalent torque after one phase open-circuited.

When there are two phases open-circuited, for instance, phases A and B, the corresponding currents are set to zero simultaneously. Thus, the rotating MMF is the sum of MMFs of the remaining three healthy phases:

$$\text{MMF}'' = \alpha^2 i''_c N_{ph} + \alpha^3 i''_d N_{ph} + \alpha^4 i''_e N_{ph} \quad (6.81)$$

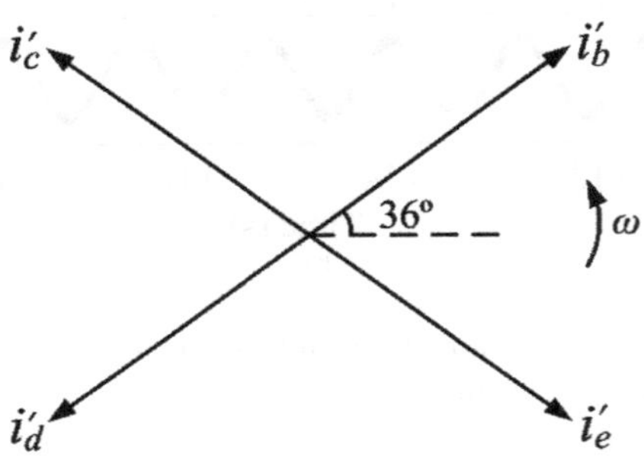

Figure 6.48 Phasor diagram of four-phase currents under remedial one-phase open-circuit brushless AC operation

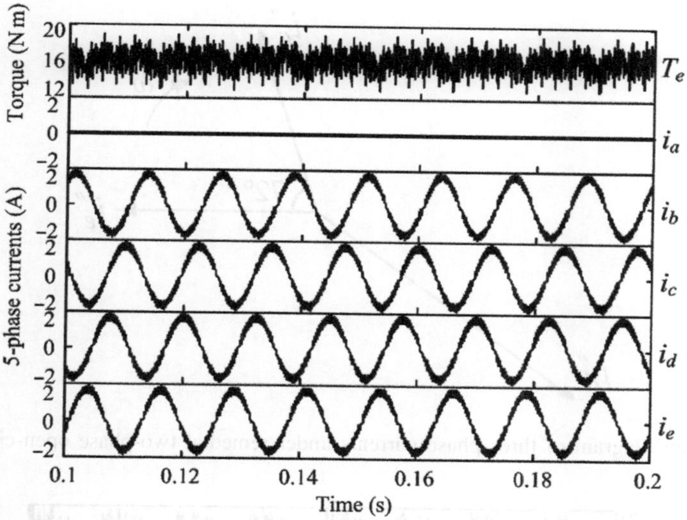

Figure 6.49 Output torque and armature current waveforms of outer-rotor FM-DSPM motor drive under remedial one-phase open-circuit brushless AC operation

Equating Eqs. (6.81) and (6.78) results in solving two equations with three variables (i_c'', i_d'', i_e''). These variables are purposely related by $i_c'' + i_d'' + i_e'' = 0$ so as to guarantee that the solution is unique and the summation of three current phasors is zero. Thus, the expressions of new phase currents can be deduced as

$$\begin{cases} i_a'' = 0 \\ i_b'' = 0 \\ i_c'' = 2.236 I_{2m} \cos(\theta + 2\pi/5) \\ i_d'' = 3.618 I_{2m} \cos(\theta - 4\pi/5) \\ i_e'' = 2.236 I_{2m} \cos\theta \end{cases} \tag{6.82}$$

The corresponding current phasors are plotted in Figure 6.50. It can be seen that the angles of phase-C and phase-E currents are regulated to those of the pre-fault phase-B and phase-A currents, respectively, while the angle of phase-D current is unchanged. The amplitudes of phase-C and phase-E currents are 223.6% of their pre-fault values, while the amplitude of phase-D current is 361.8% of its pre-fault value. The corresponding torque and current waveforms are shown in Figure 6.51. It can be found that the two-phase open-circuit operation can still maintain the output torque but suffers from a larger torque ripple.

Short-circuit faults in the FM-DSPM motor drive include the phase-to-phase winding short-circuit, inter-turn winding short-circuit, winding short-circuit at terminals, and power device short-circuit. Among them, the inter-turn winding short-circuit due to insulation failure is the most common one. When the inter-turn short-circuit occurs in one phase winding, it can be analyzed by using a transformer equivalent circuit as shown in Figure 6.52. The whole phase winding is considered as the transformer primary winding, whereas the short-circuit winding is treated as the transformer secondary winding. Thus, the circuit equations can be expressed as

$$\begin{cases} u = Ri + e = Ri + N_{\mathrm{ph}} \dfrac{d\psi}{dt} \\ e_{\mathrm{sc}} = N_{\mathrm{sc}} \dfrac{d\psi}{dt} = R_{\mathrm{sc}} i_{\mathrm{sc}} \end{cases} \tag{6.83}$$

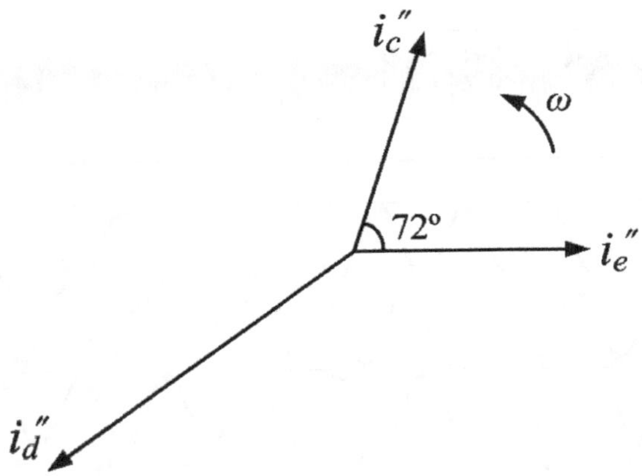

Figure 6.50 Phasor diagram of three-phase currents under remedial two-phase open-circuit brushless AC operation

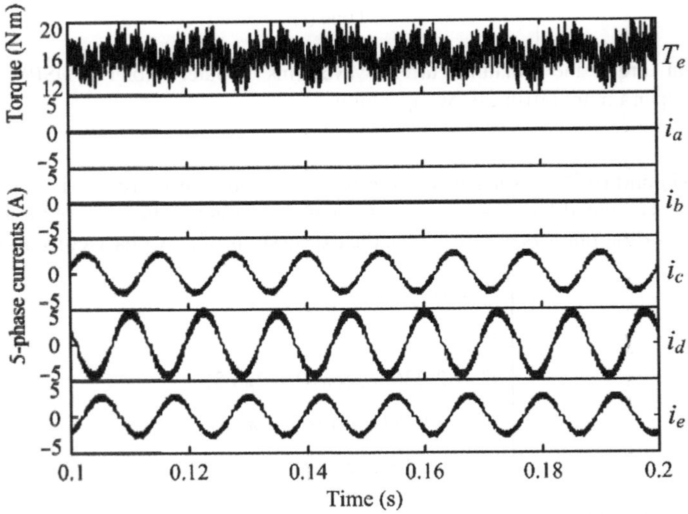

Figure 6.51 Output torque and armature current waveforms of outer-rotor FM-DSPM motor drive under remedial two-phase open-circuit BLAC operation

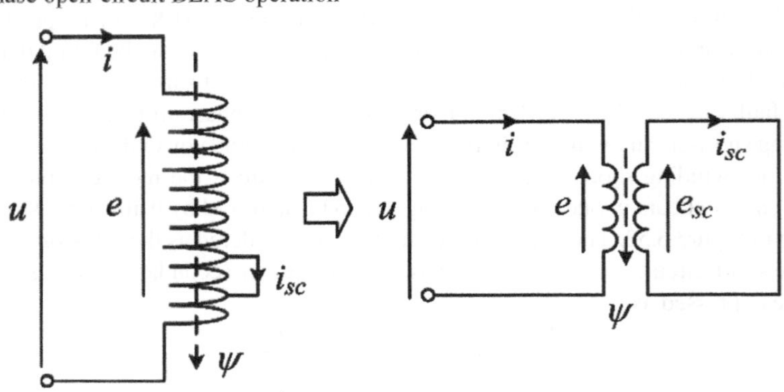

Figure 6.52 Inter-turn short-circuit and equivalent circuit

where ψ is the flux per phase, e is the induced EMF in the phase winding, e_{sc} is the induced EMF in the short-circuit winding, i_{sc} is the current in the short-circuit winding, N_{ph} is the number of turns of the phase winding, N_{sc} is the number of turns of the short-circuit winding, and R_{sc} is the resistance of the short-circuit winding. These induced EMFs and resistances can be related by

$$\begin{cases} \dfrac{e_{sc}}{e} = \dfrac{N_{sc}}{N_{ph}} \\ \dfrac{R_{sc}}{R} = \dfrac{N_{sc}}{N_{ph}} \end{cases} \tag{6.84}$$

Hence, the phase and short-circuit currents can be expressed as

$$\begin{cases} i = \dfrac{u-e}{R} \\ i_{sc} = \dfrac{e_{sc}}{R_{sc}} = \dfrac{e}{R} \end{cases} \tag{6.85}$$

Since the resistance of phase winding is usually very small and the number of turns of inter-turn short-circuit is also very few in most occasions, the corresponding influence on the phase current may be negligible. Nevertheless, the magnitude of short-circuit current is actually very large. Typically, the induced EMF, e is usually 85–95% of the terminal voltage u. Hence, substituting $e = ku$, where $k \in [0.85, 0.95]$, into Eq. (6.85) results in

$$i_{sc} = \frac{k}{1-k} i \tag{6.86}$$

which indicates that there is a large current, 5.7–19 times the phase current, circulating through the shorted turns. Consequently, it releases a large amount of heat energy that will worsen the inter-turn winding short-circuit. Therefore, it is necessary to develop a remedial operation for such short-circuit fault.

Traditionally, the remediation of short-circuit fault is to disable the faulty phase. It is equivalent to transfer the short-circuit fault problem to the open-circuit fault problem, and then employ the open-circuit remediation. However, this strategy suffers from two drawbacks: one is the lower efficiency due to the limitation of remaining healthy phases; another is the larger torque ripple due to the presence of zero-torque zone. Thus, an alternative remediation strategy is necessary to enable the faulty phase continuing operation in a special way.

In order to reduce the short-circuit current i_{sc}, it is required to decrease the induced EMF e as given by

$$e = N_{ph}\frac{d\psi}{dt} = \frac{d\psi_{PM}}{dt} + \frac{d(Li)}{dt} \tag{6.87}$$

It indicates that if the PM flux linkage, ψ_{PM} is set to zero, e and hence i_{sc} will drop significantly. On the other hand, i will increase significantly as revealed in Eq. (6.85), hence compensating the reduction of the developed torque T_e because of the vanishing of the PM torque component T_{PM}. In this machine, setting ψ_{PM} to zero means that the Al-Ni-Co PMs are fully demagnetized, which can be easily realized by applying a negative current pulse to the magnetizing winding. Thus, the FM-DSPM machine becomes working as an SR machine. In other words, when the inter-turn short-circuit fault occurs, the remediation is to change the mode of operation of the motor drive from the DSPM mode to the SR mode in such a way that the average torque can be maintained.

During the DSPM mode of this fault-tolerant operation, the average output torque per phase, T_D can be derived as

$$T_D = \frac{1}{2\pi}\left(\int_{\theta_1}^{\theta_2} I_D K_{PM} d\theta + \int_{\theta_3}^{\theta_4} (-I_D)(-K_{PM}) d\theta\right) + \frac{1}{2\pi}\left(\int_{\theta_1}^{\theta_2} \frac{1}{2} I_D^2 K_D d\theta + \int_{\theta_3}^{\theta_4} \frac{1}{2}(-I_D)^2 (-K_D) d\theta\right) = \frac{(\theta_2 - \theta_1)}{\pi} I_D K_{PM} \tag{6.88}$$

where I_D is the armature current, $\theta_i (i = 1, 2, 3, 4)$ is the turn-on or turn-off electrical rotor position, K_{PM} is the slope of variation of ψ_{PM}, and K_D is the slope of variation of L_D during the DSPM mode.

During the SR mode of this fault-tolerant operation, the average output torque per phase T_S can similarly be derived as

$$T_S = \frac{1}{2\pi}\int_{\theta_1}^{\theta_2} \left(\frac{1}{2} I_S^2 K_S\right) d\theta = \frac{(\theta_2 - \theta_1)}{4\pi} I_S^2 K_S \tag{6.89}$$

where I_S is the armature current, L_S is the self-inductance, $\theta_i (i = 1, 2)$ is the turn-on or turn-off electrical rotor position, and K_S is the slope of variation of L_S during the SR mode. In order to maintain the same output torque under these two modes, by equating Eqs. (6.88) and (6.89), the relationship between I_D and I_S can be obtained as

$$I_S = 2\sqrt{\frac{I_D K_{PM}}{K_S}} \tag{6.90}$$

where K_{PM} and K_S are readily found from finite element analysis or experimental measurement. For a given value of I_D that is adopted for the DSPM mode of operation, the required value of I_S that is deduced from Eq. (6.90) for the SR mode of operation can maintain the same average output torque under the inter-turn short-circuit fault at all phases as illustrated in Figure 6.53.

6.10 Potential Applications of Stator-PM Motor Drives in EVs

The stator-PM motor drives have high potentiality for EV application, because they can solve two fundamental problems of the existing PM brushless motor drives, including the PM synchronous motor drive and PM BLDC motor drive, as detailed below:

- There are no PMs in the rotor, thus avoiding the problem on how to firmly mount them on the high-speed rotor and hence to withstand the high centrifugal force.
- All PMs are located in the stator, thus facilitating the cooling arrangement and hence solving the thermal instability problem of PMs.

As discussed earlier, there are five major types of stator-PM motor drives for EV propulsion: DSPM, FRPM, FSPM, HEPM, and FMPM. Notice that both the HEPM and FMPM can incorporate the DSPM, FRPM, or FSPM to form individual topologies such as the HE-DSPM and FM-DSPM motor drives. These five major types of stator-PM motor drives are compared for EV propulsion in terms of their power density, torque density, efficiency, controllability, PM immunity, robustness, manufacturability, and maturity. The corresponding evaluation is summarized in Table 6.4.

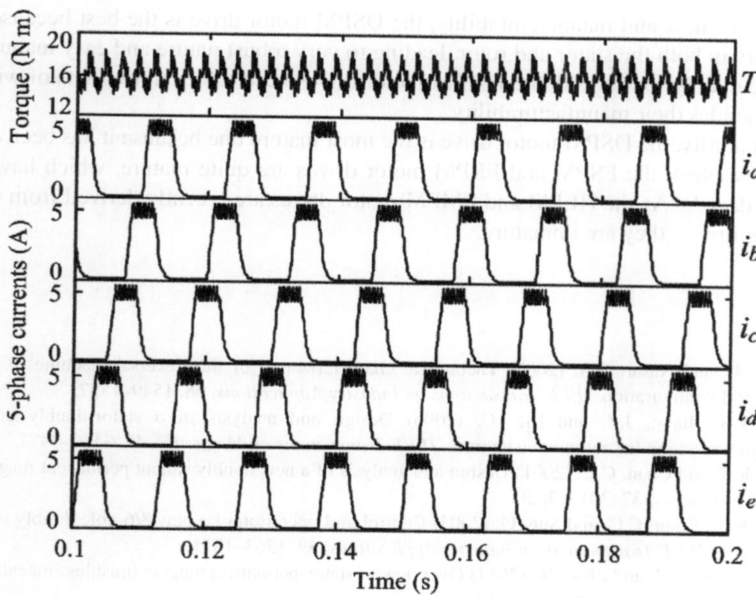

Figure 6.53 Output torque and armature current waveforms of outer-rotor FM-DSPM motor drive under remedial short-circuit operation

Table 6.4 Evaluation of stator-PM motor drives for EVs

	DSPM	FRPM	FSPM	HEPM	FMPM
Power density	Medium	Good	High	Medium	Medium
Torque density	Medium	Good	High	High	High
Efficiency	Good	Good	Good	High	High
Controllability	Medium	Medium	Medium	Superb	Superb
PM immunity	Medium	Weak	Good	Good	Good
Robustness	Strong	Medium	Medium	Medium	Medium
Manufacture	Easy	Medium	Medium	Hard	Hard
Maturity	High	Medium	Medium	Low	Low

In terms of power density and torque density, the FSPM motor drive is the best because it employs the flux-focusing arrangement. Meanwhile, the power density of the doubly-fed HEPM and FMPM is degraded because of the use of additional field or magnetizing winding. Nevertheless, the corresponding torque, especially during starting, can be temporarily boosted up by instantaneous flux strengthening. Those doubly-fed HEPM and FMPM motor drives possess the capability of air-gap flux control, which can achieve higher efficiency than those with singly-fed DSPM, FRPM, and FSPM motor drives. Concerning controllability, those doubly-fed HEPM and FMPM motor drives are much better than the singly-fed motor drives.

As the PM material is relatively delicate, their immunity to accidental demagnetization and mechanical abuse are of concern. The PM immunity of the FRPM motor drive is relatively weak because the corresponding PMs are mounted on the surfaces of stator poles that are prone to be partially demagnetized under high armature field or high temperature operation and are vulnerable to physical damage under severe vibration.

In terms of robustness and manufacturability, the DSPM motor drive is the best because it offers the highest simplicity in both the stator and rotor, leading to very robust nature and easy manufacturing. On the contrary, the doubly-fed HEPM and FMPM motor drives need to install two sets of windings in the stator, which degrades their manufacturability.

Concerning maturity, the DSPM motor drive is the most mature one because it has been developed for over two decades. Next, the FSPM and FRPM motor drives are quite mature, which have been developed for over a decade. As the HEPM and FMPM motor drives are recently derived from the singly-fed stator-PM motor drives, they are immature.

References

Boldea, I., Zhang, L. and Nasar, S.A. (2002) Theoretical characterization of flux reversal machine in low-speed servo drives-the pole-PM configuration. *IEEE Transactions on Industry Applications*, **38**, 1549–1557.

Chau, K.T., Li, Y.B., Jiang, J.Z. and Liu, C. (2006) Design and analysis of a stator-doubly-fed doubly-salient permanent-magnet machine for automotive engines. *IEEE Transactions on Magnetics*, **42**, 3470–3472.

Cheng, M., Chau, K.T. and Chan, C.C. (2001) Design and analysis of a new doubly salient permanent magnet motor. *IEEE Transactions on Magnetics*, **37**, 3012–3020.

Cheng, M., Chau, K.T., Chan, C.C. and Sun, Q. (2003) Control and operation of a new 8/6-pole doubly salient permanent magnet motor drive. *IEEE Transactions on Industry Applications*, **39**, 1363–1371.

Cheng, M., Hua, W., Zhang, J. and Zhao, W. (2011) Overview of stator-permanent magnet brushless machines. *IEEE Transactions on Industrial Electronics*, **58**, 5087–5106.

Deodhar, R.P., Andersson, S., Boldea, I. and Miller, T.J.E. (1997) The flux-reversal machine: a new brushless doubly-salient permanent-magnet machine. *IEEE Transactions on Industry Applications*, **33**, 925–934.

Gong, Y., Chau, K.T., Jiang, J.Z. *et al.* (2009) Analysis of doubly salient memory motors using Preisach theory. *IEEE Transactions on Magnetics*, **45**, 4676–4679.

Hua, W., Cheng, M. and Zhang, G. (2009) A novel hybrid excitation flux switching motor for hybrid vehicles. *IEEE Transactions on Magnetics*, **45**, 4728–4731.

Kim, T.H. and Lee, J. (2004) A study of the design for the flux reversal machine. *IEEE Transactions on Magnetics*, **40**, 2053–2055.

Kim, T.H., Won, S.H., Bong, K. and Lee, J. (2005) Reduction of cogging torque in flux-reversal machine by rotor teeth pairing. *IEEE Transactions on Magnetics*, **41**, 3964–3966.

Li, W., Chau, K.T., Gong, Y. *et al.* (2011) A new flux-mnemonic dual-magnet brushless machine. *IEEE Transactions on Magnetics*, **47**, 4223–4226.

Li, F., Chau, K.T., Liu, C. *et al.* (2012) Design and analysis of magnet proportioning for dual-memory machines. *IEEE Transactions on Applied Superconductivity*, **22**, 4905404-1–4905404-4.

Liu, C., Chau, K.T., Li, W. and Yu, C. (2009) Efficiency optimization of a permanent-magnet hybrid brushless machine using dc field current control. *IEEE Transactions on Magnetics*, **45**, 4652–4655.

Ostovic, V. (2003) Memory motor. *IEEE Industry Applications Magazine*, **9**, 52–61.

Owen, R.L., Zhu, Z.Q., Thomas, A.S. *et al.* (2010) Alternate poles wound flux-switching permanent-magnet brushless ac machines. *IEEE Transactions on Industry Applications*, **46**, 790–797.

Rauch, S.E. and Johnson, L.J. (1955) Design principles of flux-switching alternators. *AIEE Transactions*, **74**, 1261–1268.

Yu, C. and Chau, K.T. (2011a) Design, analysis and control of DC-excited memory motors. *IEEE Transactions on Energy Conversion*, **26**, 479–489.

Yu, C. and Chau, K.T. (2011b) Dual-mode operation of DC-excited memory motors under flux regulation. *IEEE Transactions on Industry Applications*, **47**, 2031–2041.

Yu, C. and Chau, K.T. (2011c) New fault-tolerant flux-mnemonic doubly-salient permanent-magnet motor drive. *IET Electric Power Applications*, **5**, 393–403.

Zhu, Z.Q., Chen, J.T., Pang, Y. *et al.* (2008) Analysis of a novel multi-tooth flux-switching PM brushless ac machine for high torque direct-drive applications. *IEEE Transactions on Magnetics*, **44**, 4313–4316.

Zhu, Z.Q., Pang, Y., Howe, D. *et al.* (2005) Analysis of electromagnetic performance of flux-switching permanent-magnet machines by nonlinear adaptive lumped parameter magnetic circuit model. *IEEE Transactions on Magnetics*, **41**, 4277–4287.

7

Magnetic-Geared Motor Drives

For modern electric vehicles (EVs), the use of in-wheel motor drives is becoming more and more attractive. In particular, it can play the role of electronic differential, namely eliminating the bulky and heavy mechanical differential. Since the wheel spins only a few hundred revolutions per minute, the in-wheel motor is either a low-speed gearless outer-rotor one or a high-speed planetary-geared inner-rotor one. The former takes the advantage of gearless operation, but its low-speed design causes bulky size and heavy weight. The latter takes the merit of reduced overall size and weight, but suffers from transmission loss, acoustic noise, and need for regular lubrication. By incorporating the magnetic gear into the electric motor, the magnetic-geared motor can simultaneously possess the advantages of high-speed motor design and low-speed output motion as well as pseudo-gearless transmission.

In this chapter, magnetic-geared (MG) motor drives, including their system configurations, magnetic gears, MG machines, power inverters, and control strategies, are discussed. Their design criteria, design examples, and potential applications are also presented.

7.1 System Configurations

The MG motor drive essentially consists of the MG machine, power inverter, and electronic controller. The core is the MG machine which basically couples the magnetic gear and electric machine together. On the basis of the types of magnetic gears, namely, the spur gear, planetary gear, and coaxial gear, and the arrangements of electric machines, namely the inner-rotor type and outer-rotor type, the MG machine can be classified into the following six types as depicted in Figure 7.1:

- Inner-rotor machine coupled with magnetic spur gear.
- Outer-rotor machine coupled with magnetic spur gear.
- Inner-rotor machine coupled with magnetic planetary gear.
- Outer-rotor machine coupled with magnetic planetary gear.
- Inner-rotor machine coupled with magnetic coaxial gear.
- Outer-rotor machine coupled with magnetic coaxial gear.

Among the aforementioned types, the outer-rotor machine coupled with magnetic planetary gear and the outer-rotor machine coupled with the magnetic coaxial gear are preferred to the others because the magnetic planetary gear and magnetic coaxial gear can directly mount onto the outer rotor of the machine. Hence, the outer rotor of the machine can share its rotating body with the sun gear of the magnetic planetary gear as shown in Figure 7.2, and it can also share its rotating body with the inner gear of the magnetic coaxial gear as shown in Figure 7.3.

Electric Vehicle Machines and Drives: Design, Analysis and Application, First Edition. K. T. Chau.

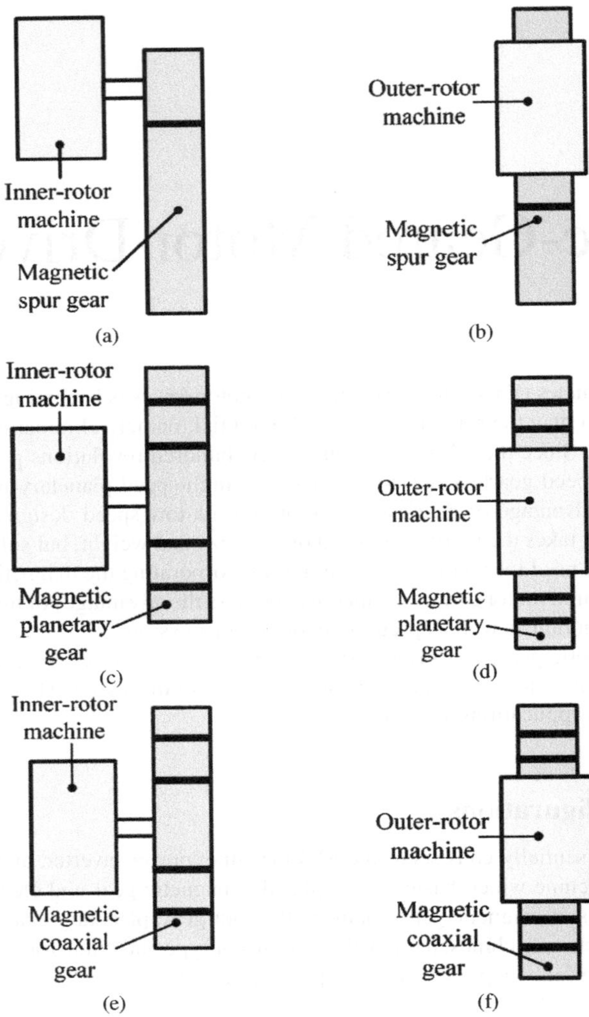

Figure 7.1 Basic configurations of MG machines: (a) inner rotor plus spur gear; (b) outer rotor plus spur gear; (c) inner rotor plus planetary gear; (d) outer rotor plus planetary gear; (e) inner rotor plus coaxial gear; and (f) outer rotor plus coaxial gear

Between the two outer-rotor configurations, the magnetic-coaxial-geared machine takes the following advantages over the magnetic-planetary-geared machine:

- All magnets of the magnetic coaxial gear contribute to torque transmission simultaneously, whereas only the pseudo-contact magnets of the magnetic planetary gear contribute to torque transmission. Hence, the magnetic-coaxial-geared machine can offer higher torque density.
- The thickness of the modulation ring of the magnetic coaxial gear is much smaller than the diameter of the planet gear of the magnetic planetary gear. Hence, the magnetic-coaxial-geared machine can be much more compact.

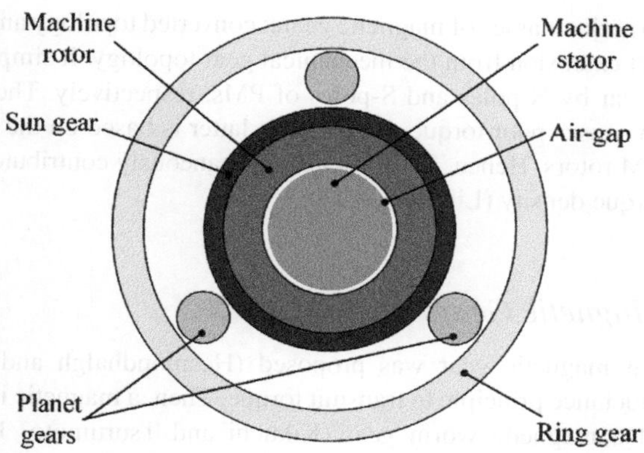

Figure 7.2 Magnetic-planetary-geared machine configuration

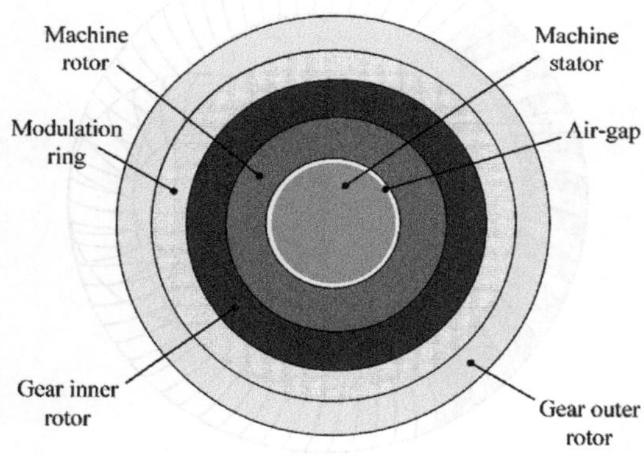

Figure 7.3 Magnetic-coaxial-geared machine configuration

7.2 Magnetic Gears

Mechanical gears are well developed and utilized to a great extent in industrial applications that a couple of mechanical gears have been adopted as a symbol of modern industrial civilization. In short, mechanical gears are employed to transmit torques or forces as well as alter the forms of motion, which include the speeds of motion and the directions of motion. Although mechanical gears work well in most kinds of applications, the corresponding contact mechanism inevitably arouses many nuisances, such as the frictional loss, audible noise, and mechanical vibration as well as the need for regular lubrication and maintenance.

With the advent of rare-earth permanent magnets (PMs), such as the neodymium-iron-boron (Nd-Fe-B) alloy, the development of magnetic gears has attracted wide attention. In fact, the idea of magnetic gears can be traced back to the beginning of the twentieth century. In 1913, a US Patent Application described an electromagnetic gear, which should be the original topology (Neuland, 1916), but almost no one was interested in it at that time. Until a magnetic spur gear was proposed in 1941 (Faus, 1941), people paid attention to magnetic gearing.

Basically, there are two major classes of magnetic gears: converted topology and field-modulated topology. The former is a direct extension from the mechanical gear topology. It simply replaces the slots and teeth of the mechanical gear by N-poles and S-poles of PMs, respectively. The low utilization of PMs is the key problem which causes poor torque density. The latter is based on the modulation of magnetic fields produced by two PM rotors. Hence, all PMs can simultaneously contribute to torque transmission, leading to offer higher torque density (Li *et al.*, 2013).

7.2.1 Converted Magnetic Gears

In 1980, a multi-element magnetic gear was proposed (Hesmondhalgh and Tipping, 1980), which employed the variable reluctance principle to transmit torque. Then, a magnetic involute gear (Tsurumoto and Kikuchi, 1987) and a magnetic worm gear (Kikuchi and Tsurumoto, 1993) were proposed as shown in Figures 7.4 and 7.5, respectively. All these magnetic gears are essentially converted from their

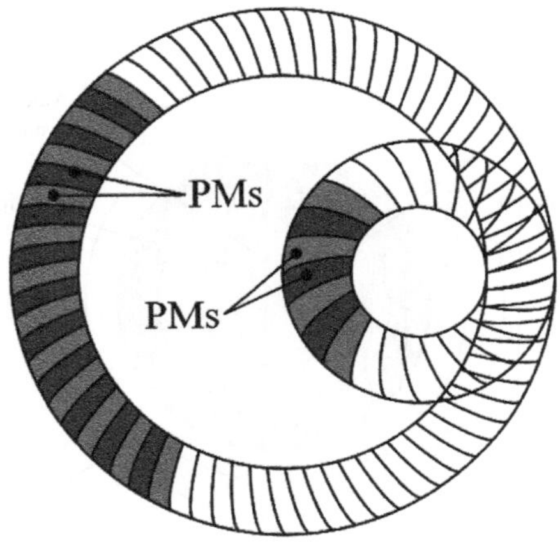

Figure 7.4 Magnetic involute gear

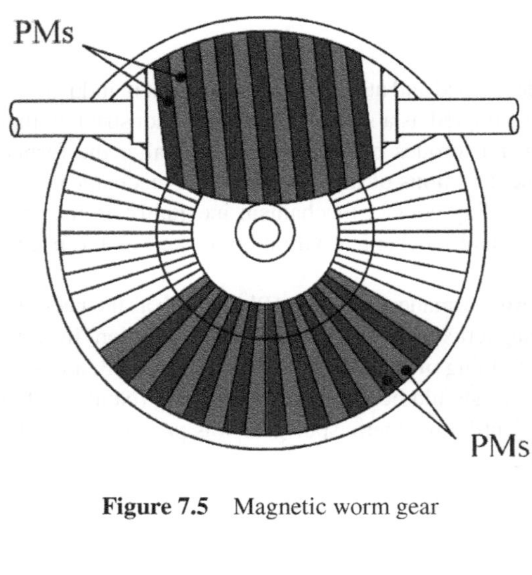

Figure 7.5 Magnetic worm gear

mechanical counterparts. Due to the low utilization of PMs, they all have a poor torque density, typically less than 2 kN m/m^3.

Rather than developing those complicated magnetic involute gear and worm gear, many researchers focus on the development of parallel-axis magnetic spur gears, which include two different magnetic couplings: radial-coupling and axial-coupling (Jorgensen, Andersen, and Rasmussen, 2005). Figure 7.6 shows two different radial-coupling topologies and one axial-coupling topology. In addition, a perpendicular-axis

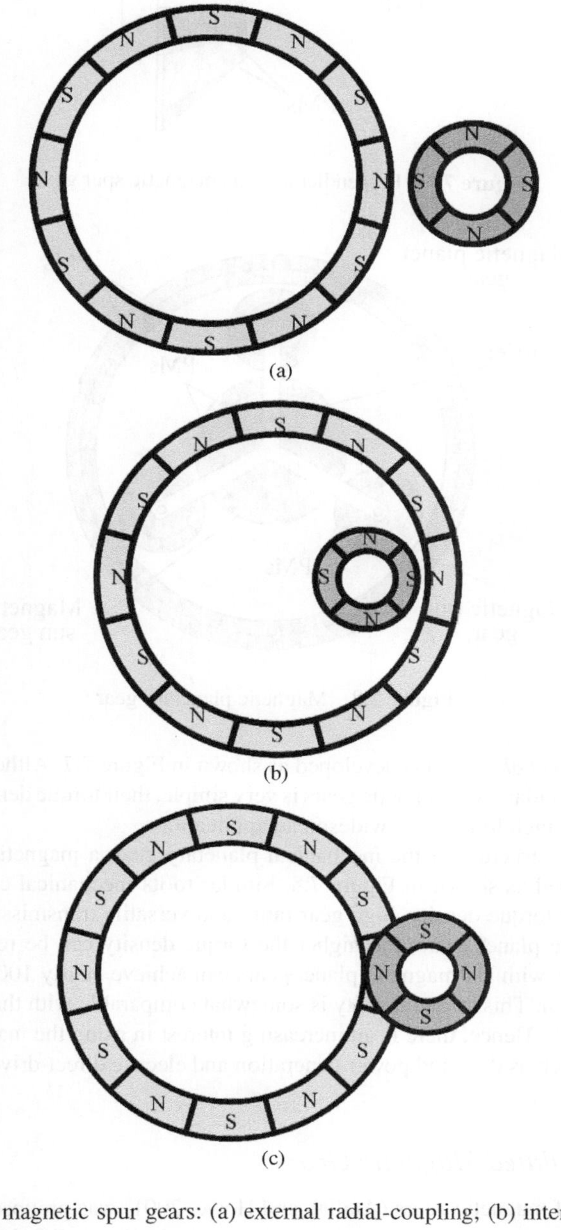

Figure 7.6 Parallel-axis magnetic spur gears: (a) external radial-coupling; (b) internal radial-coupling; and (c) axial-coupling

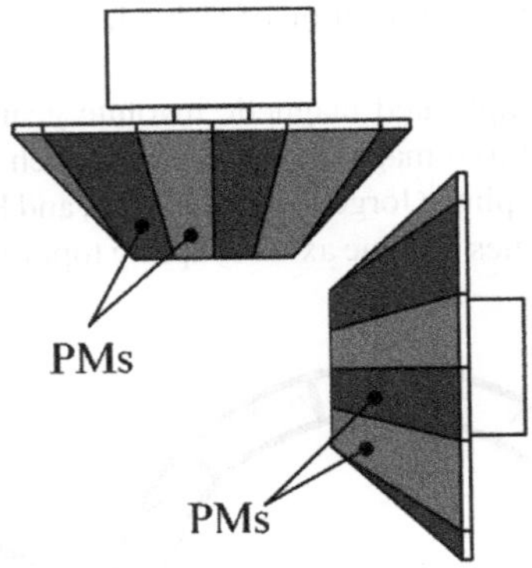

Figure 7.7 Perpendicular-axis magnetic spur gear

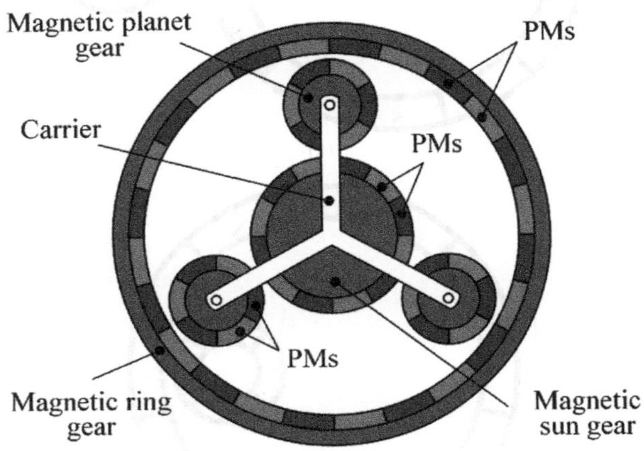

Figure 7.8 Magnetic planetary gear

magnetic spur gear (Yao *et al.*, 1996) is developed as shown in Figure 7.7. Although the configuration of parallel-axis or perpendicular-axis magnetic gears is very simple, their torque density is very low, typically less than 12 kN m/m^3, which limits their widespread applications.

With reference to the structure of the mechanical planetary gear, a magnetic planetary gear (Huang *et al.*, 2008) is developed as shown in Figure 7.8. Similar to its mechanical counterpart, the magnetic planetary gear has high torque density, high gear ratio, and versatile transmission. In general, the more the number of magnetic planet gears, the higher the torque density can be resulted. For instance, the magnetic planetary gear with six magnetic planet gears can achieve nearly 100 kN m/m^3 of shear stress on the magnetic ring gear. This torque density is somewhat comparable with that of the mechanical spur gear (100–200 kN m/m^3). Hence, there is an increasing interest in using the magnetic planetary gear for various applications, such as the wind power generation and electric direct-drive propulsion.

7.2.2 *Field-Modulated Magnetic Gears*

In 2001, a new breed of magnetic gears (Atallah and Howe, 2001) was proposed which was fundamentally different from the converted magnetic gears. As shown in Figure 7.9, it consists of three coaxial

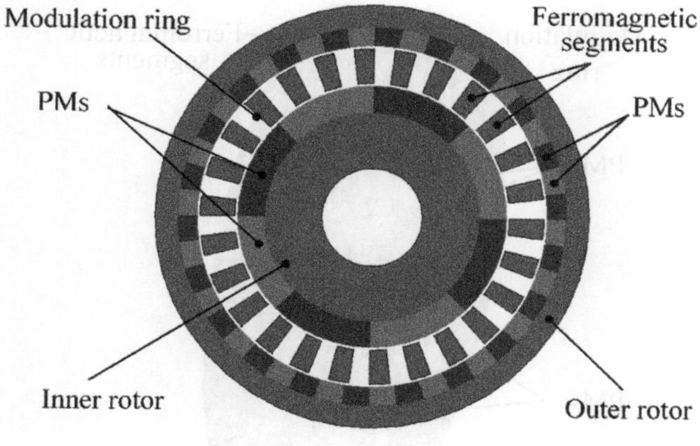

Figure 7.9 Magnetic coaxial gear

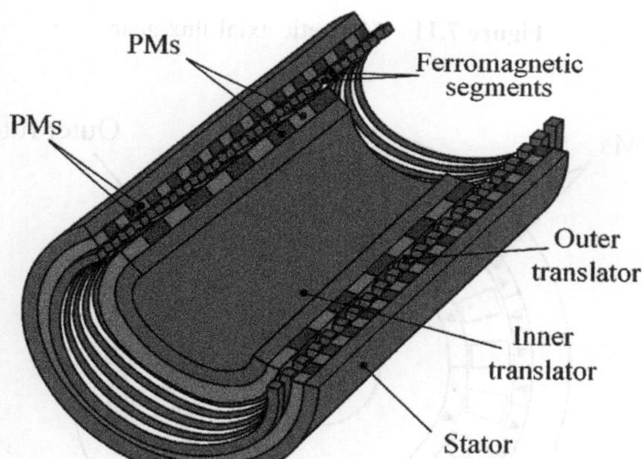

Figure 7.10 Magnetic linear gear

bodies: two of them are mounted with PMs to serve as the inner and outer rotors and the sandwiched body is a stationary modulation ring mounted with ferromagnetic segments. Its principle of operation relies on the use of ferromagnetic segments to modulate the magnetic fields produced by the two PM rotors (Atallah, Calverley, and Howe, 2004; Jian and Chau, 2009a). Because of the simultaneous contribution of all PMs to the torque transmission, this magnetic coaxial gear exhibits a high torque density, typically 50–150 kN m/m^3.

Following the advent of the magnetic coaxial gear, which is a radial-flux magnetic gear, other two morphologies have also been proposed: the magnetic linear gear (Atallah, Wang, and Howe, 2005; Li and Chau, 2012) and magnetic axial-flux gear (Mezani, Atallah, and Howe, 2006) as shown in Figures 7.10 and 7.11, respectively. With the use of Nd-Fe-B PM material, the magnetic linear gear can offer the force density exceeding 1.7 MN/m^3, which makes it promising for many kinds of linear-drive applications such as electric railway traction and wave energy harvesting (Li, Chau, and Jiang, 2011). Similarly, the magnetic axial-flux gear can achieve the torque density exceeding 70 kN m/m^3, which is particularly attractive for applications to pancake-shaped direct drives such as the in-wheel propulsion and aerospace pumps, where a hermetic isolation between the input and output shafts is required.

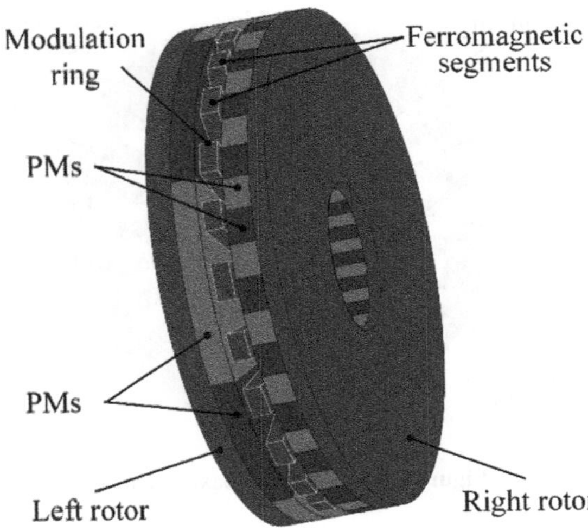

Figure 7.11 Magnetic axial-flux gear

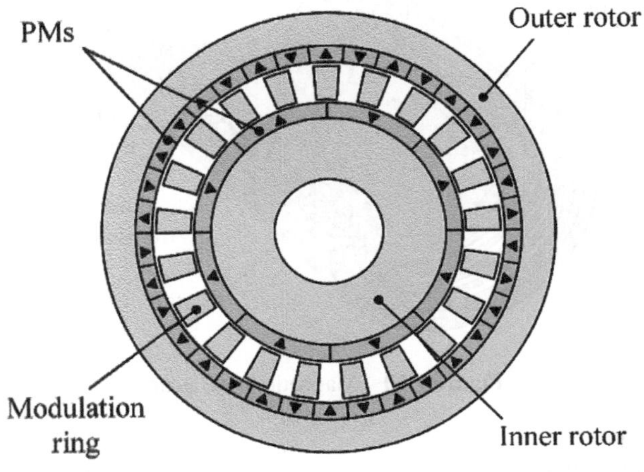

Figure 7.12 Magnetic coaxial gear with radially magnetized PMs

The first generation of magnetic coaxial gears is based on the use of radially magnetized PMs mounted on the outer surface of the inner rotor and the inner surface of the outer rotor as shown in Figure 7.12, where the arrow indicates the direction of magnetization. This radial-PM or surface-PM configuration takes the advantage of simplicity, but suffers from mechanical integrity under high-speed operation.

By magnetizing several pieces of PMs with proper orientations of magnetization to form a magnetic pole, it offers some attractive features: near-sinusoidal air-gap flux density distribution, strong magnetic field intensity, and good self-shielding magnetization (Halbach, 1980). By utilizing the concept of Halbach array of PMs, the resulting magnetic coaxial gear (Jian and Chau, 2010) is shown in Figure 7.13, which can offer higher torque density and lower torque ripple than the original one. However, the manufacturing process of Halbach-PM arrays is complicated and costly. In addition, taking into account the centrifugal force and mechanical stress, Halbach configuration is not preferable for high-speed torque transmission.

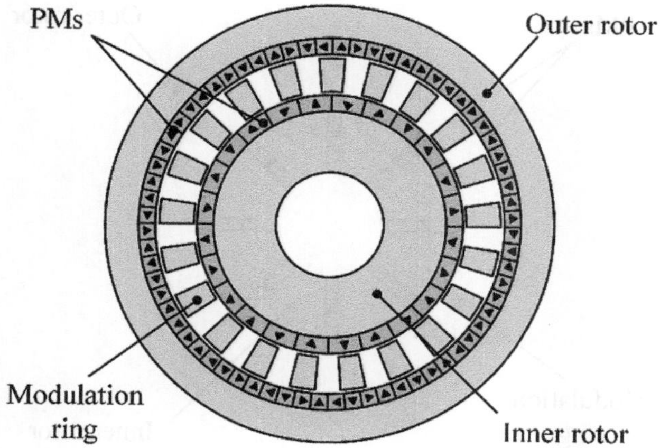

Figure 7.13 Magnetic coaxial gear with Halbach-magnetized PMs

In order to improve the mechanical integrity or robustness, the PMs can be inserted or buried in the iron core. Meanwhile, these interior PMs can be tangentially magnetized in such a way that two PM pieces are arranged to constitute a PM pole, leading to strengthen the air-gap flux density – the so-called flux-focusing or flux-concentrating arrangement. Figure 7.14a shows the magnetic coaxial gear with all PMs in the inner rotor tangentially magnetized (Rasmussen *et al.*, 2005), whereas Figure 7.14b shows the magnetic coaxial gear with all PMs in the outer rotor tangentially magnetized (Li *et al.*, 2011). Both of them adopt the flux-focusing arrangement, hence improving the flux density in the relevant air-gap (Jian *et al.*, 2009). Whether the inner rotor or outer rotor or both rotors adopting this arrangement depends on the rotational speed and hence the centrifugal force that need to withstand.

To further improve the mechanical integrity, the bipolar-magnetized PM poles (namely the N-pole and S-pole are alternating along the circumference) can be replaced by the unipolar-magnetized PM poles (namely the PM pole and iron tooth are alternating along the circumference) so that the iron tooth is induced to have an opposite magnetic polarity with that of the neighboring PM pole (Liu *et al.*, 2009). Figure 7.15 shows the configuration of a typical magnetic coaxial gear using unipolar-magnetized PMs in the outer rotor. This configuration can greatly improve the mechanical integrity and simplify the manufacturing process, while the torque density can be maintained.

The aforementioned magnetic coaxial gears are compared in terms of their torque density, torque ripple, manufacturability, mechanical integrity, and maturity. Table 7.1 summarizes the results that illustrate that the radial-PM configuration is preferred when simplicity and maturity are desired, whereas the Halbach-PM configuration is most favorable when torque density and torque ripple are of particular concern. On the other hand, the tangential PM configuration is attractive when torque density and mechanical integrity are essential, whereas the unipolar-PM configuration becomes attractive when manufacturability and mechanical integrity are important.

7.3 MG Machines

The MG machine can be considered as a machine integrated with a magnetic coaxial gear. The principle and modeling of the machine part are the same as the original machine, which will not be further discussed. The operating principle and mathematical modeling of the magnetic coaxial gear is the key to realize the desired performance, namely the high-speed machine design and low-speed output rotation, which will be described in detail.

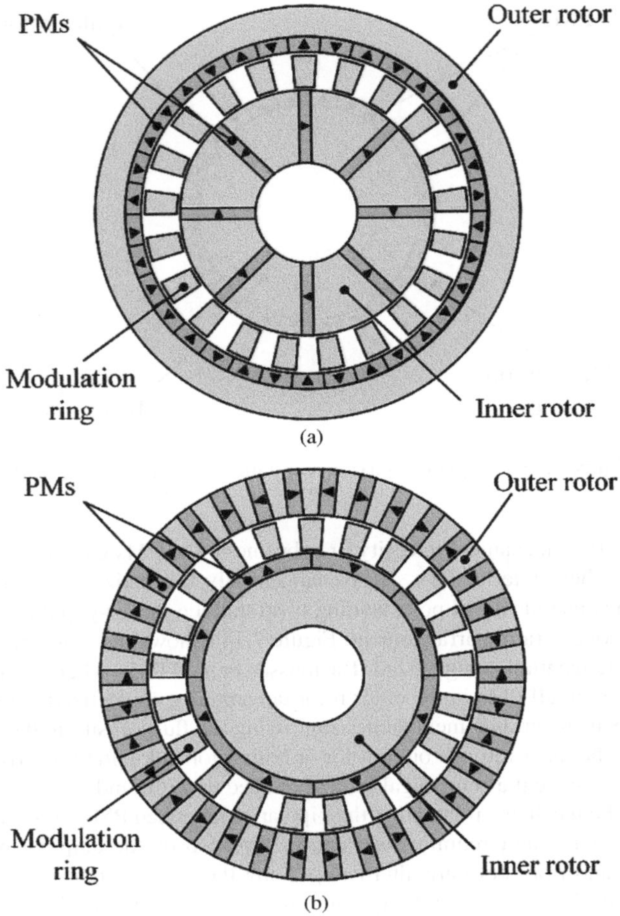

Figure 7.14 Magnetic coaxial gear with tangentially magnetized PMs: (a) inner rotor and (b) outer rotor

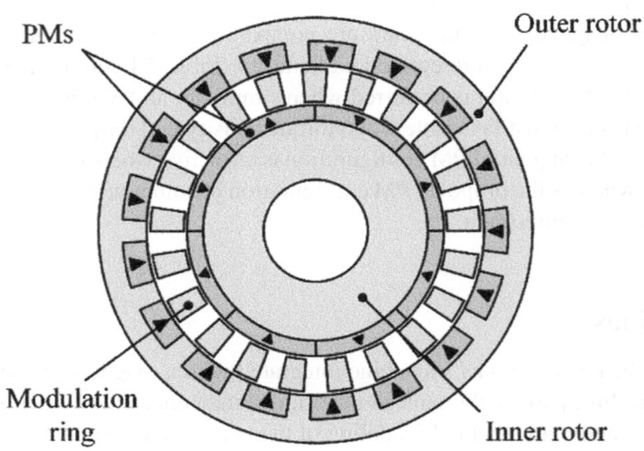

Figure 7.15 Magnetic coaxial gear with unipolar-magnetized PMs

Table 7.1 Comparison of different magnetic coaxial gear configurations

Configuration	Torque density	Torque ripple	Manufacturability	Mechanical integrity	Maturity
Radial-PM	Good	Medium	Easy	Medium	High
Halbach-PM	Excellent	Low	Difficult	Low	Low
Tangential-PM	Very good	High	Medium	High	Medium
Unipolar-PM	Good	High	Easy	High	Medium

7.3.1 *Principle of MG Machines*

Since the machine part and the magnetic coaxial gear are generally designed with no electromagnetic field couplings, the operating principle of the machine part is the same as that of the original machine. Meanwhile, the operating principle of the magnetic coaxial gear is fundamentally different from that of the machine part and plays an important role for torque transmission.

The magnetic fields in the magnetic coaxial gear are due to the PMs in both the inner and outer rotors under the influence of the stationary modulation ring. To realize its principle of operation, the modulation ring is initially removed as shown in Figure 7.16. The corresponding magnetic field distribution can be described by the Laplacian and quasi-Poissonian equations of the scalar magnetic potential (Xia, Zhu, and Howe, 2004). Detailed derivations can be found in Atallah, Calverley, and Howe (2004) and Jian (2010).

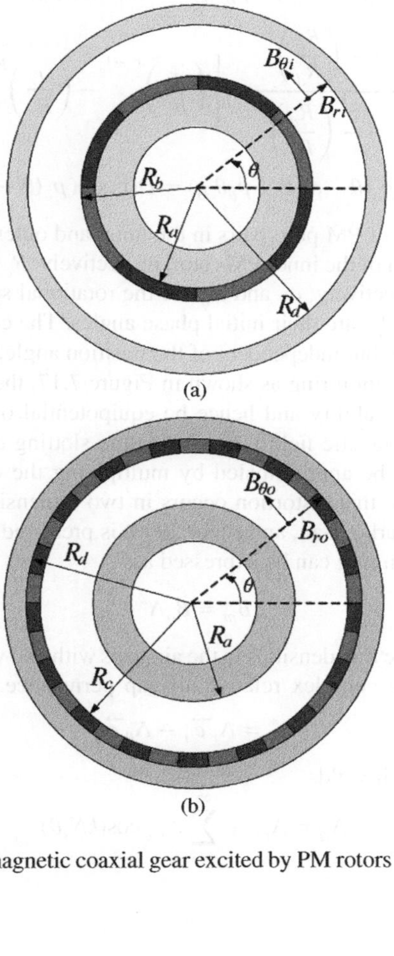

Figure 7.16 Magnetic fields of magnetic coaxial gear excited by PM rotors without modulation ring: (a) inner rotor and (b) outer rotor

The radial and tangential components of the magnetic flux density solely produced by the inner PM rotor and outer PM rotor are given by

$$B_{ri}(r) = \frac{B_r p_i}{1+p_i} \frac{1-\left(\frac{R_a}{R_b}\right)^{p_i+1}}{1-\left(\frac{R_a}{R_d}\right)^{2p_i}} \left[\left(\frac{r}{R_d}\right)^{p_i-1}\left(\frac{R_b}{R_d}\right)^{p_i+1} + \left(\frac{R_b}{r}\right)^{p_i+1}\right]$$
$$\times \cos(p_i(\theta-\omega_i t)+p_i\theta_{i0}) = A(r)\cos(p_i(\theta-\omega_i t)+p_i\theta_{i0}) \tag{7.1}$$

$$B_{\theta i}(r) = \frac{-B_r p_i}{1+p_i} \frac{1-\left(\frac{R_a}{R_b}\right)^{p_i+1}}{1-\left(\frac{R_a}{R_d}\right)^{2p_i}} \left[\left(\frac{r}{R_d}\right)^{p_i-1}\left(\frac{R_b}{R_d}\right)^{p_i+1} - \left(\frac{R_b}{r}\right)^{p_i+1}\right]$$
$$\times \sin(p_i(\theta-\omega_i t)+p_i\theta_{i0}) = B(r)\sin(p_i(\theta-\omega_i t)+p_i\theta_{i0}) \tag{7.2}$$

$$B_{ro}(r) = \frac{-B_r p_o}{1-p_o} \frac{1-\left(\frac{R_c}{R_d}\right)^{p_o-1}}{1-\left(\frac{R_a}{R_d}\right)^{2p_o}} \left[\left(\frac{r}{R_c}\right)^{p_o-1} + \left(\frac{r}{R_c}\right)^{p_o-1}\left(\frac{R_a}{r}\right)^{2p_o}\right]$$
$$\times \cos(p_o(\theta-\omega_o t)+p_o\theta_{o0}) = C(r)\cos(p_o(\theta-\omega_o t)+p_o\theta_{o0}) \tag{7.3}$$

$$B_{\theta o}(r) = \frac{B_r p_o}{1-p_o} \frac{1-\left(\frac{R_c}{R_d}\right)^{p_o-1}}{1-\left(\frac{R_a}{R_d}\right)^{2p_o}} \left[\left(\frac{r}{R_c}\right)^{p_o-1} - \left(\frac{r}{R_c}\right)^{p_o-1}\left(\frac{R_a}{r}\right)^{2p_o}\right]$$
$$\times \sin(p_o(\theta-\omega_o t)+p_o\theta_{o0}) = D(r)\sin(p_o(\theta-\omega_o t)+p_o\theta_{20}) \tag{7.4}$$

where p_i and p_o are the numbers of PM pole-pairs in the inner and outer PM rotors, respectively; R_a and R_b are the inside and outside radii of the inner PM rotor, respectively; R_c and R_d are the inside and outside radii of the outer PM rotor, respectively; ω_i and ω_o are the rotational speeds of the inner and outer PM rotors, respectively; and θ_{i0} and θ_{o0} are their initial phase angles. The coefficients *A, B, C,* and *D* are all functions of the radial distance, *r*, but independent of the position angle, θ.

Taking into account the modulation ring as shown in Figure 7.17, the ferromagnetic segments can be considered to have infinite permeability and hence be equipotential on both inner and outer air-gaps. Their distortion effect on the magnetic field is similar to the slotting effect in PM brushless machines (Wang *et al.*, 2003), which can be approximated by multiplying the original field distribution with a modulating function. Also, since this distortion occurs in two dimensions, the use of two-dimensional complex permeance function (Zarko, Ban, and Lipo, 2006) is preferred. Thus, the magnetic flux density in the presence of the modulation ring can be expressed as

$$B_m = B_o \Lambda^* \tag{7.5}$$

where B_m and B_o are the magnetic flux densities in the air-gaps with and without the stationary modulation ring, respectively and Λ^* is the complex relative air-gap permeance. In polar coordinates, it can be written as

$$\Lambda^* = \Lambda_r \vec{e}_r - \Lambda_\theta \vec{e}_\theta \tag{7.6}$$

For the inner and outer air-gaps, it yields

$$\Lambda_{ri} = \Lambda_{ri0} + \sum_{k=1}^{\infty} \Lambda_{rik} \cos(kN_s\theta) \tag{7.7}$$

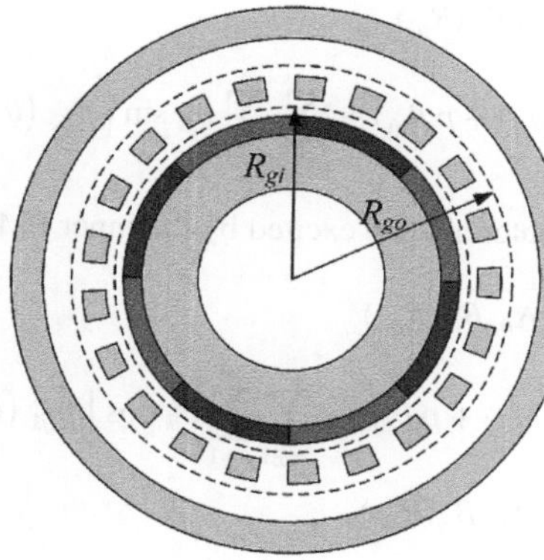

(a)

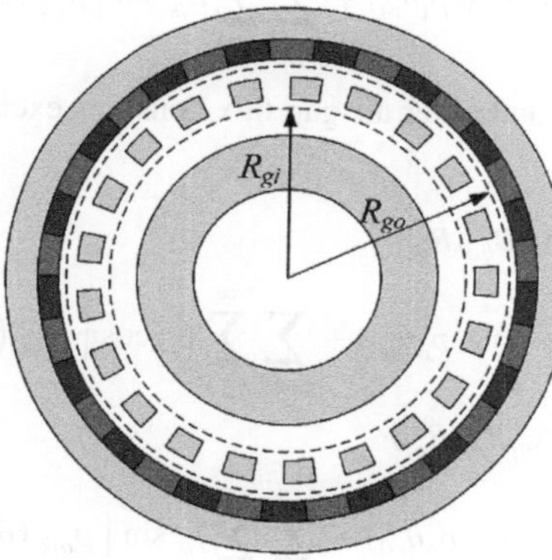

(b)

Figure 7.17 Magnetic fields of magnetic coaxial gear excited by PM rotors with modulation ring: (a) inner rotor and (b) outer rotor

$$\Lambda_{\theta i} = \sum_{k=1}^{\infty} \Lambda_{\theta ik} \sin(kN_s\theta) \tag{7.8}$$

$$\Lambda_{ro} = \Lambda_{ro0} + \sum_{k=1}^{\infty} \Lambda_{rok} \cos(kN_s\theta) \tag{7.9}$$

$$\Lambda_{\theta o} = \sum_{k=1}^{\infty} \Lambda_{\theta ok} \sin(kN_s\theta) \tag{7.10}$$

where N_s is the number of ferromagnetic segments in the modulation ring; Λ_{ri0}, Λ_{rik}, and $\Lambda_{\theta ik}$ are the Fourier coefficients of the inner air-gap permeance; and Λ_{ro0}, Λ_{rok}, and $\Lambda_{\theta ok}$ are the Fourier coefficients of the outer air-gap permeance. Thus, the modulated flux density in the inner air-gap excited by the inner PM rotor can be expressed as

$$\begin{aligned} B_{rii} &= \Lambda_{ri} B_{ri}(R_{gi}) + \Lambda_{\theta i} B_{\theta i}(R_{gi}) \\ &= a_0 \cos(p_i(\theta - \omega_i t) + p_i\theta_{i0}) + \sum_{j=\pm 1} \sum_{k=1}^{\infty} a_{jk} \cos\left[p_{ijk}\left(\theta - \omega_{ijk} t\right) + jp_i\theta_{i0}\right] \end{aligned} \tag{7.11}$$

$$B_{\theta ii} = \Lambda_{ri} B_{\theta i}(R_{gi}) - \Lambda_{\theta i} B_{ri}(R_{gi})$$
$$= b_0 \sin(p_i(\theta - \omega_i t) + p_i \theta_{i0}) + \sum_{j=\pm1} \sum_{k=1}^{\infty} b_{jk} \sin\left[p_{ijk}\left(\theta - \omega_{ijk} t\right) + jp_i \theta_{i0}\right] \tag{7.12}$$

The modulated flux density in the outer air-gap excited by the inner PM rotor can be expressed as

$$B_{roi} = \Lambda_{ro} B_{ri}(R_{go}) + \Lambda_{\theta o} B_{\theta i}(R_{go})$$
$$= c_0 \cos(p_i(\theta - \omega_i t) + p_i \theta_{i0}) + \sum_{j=\pm1} \sum_{k=1}^{\infty} c_{jk} \cos\left[p_{ijk}\left(\theta - \omega_{ijk} t\right) + jp_i \theta_{i0}\right] \tag{7.13}$$

$$B_{\theta oi} = \Lambda_{ro} B_{\theta i}(R_{go}) - \Lambda_{\theta o} B_{ri}(R_{go})$$
$$= d_0 \sin(p_i(\theta - \omega_i t) + p_i \theta_{i0}) + \sum_{j=\pm1} \sum_{k=1}^{\infty} d_{jk} \sin\left[p_{ijk}\left(\theta - \omega_{ijk} t\right) + jp_i \theta_{i0}\right] \tag{7.14}$$

Similarly, the corresponding inner and outer air-gap flux densities excited by the outer PM rotor can be expressed as

$$B_{rio} = \Lambda_{ri} B_{ro}(R_{gi}) + \Lambda_{\theta i} B_{\theta o}(R_{gi})$$
$$= e_0 \cos(p_o(\theta - \omega_o t) + p_o \theta_{o0}) + \sum_{j=\pm1} \sum_{k=1}^{\infty} e_{jk} \cos\left[p_{ojk}\left(\theta - \omega_{ojk} t\right) + jp_o \theta_{o0}\right] \tag{7.15}$$

$$B_{\theta io} = \Lambda_{ri} B_{\theta o}(R_{gi}) - \Lambda_{\theta i} B_{ro}(R_{gi})$$
$$= f_0 \sin(p_o(\theta - \omega_o t) + p_o \theta_{o0}) + \sum_{j=\pm1} \sum_{k=1}^{\infty} f_{jk} \sin\left[p_{ojk}\left(\theta - \omega_{ojk} t\right) + jp_o \theta_{o0}\right] \tag{7.16}$$

$$B_{roo} = \Lambda_{ro} B_{ro}(R_{go}) + \Lambda_{\theta o} B_{\theta o}(R_{go})$$
$$= g_0 \cos(p_o(\theta - \omega_o t) + p_o \theta_{o0}) + \sum_{j=\pm1} \sum_{k=1}^{\infty} g_{jk} \cos\left[p_{ojk}\left(\theta - \omega_{ojk} t\right) + jp_o \theta_{o0}\right] \tag{7.17}$$

$$B_{\theta oo} = \Lambda_{ro} B_{\theta o}(R_{go}) - \Lambda_{\theta o} B_{ro}(R_{go})$$
$$= h_0 \sin(p_o(\theta - \omega_o t) + p_o \theta_{o0}) + \sum_{j=\pm1} \sum_{k=1}^{\infty} h_{jk} \sin\left[p_{ojk}\left(\theta - \omega_{ojk} t\right) + jp_o \theta_{o0}\right] \tag{7.18}$$

where R_{gi} and R_{go} are the radii of the inner and outer air-gaps, respectively; $a_0, b_0, c_0, d_0, e_0, f_0, g_0,$ and h_0 as well as $a_{jk}, b_{jk}, c_{jk}, d_{jk}, e_{jk}, f_{jk}, g_{jk},$ and h_{jk} are all constant coefficients; p_{ijk} and p_{ojk} are the pole-pair numbers of the harmonic components in the inner and outer air-gaps, respectively, which are given by

$$\begin{cases} p_{ijk} = kN_s + jp_i \\ p_{ojk} = kN_s + jp_o \end{cases} \tag{7.19}$$

where $j = \pm 1$ and $k = 1, \ldots, \infty$; ω_{ijk} and ω_{ojk} are the corresponding rotational speeds as given by

$$\begin{cases} \omega_{ijk} = \dfrac{jp_i \omega_i}{kN_s + jp_i} \\ \omega_{ojk} = \dfrac{jp_o \omega_o}{kN_s + jp_o} \end{cases} \tag{7.20}$$

where ω_i and ω_o are the speeds of the inner and outer rotors, respectively. Finally, ignoring the nonlinear factors of the magnetic path, the resultant magnetic fields in the inner and outer air-gaps are obtained as

$$\begin{aligned} B_{ri} &= B_{rii} + B_{rio} \\ &= a_0 \cos(p_i(\theta - \omega_i t) + p_i\theta_{i0}) + e_0 \cos(p_o(\theta - \omega_o t) + p_o\theta_{o0}) \\ &+ \sum_{j=\pm 1} \sum_{k=1}^{\infty} \left[a_{jk} \cos\left(p_{ijk}\left(\theta - \omega_{ijk}t\right) + jp_i\theta_{i0}\right) + e_{jk} \cos(p_{ojk}(\theta - \omega_{ojk}t) + jp_o\theta_{o0})\right] \end{aligned} \tag{7.21}$$

$$\begin{aligned} B_{\theta i} &= B_{\theta ii} + B_{\theta io} \\ &= b_0 \sin(p_i(\theta - \omega_i t) + p_i\theta_{i0}) + f_0 \sin(p_o(\theta - \omega_o t) + p_o\theta_{o0}) \\ &+ \sum_{j=\pm 1} \sum_{k=1}^{\infty} \left[b_{jk} \sin\left(p_{ijk}\left(\theta - \omega_{ijk}t\right) + jp_i\theta_{i0}\right) + f_{jk} \sin(p_{ojk}(\theta - \omega_{ojk}t) + jp_o\theta_{o0})\right] \end{aligned} \tag{7.22}$$

$$\begin{aligned} B_{ro} &= B_{roi} + B_{roo} \\ &= c_0 \cos(p_i(\theta - \omega_i t) + p_i\theta_{i0}) + g_0 \cos(p_o(\theta - \omega_o t) + p_o\theta_{o0}) \\ &+ \sum_{j=\pm 1} \sum_{k=1}^{\infty} \left[c_{jk} \cos\left(p_{ijk}\left(\theta - \omega_{ijk}t\right) + jp_i\theta_{i0}\right) + g_{jk} \cos(p_{ojk}(\theta - \omega_{ojk}t) + jp_o\theta_{o0})\right] \end{aligned} \tag{7.23}$$

$$\begin{aligned} B_{\theta o} &= B_{\theta oi} + B_{\theta oo} \\ &= d_0 \sin(p_i(\theta - \omega_i t) + p_i\theta_{i0}) + h_0 \sin(p_o(\theta - \omega_o t) + p_o\theta_{o0}) \\ &+ \sum_{j=\pm 1} \sum_{k=1}^{\infty} \left[d_{jk} \sin\left(p_{ijk}\left(\theta - \omega_{ijk}t\right) + jp_i\theta_{i0}\right) + h_{jk} \sin(p_{ojk}(\theta - \omega_{ojk}t) + jp_o\theta_{o0})\right] \end{aligned} \tag{7.24}$$

In order to function as a mechanical gear, the magnetic gear should provide stable torque transmission while the rotors rotate at different speeds. To achieve this goal, the combination of (p_i, p_o, N_s) is governed by

$$N_s = p_i + p_o \tag{7.25}$$

The resulting gear ratio, G_r is given by

$$G_r = -\frac{p_o}{p_i} = -\frac{\omega_i}{\omega_o} \tag{7.26}$$

where the negative sign means that the two rotors rotate in opposite directions.

The average magnetic torque developed on the inner rotor can be determined by calculating the Maxwell stress in the inner air-gap:

$$T_{mi} = \frac{L_e R_{gi}^2}{\mu_0} \int_0^{2\pi} B_{ri} B_{\theta i} d\theta \tag{7.27}$$

where L_e is the effective axial length. From Eqs. (7.21) and (7.22), it can be found that both the radial and tangential flux densities in the inner air-gap involve infinite harmonic terms. Each term can be generally expressed as

$$b_{ri} = \delta \cos(p_m\theta + \alpha) \tag{7.28}$$

$$b_{\theta i} = \varepsilon \sin(p_n\theta + \beta) \tag{7.29}$$

where p_m and p_n are the pole-pair numbers; α and β are the phase angles; and δ and ε are constant coefficients. Thus, it yields

$$\int_0^{2\pi} b_{ri} b_{\theta i} d\theta = \int_0^{2\pi} \delta\varepsilon \cos(p_m\theta + \alpha)\sin(p_n\theta + \beta) d\theta$$
$$= \begin{cases} 0 & p_m \neq p_n \\ \delta\varepsilon\pi \sin(\beta - \alpha) & p_m = p_n \end{cases} \quad (7.30)$$

It can be found that the product term $b_{ri}b_{\theta i}$ can contribute to the average magnetic torque if it satisfies two rules:

- R1: b_{ri} and $b_{\theta i}$ have the same pole-pair number.
- R2: b_{ri} and $b_{\theta i}$ have different phase angles.

Therefore, based on Eqs. (7.19), (7.20), (7.25), and (7.26), and the rules R1 and R2, the product term, $b_{ri}b_{\theta i}$ that has contribution to the average magnetic torque can be deduced, which is denoted as the effective harmonic. Consequently, the pole-pair numbers of the effective harmonics in the inner air-gap and their contributed magnetic torques are obtained as listed in Table 7.2.

Similarly, the average magnetic torque developed on the outer rotor can be determined by calculating the Maxwell stress in the outer air-gap:

$$T_{mo} = \frac{L_e R_{go}^2}{\mu_0} \int_0^{2\pi} B_{ro} B_{\theta o} d\theta \quad (7.31)$$

Consequently, the pole-pair numbers of the effective harmonics in the outer air-gap and their contributed magnetic torques can be deduced, which are summarized in Table 7.3.

Table 7.2 Magnetic torque constitution in inner air-gap of magnetic coaxial gear

Pole-pair number	Magnetic torque contributed
$p_i = p_{o-11}$	$\frac{L_e R_{gi}^2}{\mu_0}\left[\pi\left(a_0 f_{-11} - b_0 e_{-11}\right)\sin(p_o\theta_{o0} - p_i\theta_{i0})\right]$
$p_o = p_{i-11}$	$\frac{L_e R_{gi}^2}{\mu_0}\left[\pi\left(f_0 a_{-11} - e_0 b_{-11}\right)\sin(p_o\theta_{o0} - p_i\theta_{i0})\right]$
$p_{o-1(k+1)} = p_{i1k}$ $(k = 1, \ldots, \infty)$	$\frac{L_e R_{gi}^2}{\mu_0}\left[\pi\left(a_{1k} f_{-1(k+1)} - b_{1k} e_{-1(k+1)}\right)\sin(p_o\theta_{o0} - p_i\theta_{i0})\right]$
$p_{i-1(k+1)} = p_{o1k}$ $(k = 1, \ldots, \infty)$	$\frac{L_e R_{gi}^2}{\mu_0}\left[\pi\left(f_{1k} a_{-1(k+1)} - e_{1k} b_{-1(k+1)}\right)\sin(p_o\theta_{o0} - p_i\theta_{i0})\right]$

Table 7.3 Magnetic torque constitution in outer air-gap of magnetic coaxial gear

Pole-pair number	Magnetic torque contributed
$p_i = p_{o-11}$	$\frac{L_e R_{go}^2}{\mu_0}\left[\pi\left(c_0 h_{-11} - d_0 g_{-11}\right)\sin(p_o\theta_{o0} - p_i\theta_{i0})\right]$
$p_o = p_{i-11}$	$\frac{L_e R_{go}^2}{\mu_0}\left[\pi\left(h_0 c_{-11} - g_0 d_{-11}\right)\sin(p_o\theta_{o0} - p_i\theta_{i0})\right]$
$p_{o-1(k+1)} = p_{i1k}$ $(k = 1, \ldots, \infty)$	$\frac{L_e R_{go}^2}{\mu_0}\left[\pi\left(c_{1k} h_{-1(k+1)} - d_{1k} g_{-1(k+1)}\right)\sin(p_o\theta_{o0} - p_i\theta_{i0})\right]$
$p_{i-1(k+1)} = p_{o1k}$ $(k = 1, \cdots, \infty)$	$\frac{L_e R_{go}^2}{\mu_0}\left[\pi\left(h_{1k} c_{-1(k+1)} - g_{1k} d_{-1(k+1)}\right)\sin(p_o\theta_{o0} - p_i\theta_{i0})\right]$

Both Tables 7.2 and 7.3 illustrate that the magnetic torques developed on the inner and outer rotors are sinusoidal functions of the initial phase angle difference of the two rotors. Since they are independent of time, the magnetic gear can transmit stable magnetic torque while realizing speed variation as governed by Eq. (7.26).

7.3.2 *Modeling of MG Machines*

Since the magnetic gear only provides the torque transmission function, the mathematical modeling of the MG machine depends on the machine that is integrated with the magnetic gear. While the magnetic gear inevitably utilizes the rare-earth PM material, the machine part naturally adopts the PM brushless machine. This PM brushless machine can be sinusoidal-fed (PM brushless AC machine) or rectangular-fed (PM brushless DC machine).

The MG PM brushless DC machine has been identified as the most promising candidate for in-wheel motor drive application. It takes the definite advantages of high torque density and simple control. The corresponding modeling is similar to that of a standard PM brushless DC machine as given by

$$\begin{bmatrix} u_a \\ u_b \\ u_c \end{bmatrix} = \begin{bmatrix} R & 0 & 0 \\ 0 & R & 0 \\ 0 & 0 & R \end{bmatrix} \begin{bmatrix} i_a \\ i_b \\ i_c \end{bmatrix} + \begin{bmatrix} L-M & 0 & 0 \\ 0 & L-M & 0 \\ 0 & 0 & L-M \end{bmatrix} p \begin{bmatrix} i_a \\ i_b \\ i_c \end{bmatrix} + \begin{bmatrix} e_a \\ e_b \\ e_c \end{bmatrix} \tag{7.32}$$

where $[u_a, u_b, u_c]$, $[e_a, e_b, e_c]$, and $[i_a, i_b, i_c]$ are the stator voltage vector, back electromotive force (EMF) vector, and armature current vector of the phases *A, B,* and *C*, respectively, and *R, L,* and *M* are the phase resistance, self-inductance, and mutual-inductance, respectively.

The electromagnetic torque of this PM brushless DC machine is given by

$$T_e = (e_a i_a + e_b i_b + e_c i_c)/\omega_i \tag{7.33}$$

which serves to drive the inner rotor of the magnetic coaxial gear. Hence, the motions of the inner and outer rotors are governed by

$$J_i \frac{d\omega_i}{dt} = T_e - T_{mi} - B_i \omega_i \tag{7.34}$$

$$J_o \frac{d\omega_o}{dt} = T_{mo} - T_L - B_o \omega_o \tag{7.35}$$

where J_i, J_o and B_i, B_o are the moments of inertia and viscous friction coefficients of the inner and outer rotors, respectively, and T_{mi} and T_{mo} are the developed magnetic torques on the inner and outer rotors, respectively.

7.4 Inverters for MG Motors

As mentioned earlier, the magnetic gear of the MG machine for motoring operation is essentially passive so that the power converter for the MG motor is governed by the machine itself. Again, since the natural choice of the machine for the MG motor is the PM brushless machine, the traditional inverter can readily be used.

Although the inverter can be voltage-fed or current-fed, the latter one is seldom used because it usually needs to install a large series inductance to emulate the current source. In fact, the voltage-fed inverter shown in Figure 7.18 is almost exclusively used because of its simple topology and inherent bidirectional power flow capability. Meanwhile, this voltage-fed inverter can be either voltage-controlled or current-controlled. The former takes the definite advantage of simple voltage control, whereas the latter offers the distinct advantage of direct current control. For EV propulsion, the current-controlled inverter is preferred to provide the desired current and torque responses for high-performance operation.

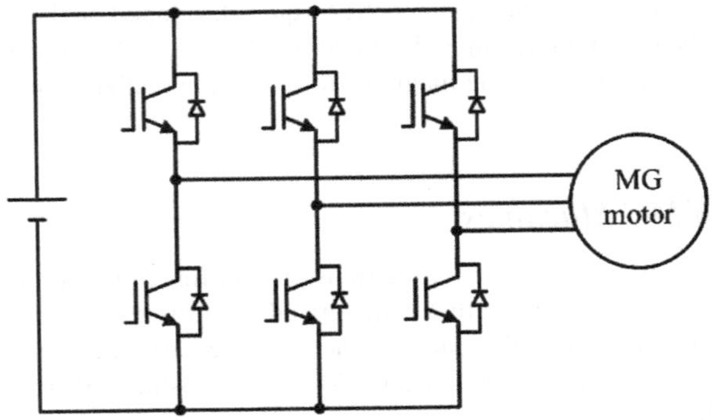

Figure 7.18 Three-phase full-bridge voltage-fed inverter

7.5 MG Motor Control

Similar to the modeling and inverter for the MG machine, the control strategy depends on the machine itself. Apart from the control strategies for the PM brushless motor drive that have been discussed in detail in Chapter 4, four emerging control strategies, namely the efficiency optimizing control, direct torque control, artificial intelligent control, and position sensorless control, are briefly discussed and compared for application to the MG motor drive (Chau, Chan, and Liu, 2008).

Different motor drives may employ different ways for efficiency optimization. When the MG motor drive adopts the PM brushless AC machine, the efficiency optimizing control strategy is particularly attractive for EVs because the EV driving range per charge is still quite limited. An improvement in efficiency for electric propulsion can significantly lengthen the driving range. Figure 7.19a depicts the control strategy in which the efficiency optimization is achieved by online tuning the d-axis armature current, I_d in such a way that the total loss of the motor drive can be minimized (Cavallaro *et al.*, 2005). The corresponding loss, P_{loss} can be represented by

$$P_{\text{loss}}(I_d, T, \omega) = P_{\text{Cu}}(I_d, T, \omega) + P_{\text{Fe}}(I_d, T, \omega) \tag{7.36}$$

where P_{Cu} is the copper loss and P_{Fe} is the iron loss for the given torque T and speed ω. As shown in Figure 7.19b, there is a unique d-axis armature current for a given operating point (T, ω) to achieve the optimal efficiency. In particular, the minimum total loss occurs at a lower d-axis armature current than that of the minimum copper loss, hence illustrating that the maximum torque per ampere control cannot maximize the overall efficiency.

Direct torque control is becoming attractive for EVs, particularly for in-wheel motor propulsion, which desires fast torque response. It takes the definite advantages that it does not rely on current control and less depends on parameters. When the MG motor drive adopts the PM brushless AC machine, the direct torque controller adjusts the torque and flux linkage independently. The controller outputs proper voltage vectors via the inverter in such a way that these two variables are forced to follow the predefined trajectories (Pascas and Weber, 2005). Figure 7.20 shows the corresponding control block diagram and its typical torque tracking performance.

Artificial intelligence-based control strategies, such as the self-tuning control, fuzzy logic control, neural network control, neuro-fuzzy control, and genetic algorithm control, are becoming attractive for motion control. Among them, the self-tuning control and fuzzy logic control are relatively mature. They are particularly attractive for the MG motor drive because they can effectively handle system nonlinearities and sensitivities to parameter variations under harsh working environment in EVs. Figure 7.21 shows the

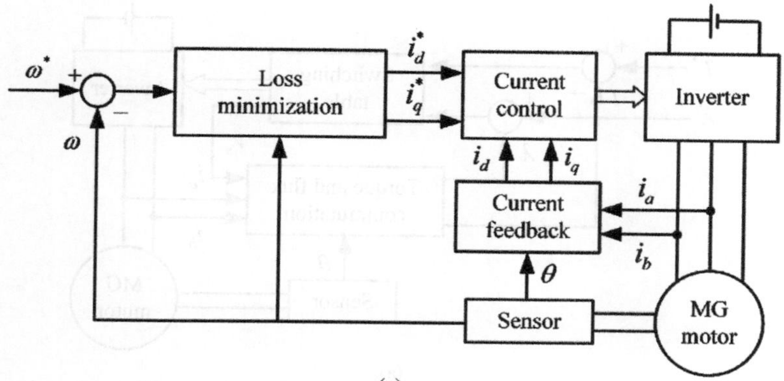

(a)

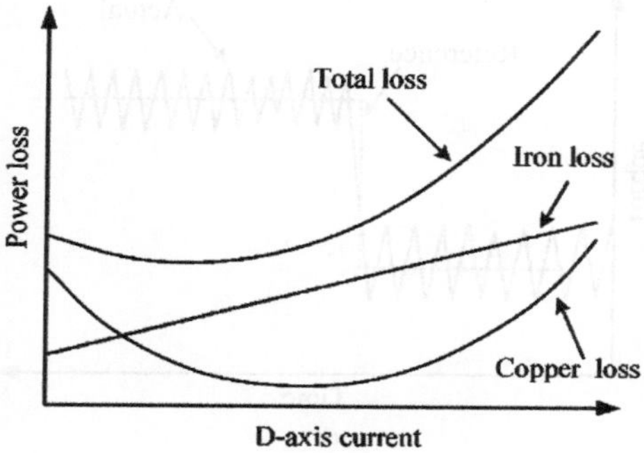

(b)

Figure 7.19 Efficiency optimizing control: (a) control block diagram and (b) loss minimization

self-tuning (ST) fuzzy proportional-integral (PI) control block diagram and its typical transient speed response as compared with that of the conventional PI control (Cheng, Sun, and Zhou, 2006).

In order to achieve high performance for EV propulsion, the MG motor drive desires accurate position feedback. In order to get rid of the costly and bulky position encoder, the use of position sensorless control is becoming attractive. There are various position estimation techniques that can mainly be classified as the motional EMF, inductance variation, and flux-linkage variation. Basically, the position information is derived by online analysis of the measurable voltages and currents of the motor. Figure 7.22 shows the corresponding control block diagram and the estimated speed with respect to the actual one (Silva, Asher, and Sumner, 2006). It should be noted that the position encoder is almost mandatory for EV motor drives because maturity and reliability are the major concerns, and the cost of position encoder is in fact insignificant compared with the total EV propulsion cost. Nevertheless, with ever-increasing concern on EV fault-tolerant operation, the position sensorless control can play an important role in fault-tolerant operation when the position encoder is malfunction or under fault condition.

7.6 Design Criteria of MG Motor Drives for EVs

In order to kick off the design of MG motor drive, the initialization of some basic dimensions of the MG machine is important. There are some design criteria for this MG machine, aiming to achieve torque density over 100 N m/l (Rasmussen *et al.*, 2009):

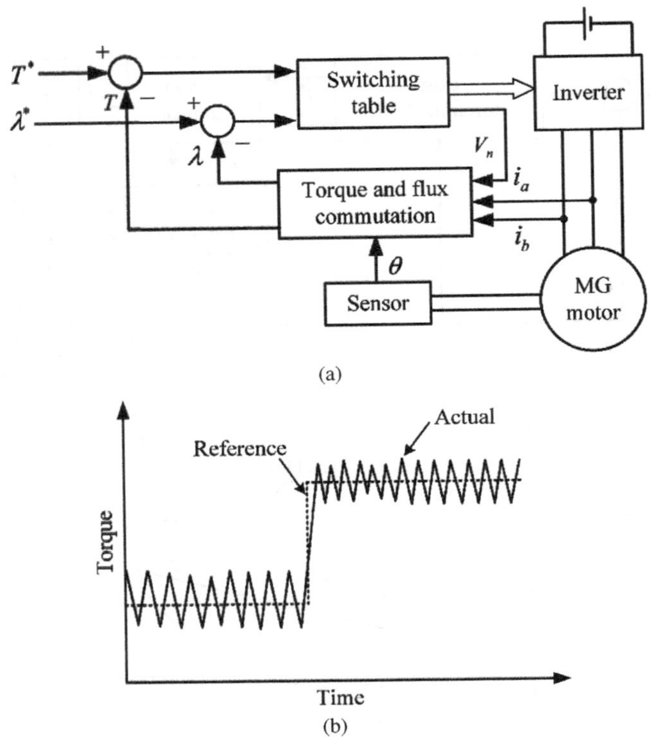

Figure 7.20 Direct torque control: (a) control block diagram and (b) torque tracking

- The PM thickness in the high-speed inner rotor is more important than that in the low-speed outer rotor in terms of the torque density. It is due to the fact that the PM pole arc in the outer rotor is so short that thicker PMs will not significantly increase the flux density in the presence of larger flux leakage; meanwhile, the PM pole arc in the inner rotor is relatively long so that thicker PMs will actually improve the flux density. While adopting thicker PMs in the inner rotor to improve the torque density, the corresponding centrifugal forces under high-speed operation are concerned.
- As the ferromagnetic segments of the modulation ring are exposed under changing magnetic field, they have to be laminated to avoid extensive iron losses. Between those laminated segments, nonmagnetic material such as plastics should be inserted in order to improve the mechanical integrity without disturbing the magnetic field.
- In order to fix the PMs on the inner and outer rotors, high-tensile-strength adhesives are usually employed. When high centrifugal forces are anticipated, a nonmagnetic retaining can is used to secure the PMs on the high-speed rotor. Nevertheless, the use of retaining can will inevitably increase the effective air-gap length and hence degrade the torque density.
- The yoke of the inner rotor of the magnetic gear, which is also the outer rotor of the electric machine part, should be thick enough to decouple the magnetic fields between the electric machine part and magnetic gear. Otherwise, the speed or torque control of the electric machine part will be affected by the magnetic gear.
- Since the machine part is generally designed to offer flux-weakening operation, the corresponding stack length can be shorter than the magnetic gear. By adopting such unequal axial lengths, there is no need to spend extra space to accommodate the end-windings of the machine part.

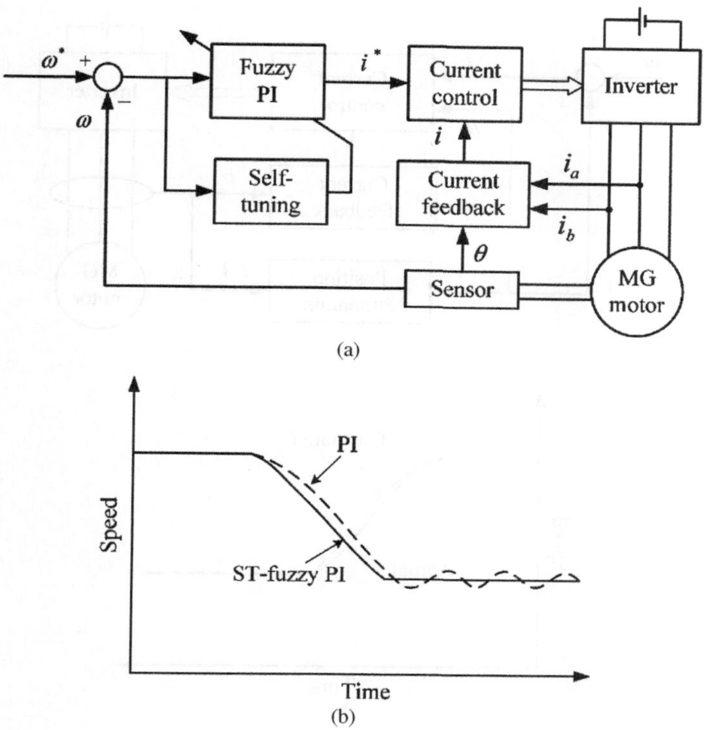

Figure 7.21 Self-tuning fuzzy PI control: (a) control block diagram and (b) speed response

- Since the MG machine generally adopts the inner-stator arrangement, the heat dissipation is a challenge. In the absence of special cooling, the current density of the armature winding should be limited to 5 A/mm^2 and the slot fill factor should not be higher than 60%.

7.7 Design Examples of MG Motor Drives for EVs

Since the magnetic gear inevitably uses PMs, the machine part of the MG motor naturally adopts the PM brushless machine. As the MG motor drive possesses the definite merit of gearing effect, the designs of two MG PM brushless motor drives for in-wheel EV propulsion are used for exemplification.

7.7.1 MG PM Brushless DC In-Wheel Motor Drive

Figure 7.23 shows the MG PM brushless DC in-wheel motor (Jian, 2010) in which the outer-rotor PM brushless DC motor is artfully integrated with the magnetic coaxial gear to achieve compact construction. It consists of four main parts: the motor stator, the motor outer rotor, which is also the gear inner rotor, the modulation ring, and the gear outer rotor. There are totally three air-gaps that separate these four parts from one another. In order to provide magnetic paths while reducing eddy current loss, the modulation ring is made up of laminated ferromagnetic segments with epoxy filled between them to enforce the structural strength for high torque transmission. Both the gear inner rotor and outer rotor are supported by bearings on the shafts to guarantee free rotations. For direct driving, the tire rim is directly mounted onto the gear outer rotor. The inner rotor is designed like a cup with PMs mounted on its inner and outer surfaces. In

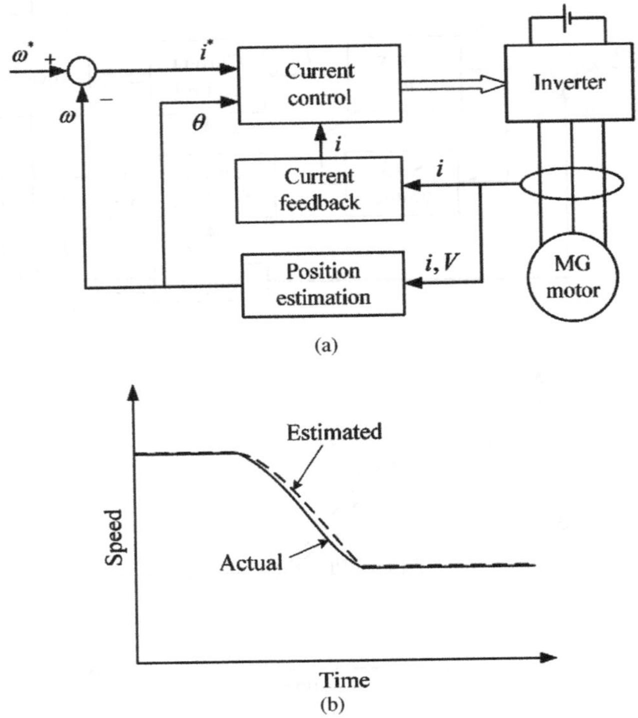

Figure 7.22 Position sensorless control: (a) control block diagram and (b) estimated speed

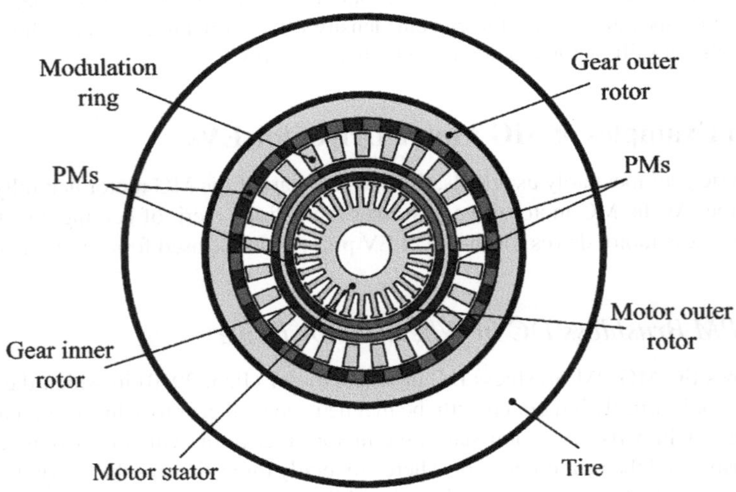

Figure 7.23 MG PM brushless DC in-wheel motor configuration

Table 7.4 Design parameters of MG PM brushless DC in-wheel motor

Rated power	5 kW
Rated frequency	220 Hz
Rated gear inner-rotor speed	4400 rpm
Rated gear outer-rotor speed	600 rpm
Number of phases	3
Number of stator slots	27
Number of inner-rotor pole-pairs	3
Number of outer-rotor pole-pairs	22
Number of ferromagnetic segments	25
Outside diameter of stator	120 mm
Inside diameter of inner rotor	121.2 mm
Outside diameter of inner rotor	142.8 mm
Inside diameter of modulation ring	144 mm
Outside diameter of modulation ring	170 mm
Inside diameter of outer rotor	172 mm
Outside diameter of outer rotor	184 mm
Axial length	100 mm
PM material	Nd-Fe-B

addition, PMs are attached on the inner surface of the gear outer rotor. All the PMs are radially magnetized, and the PM pole-pair numbers of the inner and outer rotors are 3 and 22, respectively, while the number of ferromagnetic segments is 25. Thus a gear ratio of −22 : 3 is resulted. The key design parameters are listed in Table 7.4.

In order to take into account the saturation and fringing effects, finite element analysis is employed. The electromagnetic field distributions of this MG PM brushless DC in-wheel motor under no-load and full-load conditions are shown in Figure 7.24. It can be observed that most of flux lines can pass through all three air-gaps, thus contributing to torque transmission. There are also some flux lines directly turning around the boundaries of adjacent PMs, which are useless for torque transmission and arouse power loss. Moreover, by comparing the field distributions at no load and full load, it can be observed that the armature reaction has only a slight effect on the magnetic gear. Figure 7.25 illustrates the no-load radial flux density waveforms in the inner, middle, and outer air-gaps. It can be found that the effective harmonics of the flux densities in the inner and middle air-gaps has 3 pole-pairs whereas that in the outer air-gap has 22 pole-pairs, thus verifying the modulating effect of the modulation ring.

On the basis of the inner air-gap flux density distribution, the back EMF induced in the three-phase armature winding of the stator can readily be deduced. Figure 7.26 shows the three-phase EMF waveforms when the inner rotor is rotating at a rated speed of 4400 rpm. As these back EMF waveforms are essentially trapezoidal, the PM brushless DC motor operates in the 120° conduction mode. Thus, by applying rectangular phase current with 120° conduction, and at any time two phases are conducted, the steady torque can be developed.

Consequently, the dynamic performances of this magnetic-geared PM brushless DC motor drive are assessed. Figure 7.27 shows the start-up speed responses under the motor speed command of 2000 rpm. It can be observed that the steady-state outer-rotor speed of the gear is reduced by about 7.3 times as compared with the steady-state inner-rotor speed of the gear. As they rotate in opposite directions, it well agrees with the designed gear ratio of −22/3. Moreover, Figure 7.28 shows the steady-state torque responses, which illustrate that the torque magnitude developed by the PM brushless DC motor can be successfully amplified by about 7.3 times. Notice that these torque responses are in phase but in opposite directions.

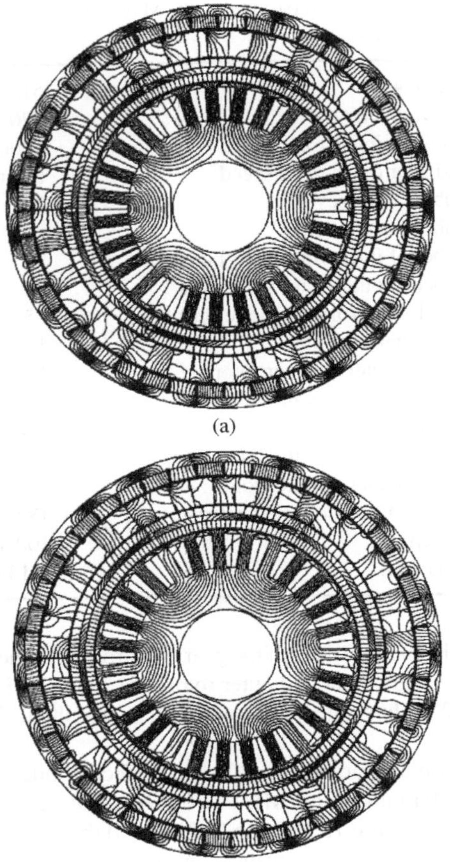

(a)

(b)

Figure 7.24 Electromagnetic field distributions of MG PM brushless DC in-wheel motor: (a) no load and (b) full load

7.7.2 MG PM Brushless AC In-Wheel Motor Drive

In order to further improve the performance of MG PM brushless in-wheel motor drives, the Halbach-array PMs can be employed to constitute the magnetic poles on both the inner and outer rotors (Jian and Chau, 2009b; Jian, 2010). The key is to utilize the unique merits of Halbach-PMs, namely the sinusoidal field distribution, strong magnetic field intensity, and self-shielding effect of magnetization. Because of the strong magnetic field intensity with sinusoidal field distribution, the use of Halbach-PMs for the MG motor can improve the torque density and reduce the cogging torque. Also, because of the self-shielding effect of magnetization, the use of Halbach-PMs can help decouple the electromagnetic fields between the inner motor field and the outer gear field. Moreover, since the magnetic field distribution in the inner air-gap is inherently sinusoidal, the motor operates well in the PM brushless AC mode than in the PM brushless DC mode.

Figure 7.29 shows the MG PM brushless AC motor for in-wheel application. It has a similar configuration as the aforementioned MG PM brushless DC motor. The key difference is that the Halbach-PMs are adopted on both the inner and outer rotors to constitute the desired numbers of magnetic poles. There are three PM pole-pairs in the motor outer rotor and three PM pole-pairs in the gear inner rotor, which

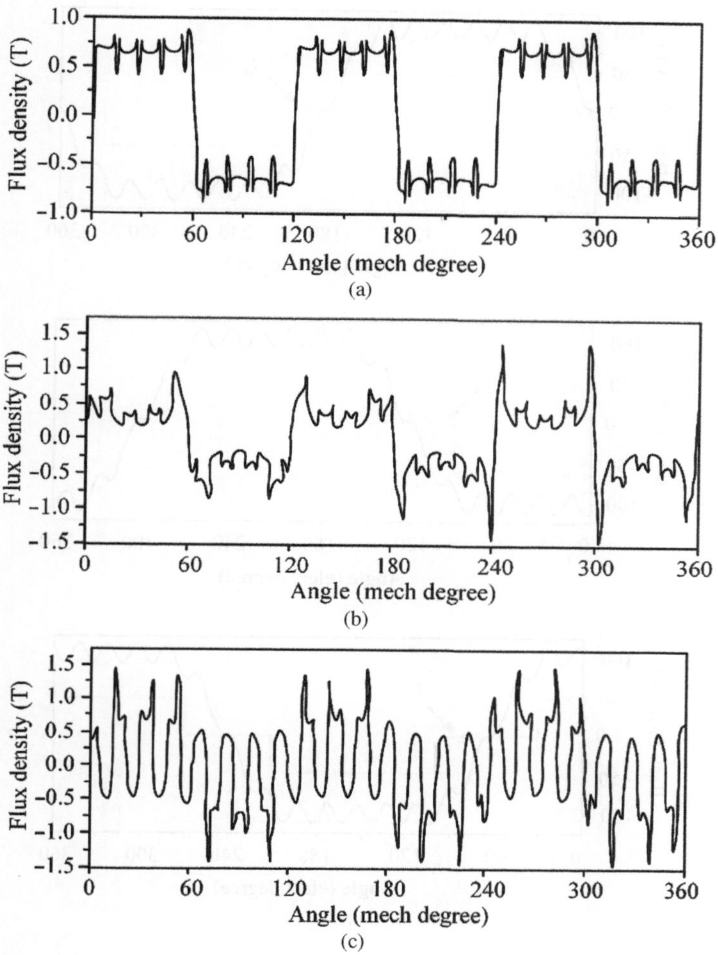

Figure 7.25 Radial flux densities of MG PM brushless DC in-wheel motor at no load: (a) inner air-gap, (b) middle air-gap, and (c) outer air-gap

are physically mounted back-to-back together. Meanwhile, there are 16 PM pole-pairs mounted on the inner surface of the gear outer rotor. According to Eq. (7.25), there are 19 ferromagnetic segments in the modulation ring. For the Halbach-PMs, the more the number of PM pieces to constitute a magnetic pole, the more the sinusoidal field distribution is resulted, but suffering from higher manufacturing difficulty. Particularly, since the PM pole arc in the gear outer rotor is generally shorter than that in the gear inner rotor or motor outer rotor, the number of PM pieces per magnetic pole in the gear outer rotor is usually smaller than that in the gear inner rotor or motor outer rotor. So, in this MG motor, there are two PM pieces per magnetic pole in the gear outer rotor, four PM pieces per magnetic pole in the gear inner rotor and also four PM pieces per magnetic pole in the motor outer rotor. Consequently, there are totally 24 PM pieces in the motor outer rotor, also 24 PM pieces in the gear inner rotor and 64 PM pieces in the gear outer rotor, which definitely causes complication in manufacture.

Besides using the Halbach-PMs with different numbers of pole-pairs in the PM rotors, the design parameters of this MG PM brushless AC in-wheel motor are similar with the previous MG PM brushless DC

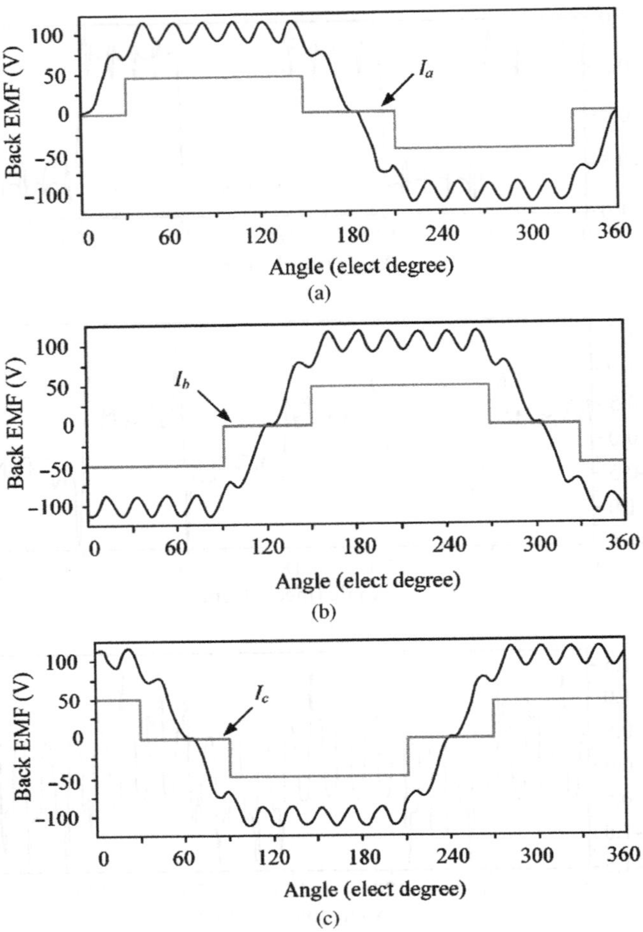

Figure 7.26 Back EMF waveforms of MG PM brushless DC in-wheel motor: (a) phase A, (b) phase B, and (c) phase C

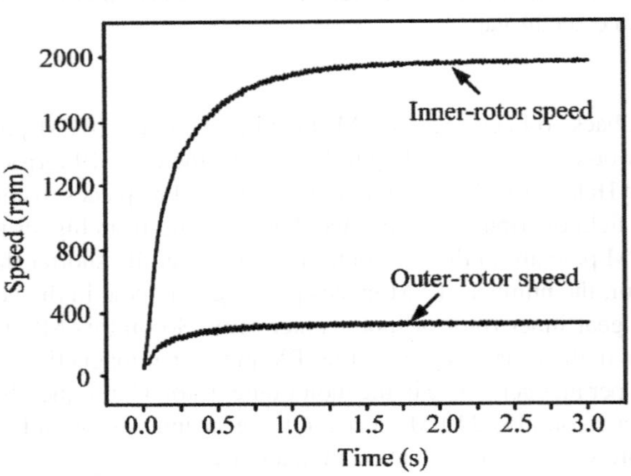

Figure 7.27 Start-up speed responses of MG PM brushless DC in-wheel motor

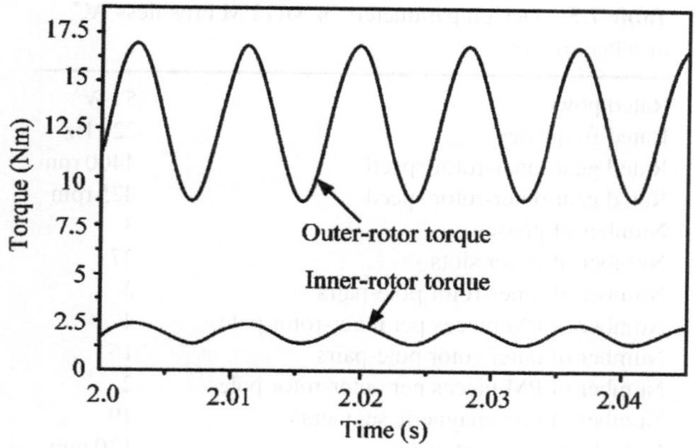

Figure 7.28 Steady-state torque responses of MG PM brushless DC in-wheel motor

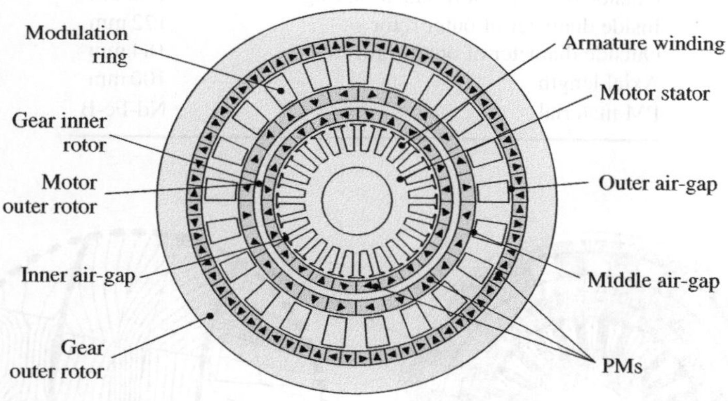

Figure 7.29 MG PM brushless AC in-wheel motor configuration with Halbach-PMs

one, which are summarized in Table 7.5. By using finite element analysis, the magnetic field distribution of this MG motor is depicted in Figure 7.30. Thanks to the self-shielding of magnetization of the Halbach-PMs, it can be observed that the magnetic field of the motor outer rotor and the magnetic field of the gear inner rotor are effectively decoupled. As there is no need to use a thick yoke between the motor outer rotor and gear inner rotor to facilitate such magnetic decoupling, the overall size, and weight of this MG motor can be further reduced.

Figure 7.31 shows the radial flux density waveform in the inner air-gap of the MG motor. It can be observed that it exhibits a near-sinusoidal waveform, which is actually due to the feature of sinusoidal field distribution of the Halbach-PMs. The corresponding harmonic spectrum is also plotted in Figure 7.32. It can be found that the effective harmonic component of three pole-pairs is highly dominant while the other ineffective harmonic components of various pole-pairs are dramatically suppressed due to the merit of sinusoidal field distribution. This can significantly reduce the cogging torque caused by the stator slots. Also, because of the sinusoidal field distribution, the back EMF waveforms are sinusoidal as depicted in Figure 7.33, which actually confirms that this MG motor operates well in the brushless AC mode than in the brushless DC mode.

Table 7.5 Design parameters of MG PM brushless AC in-wheel motor

Rated power	5 kW
Rated frequency	220 Hz
Rated gear inner-rotor speed	4400 rpm
Rated gear outer-rotor speed	825 rpm
Number of phases	3
Number of stator slots	27
Number of inner-rotor pole-pairs	3
Number of PM pieces per inner-rotor pole	4
Number of outer-rotor pole-pairs	16
Number of PM pieces per outer-rotor pole	2
Number of ferromagnetic segments	19
Outside diameter of stator	120 mm
Inside diameter of inner rotor	121.2 mm
Outside diameter of inner rotor	142.8 mm
Inside diameter of modulation ring	144 mm
Outside diameter of modulation ring	170 mm
Inside diameter of outer rotor	172 mm
Outside diameter of outer rotor	184 mm
Axial length	100 mm
PM material	Nd-Fe-B

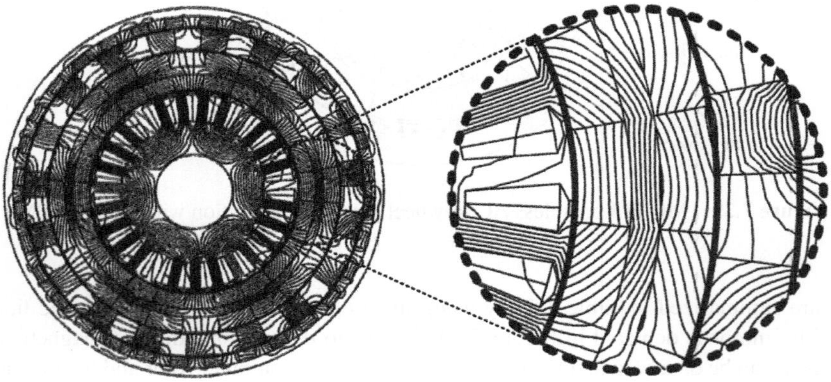

Figure 7.30 Magnetic field distribution of MG PM brushless AC in-wheel motor with Halbach-PMs

By calculating Maxwell's stress tensors in the inner and outer air-gaps, the torque–angle relationships on both the inner and outer rotors can be obtained as depicted in Figure 7.34. As expected, with the use of Halbach-PMs, the two torque–angle curves are very sinusoidal with insignificant torque ripples. It can be observed that the torque magnitude of the outer rotor is amplified by about 5.3 times as compared with that of the inner rotor, which well agrees with the designed gear ratio of $-16/3$.

Because of the self-shielding magnetization of the Halbach-PMs, the outer rotor can adopt the yokeless design. That is, the outer-rotor iron core can be replaced by nonmagnetic material such as aluminum, thereby eliminating the corresponding iron loss and further reducing the overall size and weight.

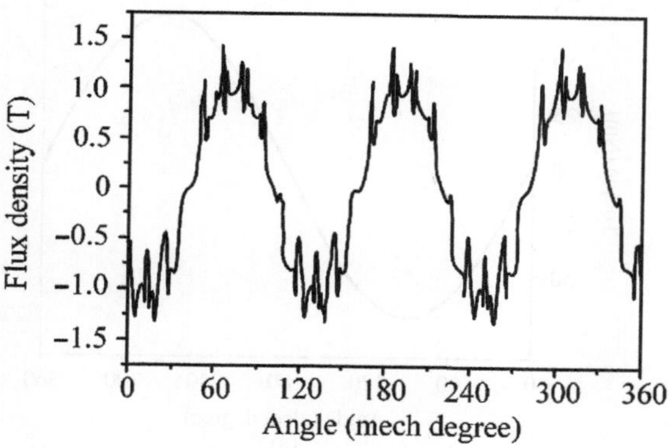

Figure 7.31 Radial flux density waveform in inner air-gap of MG PM brushless AC in-wheel motor with Halbach-PMs

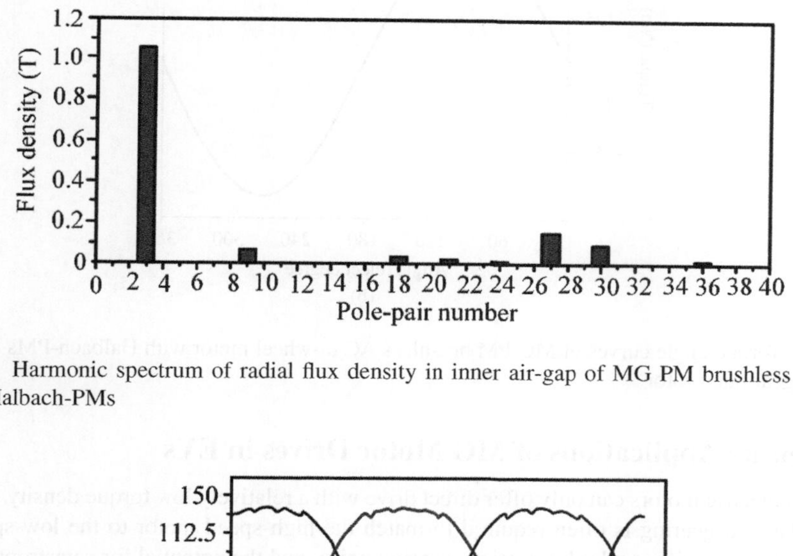

Figure 7.32 Harmonic spectrum of radial flux density in inner air-gap of MG PM brushless AC in-wheel motor with Halbach-PMs

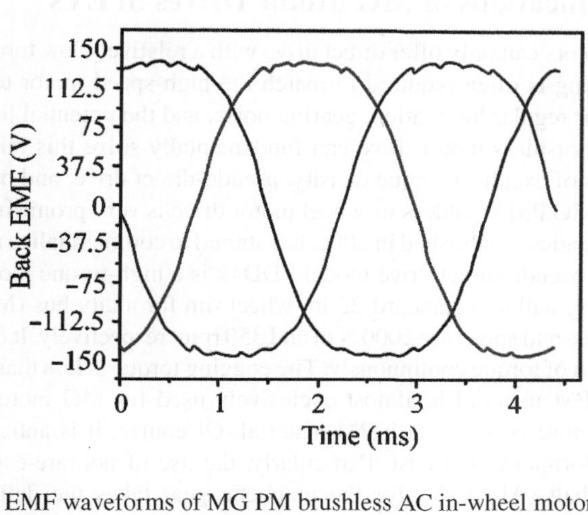

Figure 7.33 Back EMF waveforms of MG PM brushless AC in-wheel motor with Halbach-PMs

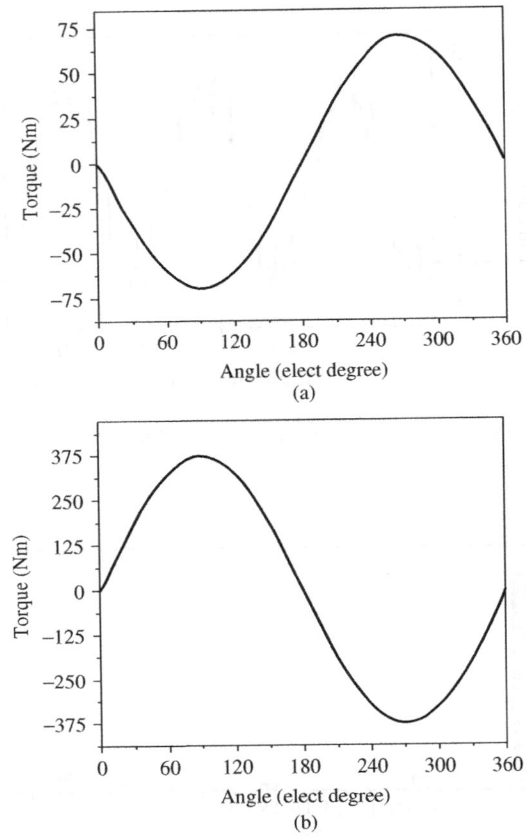

Figure 7.34 Torque-angle curves of MG PM brushless AC in-wheel motor with Halbach-PMs (a) gear inner rotor and (b) gear outer rotor

7.8 Potential Applications of MG Motor Drives in EVs

Conventional electric motors can only offer direct drive with a relatively low torque density. For compact design, mechanical gearing is often required to match the high-speed motor to the low-speed motion, which causes the need for regular lubrication, gearing noise, and the potential for catastrophic failure on overload. The MG PM brushless motor drive can fundamentally solve this mismatch problem. It possesses the definite merits of excellent torque density, pseudo-direct drive, and no need for lubrication or maintenance. Thus, the MG PM brushless in-wheel motor drive is very promising for EV propulsion.

Very recently, Magnomatics, established in 2006, has started to commercialize research on MG in-wheel motor drive for EVs. Its pseudo-direct drive model PDD4k is a high-torque motor drive rated at 65 kW, which is capable of fitting within a standard 22-in. wheel rim for a city bus (Magnomatics, 2014). The corresponding rated torque and speed are 2000 N m and 350 rpm, respectively. It can produce a peak power of 120 kW up to 4000 N m of torque continuously. The cogging torque is less than 0.5% of the rated value.

While the rare-earth PM material is almost exclusively used for MG motor drives, there is a possibility to make use of those nonrare-earth PM material. Of course, it is anticipated that there will be a trade-off between performance and cost. Particularly, the use of nonrare-earth PM material such as the aluminum-nickel-cobalt (Al-Ni-Co) for the magnetic gear takes the definite advantages of high cost-effectiveness and high tolerance for high-temperature operation (Chen *et al.*, 2014).

References

Atallah, K., Calverley, S.D. and Howe, D. (2004) Design, analysis and realization of a high-performance magnetic gear. *IEE Proceedings – Electric Power Applications*, **151**, 135–143.

Atallah, K. and Howe, D. (2001) A novel high-performance magnetic gear. *IEEE Transactions on Magnetics*, **37**, 2844–2846.

Atallah, K., Wang, J. and Howe, D. (2005) A high-performance linear magnetic gear. *Journal of Applied Physics*, **97**, 10N516-1–10N516-3.

Cavallaro, C., Tommaso, A.O.D., Miceli, R. *et al.* (2005) Efficiency enhancement of permanent-magnet synchronous motor drives by online loss minimization approaches. *IEEE Transactions on Industrial Electronics*, **52**, 1153–1160.

Chau, K.T., Chan, C.C. and Liu, C. (2008) Overview of permanent-magnet brushless drives for electric and hybrid electric vehicles. *IEEE Transactions on Industrial Electronics*, **55**, 2246–2257.

Chen, M., Chau, K.T., Li, W. and Liu, C. (2014) Cost-effectiveness comparison of coaxial magnetic gears with different magnet materials. *IEEE Transactions on Magnetics*, **50**, 7020304-1–7020304-4.

Cheng, M., Sun, Q. and Zhou, E. (2006) New self-tuning fuzzy PI control of a novel doubly salient permanent-magnet motor drive. *IEEE Transactions on Industrial Electronics*, **53**, 814–821.

Faus, H.T. (1941) Magnet gearing. US Patent 2 243 555.

Halbach, K. (1980) Design of permanent multipole magnets with oriented rare earth cobalt material. *Nuclear Instruments and Methods*, **169**, 1–10.

Hesmondhalgh, D.E. and Tipping, D. (1980) A multielement magnetic gear. *IEE Proceedings*, **127**, 129–138.

Huang, C.C., Tsai, M.C., Dorrell, D.G. and Lin, B.J. (2008) Development of a magnetic planetary gearbox. *IEEE Transactions on Magnetics*, **44**, 403–412.

Jian, L. (2010) Design, analysis and application of coaxial magnetic gears. PhD. thesis. The University of Hong Kong, Hong Kong.

Jian, L. and Chau, K.T. (2009a) Analytical calculation of magnetic field distribution in coaxial magnetic gears. *Progress in Electromagnetics Research*, **92**, 1–16.

Jian, L. and Chau, K.T. (2009b) Design and analysis of an integrated Halbach-magnetic-geared permanent-magnet motor for electric vehicles. *Journal of Asian Electric Vehicles*, **7**, 1213–1219.

Jian, L. and Chau, K.T. (2010) A coaxial magnetic gear with Halbach permanent-magnet arrays. *IEEE Transactions on Energy Conversion*, **25**, 319–328.

Jian, L., Chau, K.T., Gong, Y. *et al.* (2009) Comparison of coaxial magnetic gears with different topologies. *IEEE Transactions on Magnetics*, **45**, 4526–4529.

Jorgensen, F.T., Andersen, T.O., and Rasmussen, P.O. (2005) Two dimensional model of a permanent magnet spur gear. Proceedings of IEEE Industry Applications Conference, pp. 261–265.

Kikuchi, S. and Tsurumoto, K. (1993) Design and characteristics of a new magnetic worm gear using permanent magnet. *IEEE Transactions on Magnetics*, **29**, 2923–2925.

Li, W. and Chau, K.T. (2012) Analytical field calculation for linear tubular magnetic gears using equivalent anisotropic magnetic permeability. *Progress in Electromagnetics Research*, **127**, 155–171.

Li, X., Chau, K.T., Cheng, M. and Hua, W. (2013) Comparison of magnetic-geared permanent-magnet machines. *Progress in Electromagnetics Research*, **133**, 177–198.

Li, W., Chau, K.T. and Jiang, J.Z. (2011) Application of linear magnetic gears for pseudo-direct-drive oceanic wave energy harvesting. *IEEE Transactions on Magnetics*, **47**, 2624–2627.

Li, X., Chau, K.T., Cheng, M. *et al.* (2011) An improved coaxial magnetic gear using flux focusing. Proceedings of International Conference on Electrical Machines and Systems, pp. 1–4.

Liu, X., Chau, K.T., Jiang, J.Z. and Yu, C. (2009) Design and analysis of interior-magnet outer-rotor concentric magnetic gears. *Journal of Applied Physics*, **105**, 07F101-1–07F101-3.

Magnomatics (2014) PDD High Torque Traction Motor, Magnomatics, http://www.magnomatics.com/images/pdfs/PDD_Traction_motor_brochure.pdf (accessed September 2014).

Mezani, S., Atallah, K. and Howe, D. (2006) A high-performance axial-field magnetic gear. *Journal of Applied Physics*, **99**, 08R303-1–08R303-3.

Neuland, A.H. (1916) Apparatus for transmitting power. US Patent 1 171 351.

Pascas, M. and Weber, J. (2005) Predictive direct torque control for the PM synchronous machine. *IEEE Transactions on Industrial Electronics*, **52**, 1350–1356.

Rasmussen, P.O., Andersen, T.O., Jorgensen, F.T. and Nielsen, O. (2005) Development of a high-performance magnetic gear. *IEEE Transactions on Industry Applications*, **41**, 764–770.

Rasmussen, P.O., Mortensen, H.H., Matzen, T.N., and Jahns, T.M. (2009) Motor integrated permanent magnet gear with a wide torque-speed range. Proceedings of IEEE Energy Conversion Congress and Exposition, pp. 1510–1518.

Silva, C., Asher, G.M. and Sumner, M. (2006) Hybrid rotor position observer for wide speed-range sensorless PM motor drives including zero speed. *IEEE Transactions on Industrial Electronics*, **53**, 373–378.

Tsurumoto, K. and Kikuchi, S. (1987) A new magnetic gear using permanent magnet. *IEEE Transactions on Magnetics*, **23**, 3622–3624.

Wang, X., Li, Q., Wang, S. and Li, Q. (2003) Analytical calculation of air-gap magnetic field distribution and instantaneous characteristics of brushless DC motors. *IEEE Transactions on Energy Conversion*, **18**, 424–432.

Xia, Z.P., Zhu, Z.Q. and Howe, D. (2004) Analytical magnetic field analysis of Halbach magnetized permanent-magnet machines. *IEEE Transactions on Magnetics*, **40**, 1864–1872.

Yao, Y.D., Huang, D.R., Hsieh, C.C. *et al.* (1996) The radial magnetic coupling studies of perpendicular magnetic gears. *IEEE Transactions on Magnetics*, **32**, 5061–5063.

Zarko, D., Ban, D. and Lipo, T.A. (2006) Analytical calculation of magnetic field distribution in the slotted air gap of a surface permanent-magnet motor using complex relative air-gap permeance. *IEEE Transactions on Magnetics*, **42**, 1828–1837.

8

Vernier Permanent Magnet Motor Drives

Direct drive is becoming more and more attractive for various applications, especially electric vehicle (EV) propulsion. It takes the definite advantages of high torque, fast response, gearless operation, and zero transmission loss. Adopting direct drive for in-wheel propulsion not only offers the well-known electronic differential action, but also facilitates many advanced vehicular functions such as the antilock braking system, antislip regulation, and electronic stability program. Among the available direct-drive machines, the vernier permanent magnet (PM) machine inherently offers high torque at low speeds, thus eliminating the use of mechanical gear or even magnetic gear to amplify the torque for low-speed operation.

In this chapter, various vernier permanent magnet (VPM) motor drives, including their system configurations, machine topologies, power inverters, and control strategies, are discussed for EV propulsion. The corresponding design criteria, design examples, and potential applications are also given.

8.1 System Configurations

Essentially, the VPM machine is a kind of synchronous machine so that the configuration of the VPM motor drive is similar to that of the PM brushless motor drive, which consists of an electric motor, an electronic controller, a power inverter, and an optional transmission gear. Because of the inherent low-speed high-torque operation of the VPM machine, the VPM motor drive is preferred to serve as a direct drive for in-wheel propulsion, thus eliminating the use of transmission gear. Additionally, the use of in-wheel propulsion can eliminate the bulky, heavy, and costly mechanical differential. It should be noted that the driveline transmission losses of the transmission gear and mechanical differential can be up to 20% of the total power generated by the motor (Jones, 2007).

Figure 8.1 depicts the system configuration of the in-wheel direct drive using the VPM motor drive for EV propulsion. The VPM motor drive consists of an electronic controller that functions to command the system operation, two low-speed high-torque VPM machines, which can be directly mounted inside the wheels and hence called the in-wheel motors or hub motors, and two inverters, which can independently control the two machines to perform the desired traction control such as the electronic differential. Because of the absence of transmission gear and mechanical differential, the transmission losses are almost eliminated and the operating efficiency is significantly improved. Since regenerative braking can take place in each driving wheel, the corresponding energy recovery becomes more effective, which can extend the EV driving range. Moreover, because of the elimination of bulky and heavy mechanical differential, additional batteries can be installed which helps further improve the driving range per charge (Jain and Williamson, 2009).

Electric Vehicle Machines and Drives: Design, Analysis and Application, First Edition. K. T. Chau.

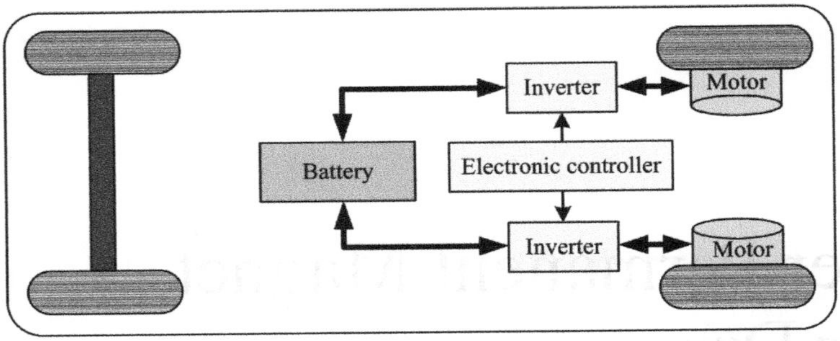

Figure 8.1 Vernier PM in-wheel motor drive using dual inverters

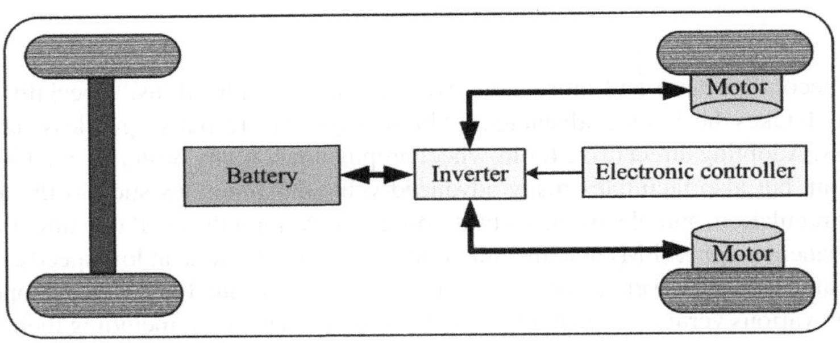

Figure 8.2 Vernier PM in-wheel motor drive using single inverter

In general, a two-wheel drive EV needs two machines and two inverters, whereas a four-wheel drive EV requires four machines and four inverters. The corresponding electronic controller is a single control unit to command the switching operation of all inverters. In order to reduce the inverter cost, a single inverter, such as the five-leg inverter or nine-switch inverter, can be employed to supersede two standard three-phase three-leg inverters, hence independently controlling the two VPM machines. Figure 8.2 depicts the system configuration of a VPM in-wheel motor drive using a single inverter, which can offer the merits of lower cost and space saving. Nevertheless, the use of this single inverter to independently control two machines suffers from the drawbacks that the pulse-width modulation (PWM) scheme needs to be modified and the cooling arrangement is more demanding.

8.2 Vernier PM Machines

It is well known that the vernier reluctance machine can inherently offer low-speed high-torque operation, while avoiding the increase of the number of armature winding pole-pairs. In order to improve the power density, the vernier reluctance machine usually incorporates PMs to provide the field excitation, thus becoming the VPM machine.

8.2.1 Vernier PM versus Magnetic-Geared PM

As discussed in Chapter 7, the magnetic-geared (MG) machine is an artful integration of the magnetic coaxial gear and PM brushless machine, which can simultaneously achieve the low-speed requirement for direct drive and high-speed requirement for machine design. Naturally, an outer-rotor PM brushless machine can be combined with the magnetic coaxial gear to share a common rotor – namely the inner rotor of the gear and the outer rotor of the machine – to form the MG machine. Since this topology involves three air-gaps and two rotors as shown in Figure 8.3, the corresponding alignment and precision requirements are so complicated and very demanding.

Rather than using the outer rotor of the PM brushless machine to serve as the physical input of the magnetic coaxial gear, the high-speed rotating field created by the armature winding in the stator can directly serve this purpose. The resulting MG machine involves only two air-gaps as shown in Figure 8.4 in which a stationary armature is employed to replace the high-speed inner rotor of the magnetic coaxial gear, hence reducing the overall size and weight (Wang *et al.*, 2009). Since the stator is stationary, there is no need to use an air-gap to separate the stationary armature and the stationary modulation ring. By further removing this air-gap, the MG machine involving only one air-gap is resulted as shown in Figure 8.5. When the number of ferromagnetic segments in the modulation ring is an integer multiple of the number of stator teeth, the MG machine with one air-gap becomes the VPM machine with all PMs located in the rotor (Li, Chau, and Li, 2011; Li *et al.*, 2013).

8.2.2 *Structure of Vernier PM Machines*

The VPM machine possesses three types of structures: the rotor-PM type, where all PMs are mounted on the rotor; the stator-PM type, where all PMs are mounted on the stator, and the all-PM type, where PMs are mounted on both the rotor and stator. As the rotor-PM VPM machine is most mature, it is simply

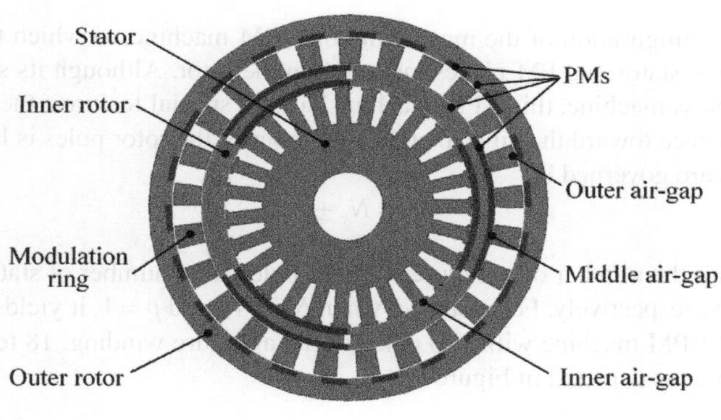

Figure 8.3 MG machine with three air-gaps

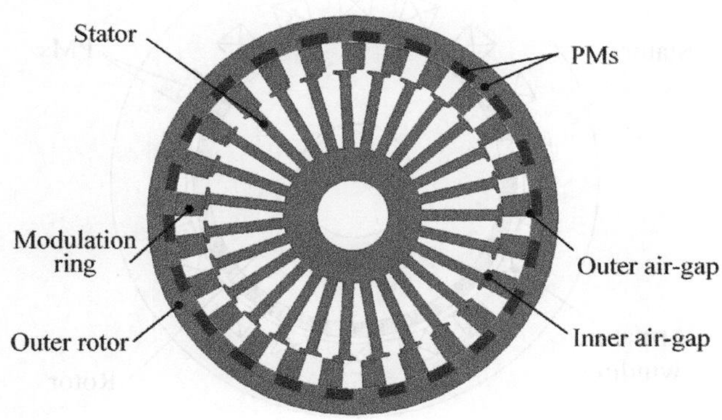

Figure 8.4 MG machine with two air-gaps

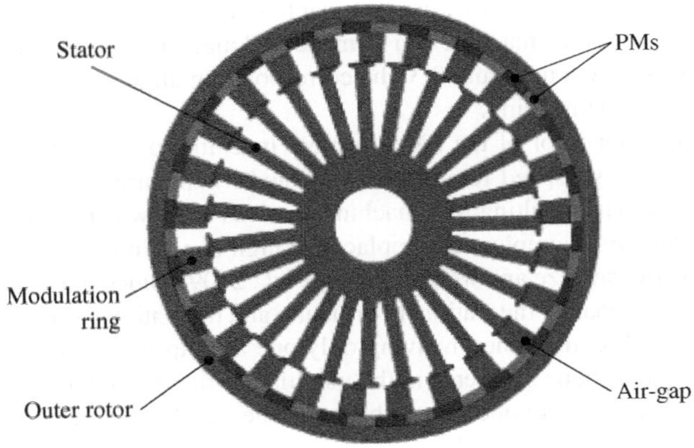

Figure 8.5 MG machine with one air-gap

called the VPM machine. Unless otherwise stated, all discussions on the VPM machine actually refer to the rotor-PM VPM machine.

8.2.2.1 VPM Machine with PMs in Rotor

Figure 8.6 shows the configuration of the most common VPM machine, in which there are three-phase armature winding in the stator and PM pieces mounted on the rotor. Although its structure is similar to an ordinary PM brushless machine, this VPM machine has two special features: the stator has uniformly pitched teeth on its surface toward the air-gap and the number of the rotor poles is large (Toba and Lipo, 1999). These features are governed by

$$N_r = N_s \pm p \tag{8.1}$$

where p, N_s, and N_r are the number of armature winding pole-pairs, number of stator teeth, and number of rotor PM pole-pairs, respectively. For instance, when $N_s = 18$ and $p = 1$, it yields $N_r = 17$, which is a three-phase rotor-PM VPM machine with two poles of the armature winding, 18 teeth in the stator, and 34 PM poles in the rotor as depicted in Figure 8.6.

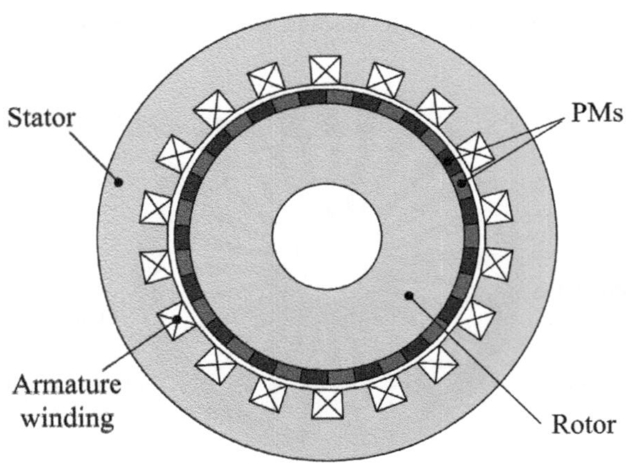

Figure 8.6 Rotor-PM VPM machine configuration

In essence, the PMs create a multipole magnetomotive force (MMF) field, which is modulated by the permeance variation of teeth. The number of PM pole-pairs and the number of iron teeth are similar but not equal so that the armature flux has a relatively low pole number. The behavior of this magnetic circuit resembles the action of the vernier gauge, where there are patterns of alignment and misalignment. When the rotor rotates one PM pole pitch, the flux linkage reverses the polarity. The speed of rotation of the alignment pattern is thus much higher than that of the rotor itself. Hence, the electromotive force (EMF) induced in the armature winding, which is proportional to the rate of change of flux linkage, is of high frequency and high magnitude. This speed reduction from the armature field speed to the rotor speed is called the magnetic gearing effect, which works as a mechanical gear to reduce the speed or amplify the torque. Therefore, the VPM machine has the inherent feature of low-speed high-torque operation.

For the stator, there are two kinds of slot-and-tooth arrangements: split-pole and open-slot (Toba and Lipo, 1999). As shown in Figure 8.7, the split-pole arrangement takes the advantage of offering a large number of teeth, but suffers from the problem of dead spaces between the teeth. On the other hand, the open-slot arrangement fully utilizes its slots to accommodate the coils, but lacks the flexibility to change the number of teeth.

For the rotor, there are two main types of PM arrangements: surface-magnet and inset-magnet (Toba and Lipo, 1999). As shown in Figure 8.8, the surface-magnet arrangement has PM pole pieces with alternating N and S polarities mounted on the rotor surface, which possesses the merit of simple mounting, whereas the inset-magnet arrangement has PM pole pieces with the same N or S polarity inset onto the rotor surface, which takes the advantage of physical robustness. The most critical issue between them is the effect of the location of PMs on the phase inductance. With the surface-magnet rotor, the thickness of PMs virtually enlarges the air-gap so that the machine exhibits a relatively low phase inductance. On the contrary, when adopting the inset-magnet rotor, the salient iron core of the rotor directly faces the air-gap so that the phase inductance becomes relatively high. Such high inductance will degrade the power factor and cause difficulty in current control.

8.2.2.2 VPM Machine with PMs in Stator

The VPM machine with PMs located in the stator is commonly termed the vernier hybrid machine (Spooner and Haydock, 2003). This stator-PM VPM machine operates in a similar manner to the rotor-PM

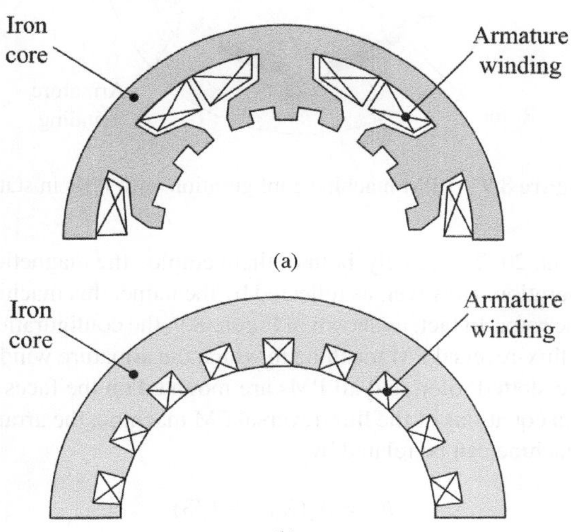

Figure 8.7 Stator arrangements of rotor-PM VPM machine: (a) split-pole and (b) open-slot

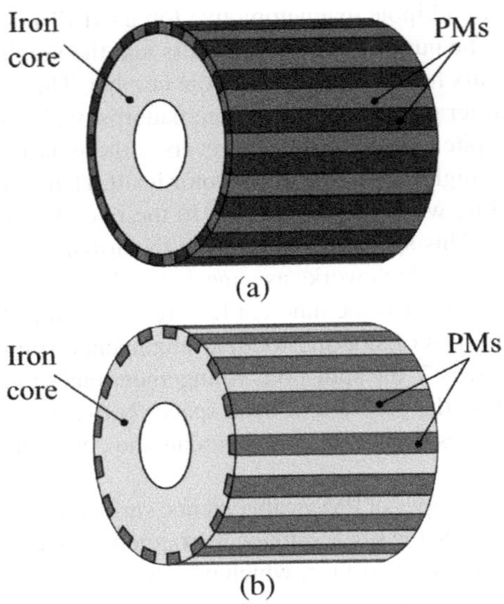

Figure 8.8 Rotor arrangements of rotor-PM VPM machine: (a) surface-magnet and (b) inset-magnet

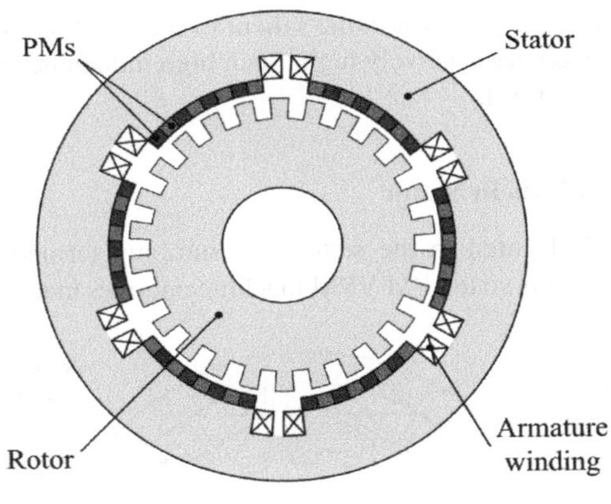

Figure 8.9 VPM machine configuration with PMs in stator

VPM machine (Li and Chau, 2012) – namely, both of them employ the magnetic gearing effect to achieve low-speed high-torque operation. However, as reflected by the name, this machine installs its PM poles in the stator, rather than in the rotor. In fact, as shown in Figure 8.9, the configuration of this VPM machine is very similar to that of the flux-reversal PM machine in which the armature winding is wound on the stator iron protrusions facing the slotted rotor, and all PMs are mounted on the faces of those iron protrusions. Borrowing from the design equations of the flux-reversal PM machine, the arrangement of teeth and PMs of this stator-PM VPM machine can be related by

$$N_r = N_s(N_{\mathrm{PM}} + 1/3) \tag{8.2}$$

where N_s, N_r, and N_{PM} are the number of stator teeth, number of rotor teeth, and number of PM pole-pairs per stator tooth, respectively. For instance, when $N_s = 6$ and $N_{\mathrm{PM}} = 4$, it yields $N_r = 26$, which is a three-phase stator-PM VPM machine having 6 stator teeth with 8 PM poles per stator tooth and 26 rotor teeth, as depicted in Figure 8.9.

This stator-PM VPM machine preferably adopts the three-phase armature winding to alleviate the undesirable cogging torque. The phase coils can be arranged with alternating polarity around the machine so that the armature flux of that phase passes only through those coils. Alternatively, the phase coils can be arranged with the same polarity so that their armature flux passes from the stator to the rotor and returns via the other phases (Spooner and Haydock, 2003).

Because of the magnetic gearing effect, the operating frequency of this machine is inherently high even for a modest rotor speed. Thus, this machine is generally dictated by the corresponding iron and eddy-current losses rather than the mechanical stress to limit its practical maximum speed.

8.2.2.3 VPM Machine with PMs in Both Rotor and Stator

Figure 8.10 shows a VPM machine with PMs located in both the rotor and stator, the so-called all-PM VPM machine. The three-phase armature winding is wound in the stator. There are small slots chipped on both the rotor and stator at regular intervals in which the PMs located in these slots are magnetized in the same direction, namely the PMs in the stator facing the air-gap are N poles while the PMs in the rotor facing the air-gap are S poles. The number of winding pole-pairs and the number of PM pieces are governed by (Hosoya and Shimomura, 2011)

$$\begin{cases} N_s = 6aqp \\ N_r = N_s \pm p \end{cases} \tag{8.3}$$

where a is a positive integer, q is the number of slots per pole per phase of the armature winding, p is the number of pole-pairs of the armature winding, N_s is the number of PM pieces in the stator, and N_r is the number of PM pieces in the rotor. For instance, when $N_s = 18$, it yields $a=1$, $q=1$, $p=3$, and $N_r = 15$, which is a three-phase all-PM VPM machine with six poles of the armature winding, 18 PM pieces in the stator, and 15 PM pieces in the rotor as depicted in Figure 8.10.

This all-PM VPM machine employs the magnetic gearing effect to achieve low-speed high-torque operation, which is similar to the rotor-PM or stator-PM VPM machine. However, the design of this VPM machine is relatively complex because it is susceptible to magnet saturation in the iron core. By carefully tuning both the electric loading and stator inside diameter, the torque per current density can be optimized while the magnetic saturation can also be alleviated (Tasaki *et al.*, 2012).

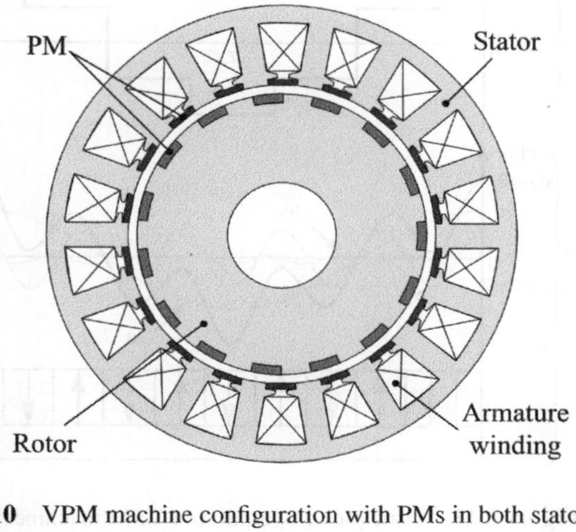

Figure 8.10 VPM machine configuration with PMs in both stator and rotor

8.2.3 Principle of Vernier PM Machines

The principle of operation of the VPM machine, the rotor-PM one, is mainly based on the interaction of the armature current field generated by the stator and the PM field generated by the rotor. This machine is generally wound with the three-phase distributed winding and driven by a three-phase supply. Figure 8.11 illustrates the principle of operation of a VPM machine in which there are six stator teeth, five rotor PM pole-pairs, and one armature winding pole-pair (Kataoka *et al.*, 2013). It can be observed that the fifth harmonic component of the rotating magnetic field is produced by the MMF of the three-phase armature winding and the air-gap permeance variation. The rotor rotates synchronously with this fifth harmonic rotating magnetic field. Thus, the rotor speed is stepped down to one-fifth of the speed of the rotating magnetic field and the torque is stepped up accordingly, hence working as a motor with magnetic gearing. Consequently, the VPM machine can offer very low-speed operation by properly increasing the number of PM poles in the rotor as governed by

$$n_r = \left(\frac{f}{p}\right)\left(\frac{p}{N_r}\right) = \frac{f}{N_r} \tag{8.4}$$

where n_r is the synchronous speed in rev/s and f is the supply frequency in Hz.

The detailed torque analysis of the VPM machine can be found in Toba and Lipo (2000). In order to illustrate how to produce the desired steady torque, the following assumptions are made:

- The relative permeability of iron core is infinite.
- The relative permeability of PM material is the same as the air-gap so that the PM can be considered as a virtual air-gap.
- The air-gap MMF, permeance, and flux density vary only along the circumference direction, and are uniform in the radial and axial directions.

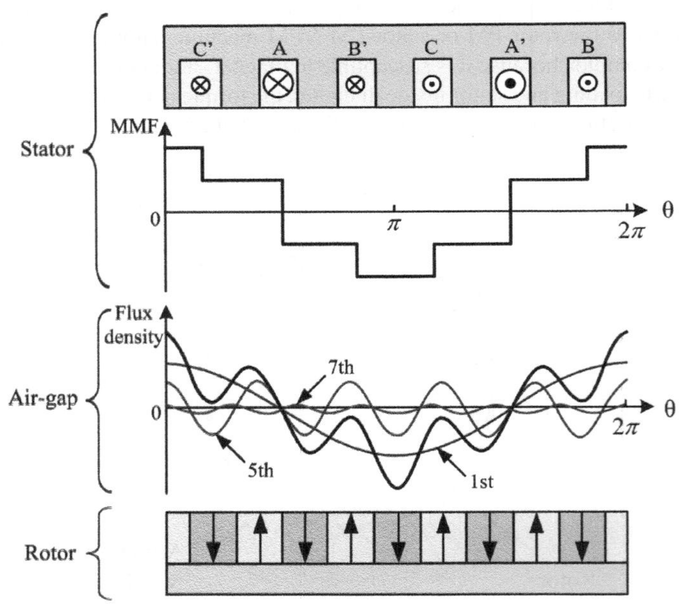

Figure 8.11 Principle of operation of VPM machines

The armature winding of the stator is fed by three-phase balanced currents, which are expressed as:

$$\begin{cases} I_a = \sqrt{2}I\cos(\omega t - \alpha) \\ I_b = \sqrt{2}I\cos\left(\omega t - \alpha - \frac{2\pi}{3}\right) \\ I_c = \sqrt{2}I\cos\left(\omega t - \alpha - \frac{4\pi}{3}\right) \end{cases} \tag{8.5}$$

where I is the phase current, ω is the angular frequency, and α is an arbitrary phase angle. The corresponding armature current MMF, F_a established in the air-gap can be expressed as

$$F_a(\theta_s, t) = \frac{3F_{a1}}{2}\left[\sum_{n=1,7,13,\cdots} \frac{k_{dn}k_{pn}}{n}\cos\left(np\theta_s - (\omega t - \alpha)\right) + \sum_{n=5,11,17,\cdots} \frac{k_{dn}k_{pn}}{n}\cos\left(np\theta_s + (\omega t - \alpha)\right)\right] \tag{8.6}$$

where n denotes the order of harmonics, θ_s is the angle of the stator with respect to the current vector of phase A, k_{dn} and k_{pn} are respectively the nth harmonic distribution factor and pitch factor, and F_{a1} is the fundamental component of the armature current MMF, which is given by

$$F_{a1} = \frac{2\sqrt{2}N_{\text{ph}}I}{\pi p} \tag{8.7}$$

where N_{ph} is the number of turns of the armature winding per phase. Meanwhile, the distribution factor, k_{dn} and pitch factor, k_{pn} are given by

$$k_{dn} = \frac{\sin(n\pi/6)}{q\sin(n\pi/6q)} \tag{8.8}$$

$$k_{pn} = \sin(n\pi\beta/2) \tag{8.9}$$

where q is the number of slots per pole per phase and β is the ratio of the coil span to the pole pitch.

The PM MMF generated by the PM rotor during standstill can be expressed as

$$F_{\text{PM}}(\theta_s) = \sum_{\text{odd } n}^{\infty} \frac{F_{\text{PM1}}}{n}\cos(nN_r\theta_r) = \sum_{\text{odd } n}^{\infty} \frac{F_{\text{PM1}}}{n}\cos(nN_r(\theta_s - \theta_m)) \tag{8.10}$$

where θ_r is the angle of the rotor with respect to the PM pole center, θ_m is the rotor position equal to $(\theta_s - \theta_r)$, and F_{PM1} is the fundamental component of the PM MMF.

Because of the slotting effect, the permeance coefficient in the air-gap in the radial direction can be expressed as (Rezzoug and Zaïm, 2013)

$$P(\theta_s) = P_0 + (-1)^{\varepsilon}\sum_{m=1}^{\infty} P_m\cos(mN_s\theta_s) \tag{8.11}$$

where P_m is the amplitude of the permeance coefficient of the mth harmonic and ε is the number of slots chorded for the short-pitch winding. Thus, the air-gap flux density B_{PM} due to the PM rotor is the product of $F_{\text{PM}}(\theta_s)$ and $P(\theta_s)$. Neglecting the higher order components, B_{PM} can be written as (Toba and Lipo, 2000)

$$\begin{aligned} B_{\text{PM}} &= [F_{\text{PM1}}\cos(N_r(\theta_s - \theta_m))]\,[P_0 + (-1)^{\varepsilon}P_1\cos(N_s\theta_s)] \\ &= (-1)^{\varepsilon}B_{\text{PM1}}\cos[(N_r - N_s)\theta_s + N_r\theta_m] + B_{\text{PM}h}\cos(N_r(\theta_s - \theta_m)) \end{aligned} \tag{8.12}$$

where B_{PM1} and $B_{\text{PM}h}$ are defined as

$$B_{\text{PM1}} = \frac{F_{\text{PM1}}P_1}{2} \tag{8.13}$$

$$B_{\text{PM}h} = F_{\text{PM1}}P_0 \tag{8.14}$$

From Eq. (8.12), it can be found that the first term has the same spatial period as the fundamental component of the armature current MMF and the second term has the same spatial period as one of the slot harmonics of the armature current MMF. It can also be deduced that the rotational directions of the first and second terms will be the same if N_r is chosen to be $(N_s + p)$, whereas they become opposite to one another when N_r is equal to $(N_s - p)$.

In general, the higher order components of the slot harmonics of the armature current MMF rotate in the same directions with the fundamental component, and vice versa. Thus, when the fundamental component of the armature current MMF is synchronous to the first-term component of air-gap flux density of PMs, which is usually the case, the second-term component of air-gap flux density of PMs also becomes synchronous to its corresponding slot harmonic component of the armature current MMF so that the interaction of these two harmonic components yields a steady torque. Consequently, considering only the fundamental and slot harmonics, the armature current MMF given by Eq. (8.6) can be rewritten as

$$\begin{aligned} F_a = &\frac{3k_{d1}k_{p1}F_{a1}}{2}[\cos(p\theta_s - (\omega t - \alpha)) \\ &+ \frac{(-1)^{\varepsilon}}{N_s/p - 1}\cos((N_s - p)\theta_s + (\omega t - \alpha)) \\ &+ \frac{(-1)^{\varepsilon-1}}{N_s/p + 1}\cos((N_s + p)\theta_s - (\omega t - \alpha))] \end{aligned} \tag{8.15}$$

Assuming that all the field energy is stored in the air-gap and the PM (which can be considered as a virtual air-gap), the developed torque T can be expressed as (Toba and Lipo, 2000)

$$T = \frac{p\tau\, l}{\pi}\int_0^{2\pi} P(F_a + F_{\text{PM}})\frac{\partial F_{\text{PM}}}{\partial\theta_m}\, d\theta_s \tag{8.16}$$

where τ is the pole pitch of the armature winding and l is the stack length of the iron core. Substituting Eqs. (8.10)–(8.15) into Eq. (8.16), the developed torque can be deduced as

$$T = \frac{3}{2}p\tau\, l\, N_r k_{d1}k_{p1}F_{a1}\left(B_{\text{PM1}} \mp \frac{B_{\text{PM}h}}{N_r/p}\right)\sin((\omega t - \alpha) \mp N_r\theta_m) \tag{8.17}$$

where the order of signs agrees with that in Eq. (8.1). It can be observed that the torque increases when $N_r = (N_s - p)$ and decreases when $N_r = (N_s + p)$. From Eqs. (8.13) and (8.14), it is evident that the harmonic torque component has a greater impact than the fundamental torque component (Ishizaki *et al.*, 1995; Llibre and Matt, 1998). It is obvious that a steady torque is resulted by setting the armature current frequency as

$$\omega = \pm N_r\omega_m \tag{8.18}$$

where the order of the sign is the same as that in Eq. (8.1), and $\omega_m = (\theta_m/t)$ is the angular speed of the rotor rotation. Substituting Eqs. (8.7), (8.18), and $\alpha = \pm\pi/2$ into Eq. (8.17), the maximum steady torque T_m is resulted as given by

$$T_m = \frac{3\sqrt{2}}{\pi}\tau\, l\, N_r k_{d1}k_{p1}N_{\text{ph}}I\left(B_{\text{PM1}} \mp \frac{B_{\text{PM}h}}{N_r/p}\right) \tag{8.19}$$

8.2.4 Modeling of Vernier PM Machines

Essentially, the principle of operation of the VPM machine is similar to that of the PM synchronous machine, whereas the torque production is based on the harmonic component rather than the fundamental component of the rotating magnetic field. Thus, the equivalent circuit of the VPM machine is similar to that of the PM synchronous machine as shown in Figure 8.12, where V is the supply voltage, E_o is the back EMF, I is the armature current, R_a is the armature resistance, and X_s is the synchronous reactance. The corresponding phasor diagram on the δ–γ axes is depicted in Figure 8.13, where ϕ is the power factor angle and $N_r\delta_L$ is the load or power angle (Kakihata *et al.*, 2012).

The voltage equation of the VPM machine in the steady state on the δ- and γ-axes can be expressed as

$$V\cos(N_r\delta_L) = E_o + IX_s\sin(\phi - N_r\delta_L) + IR_a\cos(\phi - N_r\delta_L) \quad (8.20)$$

$$V\sin(N_r\delta_L) = IX_s\cos(\phi - N_r\delta_L) - IR_a\sin(\phi - N_r\delta_L) \quad (8.21)$$

Transforming into the currents on the δ- and γ-axes, Eqs. (8.20) and (8.21) can be rewritten as (Kakihata *et al.*, 2012)

$$I_\delta = I\cos(\phi - N_r\delta_L) = \frac{V}{Z_s}\cos(\theta_s - N_r\delta_L) - \frac{E_o}{Z_s}\cos\theta_s \quad (8.22)$$

$$-I_\gamma = I\sin(\phi - N_r\delta_L) = \frac{V}{Z_s}\sin(\theta_s - N_r\delta_L) - \frac{E_o}{Z_s}\sin\theta_s \quad (8.23)$$

where $Z_s = \sqrt{R_a^2 + X_s^2}$ is the synchronous impedance, $\theta_s = \tan^{-1}(X_s/R_a)$ is the synchronous impedance angle, I_δ is the armature current along the δ-axis, and I_γ is the armature current along the γ-axis.

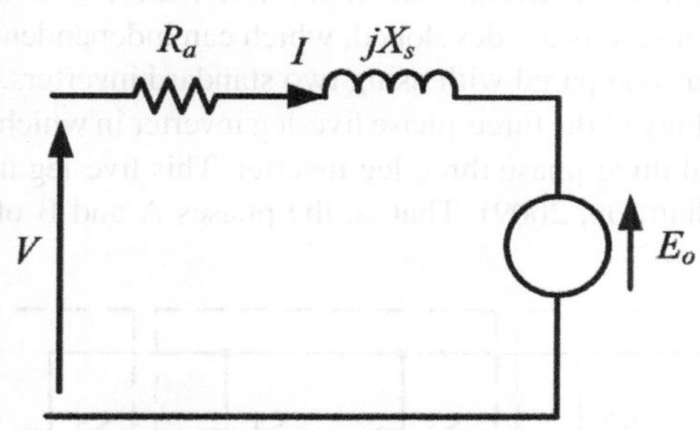

Figure 8.12 Equivalent circuit of VPM machine

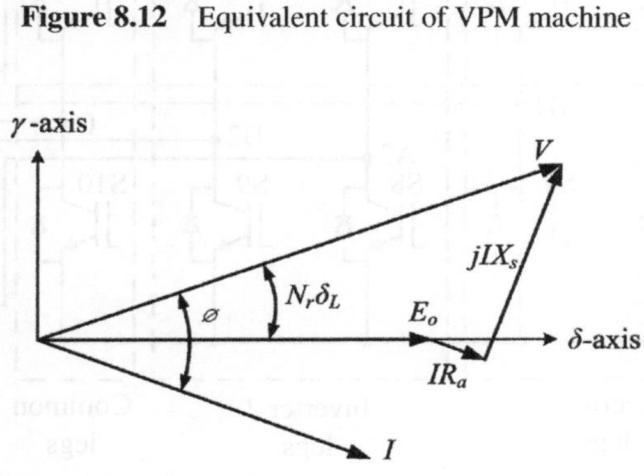

Figure 8.13 Phasor diagram of VPM machine

By multiplying E_o to Eq. (8.22), the output power P_o can be obtained as

$$P_o = 3\left[\frac{VE_o}{Z_s}\cos\left(\theta_s - N_r\delta_L\right) - \frac{E_o^2}{Z_s}\cos\theta_s\right] \tag{8.24}$$

On neglecting the armature resistance, it yields $\theta_s = 90°$ so that the output power can be simplified as

$$P_o = 3\left[\frac{VE_o}{X_s}\sin\left(N_r\delta_L\right)\right] \tag{8.25}$$

which has the same form as that of a three-phase synchronous machine. Since the angular speed of the rotor is equal to the angular supply frequency ω divided by N_r times, the output torque T can be easily deduced from Eq. (8.24) as

$$T = \frac{P_o}{\omega/N_r} = 3\frac{N_r}{\omega}\left[\frac{VE_o}{Z_s}\cos\left(\theta_s - N_r\delta_L\right) - \frac{E_o^2}{Z_s}\cos\theta_s\right] \tag{8.26}$$

8.3 Inverters for Vernier PM Motors

Since the VPM machine is a kind of synchronous machine, the PWM inverter that has been developed for the PM synchronous motor drive can be utilized for the VPM motor drive. In general, the three-phase three-leg inverter topology and the corresponding PWM switching scheme are adopted.

As mentioned earlier, the VPM motor drive is mainly applied to in-wheel direct drive for EV propulsion because of its unique low-speed high-torque feature. For a two-wheel drive EV, the propulsion system desires two VPM motors that are separately fed by two inverters, leading to increase the inverter cost and overall size as compared to the conventional single-motor drive with single inverter. To overcome this drawback, various single inverters are developed, which can independently control two motors while requiring less power devices as compared with using two standard inverters.

Figure 8.14 shows the topology of the three-phase five-leg inverter in which there are two additional legs as compared with a traditional three-phase three-leg inverter. This five-leg inverter can drive two motors independently (Jain and Williamson, 2009). That is, the phases A and B of both motors are connected

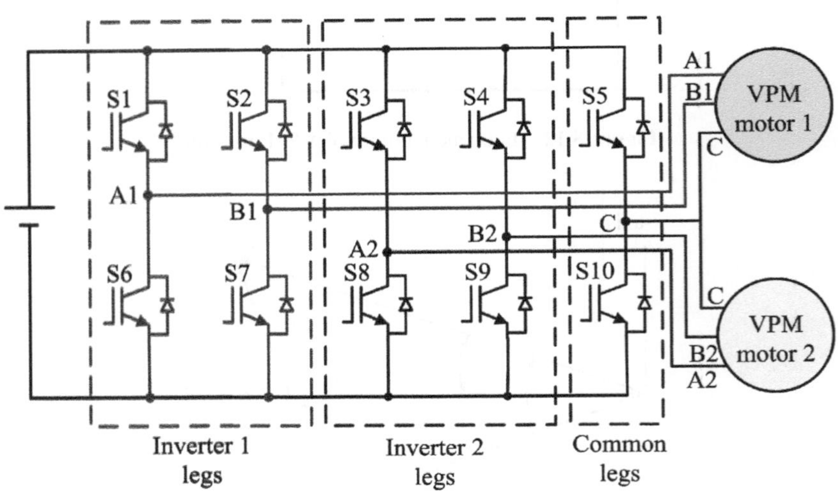

Figure 8.14 Five-leg inverter for dual VPM machines

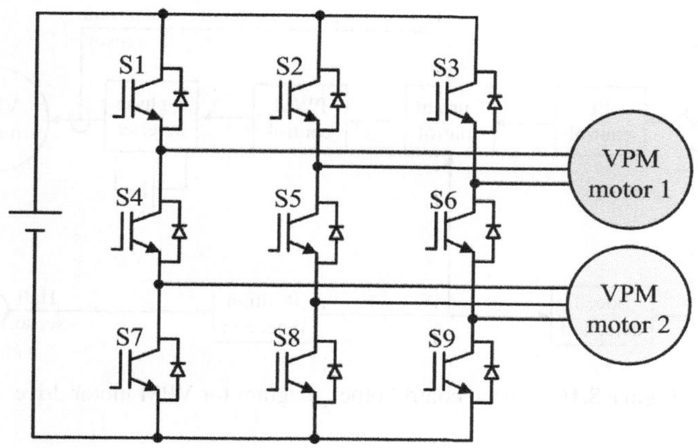

Figure 8.15 Nine-switch inverter for dual VPM machines

to each leg separately, whereas the phase C of both machines is connected to a common leg. Since this arrangement causes a difference in the switching pattern of phase C in both machines, the standard PWM switching schemes for the three-leg inverter are no longer applicable. In order to perform independent torque control of both motors, a specific PWM switching scheme known as the two-arm modulation is developed (Enokijima, Oka, and Matsuse, 2011). However, the corresponding voltage utilization factor is relatively low.

Meanwhile, the nine-switch inverter is also proposed to independently control two motors (Kominami and Fujimoto, 2007). Figure 8.15 shows the structure of this nine-switch inverter. It can be considered as an interleaved topology of an upper-deck inverter, which consists of six switches, S1–S6, and a lower-deck inverter, which also consists of six switches, S4–S9, with three switches S4–S6 representing a common branch. A specific PWM switching scheme is also developed, using both positive and negative carrier signals, to perform independent control of two motors. Since the corresponding sum of modulation index of two outputs should be less than or equal to one, the voltage amplitude of outputs is generally smaller than that of using two separate inverters.

Although the above-mentioned inverter topologies provide the advantages of cost-effectiveness and compact size, the main challenges are the control complexity and demanding cooling arrangement for such inverters. As one of the main concerns for EV motor drives is the system reliability, the EV makers currently prefer to use two traditional three-leg inverters, rather than the five-leg or nine-switch inverter, for dual-motor propulsion.

8.4 Vernier PM Motor Control

Since the VPM machine is essentially a PM synchronous machine, the control strategy of the VPM motor drive is virtually the same as that of the traditional PM synchronous motor drive. As depicted in Figure 8.16, the system control consists of dual control loops, namely the inner current loop with a hysteresis current regulator and the outer speed loop with a proportional-integral (PI) regulator.

There are many possible control varieties. The current controller can implement various control strategies for both constant-torque and constant-power operations, including the field-oriented control, direct torque control, and efficiency-optimizing control. Meanwhile, the PWM controller can implement various PWM switching schemes, including the hysteresis current control and space-vector PWM. The PI controller can also incorporate various sophisticated algorithms such as the fuzzy logic control, neural network control, adaptive control, and sliding-mode control. Of course, the Hall sensor can be eliminated by adopting the position sensorless control technique.

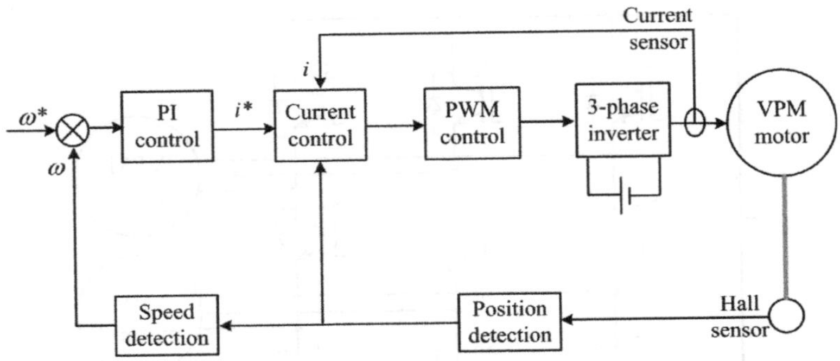

Figure 8.16 Speed control block diagram for VPM motor drive

8.5 Design Criteria of Vernier PM Motor Drives for EVs

As the VPM machine is a kind of PM brushless machine, the corresponding main dimensions can be initialized or estimated by using the standard power equation of the PM brushless machine. On the basis of these initial dimensions, finite element analysis of the machine will be conducted. Consequently, the machine dimensions will be iteratively modified until the machine performance can satisfy the requirements and constraints.

The selection criteria for the number of stator teeth N_s, number of armature winding pole-pairs p, and number of rotor pole-pairs N_r of the VPM machine are governed by Eq. (8.1). In general, the number of phases is selected as three so that the rotating field can readily be produced by using the standard PWM inverter and standard winding connection. Moreover, the selection criteria for the stator tooth width, stator slot height, stator yoke thickness, and rotor yoke thickness of the VPM machine are summarized as follows:

- The armature winding current density at the rated load should not be higher than 6 A/mm^2; otherwise, a special cooling requirement is desired.
- The magnetic flux density of the stator teeth, stator yoke, and rotor yoke should not be larger than 1.7 T.
- The slot fill factor in the stator should be about 55% to facilitate the convenience of winding.
- Because of the magnetic gearing effect, the iron loss should be limited to a reasonable value, particularly at high-speed operation.
- The mutual inductances among phase windings should be minimum.
- The cogging torque and torque ripple should be minimum.

In order to estimate the dimensions of PM pieces used for the VPM machine (Li, 2012), namely the PM thickness h_{PM} and PM surface area S_{PM}, the magnetic circuit approach is generally adopted. Assuming that the total magnetic potential drop of the external magnetic circuit is F and the magnetic potential drop in the air-gap is F_δ, they are related by

$$F = k_s F_\delta = k_s k_\delta \delta \, H_\delta \tag{8.27}$$

where k_s is the saturation coefficient, k_δ is the air-gap coefficient, δ is the air-gap length, and H_δ is the air-gap magnetic field intensity. When the PM operates at the magnetic flux density B_{PM} and magnetic field intensity H_{PM}, it yields

$$\begin{cases} H_{PM}\, h_{PM} &= F \\ B_{PM} S_{PM} &= \psi_{PM} = \sigma \psi_\delta \end{cases} \tag{8.28}$$

where ψ_{PM} is the PM flux, ψ_δ is the air-gap flux, and σ is the magnetic leakage factor. Substituting Eq. (8.27) into Eq. (8.28) yields

$$\begin{cases} k_s k_\delta \delta\, H_\delta = H_{PM} h_{PM} \\ \sigma B_\delta S_\delta = \sigma \mu_o H_\delta S_\delta = B_{PM} S_{PM} \end{cases} \tag{8.29}$$

where B_δ is the air-gap flux density and S_δ is the air-gap area. Multiplying the two equations in Eq. (8.29) yields

$$\sigma \mu_o k_s k_\delta \delta\, S_\delta H_\delta^2 = B_{PM} H_{PM} h_{PM} S_{PM} \tag{8.30}$$

So, the PM volume can be obtained as

$$V_{PM} = S_{PM} h_{PM} = \frac{\sigma \mu_0 k_s k_\delta \delta\, S_\delta H_\delta^2}{B_{PM} H_{PM}} \tag{8.31}$$

which is determined by the product of B_{PM} and H_{PM}. That is, when the operating point is designed to achieve the maximum magnetic energy product, the required PM volume is minimum. Thus, $B_{PM} = B_r/2$ and $H_{PM} = -H_c/2$ are selected so that the PM volume can be minimized as given by

$$V_{PM} = \frac{4\sigma \mu_0 k_s k_\delta \delta\, S_\delta {H_\delta}^2}{B_r H_c} \tag{8.32}$$

where B_r and H_c are the PM remanence and coercive force, respectively. Dividing the two equations in Eq. (8.29) yields

$$\frac{k_s k_\delta}{\sigma \mu_0} \frac{\delta}{S_\delta} = \frac{H_{PM}}{B_{PM}} \frac{h_{PM}}{S_{PM}} \tag{8.33}$$

At the maximum magnetic energy product, Eq. (8.33) can be rewritten as

$$\frac{h_{PM}}{S_{PM}} = \frac{k_s k_\delta}{\sigma \mu_0} \frac{B_{PM}}{H_{PM}} \frac{\delta}{S_\delta} = \frac{k_s k_\delta}{\sigma \mu_0} \frac{B_r}{H_c} \frac{\delta}{S_\delta} \tag{8.34}$$

Using Eqs. (8.32) and (8.34), the PM thickness can be deduced as

$$h_{PM} = \frac{2 k_s k_\delta B_\delta \delta}{\mu_0 H_c} \tag{8.35}$$

Hence, the PM area can also be deduced as

$$S_{PM} = \frac{2\sigma S_\delta B_\delta}{B_r} \tag{8.36}$$

It should be noted that the PM sizing is so important that it directly affects the overall machine performance and cost. Therefore, after the initialization PM dimensions by using the above-mentioned design equations, it needs to iteratively fine-tune the dimensions to achieve the desired performance and cost by using finite element analysis.

The VPM machine is a kind of PM brushless machine. Thus, similar to the PM synchronous machine, the VPM machine generally adopts the distributed winding arrangement where the armature coil of one phase is wound over several slots (Kakihata *et al.*, 2012). In terms of the magnetic field intensity and utilization of magnetic flux, this winding arrangement is superior to the concentrated winding. However, the distributed winding requires longer end-windings, which results in more copper used and higher copper loss.

On the other hand, similar to the PM brushless DC machine, the VPM machine can also adopt the concentrated winding arrangement where the armature coil of one phase is wound around one slot (Yang *et al.*, 2013). Compared with the distributed winding, the concentrated winding takes the advantages of requiring less copper and being easier for winding assembly, leading to a reduction in the overall manufacturing cost. In addition, the phase resistance of the concentrated winding machine tends to be lower than that of the distributed winding machine, thus improving the machine efficiency.

8.6 Design Examples of Vernier PM Motor Drives for EVs

As the VPM machine possesses the unique advantage of low-speed high-torque operation, the VPM motor drive is particularly beneficial for in-wheel EV direct drive, which generally prefers the outer-rotor topology. So, two outer-rotor VPM motor drives, one with the basic rotor-PM VPM machine and another with the flux-controllable rotor-PM VPM machine, are used for exemplification.

8.6.1 Outer-Rotor Vernier PM Motor Drive

Figure 8.17 shows the machine structure of the outer-rotor VPM motor drive for in-wheel EV propulsion (Li *et al.*, 2010b). The key is to adopt the split-pole arrangement, dubbed the flux-modulation pole (FMP), in the outer part of the inner stator to modulate the high-speed rotating field of the armature winding and the low-speed rotating field of the PM outer rotor (Li *et al.*, 2010a; Li, 2012). The corresponding armature winding is based on the three-phase concentrated winding connection in which the coil pitch is set equal to the slot pitch.

Similar to other rotor-PM VPM machines, the stator and rotor designs are governed by Eq. (8.1), namely $N_r = (N_s \pm p)$, where N_s is the number of FMPs in the stator, N_r is the number of PM pole-pairs in the rotor, and p is the number of armature winding pole-pairs. In general, $N_r = (N_s - p)$ is adopted. Consequently, the gear ratio G_r is given by

$$G_r = \frac{ip + jN_s}{ip} \tag{8.37}$$

where $i = 1, 3, \ldots$ and $j = 0, \pm 1, \pm 2, \ldots$ In this design, the combination of $i = 1$ and $j = -1$ is selected because it can provide the highest asynchronous space harmonic. There are nine slots in the inner stator to accommodate the three-phase armature winding having six poles ($p = 3$). Each stator tooth is split into three FMPs, resulting in totally 27 FMPs ($N_s = 27$). From Eq. (8.1), it can be deduced that there are 24 PM pole-pairs ($N_r = 24$) – that is, there are 48 PM poles mounting on the outer rotor. Meanwhile, from Eq. (8.37), it yields $G_r = -8 : 1$, which indicates that the rotor speed is only one-eighth of that in the conventional machine with the same number of armature winding pole-pairs, but rotating in the opposite direction. Therefore, when the supply frequency is 60 Hz and the corresponding rotating field speed in the stator is 1200 rpm, the outer rotor speed is scaled down by a factor of 8 to become 150 rpm.

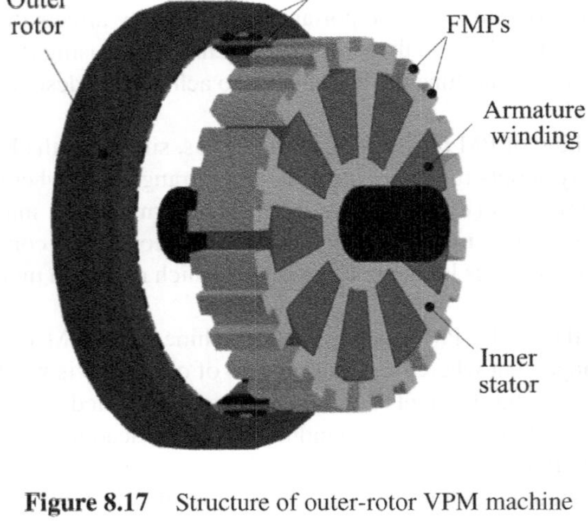

Figure 8.17 Structure of outer-rotor VPM machine

Compared with the available VPM machine designs, this machine possesses the following merits:

- The outer-rotor arrangement inherently provides a large diameter to accommodate a large number of PM poles, and enable full utilization of the inner space in the stator to accommodate the armature winding.
- The outer rotor facilitates direct coupling with the rim of the tire, thus alleviating the bearing requirements and improving the mechanical integrity.
- The inner-stator arrangement with FMPs enables to adopt compact armature winding, thus avoiding the problem of using the drum winding (Toba and Lipo, 1999).
- The armature winding adopts the concentrated winding arrangement with the coil pitch equal to the slot pitch, which can reduce the end-windings; hence, saving the copper material thereby reducing the copper loss.

Table 8.1 summarizes the key design data of this outer-rotor VPM machine. Consequently, the magnetic field distributions of this machine can be calculated by using finite element analysis. As shown in Figure 8.18, it can be found that as the outer rotor rotates through mechanical 7.5° in the anticlockwise direction, it corresponds to rotate through electrical 180°. It can also be observed that the flux lines per stator tooth can pass through the FMPs separately, hence verifying the desired flux modulating function. Meanwhile, the flux lines migrate from one stator tooth to another in the clockwise direction, which verifies that the direction of rotation of the rotor is opposite to that of the stator rotating field.

Moreover, the air-gap flux density of this machine is analyzed as depicted in Figure 8.19. It can be seen that it exhibits 24 pole-pairs in the air-gap within 360°, which corresponds to three pole-pairs of the stator rotating field, thus well agreeing with the principle of magnetic gearing. Then, the no-load EMF waveform at the rated speed can be deduced as shown in Figure 8.20. It can be found that the no-load EMF waveform is symmetrical and sinusoidal.

Furthermore, the developed torque waveform of the motor drive at the rated condition is simulated as depicted in Figure 8.21. It can be found that the average torque of this machine can achieve 113 N m, which is sufficiently high for in-wheel EV direct drive. Meanwhile, the corresponding torque ripple is about 11.8%, which is very acceptable for EV propulsion. Such torque ripple is mainly due to the cogging torque of this machine.

Table 8.1 Key design data of outer-rotor VPM machine

Rated power	2.3 kW
Rated torque	110 N m
Rated speed	200 rpm
Number of phases	3
Number of stator pole-pairs	3
Number of stator FMPs	27
Number of rotor PM pole-pairs	24
Rotor outside diameter	240 mm
Rotor inside diameter	203.2 mm
Stator outside diameter	202 mm
Air-gap length	0.6 mm
Axial length	120 mm
PM material	Nd-Fe-B

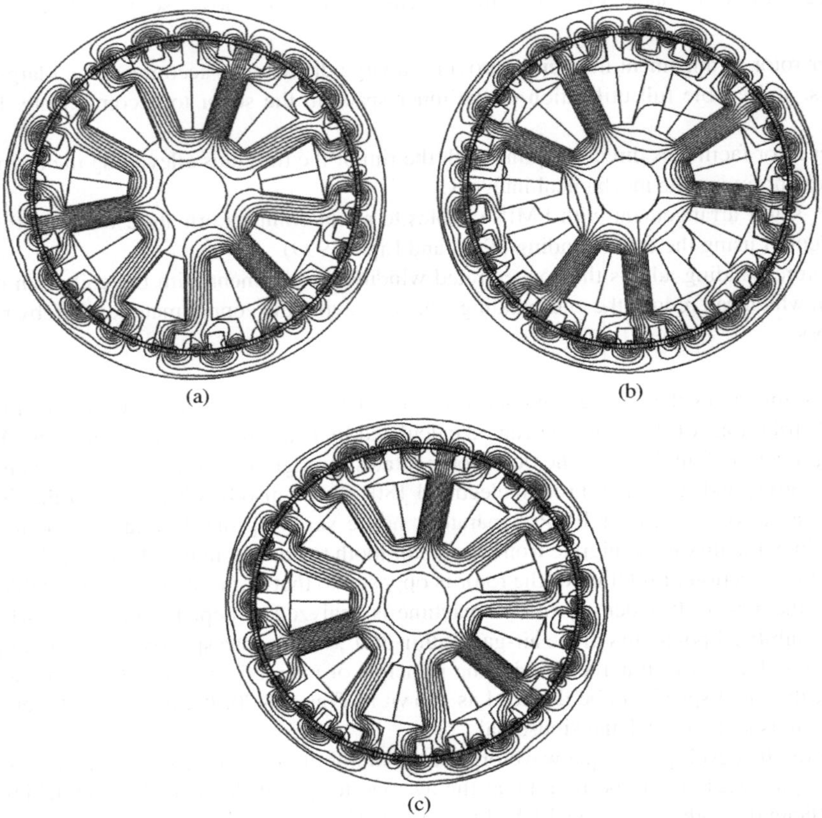

Figure 8.18 Magnetic field distributions of outer-rotor VPM machine at different rotor positions: (a) 0°, (b) 90°, and (c) 180°

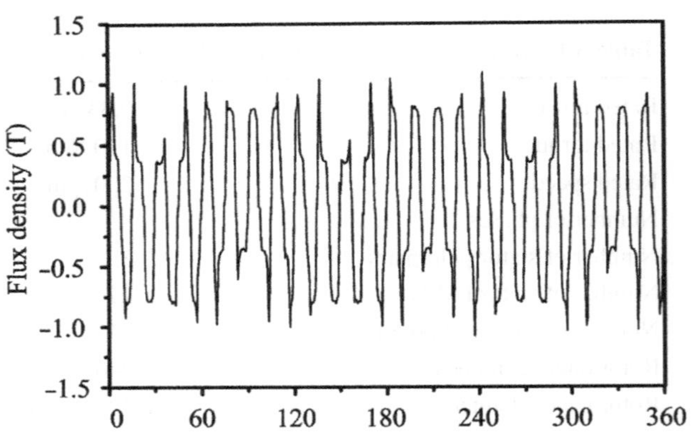

Figure 8.19 Air-gap flux density waveform of outer-rotor VPM machine

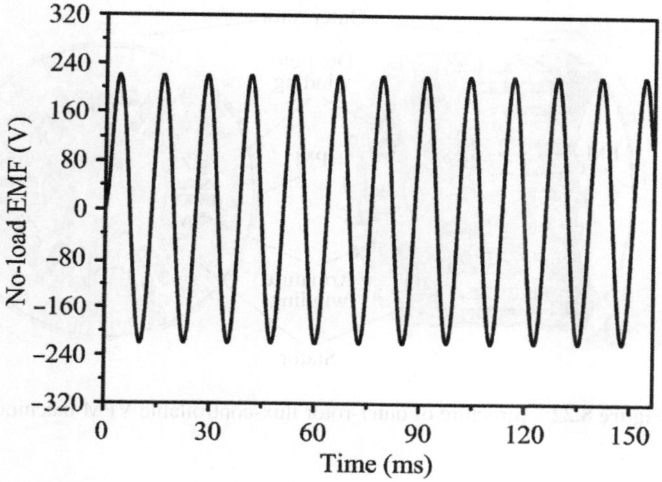

Figure 8.20 No-load EMF waveform at rated speed of outer-rotor VPM machine

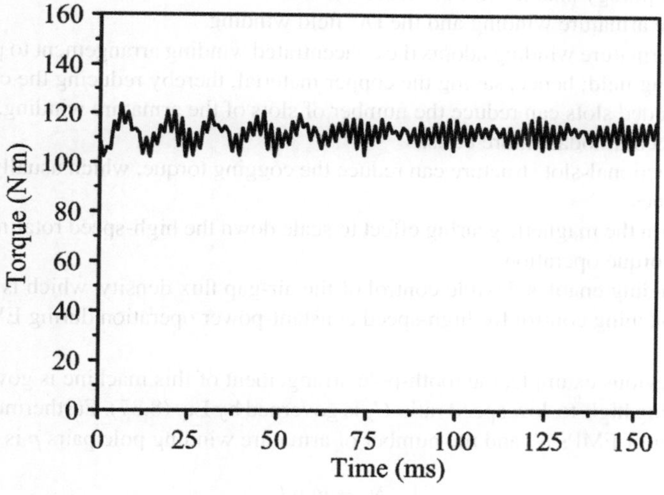

Figure 8.21 Torque waveform of outer-rotor VPM machine

8.6.2 *Outer-Rotor Flux-Controllable Vernier PM Motor Drive*

While low-speed high-torque operation is important for in-wheel EV direct drive, high-speed constant-power operation is equally important. Rather than relying on field-oriented control to tune the *d*-axis armature current for flux weakening, the use of DC field winding to tune the air-gap flux is more straightforward and effective.

Figure 8.22 shows the configuration of the outer-rotor flux-controllable VPM machine, which incorporates 22 pole-pair PMs mounted on the outer rotor, 24 FMPs located in the stator, and 6 embedded slots accommodating both the armature winding and DC field winding in the stator (Liu, Zhong, and Chau, 2011). The unique features are summarized as follows:

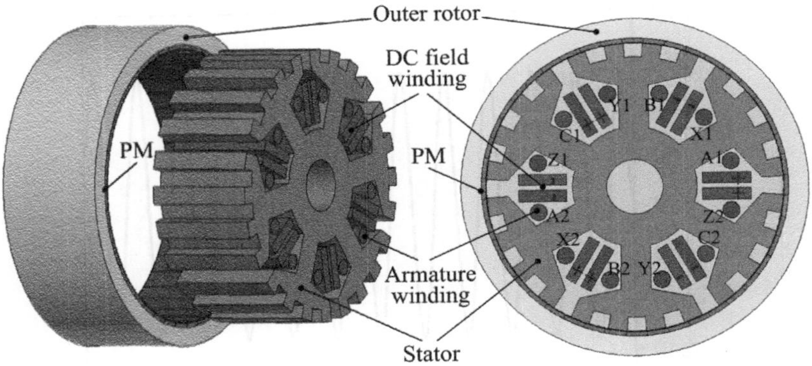

Figure 8.22 Structure of outer-rotor flux-controllable VPM machine

- The outer-rotor structure facilitates mounting with the rim of the tire for direct drive.
- The outer-rotor configuration inherently offers a large diameter to accommodate a large number of PM poles, which is desirable for VPM machines.
- The outer-rotor topology enables full utilization of the inner space to accommodate the FMPs as well as the three-phase armature winding and the DC field winding.
- The three-phase armature winding adopts the concentrated winding arrangement to produce the desired high-speed rotating field; hence, saving the copper material, thereby reducing the copper loss.
- The use of embedded slots can reduce the number of slots of the armature winding, and hence enlarge the slot area to accommodate more coils.
- The multipole fractional-slot structure can reduce the cogging torque, which usually occurs at conventional PM machines.
- The FMPs perform the magnetic gearing effect to scale down the high-speed rotating field, resulting in low-speed high-torque operation.
- The DC field winding enables flexible control of the air-gap flux density, which is highly desirable to achieve flux-weakening control for high-speed constant-power operation during EV cruising.

Similar to the previous example, the tooth-pole arrangement of this machine is governed by Eq. (8.1), and the corresponding high-to-low speed ratio G_r is governed by Eq. (8.37). Furthermore, the relationship between the number of FMPs N_s and the number of armature winding pole pairs p is given by

$$N_s = m\,p\,k \tag{8.38}$$

where m is the number of winding phases and $k = 2, 3, 4, \ldots$

In this machine design, the parameters $m=3$, $p=2$, and $k=4$ are selected. According to Eq. (8.38), it yields $N_s = 24$. Based on Eq. (8.1), it yields $N_r = 22$. Thus, it is so-called the three-phase four-pole outer-rotor flux-controllable VPM machine having 24 FMPs in the stator and 22 PM pole-pairs in the rotor. Substituting $i=1$ and $j=-1$ into Eq. (8.37) to obtain the largest space harmonic component, it yields $G_r = -11 : 1$, which indicates that the rotor speed is only one-eleventh of that in the conventional machine with the same number of armature winding pole-pairs, but rotating in the opposite direction. Namely, when the speed of rotating field in the stator is 2200 rpm, the outer rotor speed is scaled down by a factor of 11 to become 200 rpm. The key design data are summarized in Table 8.2.

The operation principle and control strategy of this machine are essentially the same as the previous VPM machine, except that the DC field winding is independently fed and controlled. The corresponding power circuit for DC field current control is basically a standard H-bridge converter as shown in

Table 8.2 Key design data of outer-rotor flux-controllable VPM machine

Rated power	2 kW
Rated torque	90 N m
Rated speed	200 rpm
Speed range	0–1000 rpm
Number of phases	3
Number of stator pole-pairs	2
Number of stator FMPs	24
Number of rotor PM pole-pairs	22
Number of DC field winding poles	6
Rotor outside diameter	246 mm
Rotor inside diameter	211.2 mm
Stator outside diameter	210 mm
Air-gap length	0.6 mm
Axial length	80 mm
PM material	Nd-Fe-B

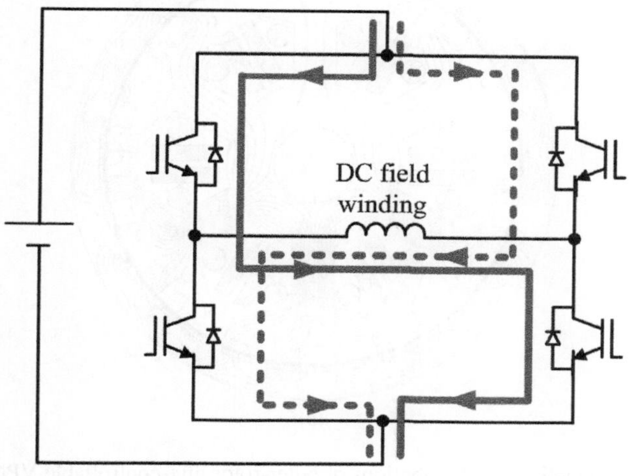

Figure 8.23 DC field current control of outer-rotor flux-controllable VPM machine

Figure 8.23. Hence, both the magnitude and direction of the DC field current can be easily regulated by adjusting the turn-on and turn-off times of the power switches.

Machine performances are analyzed by finite element analysis. First, the no-load magnetic field distributions of the machine without and with DC field control are shown in Figure 8.24. It can be seen that the flux lines are effectively modulated by the FMPs on the stator and go through the PM poles, hence verifying the desired flux modulation effect. In addition, with the DC field control using 5 A/mm^2, it can be observed that although some parts of the flux density are somewhat strengthened by the DC field, these parts are much less than that with flux weakening. Thus, the overall effect of the air-gap flux density is weakened when applying the DC field. Consequently, the corresponding flux linkage waveforms are plotted in Figure 8.25, which illustrates that the DC field control can effectively weaken the flux linkage.

In order to assess the constant-torque operation performance of this machine, the developed torque waveform at the rated condition is simulated as shown in Figure 8.26. Meanwhile, the cogging torque

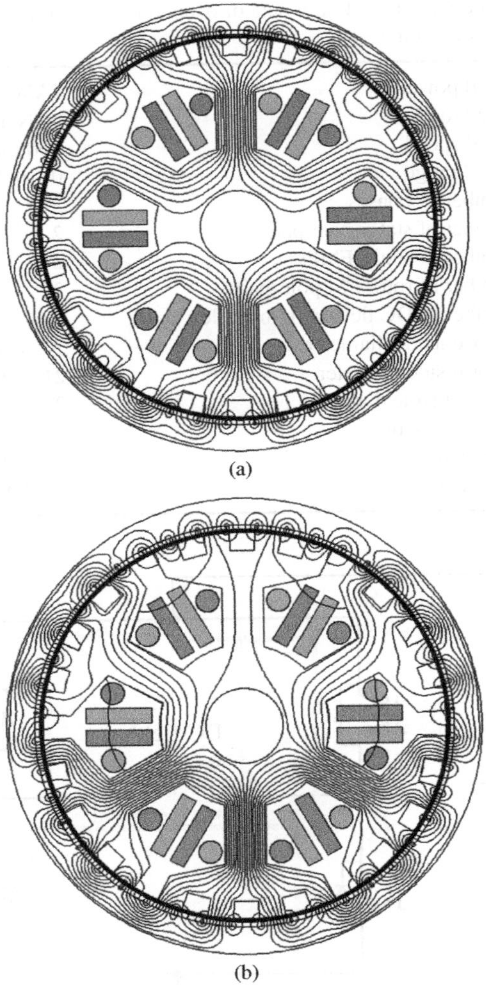

Figure 8.24 No-load magnetic field distributions of outer-rotor flux-controllable VPM machine: (a) without DC field control and (b) with DC field control

waveform is also plotted to illustrate its effect on the developed torque. It can be observed that the steady-state torque can reach 90 N m, which verifies the desired torque capability for constant-torque operation. It can also be found that the corresponding torque ripple is about 12% of the average torque, which is mainly caused by the cogging torque. Such torque ripple is very acceptable for EV propulsion.

In order to assess the constant-power operation performance of this machine, the back EMF waveforms at 200 and 1000 rpm under no flux-weakening control are simulated as shown in Figure 8.27. As expected, the back EMF amplitude increases significantly with the speed, which deteriorates the constant-power operation beyond the rated speed. Then, by applying the DC field control of 5 A/mm^2 to realize flux-weakening control, the EMF amplitude at 1000 rpm as illustrated in Figure 8.28 can be maintained at the same level as that at the rated speed. Hence, it confirms that this machine can offer the desired wide speed range of constant-power operation.

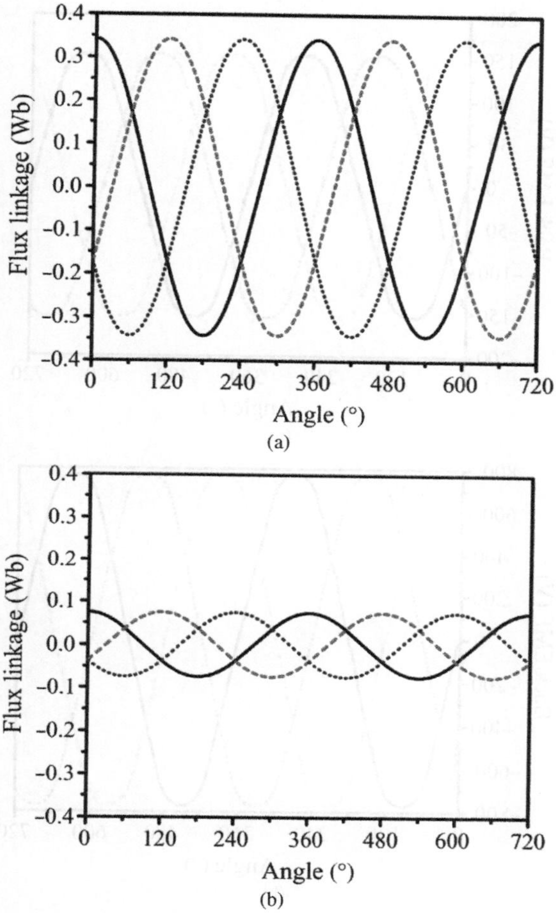

Figure 8.25 Flux linkage waveforms of outer-rotor flux-controllable VPM machine: (a) without DC field control and (b) with DC field control

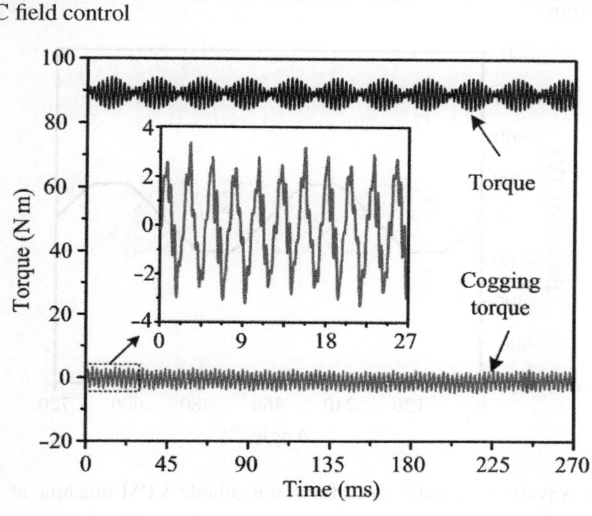

Figure 8.26 Torque waveforms of outer-rotor flux-controllable VPM machine

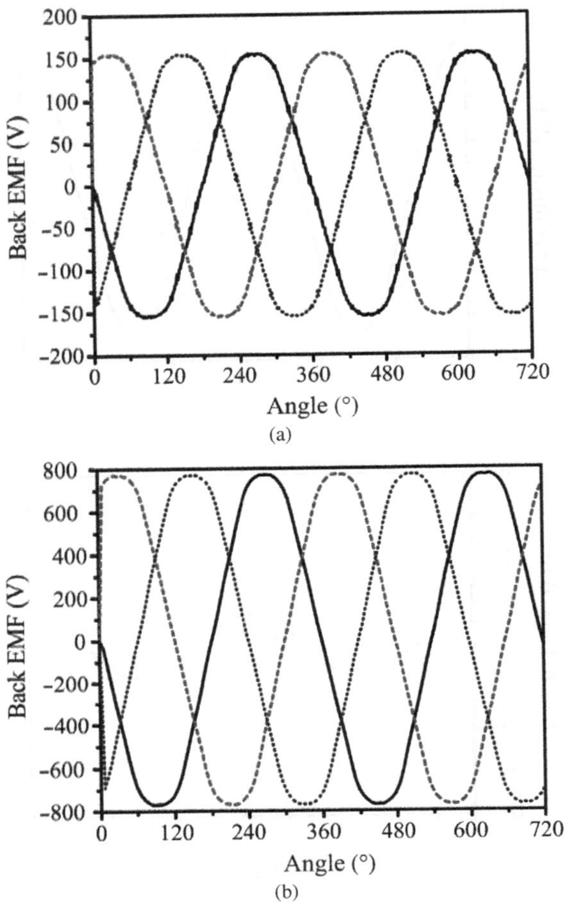

Figure 8.27 Back EMF waveforms of outer-rotor flux-controllable VPM machine without DC field control: (a) 200 rpm and (b) 1000 rpm

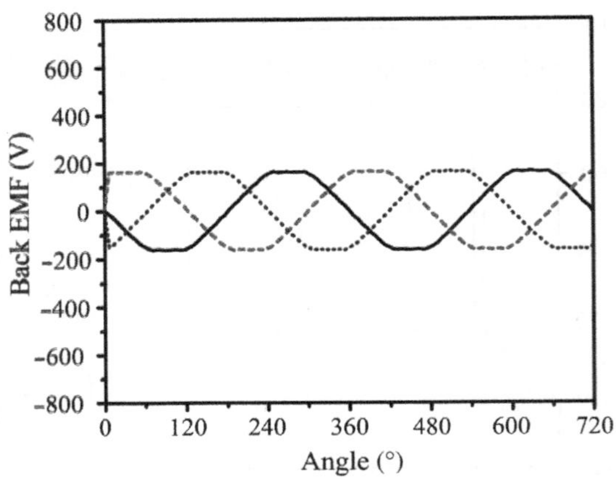

Figure 8.28 Back EMF waveform of outer-rotor flux-controllable VPM machine at 1000 rpm with DC field control

8.7 Potential Applications of Vernier PM Motor Drives in EVs

The application of VPM motor drives to EVs is rare. There was an example of using a VPM machine developed by a French company ERNEO to drive a light EV. This VPM machine adopted the conventional inner-rotor topology, and was designed to offer the steady torque of 33 N m and peak torque of 125 N m for over 1 minute (Rezzoug and Zaïm, 2013).

Although the VPM machine possesses the definite merit of low-speed high-torque operation, the use of conventional inner-rotor structure has limited its realistic application because it still needs a mechanism to transmit the rotor torque to the wheel. If the outer-rotor structure is adopted, this mechanism will be totally eliminated since the outer rotor can be directly coupled with the tire rim. Additionally, the inner space of the outer-rotor structure can be fully utilized to accommodate the flux modulation poles, the armature winding and the optional DC field winding.

Currently, the development of VPM machines for EV propulsion is much slower that they deserve. One main reason is due to the fact that the giant EV makers do not prefer to use the in-wheel direct drive for EV propulsion. When using the single-motor propulsion system, the use of PM synchronous machine with fixed gearing is generally sufficient to provide the desired low-speed high-torque operation. Nevertheless, with the ever-increasing demand on higher performance and better stability, the in-wheel direct drive will be popular for EV propulsion. The VPM motor drive will play an essential role in future EVs.

References

Enokijima, H., Oka, K. and Matsuse, K. (2011) Independent position control of two permanent magnet synchronous motor drives fed by a five-leg inverter. *Journal of International Council on Electrical Engineering*, **1**, 400–404.

Hosoya, R., and Shimomura, S. (2011) Apply to in-wheel machine of permanent magnet vernier machine using NdFeB bonded magnet – fundamental study. Proceedings of IEEE 7th International Power Electronics – ECCE Asia, pp. 2208–2215.

Ishizaki, A., Tanaka, T., Takasaki, K., and Nishikata, S. (1995) Theory and optimum design of PM vernier motor. Proceedings of IEE International Conference on Electrical Machines and Drives, pp. 208–212.

Jain, M., and Williamson, S.S. (2009) Suitability analysis of in-wheel motor direct drives for electric and hybrid electric vehicles. Proceedings of Electrical Power 0026 Energy Conference, pp. 1–5.

Jones, W.D. (2007) Putting electricity where the rubber meets the road. *IEEE Spectrum*, **44**, 18–20.

Kakihata, H., Kataoka, Y., Takayama, M. et al. (2012) Design of surface permanent magnet-type vernier motor. Proceedings of 15th International Conference on Electrical Machines and Systems, pp. 1–6.

Kataoka, Y., Takayama, M., Matsushima, Y. and Anazawa, Y. (2013) Comparison of three magnet array-type rotors in surface permanent magnet-type vernier motor. *Journal of International Conference on Electrical Machines and Systems*, **2**, 1–8.

Kominami, T., and Fujimoto, Y. (2007) A novel nine-switch inverter for independent control of two three-phase loads. Proceedings of IEEE Industry Applications Society Annual Conference, pp. 2346–2350.

Li, J. (2012) Design, analysis and control of permanent-magnet vernier machines. PhD Thesis. The University of Hong Kong.

Li, J. and Chau, K.T. (2012) Performance and cost comparison of permanent-magnet vernier machines. *IEEE Transactions on Applied Superconductivity*, **22**, 5202304:1–5202304:4.

Li, J., Chau, K.T., Jiang, J.Z. *et al.* (2010a) A new efficient permanent-magnet vernier machine for wind power generation. *IEEE Transactions on Magnetics*, **45**, 1475–1478.

Li, J., Wu, D., Zhang, X., and Gao, S. (2010b) A new permanent-magnet vernier in-wheel motor for electric vehicles. Proceedings of IEEE Vehicle Power and Propulsion Conference, pp. 1–6.

Li, X., Chau, K.T., Cheng, M. and Hua, W. (2013) Comparison of magnetic-geared permanent-magnet machines. *Progress In Electromagnetics Research*, **133**, 177–198.

Li, J., Chau, K.T. and Li, W. (2011) Harmonic analysis and comparison of permanent magnet vernier and magnetic-geared machines. *IEEE Transactions on Magnetics*, **47**, 3649–3652.

Liu, C., Zhong, J. and Chau, K.T. (2011) A novel flux-controllable vernier permanent-magnet machine. *IEEE Transactions on Magnetics*, **47**, 4238–4241.

Llibre, J.-F. and Matt, D. (1998) Harmonic study of the effort in the vernier reluctance magnet machine. Proceedings of International Conference on Electrical Machines, pp. 1664–1669.

Rezzoug, A. and Zaïm, M.E.-H. (2013) *Non-conventional Electrical Machines*, Wiley-ISTE.

Spooner, E. and Haydock, L. (2003) Vernier hybrid machines. *IEE Proceedings – Electric Power Applications*, **150**, 655–662.

Tasaki, Y., Kashitani, Y., Hosaya, R., and Shimomura, S. (2012) Design of the vernier machine with permanent magnets on both stator and rotor side. Proceedings of IEEE 7th International Power Electronics and Motion Control Conference – ECCE Asia, pp. 302–309.

Toba, A. and Lipo, T.A. (1999) Novel dual-excitation permanent magnet vernier machine. Proceedings of Industry Applications Conference, pp. 2539–544.

Toba, A. and Lipo, T.A. (2000) Generic torque-maximizing design methodology of surface permanent-magnet vernier machine. *IEEE Transactions on Industry Applications*, **36**, 1539–1546.

Wang, L.L., Shen, J.X., Luk, P.C.K. *et al.* (2009) Development of a magnetic-geared permanent magnet brushless motor. *IEEE Transactions on Magnetics*, **45**, 4578–4581.

Yang, J., Liu, G., Zhao, W. *et al.* (2013) Quantitative comparison for fractional-slot concentrated-winding configurations of permanent-magnet vernier machines. *IEEE Transactions on Magnetics*, **49**, 3826–3829.

9

Advanced Magnetless Motor Drives

Starting from the past decade, the permanent-magnet (PM) brushless motor drives have been superseding the induction motor drives for electric vehicle (EV) propulsion because of their outstanding torque density, power density, and operating efficiency. However, the supply of PM materials, especially rare-earth elements, is very limited and fluctuating that the corresponding market prices are soaring and volatile. Thus, in recent years, the development of advanced magnetless motor drives has become more and more attractive. Although these magnetless machines generally suffer from lower torque density, they can offer the definite merit of high cost-effectiveness. Meanwhile, various magnetless machine topologies have been developed for EV application.

In this chapter, advanced magnetless motor drives, including the synchronous reluctance (SynR), doubly-salient DC (DSDC), flux-switching DC (FSDC), vernier reluctance (VR), and doubly-fed vernier reluctance (DFVR) magnetless topologies as well as the axial-flux morphology, are presented. The corresponding design criteria, design examples, and potential applications for EV propulsion are also given.

9.1 What Is Advanced Magnetless?

With ever-increasing popularity of PM machines for EVs, the demand of neodymium-iron-boron (Nd-Fe-B) PM material is drastically soaring. The price of the raw material neodymium is the determining factor in pricing the Nd-Fe-B PM. As shown in Figure 9.1, it increased drastically between July 2009 and July 2011, and currently settles at several times of the initial price (Supermagnete, 2014). Both the absolute value and volatility of the neodymium price severely add uncertainty to the development of PM machines, and stimulate the research of advanced magnetless machines.

Initially, the family of magnetless machines is coined, aiming to distinguish it from the family of PM machines. Conceptually, the induction machine and switched reluctance (SR) machine are a kind of magnetless machines because they do not equip with any PMs. Since these two kinds of machines have been well developed and mature, they form their individual families. Therefore, the adjective "advanced" is incorporated to exclude them from those magnetless machines that are recently developed or relatively immature.

Focusing on those advanced magnetless machines that are viable for EV propulsion, five major types are identified:

- SynR machine
- DSDC machine

Electric Vehicle Machines and Drives: Design, Analysis and Application, First Edition. K. T. Chau.

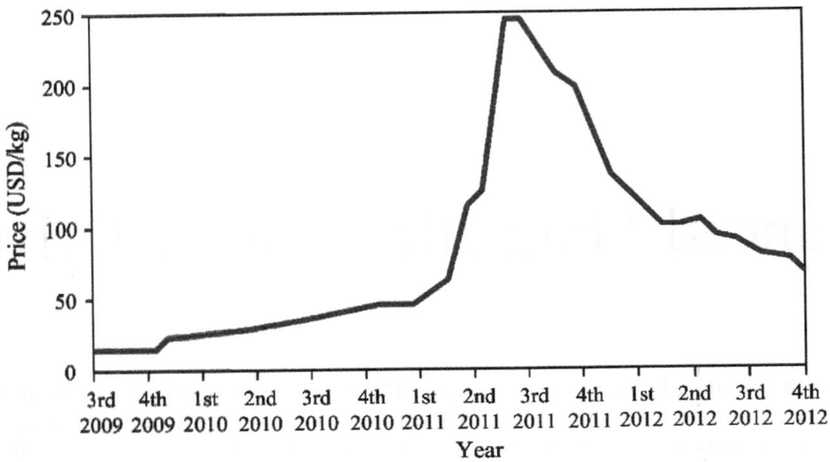

Figure 9.1 Price of neodymium

- FSDC machine
- VR machine
- DFVR machine

Meanwhile, both the radial-flux (RF) and axial-flux machine morphologies of the aforementioned five magnetless machine topologies are suitable for EV propulsion. Particularly, those axial-flux morphologies are attractive for the in-wheel motor drive application.

9.2 System Configurations

Similar to other motor drives, the advanced magnetless motor drives consist of four main components: magnetless machine, converter, sensor, and controller. There are different system configurations of the advanced magnetless motor drives, depending on the type of magnetless machines and the associated power converter.

Among them, the SynR machine and VR machine are singly fed from one power converter, whereas the others are doubly fed from two power converters. The system configuration of those singly-fed advanced magnetless motor drives is depicted in Figure 9.2 in which the armature winding in the stator is solely

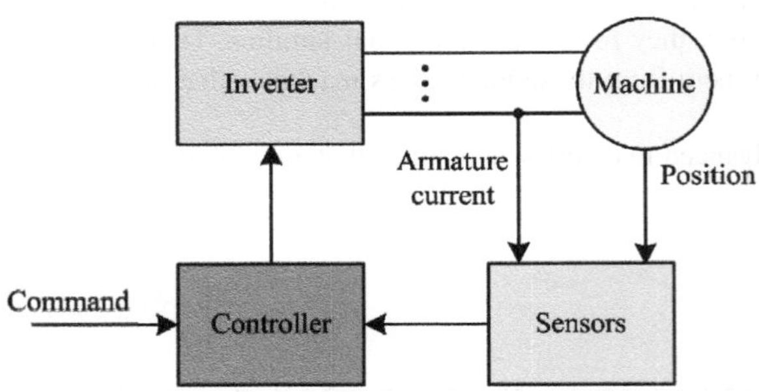

Figure 9.2 Configuration of singly-fed advanced magnetless motor drive

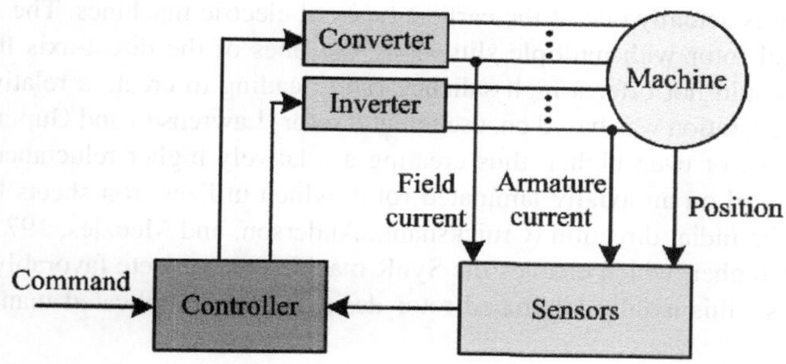

Figure 9.3 Configuration of doubly-fed advanced magnetless motor drive

fed by a power inverter. Meanwhile, the system configuration of those doubly-fed advanced magnetless motor drives is depicted in Figure 9.3 in which the armature winding is fed by an inverter while the field winding is fed by a converter. It should be noted that all windings of the advanced magnetless machines are located in the stator, and their rotors are all free from PMs or copper windings.

Between the singly-fed and doubly-fed motor drives, the former has the definite merit of simplicity, while the latter has additional controllability to improve the operating performances. Of course, both the machine topology and converter circuitry of the doubly-fed advanced magnetless motor drive are more complicated and costly.

9.3 Synchronous Reluctance Motor Drives

The PM synchronous machine has been accepted to provide outstanding power density and operating efficiency. However, it suffers from the drawbacks of high PM material cost, accidental demagnetization of PMs, and thermal instability of PMs. The SynR machine relies on reluctance torque, rather than the reaction torque, which is dominant in the cylindrical wound-rotor synchronous machine and the surface-mounted PM synchronous machine.

Compared with the SR machine, whose operation is also based on the principle of minimum reluctance, the SynR machine operates with the rotating field. This rotating field enables the machine offering smooth torque and stable operation down to low speeds, both of which are difficult to achieve in the SR machine. In addition, unlike the SR machine, the SynR machine is completely compatible with the stators and inverters of other AC motor drives.

The system configuration of SynR motor drive for EV propulsion is similar to that of the PM synchronous motor drive. It consists of the SynR motor, PWM (pulse-width modulation) inverter, electronic controller, sensor, and the optional mechanical transmission. There are a number of advantages associated with this system:

- The SynR machine does not require field windings or PMs in the rotor, hence offering high mechanical integrity to withstand high-speed operation.
- Since the SynR machine does not need expensive PM material, it is much less costly than the PM synchronous machine.
- As the SynR machine eliminates the accidental demagnetization of PMs, the motor drive can offer very high-current operation.
- As the SynR machine also eliminates the thermal instability of PMs, the motor drive can allow for working in high-temperature environment.

The SynR machine is actually one of the earliest types of electric machines. The first generation was based on a cylindrical rotor with multiple slits along the lines of the direct-axis flux (Kostko, 1923). This rotor structure could not offer a high saliency ratio, leading to create a relatively low reluctance torque. The second generation was based on a segmental rotor (Lawrenson and Gupta, 1967). Its saliency ratio could achieve five or even higher, thus creating a relatively higher reluctance torque. The latest third generation is based on an axially laminated rotor, which utilizes iron sheets bent into a U-shape and then stacked in the radial direction (Cruickshank, Anderson, and Menzies, 1971). Its saliency ratio can achieve seven or higher, which enables the SynR machine to compete favorably with the induction machine. Nevertheless, this axially laminated rotor desires more complicated manufacture and higher manufacturing cost.

A modern axially laminated SynR machine is shown in Figure 9.4. Its rotor is constructed of thin laminations that are bent into a semicircular shape. These iron segments are separated by insulation material, such as air or plastic. By selecting the ratio between the width of each iron segment and the width of each insulation material, the saliency ratio of this SynR machine can be optimized to achieve 10 or higher (Matsuo and Lipo, 1994).

Since the armature winding of the stator in the SynR machine is sinusoidally distributed, the air-gap flux harmonics are negligible. Hence, the equations that describe the behavior of the SynR machine can be derived from the conventional equations describing the conventional wound-field synchronous machine. Therefore, by eliminating the field excitation, the SynR machine can readily be modeled based on the *d–q* axis transformation as given by

$$v_{ds} = R_s i_{ds} + \frac{d\lambda_{ds}}{dt} - \omega_r \lambda_{qs} \tag{9.1}$$

$$v_{qs} = R_s i_{qs} + \frac{d\lambda_{qs}}{dt} + \omega_r \lambda_{ds} \tag{9.2}$$

where v_{ds} and v_{qs} are the *d–q* components of the stator voltage, i_{ds} and i_{qs} are the *d–q* components of the stator current, λ_{ds} and λ_{qs} are the *d–q* components of stator flux, R_s is the stator winding resistance, ω_r is the rotor speed, and magnetic saturation or iron loss is neglected. The saliency ratio is defined as

$$\kappa = \frac{L_d}{L_q} \tag{9.3}$$

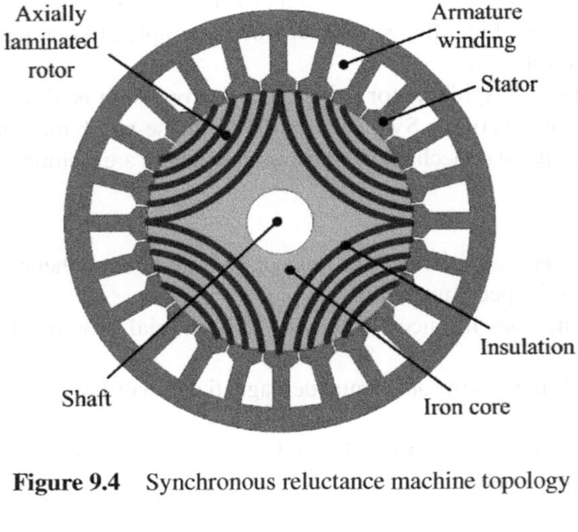

Figure 9.4 Synchronous reluctance machine topology

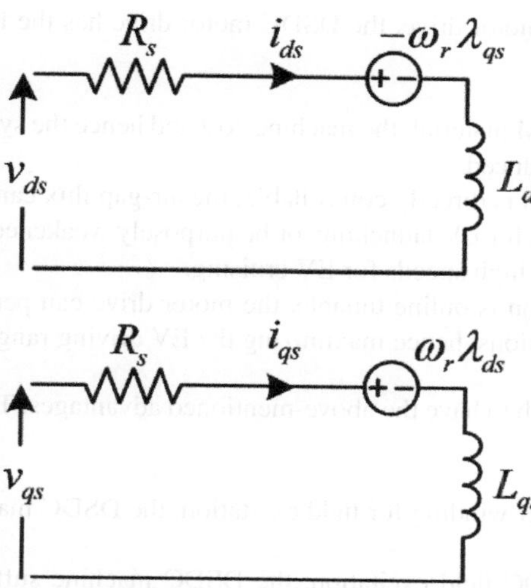

Figure 9.5 Equivalent circuit of synchronous reluctance machine

The stator flux components can be expressed as

$$\lambda_{ds} = (L_{md} + L_{ls})i_{ds} = L_{ds}i_{ds} \tag{9.4}$$

$$\lambda_{qs} = (L_{mq} + L_{ls})i_{qs} = L_{qs}i_{qs} \tag{9.5}$$

where L_{md} and L_{mq} are, respectively, the d-axis and q-axis magnetizing inductances, L_{ls} is the stator leakage inductance, and L_{ds} and L_{qs} are, respectively, the d-axis and q-axis stator inductances. Substituting Eqs. (9.4) and (9.5) into Eqs. (9.1) and (9.2), the equivalent circuit of the SynR machine can be deduced as shown in Figure 9.5. Consequently, the developed torque can be expressed as

$$T = \frac{3}{2}p\,(L_{ds} - L_{qs})\,i_{ds}i_{qs} \tag{9.6}$$

where p is the number of pole-pairs. It is obvious that the larger the value of κ, the higher the developed torque is resulted. Moreover, the saliency ratio of seven can enable the SynR machine operating at the power factor of about 0.8, which is comparable to the induction machine for EV propulsion.

Since the control of the SynR machine is similar to that of the PM synchronous machine, the control strategies that have been developed for the PM synchronous motor drive are applicable to the SynR motor drive. Such strategies include the field-oriented control (Xu *et al.*, 1991), direct torque control (Morales-Caporal and Pacas, 2008), efficiency optimizing control (Lee, Kang, and Sul, 1999), and sensorless control (Agarlita, Boldea, and Blaabjerg, 2012).

9.4 Doubly-Salient DC Motor Drives

The doubly-salient permanent-magnet (DSPM) machine is a kind of stator-PM brushless machines, which combines the advantageous features of both the PM brushless machine and SR machine. However, similar to the PM brushless machine, the DSPM machine suffers from uncontrollable PM flux for high-speed operation and high PM material cost. By directly replacing the PM excitation with the DC field winding, the DSDC machine is resulted, which consists of two types of stator windings: the polyphase armature winding and the DC field winding (Fan and Chau, 2008). Since the stator is doubly fed, this DSDC machine is also termed the stator-doubly-fed doubly-salient machine.

Compared with the DSPM motor drive, the DSDC motor drive has the following advantages that are viable for EV propulsion:

- Because of the absence of PM material, the machine cost and hence the system cost of the DSDC motor drive can be significantly reduced.
- Since the DC field excitation is directly controllable, the air-gap flux can be temporarily strengthened to boost up the motor torque for EV launching, or be purposely weakened to enable the motor offering constant-power operation at high speeds for EV cruising.
- While the DC field excitation is online tunable, the motor drive can perform efficiency optimization under various driving conditions, hence maximizing the EV driving range per charge.

Of course, there is no free lunch to have the above-mentioned advantages. The DSDC motor drive has the following drawbacks:

- Because of the use of copper winding for field excitation, the DSDC machine suffers from larger size and heavier weight.
- Because of the continual DC field excitation, the DSDC machine suffers from higher copper loss, leading to degrade the system efficiency.

Figure 9.6 depicts a three-phase 12/8-pole DSDC machine, which consists of both the three-phase concentrated armature winding and DC field winding in the stator, and simple iron core in the rotor. Both the stator and rotor have salient poles, but with different numbers. For instance, there are 12 salient poles in the stator and 8 salient poles in the rotor. Because there are no PMs, no brushes, and no windings in the rotor, it offers a very robust structure and the capability to run at very high speeds.

The design of the pole-pair arrangement of the DSDC machine is the same as that of the DSPM machine, which is governed by the following criteria:

$$\begin{cases} N_s = 2mk \\ N_r = N_s \pm 2k \end{cases} \tag{9.7}$$

$$\omega = \frac{60f}{N_r} \tag{9.8}$$

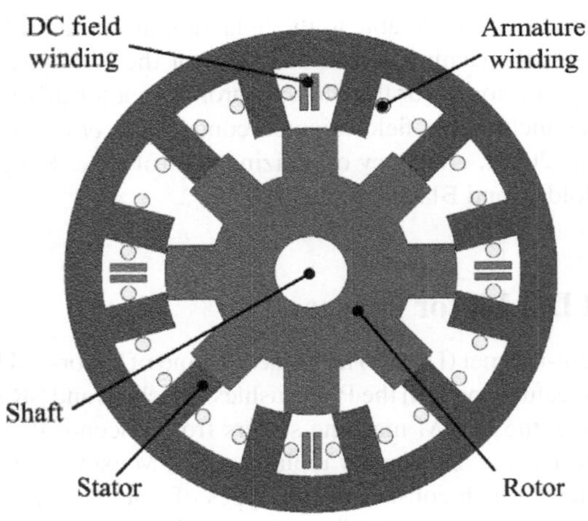

Figure 9.6 Structure of doubly-salient DC machine

where N_s the number of stator poles, N_r is the number of rotor poles, m is the number of phases, k is any integer, ω is the rotor speed, and f is the commutating frequency of any phase. In order to reduce the commutating frequency and hence the switching loss of power switches as well as the iron losses in poles and yokes, there should be as few rotor poles selected as possible. For the same rotor speed, the number of rotor poles is usually less than that of stator poles. Meanwhile, the number of phase numbers should be equal to or greater than three to enable the machine capable of starting itself in either the forward or reverse direction. Consequently, $N_s/N_r = 6/4, 8/6$, and $12/8$ are the three most viable configurations of the DSDC machine. Among them, the 6/4-pole and 12/8-pole are three-phase, whereas the 8/6-pole one is a four-phase configuration that certainly needs more power devices. When compared with the three-phase 6/4-pole one, the three-phase 12/8-pole machine possesses shorter flux paths in the yoke, resulting in lower iron losses. Also, because the flux per pole in the 12/8-pole machine is half of that in the 6/4-pole machine, the width of both the stator yoke and teeth is almost one-half of that of the 6/4-pole one. This enables a larger inner stator diameter and hence a larger rotor diameter, leading to achieve higher torque density. Moreover, a shorter width of stator teeth results in shorter end-windings, leading to reduce the winding resistance and save copper material used. Hence, the 12/8-pole DSDC machine can offer higher efficiency than the 6/4-pole one.

The operating waveforms of the DSDC machine are shown in Figure 9.7, which are similar to that of the DSPM machine. Since the flux linkage waveform is essentially trapezoidal, the brushless DC (BLDC) operation mode is generally adopted. When the flux linkage, ψ_{DSDC} is increasing and the no-load electromotive force (EMF) becomes positive, a positive armature current i_{DSDC} is applied to produce a positive torque T_{DSDC}. Similarly, a negative armature current is applied when the flux linkage is decreasing and the no-load EMF is negative, hence also producing a positive torque. Each phase performs 120° conduction. The resulting electromagnetic torque T_{DSDC} can be expressed as

$$\begin{aligned} T_{\mathrm{DSDC}} &= \frac{1}{2\pi}\int_0^{2\pi}\left(i_{\mathrm{DSDC}}\frac{d\psi_{\mathrm{DSDC}}}{d\theta} + \frac{1}{2}i_{\mathrm{DSDC}}^2\frac{dL_{\mathrm{DSDC}}}{d\theta}\right)d\theta \\ &= T_{\mathrm{DC}} + T_R \end{aligned} \tag{9.9}$$

where L_{DSDC} is the self-inductance, T_{DC} is the DC field torque component, and T_R is the reluctance torque component. In essence, the developed torque is mainly contributed by the DC field torque component, whereas the reluctance torque component is small and pulsates with a zero average value.

It should be noted that during high-speed operation, the positive current may not be able to reach its steady-state value before the armature winding is switched off at the proper rotor position angle, whereas

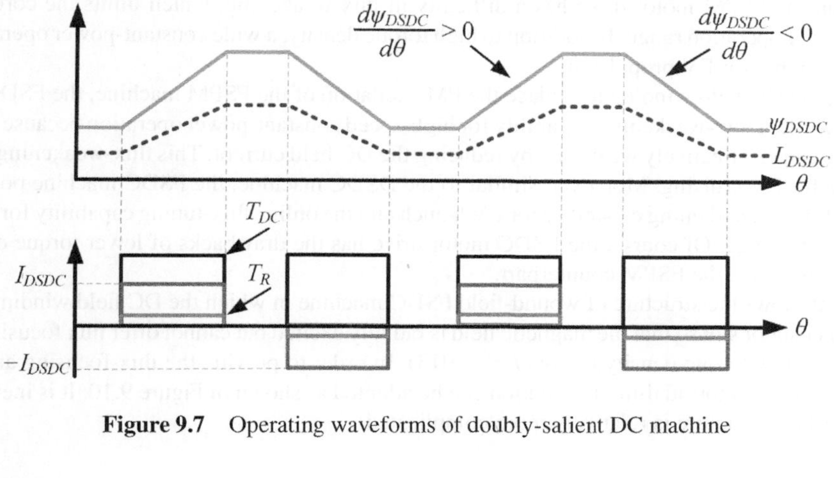

Figure 9.7 Operating waveforms of doubly-salient DC machine

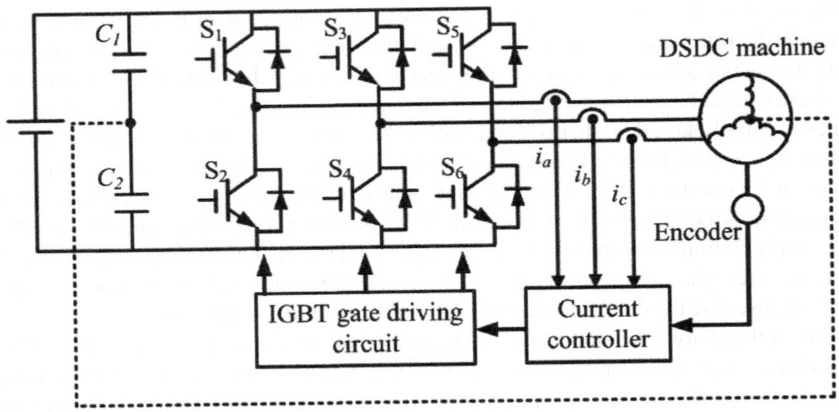

Figure 9.8 Half-bridge bipolar converter for DSDC machine

the negative current may quickly reach its steady-state value. Thus, the positive and negative currents need to be limited by using angle position control and current chopping control, respectively.

To feed the DSDC machine, a bipolar converter topology is required to offer bidirectional current operation. There are two possible converter topologies in which the phase current can be controlled independently for bidirectional operation: full-bridge converter and half-bridge converter with split capacitors. The half-bridge converter is usually adopted as shown in Figure 9.8 because it can minimize the number of power devices. A connection between the midpoint of the split capacitors and the neutral of machine windings is optionally required to accommodate the transient current during the commutation period.

9.5 Flux-Switching DC Motor Drives

Similar to the DSDC machine, the FSDC machine is a kind of doubly-salient brushless machines with both the armature winding and DC field winding in the stator. Differing from the DSDC machine, the FSDC machine allows for bipolar flux linkage and flux-focusing arrangement, leading to offer higher torque density.

With the flux-focusing arrangement in the stator, the flux-switching permanent magnet (FSPM) machine can offer high air-gap flux density and hence high torque density. However, the PM material cost and thermal instability of the FSPM machine are of concern. In addition, because of its severe magnetic saturation in the stator, the FSPM motor drive has a difficulty in flux weakening, which limits the corresponding constant-power operation range. In addition to high torque density, a wide constant-power operation range is highly desirable for EV propulsion.

By using the DC field winding to replace the PM excitation of the FSPM machine, the FSDC machine can offer excellent flux-weakening capability for high-speed constant-power operation because the air-gap flux density can be effectively weakened by reducing the DC field current. This flux-weakening capability is desirable for EV cruising. Moreover, similar to the DSDC machine, the FSDC machine possesses the temporary flux-strengthening capability for EV launch and the online flux-tuning capability for improving the EV driving range. Of course, the FSDC motor drive has the drawbacks of lower torque density and lower efficiency than the FSPM counterpart.

Figure 9.9 shows the structure of wound-field FSDC machine in which the DC field winding is wound in every other stator slot so that the magnetic field is radially excited but cannot offer flux focusing, leading to a relatively low torque density (Tang *et al.*, 2013). In order to provide the flux-focusing arrangement (Tang *et al.*, 2012), a toroidal-field excitation can be adopted as shown in Figure 9.10. It is inevitable that the toroidal-field structure is relatively more complicated.

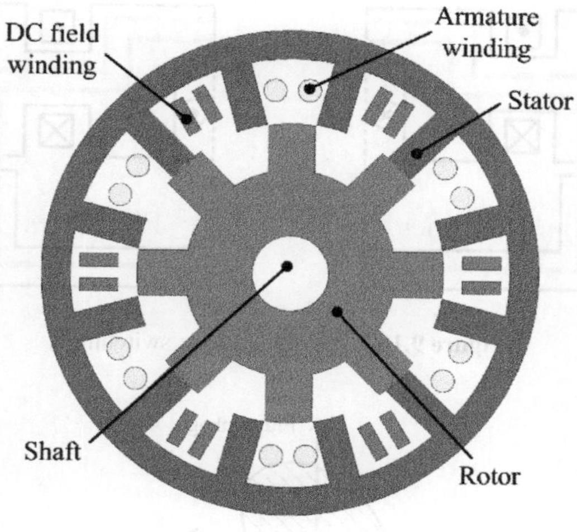

Figure 9.9 Structure of wound-field flux-switching DC machine

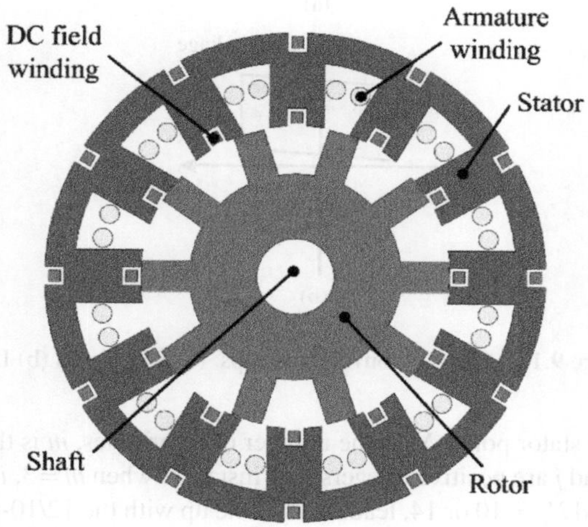

Figure 9.10 Structure of toroidal-field flux-switching DC machine

Differing from the DSDC machine, the FSDC machine exhibits bipolar flux linkage. That is, the corresponding polarity of the flux linked by the armature winding switches as the rotor rotates from one alignment to another alignment of stator and rotor poles. This flux-switching principle is illustrated in Figure 9.11. Moreover, the existence of bipolar flux linkage enables the FSDC machine to operate in all the four quadrants of the energy-conversion plane, the flux linkage versus the magnetomotive force (MMF), which actually distinguishes it from the DSDC machine as depicted in Figure 9.12.

On the basis of the use of toroidal-field winding, the structure of FSDC machine can be designed as governed by

$$\begin{cases} N_s = 2mi \\ N_r = N_s \pm 2j \end{cases} \tag{9.10}$$

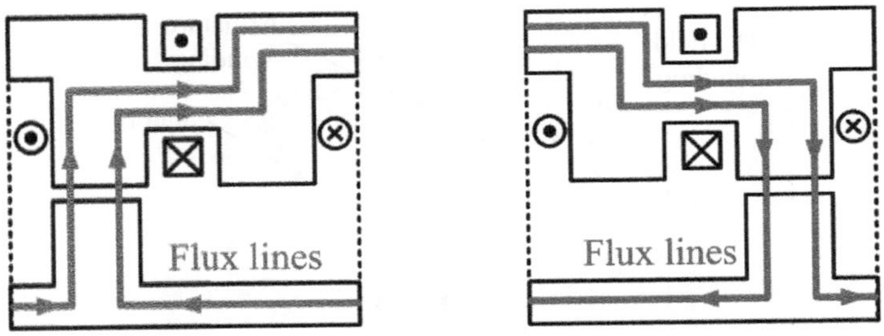

Figure 9.11 Principle of flux switching

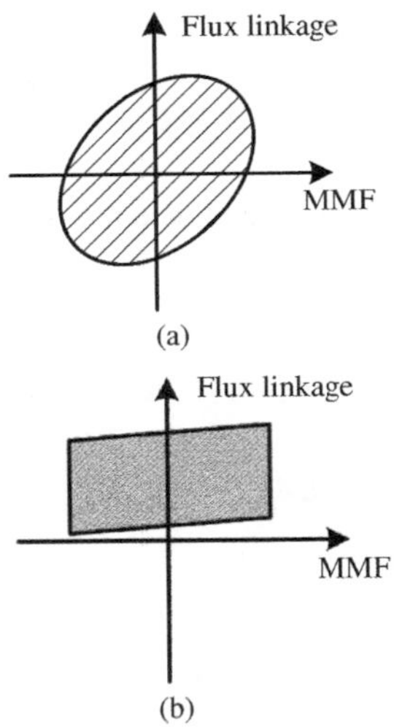

Figure 9.12 Energy-conversion loops: (a) FSDC and (b) DSDC

where N_s is the number of stator poles, N_r is the number of rotor poles, m is the number of phases of the armature winding, and i and j are positive integers. For instance, when $m=3$, $i=2$, and $j=1$ are selected, it ends up with $N_s = 12$ and $N_r = 10$ or 14, leading to come up with the 12/10-pole and 12/14-pole FSDC machine topologies.

Differing from the DSDC machine that it usually operates at the BLDC mode, the FSDC machine can operate at either the BLDC mode or brushless AC (BLAC) mode, depending on the shape of the flux linkage waveform (Lee *et al.*, 2014b). That is, when the armature adopts the concentrated winding arrangement, the flux linkage is essentially trapezoidal, which favors the BLDC operation mode; whereas when the armature adopts the distributed winding arrangement, the flux linkage is essentially sinusoidal, which favors the BLAC operation mode.

Figure 9.13 shows the operating waveforms of the FSDC machine in which the BLDC operation mode is employed. When the flux linkage ψ_{FSDC} is increasing, a positive armature current i_{FSDC} is applied to produce a positive torque T_{FSDC}. Also, when the flux linkage is decreasing, a negative armature current is applied hence producing a positive torque. Each phase performs 120° conduction. Hence, the electromagnetic torque T_{FSDC} under the BLDC operation mode can be expressed as

$$
\begin{aligned}
T_{\text{FSDC}} &= \frac{1}{2\pi}\int_0^{2\pi}\left(i_{\text{FSDC}}\frac{d\psi_{\text{FSDC}}}{d\theta} + \frac{1}{2}i_{\text{FSDC}}^2\frac{dL_{\text{FSDC}}}{d\theta}\right)d\theta \\
&= T_{\text{DC}} + T_R
\end{aligned}
\tag{9.11}
$$

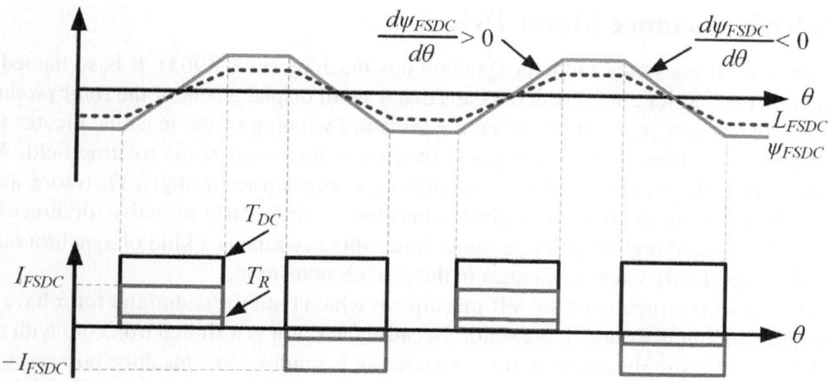

Figure 9.13 Operating waveforms of FSDC machine at BLDC mode

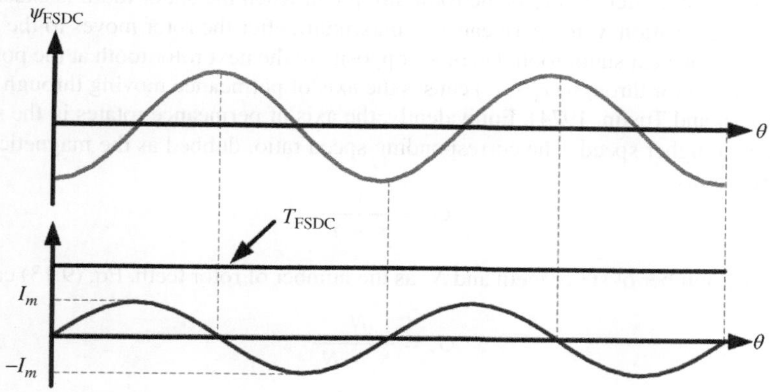

Figure 9.14 Operating waveforms of FSDC machine at BLAC mode

where L_{FSDC} is the self-inductance, T_{DC} is the DC field torque component, and T_R is the reluctance torque component. This DC field torque component is dominant and serves as the source of main torque, whereas the reluctance torque component is small and actually the source of torque ripple.

Figure 9.14 shows the operating waveforms of the FSDC machine operating at the BLAC mode. When the flux linkage ψ_{FSDC} is increasing, a positive armature current i_{FSDC} is applied to produce a positive torque; meanwhile, when the flux linkage is decreasing, a negative armature current is also applied to produce a positive torque. Each phase has 180° conduction angle. Hence, the electromagnetic torque T_{FSDC} under the BLAC operation mode can be expressed as

$$T_{FSDC} = \frac{3}{\omega}\frac{1}{2\pi}\int_0^{2\pi}((E_m \sin\theta)(I_m \sin\theta))d\theta$$
$$= \frac{3E_m I_m}{2\omega} \quad (9.12)$$

where I_m is the amplitude of the phase current, E_m is the amplitude of the fundamental component of the back EMF, and ω is the electrical angular velocity.

The power converter requirements and speed control strategies of the FSDC motor drive are essentially the same as the DSDC motor drive. Compared with the FSPM motor drive, the FSDC motor drive desires an additional DC–DC converter to control the DC field current while enjoying the flexibility to perform flux control.

9.6 Vernier Reluctance Motor Drives

The VR machine is an unexcited inductor synchronous machine (Lee, 1963). It is so named because it operates based on the vernier effect, which means that a small displacement of the rotor produces a large displacement of the axis of permeance. When the armature winding of the machine creates the rotating magnetic field, the rotor runs synchronously at a fraction of the speed of the rotating field. As the rotor speed steps down from the rotating field, the machine torque steps up accordingly. Therefore, as the vernier effect works as the magnetic gearing effect, the VR machine is particularly attractive for direct-drive applications desiring low-speed high-torque operation. Since this machine is a kind of synchronous machine, the torque is developed only when it operates in the synchronous mode.

Figure 9.15 shows the structure of the VR machine in which both the stator and rotor have open slots, and the armature winding is wound in the stator. Because the stator is a slotted iron core with an armature winding and the rotor is simply a slotted iron core without a winding, this machine possesses the definite merits of low cost, robust structure, and heat resistance. The slot pitch of the rotor is slightly different from that of the stator in such a way that there are some stator teeth opposite to rotor teeth (maximum permeance) and some stator teeth opposite to rotor slots (minimum permeance). As illustrated in Figure 9.16, where τ_s is the stator slot pitch and τ_r is the rotor slot pitch, when the stator tooth is exactly opposite to the rotor tooth at the position X, the permeance is maximum; after the rotor moves to the right by a displacement $(\tau_s - \tau_r)$, the next stator tooth becomes opposite to the next rotor tooth at the position Y. Thus, the movement of the rotor through $(\tau_s - \tau_r)$ causes the axis of permeance moving through τ_s in the same direction (Mukherji and Tustin, 1974). Equivalently, the axis of permeance rotates in the same direction as the rotor but at a higher speed. The corresponding speed ratio, dubbed as the magnetic gearing ratio, G_r, can be expressed as

$$G_r = \frac{\tau_s}{\tau_s - \tau_r} \tag{9.13}$$

Denoting N_s as the number of stator teeth and N_r as the number of rotor teeth, Eq. (9.13) can be rewritten as

$$G_r = \frac{N_r}{N_r - N_s} \tag{9.14}$$

It should be noted that when the rotor slot pitch is greater than the stator slot pitch $(\tau_r > \tau_s)$, the resulting G_r is negative, which indicates that the rotor rotates in the direction opposite to that of the rotating field. In

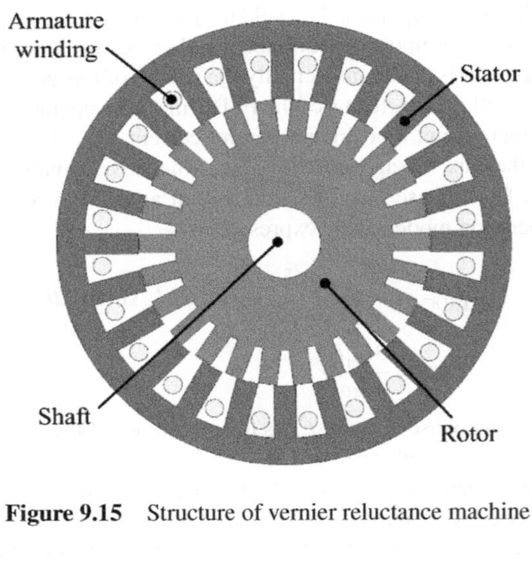

Figure 9.15 Structure of vernier reluctance machine

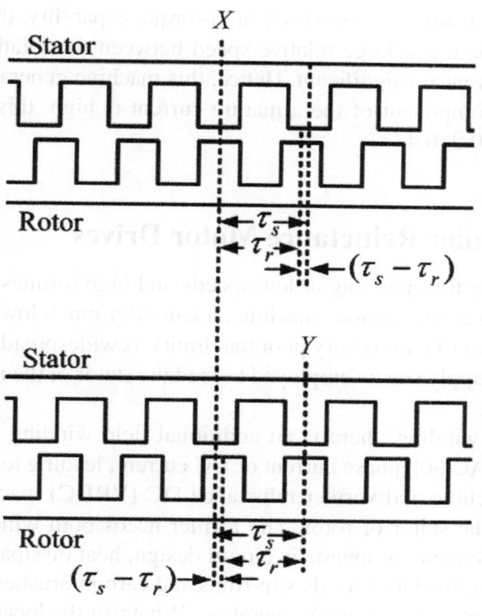

Figure 9.16 Principle of operation of vernier reluctance machine

general, it is more preferable to have the stator slot pitch greater than the rotor slot pitch ($\tau_s > \tau_r$) because it can provide more room to accommodate the stator winding.

In the VR machine, the air-gap permeance waveform should have the same number of poles as the stator MMF waveform. Thus, the relationship between the number of stator teeth N_s and the number of rotor teeth N_r is governed by

$$N_s = N_r \pm 2p \tag{9.15}$$

where p is the number of pole-pairs of the rotating magnetic field produced by the armature winding. When adopting the three-phase armature winding, the number of stator teeth can be expressed as

$$N_s = 6apq \tag{9.16}$$

where a is a positive integer and q is the number of slots per pole per phase. For instance, when $N_s = 24$ and $p = 1$ are selected, it ends up with $q = 4$, $a = 1$, and $N_r = 22$ or 26. As mentioned earlier, it is more preferable to have $\tau_s > \tau_r$ so that $N_r = 26$ is preferred. That is, it is a three-phase two-pole VR machine with 24 stator teeth and 26 rotor teeth as shown in Figure 9.15.

When the VR machine with p pole-pairs is fed by three-phase supply with frequency f, the rotating field speed or usually called synchronous speed ω_s in rpm is given by

$$\omega_s = \frac{60f}{p} \tag{9.17}$$

Substituting Eqs. (9.14) and (9.15) in Eq. (9.17), the rotor speed ω_r in rpm can be obtained as

$$\omega_r = \frac{120f}{N_r} \tag{9.18}$$

It indicates that the rotor speed is independent of the number of poles of the armature winding. Thus, the rotor speed is not affected by the number of poles but depends on the number of rotor teeth.

While offering the unique feature of low-speed high-torque capability, the VR machine suffers from two problems. First, since there is a large relative speed between the rotating field and rotor, the rotor core losses and stray load losses are significant. Hence, this machine generally exhibits lower efficiency. Second, since the reactive component of the armature current is high, this machine usually has a low power factor, in the order of 0.2–0.4.

9.7 Doubly-Fed Vernier Reluctance Motor Drives

The VR machine is attractive for operating at low speeds and high torques, particularly for direct-drive application. As it operates as a synchronous machine, it can offer much lower torque ripples than the SR counterpart. However, its power factor is very poor that limits its widespread application. In order to solve this problem, an additional supply can be employed to feed the stator of the machine, the so-called DFVR machine.

Apart from the armature winding, there is an additional field winding in the DFVR machine. This field winding can be fed by AC polyphase current or DC current, leading to create two members: vernier reluctance AC (VRAC) machine and vernier reluctance DC (VRDC) machine. Theoretically, the field winding can be located in the stator or rotor. The former needs both windings properly wound in the stator, which desires more demanding magnetic circuit design, heat dissipation arrangement, and manufacturing process. In contrast, the latter needs slip-rings and carbon brushes to access the rotor circuitry, which is undesirable for maintenance-free EV operation. Whatever the location of field winding adopted, because of the variation of the air-gap permeance and the difference between the pole-pair numbers of the armature and field windings, the DFVR machine can run synchronously at low speeds with improved power factor.

Figure 9.17 shows the structure of the VRAC machine in which there are N_s teeth in the stator and N_r teeth in the rotor. The armature winding is a traditional three-phase distributed winding with p_a pole-pairs, and the field winding is another three-phase distributed winding with p_f pole-pairs (Taibi, Tounzi, and Piriou, 2006). Both windings are located in the stator.

Assuming that the permeability of iron is infinite, the magnetic energy or coenergy, W_e of this VRAC machine is stored only in the air-gap and can be expressed as

$$W_e = \frac{1}{2}\int_0^{2\pi} (F_a + F_f)^2 P(\theta_s, \theta) d\theta_s \tag{9.19}$$

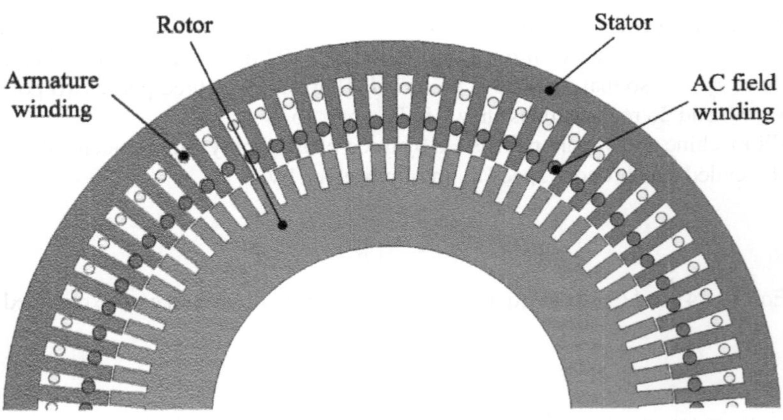

Figure 9.17 Structure of vernier reluctance AC machine

where F_a and F_f are the MMF created by the armature and field windings, respectively, and P is the air-gap permeance per angle, θ_s is the position of a given point in the air-gap with respect to the stator frame, and θ is the angle between the rotor and stator frames.

Since the armature and field windings are traditional three-phase distributed ones, their MMFs can be written as (Tounzi, Ramdane, and Zaim, 2008)

$$F_a = \sum_{i=0}^{\infty} \frac{3}{\pi} \frac{Z_a I_{am}}{(2i+1)} (-1)^i K_{aw}^i \cos(\omega_a t - (2i+1)p_a\theta_s) \tag{9.20}$$

$$F_f = \sum_{i=0}^{\infty} \frac{3}{\pi} \frac{Z_f I_{fm}}{(2i+1)} (-1)^i K_{fw}^i \cos(\omega_f t - (2i+1)p_f\theta_s) \tag{9.21}$$

where K_{aw}^i and K_{fw}^i are their winding coefficients, Z_a and Z_f are their conductor numbers per slot, I_{am} and I_{fm} are their current magnitudes, and ω_a and ω_f are their angular frequencies. Meanwhile, the air-gap permeance can be written as

$$\begin{aligned} P(\theta_s, \theta) = P_0 + \sum_{j=1}^{\infty} P_{js} \cos(jN_s\theta_s) + \sum_{k=1}^{\infty} P_{kr} \cos(kN_r(\theta_s - \theta)) \\ + \frac{1}{2} \sum_{j=1}^{\infty} \sum_{k=1}^{\infty} P_{js-kr} \cos(jN_s\theta_s + \varepsilon kN_r(\theta_s - \theta)) \end{aligned} \tag{9.22}$$

where P_0, P_{js}, P_{kr}, and P_{js-kr} are coefficients depending on the geometric characteristics and $\varepsilon = \pm 1$. By substituting Eqs. (9.20)–(9.22) in Eq. (9.19), the developed torque can be obtained as

$$T = \left. \frac{\partial W_e}{\partial \theta} \right|_{\text{constant current}} \tag{9.23}$$

In order to yield the developed torque, the numbers of stator and rotor teeth as well as the numbers of armature and field pole-pairs are governed by

$$|N_s + \varepsilon\, N_r| = |p_a + \varepsilon\, p_f| \tag{9.24}$$

In order to develop the torque using only the interaction between the fields of armature and field windings, the numbers of stator and rotor teeth as well as the numbers of armature and field pole-pairs are governed by

$$|N_s + \varepsilon\, N_r| \neq 2p_a \neq 2p_f \tag{9.25}$$

$$|N_s + \varepsilon\, N_r| \neq p_a \neq p_f \tag{9.26}$$

Hence, the developed torque can be written as

$$T = \frac{4}{\pi} N_r F_{am} F_{fm} P_{1s-1f} \sin((\omega_a + \varepsilon\omega_f)t + \varepsilon N_r(\theta - \theta_0)) \tag{9.27}$$

where F_{am} and F_{fm} are the magnitudes of the armature MMF and field MMF, respectively, and θ_0 is a constant angle. This torque is constant when the rotor rotates at the synchronous speed ω_r as given by

$$\omega_r = \left| \frac{\omega_a + \varepsilon\, \omega_f}{\varepsilon\, N_r} \right| \tag{9.28}$$

Thus, the machine operates in a manner similar to that of a traditional synchronous machine.

From the desired synchronous speed, the armature winding frequency ($f_a = \omega_a/2\pi$), the field winding frequency ($f_f = \omega_f/2\pi$), the number of rotor teeth can be deduced from Eq. (9.28). For instance, when $\omega_r = 50$ rpm, $f_a = 50$ Hz, and $f_f = 8.33$ Hz are selected, it ends up with $N_r = 70$; then, using Eqs. (9.24)–(9.26), $N_s = 72$, $p_a = 6$, and $p_f = 4$ can be resulted. It should be noted that there are many possible solutions satisfying the design equations. The rule of thumb is to minimize the difference between the numbers of stator and rotor teeth so as to get a smooth sinusoidal air-gap permeance.

When the field winding of this VRAC machine is connected in series to provide the same number of pole-pairs and then fed by DC field current, it becomes the VRDC machine. The corresponding synchronous speed can be deduced from Eq. (9.28) as given by

$$\omega_r = \frac{\omega_a}{N_r} \tag{9.29}$$

For example, when the machine having 70 rotor teeth is supplied with a frequency of 50 Hz, the operating rotor speed is 43 rpm.

Figure 9.18 shows another viable structure of VRDC machine, also dubbed as the dual-winding reluctance machine, in which the armature winding and DC field winding are wound with the nonoverlapping concentrated winding arrangement (Fukami *et al.*, 2010). The use of concentrated winding arrangement can simplify the winding structure as well as reduce the end-winding and the amount of copper material used. There are p_a pole-pairs of the armature winding, p_f pole-pairs of the DC field winding, and p_r rotor pole-pairs, which is equivalent to $2p_r$ rotor teeth.

The key of operation is to make use of the reluctance rotor to modulate the stationary DC field, and to magnetically couple with the rotating armature field. That is, when the DC field winding is excited, a stationary field with p_f pole-pairs is created in the air-gap. This field is modulated by the reluctance rotor to form a rotating field with p_a pole-pairs, which then couples magnetically with the rotating armature field with p_a pole-pairs. Hence, the machine operates as a synchronous machine with $(p_a + p_f)$ pole-pairs. To enable this operation, the pole-pair numbers of the armature winding, field winding, and rotor are related by

$$p_r = \frac{p_a + p_f}{2} \tag{9.30}$$

Hence, the number of rotor teeth, N_r is given by

$$N_r = 2p_r = p_a + p_f \tag{9.31}$$

Since each stator tooth is wound as a field pole, the number of stator teeth N_s is given by

$$N_s = 2p_f \tag{9.32}$$

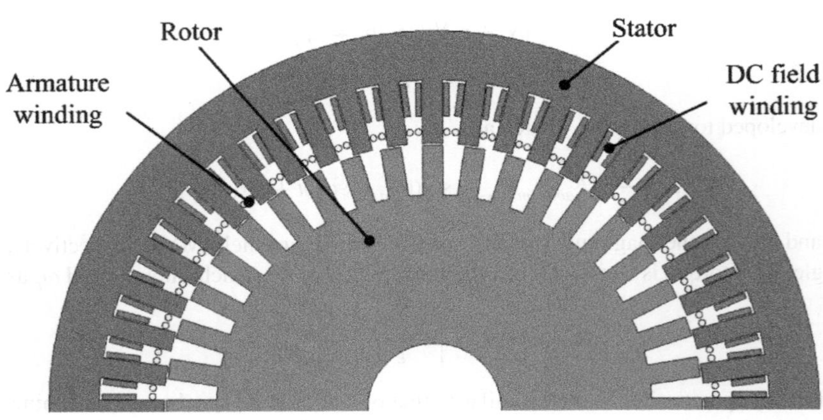

Figure 9.18 Structure of vernier reluctance DC machine

In order to magnetically couple the modulated DC field and the armature field, the relationship between the pole-pair numbers of the armature and field windings is given by

$$p_f = K\, p_a \tag{9.33}$$

where K is the number of stator teeth per armature pole. Consequently, the rotor speed can be written as

$$\omega_r = \frac{\omega_a}{N_r} = \frac{2\omega_a}{p_a + p_f} \tag{9.34}$$

which indicates that the machine operates as a synchronous machine with $(p_a + p_f)$ pole-pairs. Equivalently, it offers the magnetic gearing ratio equal to the number of rotor teeth as given by

$$G_r = N_r \tag{9.35}$$

Therefore, this dual-winding reluctance machine is a kind of VRDC machine, but with more specific design and operation.

On the basis of the desired rotor speed and armature winding frequency, the number of rotor teeth can be calculated by using Eq. (9.34). Then, the numbers of armature and field winding pole-pairs as well as the number of stator teeth can be deduced from Eqs. (9.31)–(9.33). For instance, when $\omega_r = 75$ rpm and $f_a = 50$ Hz are selected, it ends up with $N_r = 40$ and then, $p_a = 16$, $p_f = 24$, and $N_s = 48$.

9.8 Axial-Flux Magnetless Motor Drives

On the basis of the direction of magnetic flux paths and location of current-carrying conductors, all machine topologies can generally be represented as RF, axial-flux, and linear-flux morphologies. Namely, all machine topologies, such as induction, synchronous, SR, and advanced magnetless machines, can be designed with the axial-flux morphology. For instance, the DSDC machine topology can have different morphological representations as shown in Figure 9.19.

While most machines are developed in the RF morphology, where the air-gap flux is radially directed, the axial-flux machine exhibits an axially directed air-gap flux. The corresponding stator usually has a ring structure and the rotor is a disc, with almost the same active inside and outside diameters. The radial length from the stator inside diameter to outside diameter is the active part of the machine to develop the torque. The axial length depends on the flux density in the stator and rotor yokes. Since both the stator and rotor can be utilized with a proper yoke design, the axial-flux machine generally offers higher torque density and higher power density than its RF counterpart. As the number of poles increases, the radial active part of the machine remains unchanged. Hence, the axial-flux machine in the pancake shape is particularly advantageous for low-speed high-torque motor drives, such as the direct-drive in-wheel motor (Profumo, Zhang, and Tenconi, 1997).

The basic axial-flux magnetless (AFM) machine is the single-stator single-rotor structure as depicted in Figure 9.20. The major problem of this design is the large axial force exerted on the stator by the rotor. This axial force can be alleviated by adopting the single-stator double-rotor structure or the double-stator single-rotor structure as depicted in Figures 9.21 and 9.22, respectively. That is, the axial force between one pair of rotor and stator can be canceled out by the axial force between another pair of rotor and stator.

The single-stator double-rotor structure allows two different arrangements that result in different flux distributions as shown in Figure 9.23. First, the armature windings in the stator are designed in such a way that the stator current flows in reverse directions in each of the two back-to-back stator slots. Since the end-winding length is almost equal to the stator yoke axial length, it can save copper material and reduce the copper loss. However, this arrangement needs a large stator yoke size. Second, the two identical armature windings in the stator are connected in such a way that the stator current flows in the same direction in each of the two back-to-back stator slots. There is only one main flux that links the stator windings and two rotors. Thus, it has a small stator yoke or may be even yokeless, which can reduce the core material and iron loss, but suffers from long end-windings which result in high copper loss.

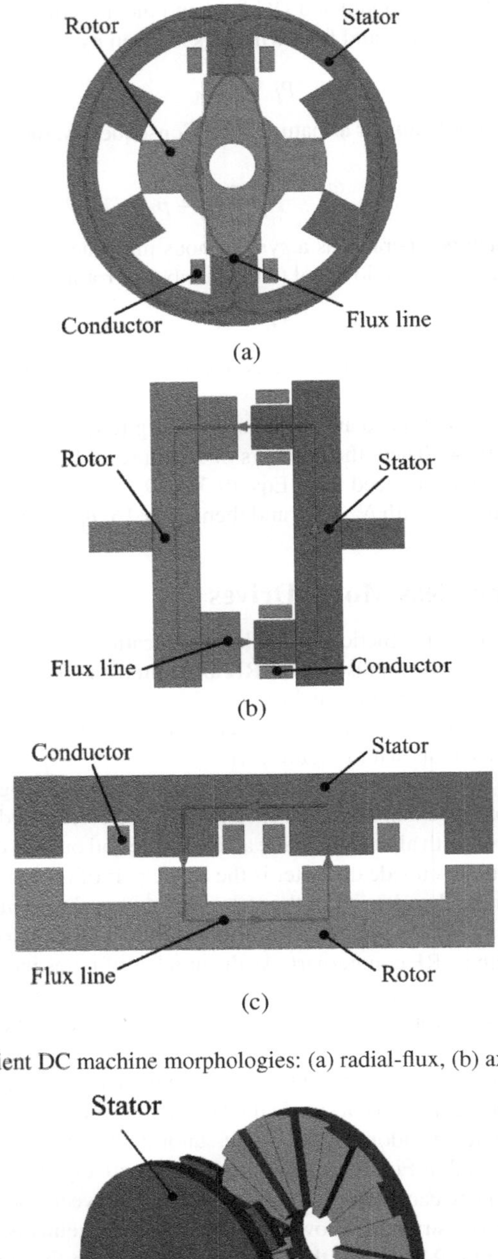

Figure 9.19 Doubly-salient DC machine morphologies: (a) radial-flux, (b) axial-flux, and (c) linear-flux

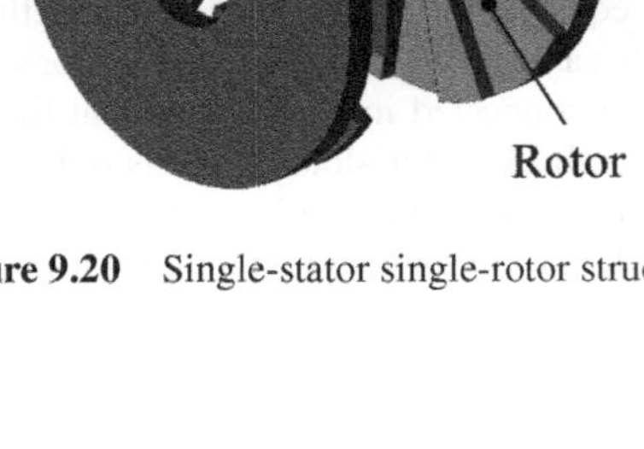

Figure 9.20 Single-stator single-rotor structure

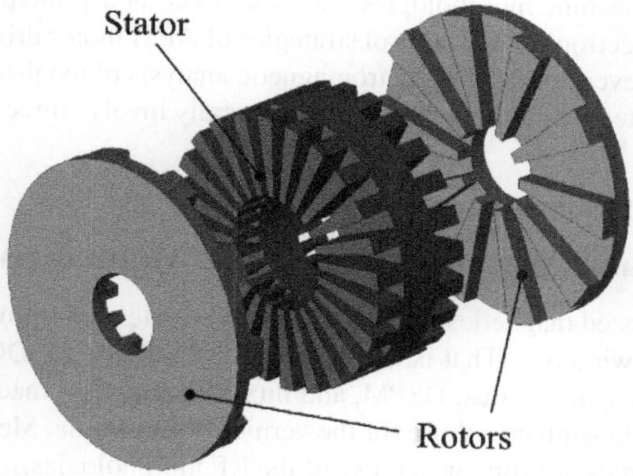

Figure 9.21 Single-stator double-rotor structure

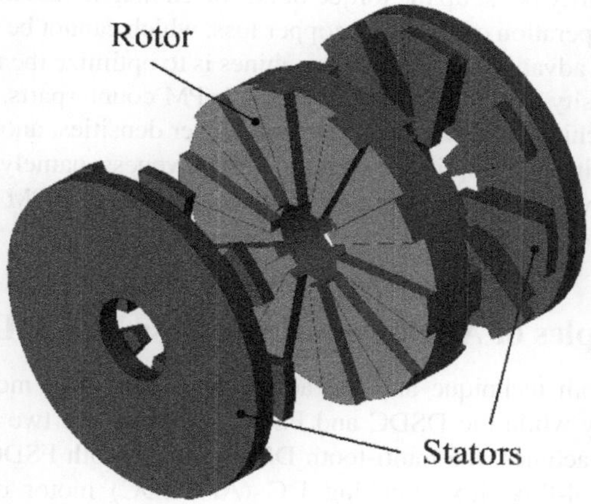

Figure 9.22 Double-stator single-rotor structure

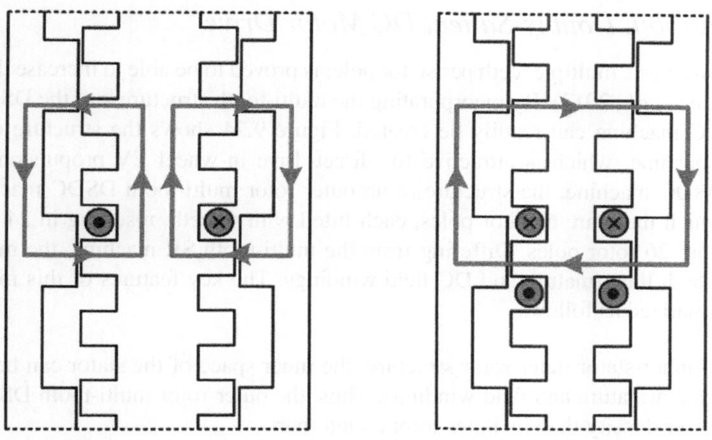

Figure 9.23 Flux arrangements of single-stator double-rotor structure

Since the difference of machine morphologies does not affect their principle of operation, the mathematical modeling, power electronics, and control strategies of AFM motor drives are essentially the same as their RF counterparts. Nevertheless, the electromagnetic analysis of axial-flux machines is much more difficult than their RF counterparts because the former generally involve three-dimensional finite element analysis.

9.9 Design Criteria of Advanced Magnetless Motor Drives for EVs

The design criteria of advanced magnetless motor drives can be extended from their PM counterparts with the PMs replaced by field windings. That is, the criteria for the SynR, DSDC, and FSDC machines are similar to those for the PM synchronous, DSPM, and flux-switching PM machines, respectively, and the VR and DFVR machines are similar to those for the vernier PM machine. Meanwhile, the design criteria of the axial-flux morphologies are similar to those of the RF morphologies.

Because of the absence of high-energy PM material for field excitation, the advanced magnetless machines generally suffer from lower torque density than their PM counterparts. Although the increase of field current can temporarily boost up the torque of advanced magnetless machines, such as the DSDC and FSDC machines, this operation causes high copper loss, which cannot be sustainable. Therefore, one of the key design criteria of advanced magnetless machines is to optimize the machine dimensions, hence maximizing the torque density to be comparable with their PM counterparts. On the other hand, instead of competing with PM machines in terms of torque and power densities, another key design criterion of advanced magnetless machines is to optimize their cost-effectiveness, namely the torque per cost (N m/$) or power per cost (kW/$), which enables them favorably competing with PM machines for application to low-cost EVs.

9.10 Design Examples of Advanced Magnetless Motor Drives for EVs

Since the use of multi-tooth technique and the adoption of axial-flux morphology can significantly improve the torque density while the DSDC and FSDC machines are two most promising topologies of advanced magnetless machines, the multi-tooth DSDC, multi-tooth FSDC, axial-flux doubly-salient DC (AF-DSDC), and axial-flux flux-switching DC (AF-FSDC) motor drives are used for design exemplification.

9.10.1 Multi-tooth Doubly-Salient DC Motor Drive

The multi-tooth structure, multiple teeth per stator pole, is proved to be able to increase the torque density of SR machine (Lee *et al.*, 2013). By incorporating the multi-tooth structure into the DSDC machine, the multi-tooth DSDC machine can readily be created. Figure 9.24 shows the structure of an outer-rotor multi-tooth SR machine, which is attractive for direct-drive in-wheel EV propulsion. Extending this concept to the DSDC machine, the structure of an outer-rotor multi-tooth DSDC machine is shown in Figure 9.25 in which there are 6 stator poles, each fitted with 4 teeth, resulting in 24 equivalent stator poles, and there are 26 rotor poles. Differing from the multi-tooth SR machine, the multi-tooth DSDC machine installs both the armature and DC field windings. The key features of this multi-tooth DSDC machine are summarized as follows:

- Because of the inner-stator outer-rotor structure, the inner space of the stator can be fully utilized to accommodate the armature and field windings. Thus, the outer-rotor multi-tooth DSDC machine can offer higher power density than its inner-rotor counterpart.
- With the use of multi-tooth structure, the torque density can be further improved. Thus, the multi-tooth DSDC machine can offer higher torque density than the DSDC machine.

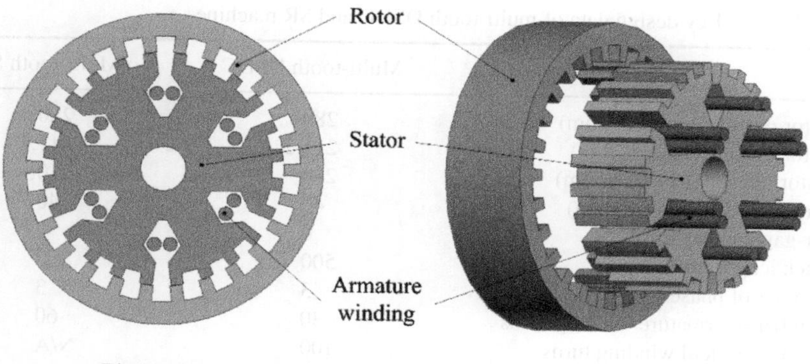

Figure 9.24 Structure of outer-rotor multi-tooth SR machine

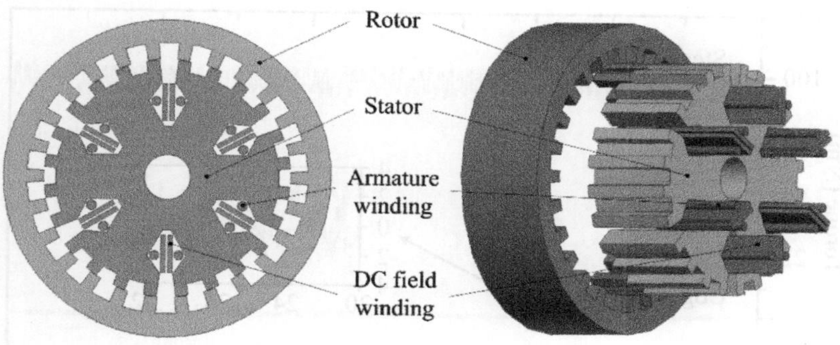

Figure 9.25 Structure of outer-rotor multi-tooth DSDC machine

- With the use of DC field winding, both the inductance-rising and inductance-falling zones can be utilized for torque production. Thus, the multi-tooth DSDC machine can offer higher torque density and lower torque ripple than the multi-tooth SR machine.

Since the multi-tooth DSDC machine is extended from the multi-tooth SR machine, its design equations involve the design criteria of both the multi-tooth SR machine and the DSDC machine, which are governed by

$$\begin{cases} N_{sp} = 2mk \\ N_{se} = N_{sp} N_{st} \\ N_r = N_{se} + 2k \end{cases} \tag{9.36}$$

where N_{sp} is the number of stator poles, N_{st} is the number of stator teeth per pole, N_{se} is the equivalent number of stator poles, N_r is the number of rotor poles, m is the number of phases, and k is an integer. For instance, by selecting $N_{sp} = 6$, $N_{st} = 4$, $m=3$, and $k=1$, it ends up with $N_{se} = 24$ and $N_r = 26$ as depicted in Figure 9.25. Then, the machine dimensions, namely the stator diameter, rotor diameter, core length, pole arc, pole height, and yoke thickness, are designed in such a way that the magnetic saturation and hence the core losses can be minimized.

As the principle of operation of the multi-tooth SR machine is the same as that of the traditional SR machine, the principle of operation of the multi-tooth DSDC machine is also the same as that of the DSDC machine. The key design data of this multi-tooth DSDC machine are listed in Table 9.1, where similar data of the multi-tooth SR machine are also listed for comparison.

Table 9.1 Key design data of multi-tooth DSDC and SR machines

	Multi-tooth DSDC	Multi-tooth SR
Rotor outside diameter (mm)	280	280
Rotor inside diameter (mm)	211.2	211.2
Stator outside diameter (mm)	210	210
Stator inside diameter (mm)	40	40
Air-gap length (mm)	0.6	0.6
Stack length (mm)	500	500
Number of phases	3	3
Number of armature winding turns	30	60
Number of field winding turns	100	N/A

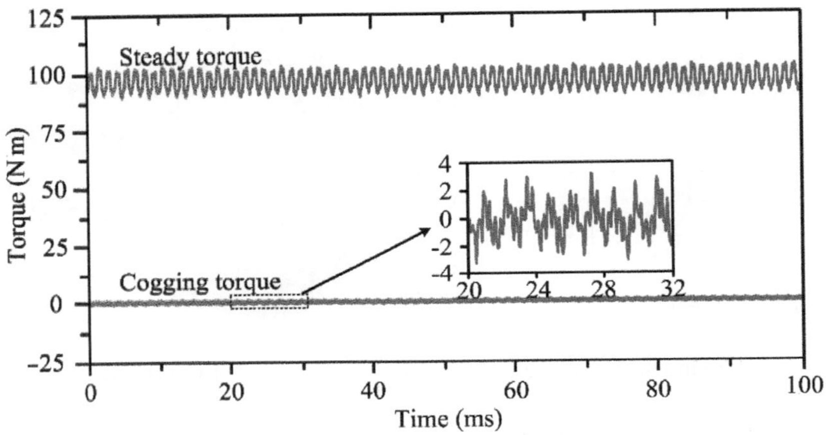

Figure 9.26 Torque waveforms of multi-tooth DSDC machine

By using finite element analysis, the output torque waveforms of this multi-tooth DSDC machine are simulated. As shown in Figure 9.26, the rated torque is 92.3 N m, and the rated cogging torque is 6.4 N m. The corresponding torque ripple is 19.6%, which is comparable with the normal DSDC machine. It should be noted that the cogging torque is the detent torque due to the interaction between the stator and rotor teeth in the presence of field excitation. Therefore, it can be considered as the torque at no-load condition. Meanwhile, the torque ripple is a measure of torque fluctuation, which is defined as the ratio of the peak-to-peak difference to the average value.

The overall performances of this multi-tooth DSDC machine are compared with those of the multi-tooth SR machine as listed in Table 9.2. It can be found that the multi-tooth DSDC machine can significantly improve the power density and torque density by about 2.6 times and suppress the torque ripple by 3.3 times as compared with its multi-tooth SR counterpart.

9.10.2 Multi-tooth Flux-Switching DC Motor Drive

Figure 9.27 shows the structure of outer-rotor multi-tooth FSDC machine in which there are 6 stator poles, each fitted with 4 teeth, resulting in 24 equivalent stator poles, and there are 20 rotor poles. When compared with the multi-tooth DSDC machine, the multi-tooth FSDC machine possesses some advantageous features:

Table 9.2 Key performance data of multi-tooth DSDC and SR machines

	Multi-tooth DSDC	Multi-tooth SR
Rated power (kW)	4.8	1.8
Rated speed (rpm)	500	500
Rated armature current (A/mm^2)	5	5
Rated field excitation (A-turn)	500	N/A
Rated torque (N m)	92.3	35.2
Cogging torque (N m)	6.4	N/A
Torque ripple (%)	19.6	64.3

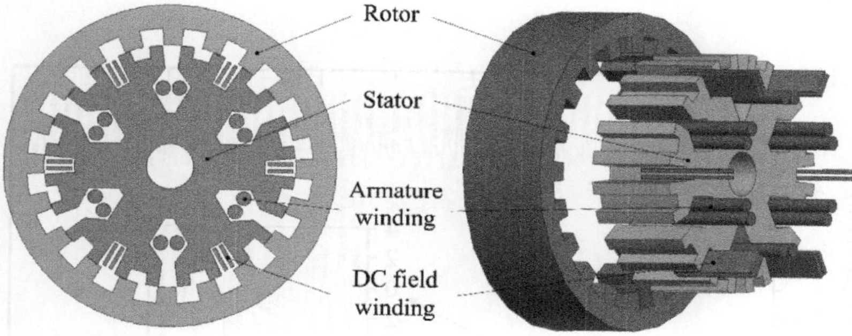

Figure 9.27 Structure of outer-rotor multi-tooth FSDC machine

- Both the armature and DC field windings of the multi-tooth DSDC machine are located in the same slot, whereas the windings of the multi-tooth FSDC machine are located in different slots. Thus, the multi-tooth FSDC machine can accommodate more winding turns, and provide easier winding process.
- The flux linkage variation of the multi-tooth DSDC machine is unipolar, whereas the flux linkage variation of the multi-tooth FSDC machine is bipolar. Hence, the multi-tooth FSDC machine can fully utilize the iron core, thereby offering higher torque density and higher power density.

The design equations of the multi-tooth FSDC machine involve the design criteria of both the multi-tooth SR machine and the FSDC machine, which are governed by

$$\begin{cases} N_{sp} = 2mi \\ N_{se} = N_{sp} N_{st} \\ N_r = N_{se} - N_{sp} \pm 2j \end{cases} \tag{9.37}$$

where N_{sp} is the number of stator poles, N_{st} is the number of stator teeth per pole, N_{se} is the equivalent number of stator poles, N_r is the number of rotor poles, m is the number of phases, and i and j are integers. For instance, by selecting $N_{sp} = 6$, $N_{st} = 4$, $m = 3$, $i = 1$, and $j = 1$, it ends up with $N_{se} = 24$ and $N_r = 20$ as shown in Figure 9.27. Then, the machine dimensions, namely the stator diameter, rotor diameter, core length, pole arc, pole height, and yoke thickness, are designed in such a way that the magnetic saturation and hence the core losses can be minimized. The key design data of this multi-tooth FSDC machine are listed in Table 9.3, in which similar data of the multi-tooth SR machine are also listed for comparison.

The principle of operation of the multi-tooth FSDC machine is the same as that of the FSDC machine. On the basis of finite element analysis, the output torque waveforms of the multi-tooth FSDC machine

Table 9.3 Key design data of multi-tooth FSDC and SR machines

	Multi-tooth FSDC	Multi-tooth SR
Rotor outside diameter (mm)	280	280
Rotor inside diameter (mm)	211.2	211.2
Stator outside diameter (mm)	210	210
Stator inside diameter (mm)	40	40
Air-gap length (mm)	0.6	0.6
Stack length (mm)	500	500
Number of phases	3	3
Number of armature winding turns	60	60
Number of field winding turns	80	N/A

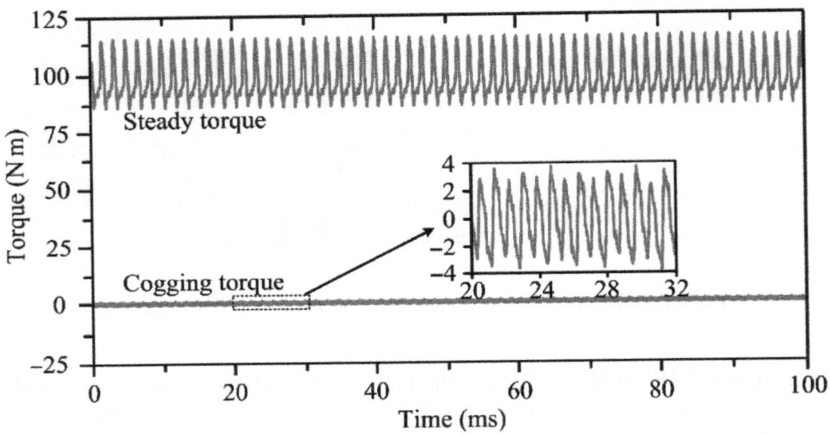

Figure 9.28 Torque waveforms of multi-tooth FSDC machine

are simulated as shown in Figure 9.28. It can be observed that the rated torque is 105.8 N m, while the rated cogging torque is 7.8 N m. The corresponding torque ripple is 31.6%, which is comparable with the normal FSDC machine. The overall performances of this multi-tooth FSDC machine are compared with those of the multi-tooth SR machine as listed in Table 9.4. It confirms that the multi-tooth FSDC machine can offer much higher power density and torque density as well as much lower torque ripple than its multi-tooth SR counterpart. Moreover, by comparing Tables 9.2 and 9.4, it can be found that the multi-tooth FSDC machine can offer higher power density and higher torque density, while suffering from higher cogging torque and higher torque ripple than the multi-tooth DSDC counterpart.

9.10.3 Axial-Flux Doubly-Salient DC Motor Drive

An AF-DSDC machine is designed for EV in-wheel motor drive. The specifications of this in-wheel motor drive are listed in Table 9.5. Figure 9.29 shows the topology of this AF-DSDC machine, which adopts the double-stator single-rotor structure. It installs with both the armature winding and DC field winding in each of the stator, and both windings adopt the concentrated winding arrangement (Lee, Liu, and Chau, 2014). On the basis of this structure with proper dimensions, the machine is able to mount its

Table 9.4 Key performance data of multi-tooth FSDC and SR machines

	Multi-tooth FSDC	Multi-tooth SR
Rated power (kW)	5.5	1.8
Rated speed (rpm)	500	500
Rated armature current (A/mm^2)	5	5
Rated field excitation (A-turn)	500	N/A
Rated torque (N m)	105.8	35.2
Cogging torque (N m)	7.8	N/A
Torque ripple (%)	31.6	64.3

Table 9.5 Specifications of in-wheel motor drive

Peak DC voltage	360 V
Rated power	4.7 kW
Rated torque	150 N m
Constant-torque operation	0–300 rpm
Constant-power operation	300–900 rpm
Peak torque	130% for 10 s
Wheel dimension	195/65 R15

rotor to the tire directly so that the in-wheel direct drive can be achieved as shown in Figure 9.30. The machine is designed based on a four-phase topology. The corresponding design equations are the same as the radial-flux doubly-salient DC (RF-DSDC) machine. Thus, the pole arrangements of this AF-DSDC machine are selected as $N_s = 8$ and $N_r = 10$, resulting in 8 poles in each stator and 10 poles on each side of the rotor. The key design data are tabulated in Table 9.6.

Since each of the two sets of windings is installed with their magnetic axes parallel to each other and with opposite directions, the magnetic flux generated by the DC field winding is virtually divided into two equal paths in the rotor yoke as depicted in Figure 9.31. Hence, the attractive force between the stator 1 and rotor can be canceled out with the attractive force between the stator 2 and rotor, thus solving the key drawback that normally occurs at the single-stator single-rotor axial-flux machines. In addition, in order to ease the operation complexity, the sets of windings in the two stators are purposely connected in series.

The power electronics for this AF-DSDC machine is the same as the RF-DSDC machine. That is, the full-bridge inverter is equipped to feed the armature windings while providing the independent phase control capability. Meanwhile, the full-bridge DC–DC converter is equipped to control the DC field windings while providing bidirectional DC field current to facilitate the flux strengthening and weakening capabilities.

Similar to the RF-DSDC machine, the AF-DSDC machine normally operates at the BLDC mode. Moreover, it normally operates at the bipolar conduction scheme in order to develop torque during both inductance-rising and inductance-falling zones. Nevertheless, it can operate at the unipolar conduction scheme when the DC field winding is under open-circuit fault. In fact, the DSDC motor drive operating at the BLDC mode under the unipolar conduction scheme is equivalent to an SR motor drive. To be more specific, the armature current at the unipolar conduction scheme needs to be boosted up so that the resulting machine torque should be comparable with that offered at the bipolar conduction scheme, leading to suffer from degraded efficiency during this fault-tolerant operation. Since this unipolar conduction scheme utilizes only half of the torque-producing period, the corresponding torque ripple is also worse than that at the bipolar conduction scheme.

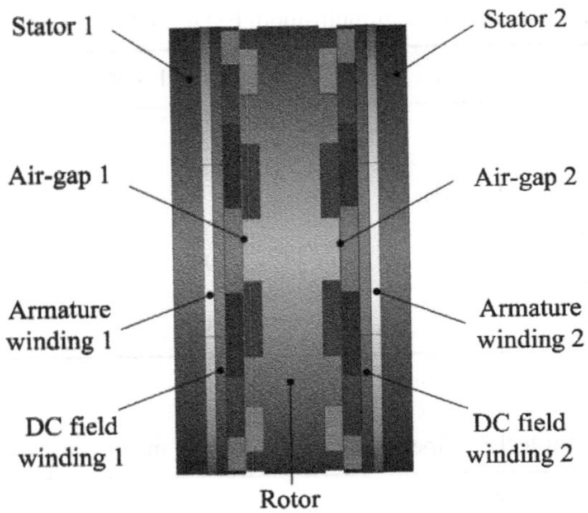

(a)

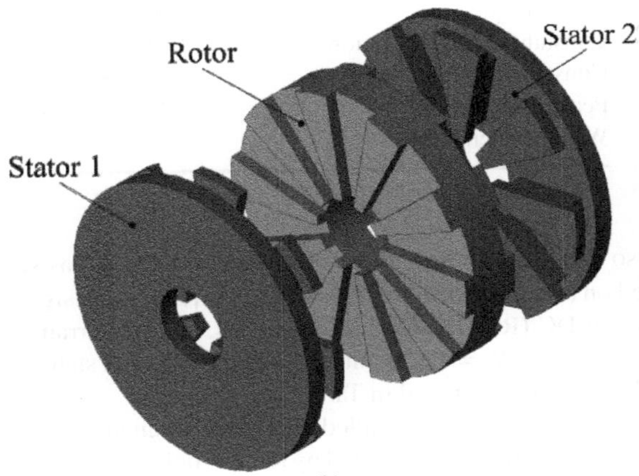

(b)

Figure 9.29 Axial-flux DSDC machine: (a) topology and (b) structure

Differing from the electromagnetic field analysis of RF machines, the analysis of axial-flux machines generally involves three-dimensional finite element analysis. Figure 9.32 shows the no-load magnetic field distribution of the AF-DSDC machine. It can be found that the magnetic flux in the rotor yoke is divided into two symmetrical paths passing through the two stators, illustrating that the attractive forces of the two stators exerted on the rotor are canceled out. Then, the flux density waveforms at both air-gaps under the no-load condition are shown in Figure 9.33, where the DC field excitation is set at the rated current density of 5 A/mm^2. It can be found that both air-gap flux density waveforms are almost the same, confirming that the two stators can develop almost the same torque.

The torque waveform of the AF-DSDC machine under the rated armature current of 5 A/mm^2 and rated DC field excitation of 5 A/mm^2 is shown in Figure 9.34. It can be observed that the developed torque can achieve an average value of 154.2 N m, which can satisfy the desired specification of rated torque listed in Table 9.5. Since the DC field winding can be independently controlled, the air-gap flux density can be flexibly regulated. When the DC field excitation is temporarily strengthened to 10 A/mm^2, the developed torque can achieve an average value of 203.2 N m as shown in Figure 9.35, which is very useful for temporarily boosting the torque for EV launch or overtaking.

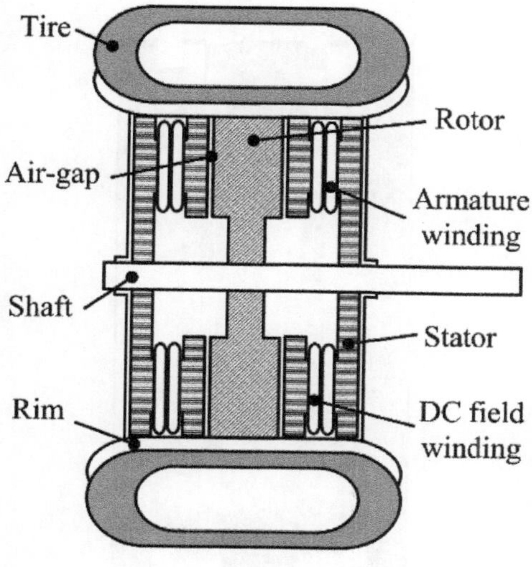

Figure 9.30 Structure of in-wheel direct drive using AF-DSDC machine

Table 9.6 Key design data of AF-DSDC machine

Radial outside diameter	381 mm
Radial inside diameter	100 mm
Axial stack length	195 mm
Stator inside diameter	100 mm
Length of each air-gap	0.5 mm
Number of poles on each stator face	8
Number of poles on each rotor face	10
Number of armature phases	4
Number of turns per armature coil	50
Number of turns per field coil	50

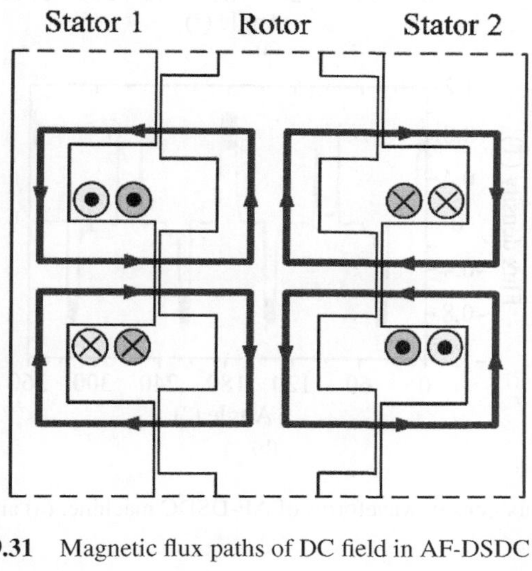

Figure 9.31 Magnetic flux paths of DC field in AF-DSDC machine

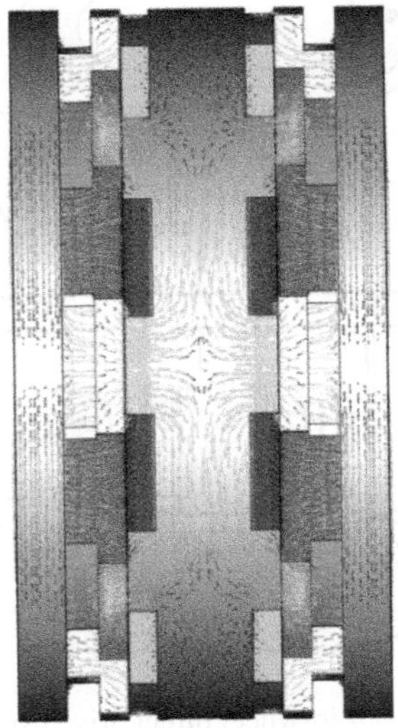

Figure 9.32 No-load magnetic field distribution of AF-DSDC machine

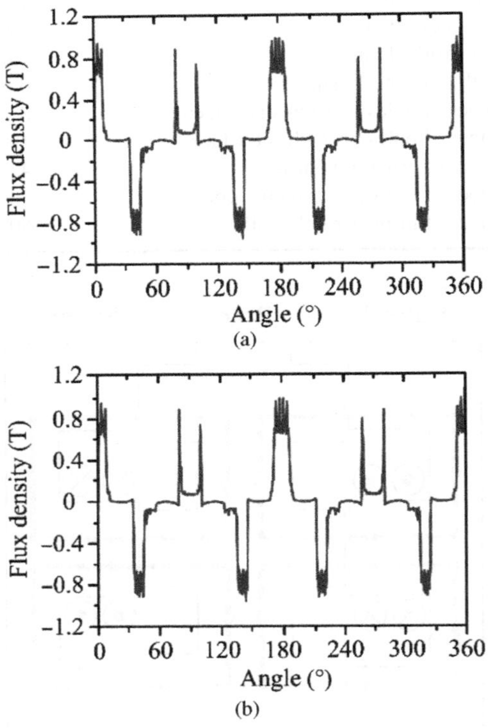

Figure 9.33 Air-gap flux density waveforms of AF-DSDC machine: (a) air-gap 1 and (b) air-gap 2

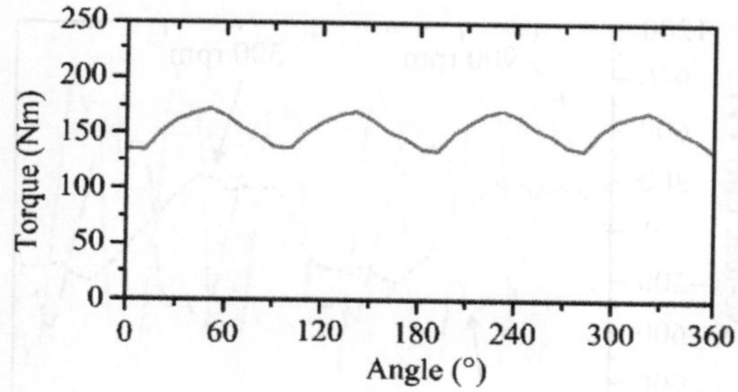

Figure 9.34 Torque waveform of AF-DSDC machine under rated DC field excitation

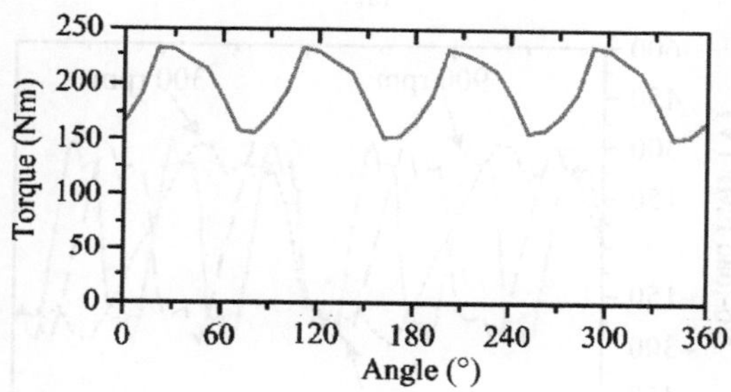

Figure 9.35 Torque waveform of AF-DSDC machine under strengthened DC field excitation

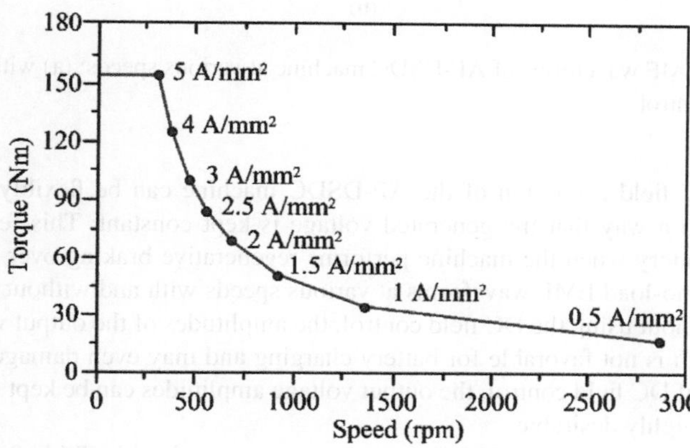

Figure 9.36 Torque–speed capability of AF-DSDC machine

Differing from the DSPM machine, the DSDC machine possesses the definite merits that the field excitation can be flexibly controlled, hence achieving high flux-weakening capability for wide speed constant-power operation. The torque–speed capability of the AF-DSDC machine is simulated as shown in Figure 9.36 in which the DC field current is weakened with the increase of speed. It confirms that the machine can maintain constant-power operation over a very wide speed range. Particularly, with the DC field excitation of 1.5 A/mm^2, the operating speed is 907 rpm, which can fully cover the desired constant-power speed range as listed in Table 9.5.

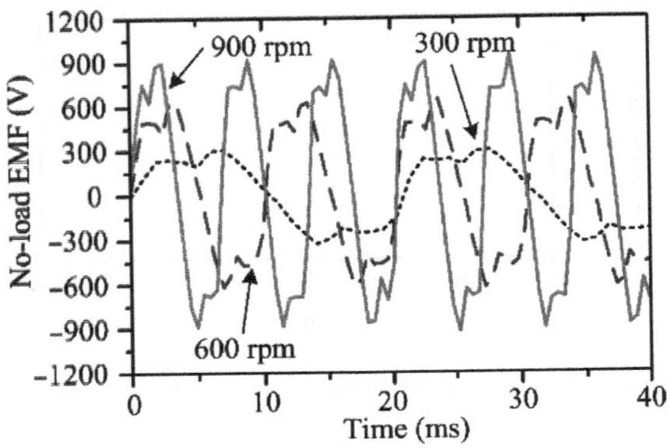

(a)

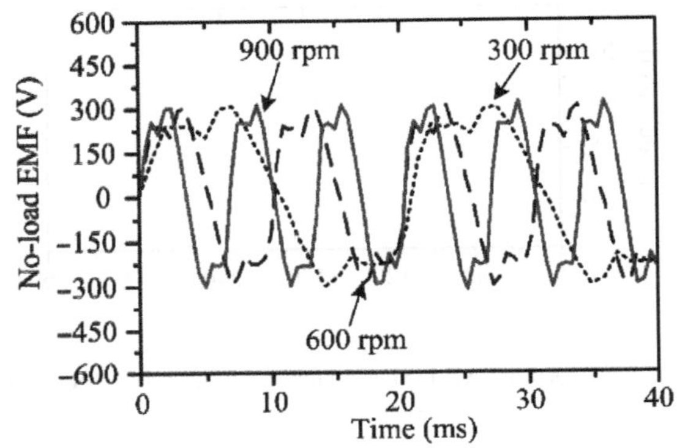

(b)

Figure 9.37 No-load EMF waveforms of AF-DSDC machine at various speeds: (a) without DC field control and (b) with DC field control

Furthermore, the DC field excitation of the AF-DSDC machine can be flexibly tuned to vary with the rotor speed in such a way that the generated voltage is kept constant. This feature is particularly useful to charge the battery when the machine performs regenerative braking over a wide speed range. Figure 9.37 shows the no-load EMF waveforms at various speeds with and without employing DC field control. Obviously, without using the DC field control, the amplitudes of the output voltage vary with the operating speeds, which is not favorable for battery charging and may even damage the battery. On the contrary, with the use of DC field control, the output voltage amplitudes can be kept constant over a wide speed range, which is highly desirable.

The overall performances of the AF-DSDC machine are summarized in Table 9.7. In order to justify the use of axial-flux morphology, the performances of the AF-DSDC machine are compared with three RF counterparts, namely the RF-SR, RF-DSDC, and RF-DSPM machines. To have a fair comparison, all machines adopt the same dimensions, namely the radial outside diameter, radial inside diameter, axial stack length, and air-gap length. In addition, the winding factors and current densities are selected to be the same. On the basis of finite element analysis, their key performances are calculated as summarized in Table 9.8. As expected, since the RF-SR machine utilizes only half of its torque-producing period, the corresponding gravimetric and volumetric rated torque densities are the worst. Meanwhile, by utilizing all the torque-producing periods, the RF-DSDC machine can offer better torque performances than the RF-SR machine, but still much lower than those of the RF-DSPM machine. With the use of axial-flux morphology, the AF-DSDC machine can offer much higher torque densities than its RF-DSDC counterpart, and become

Table 9.7 Overall performances of AF-DSDC machine

Rated power	4.8 kW
Power density	35.4 W/kg
Operating frequency at base speed	50 Hz
No-load EMF of armature winding 1	153 V
No-load EMF of armature winding 2	153 V
Rated DC field excitation	5 A/mm^2
Rated torque	154.2 N m
Torque ripple at rated torque	27.4%
Boosted DC field excitation	10 A/mm^2
Boosted torque	203.2 N m

Table 9.8 Performance comparison of axial-flux and radial-flux machines

	RF-SR	RF-DSDC	RF-DSPM	AF-DSDC
Rated torque/mass (N m/kg)	0.51	0.61	1.21	1.13
Rated torque/volume (kN m/m^3)	4.12	4.85	9.51	8.97
Material cost (USD)	208.4	209.8	411.9	239.5
Rated torque/cost (N m/USD)	0.34	0.41	0.39	0.65

comparable with those of the RF-DSPM machine. Therefore, it illustrates that the advanced magnetless machine can offer the torque densities in par with its PM counterpart. Although the RF-DSPM machine can still offer slightly higher torque densities than the AF-DSDC machine, the corresponding material cost is much higher because of the use of expensive PM material. To assess the cost-effectiveness of these machines, the rated torque per unit cost is calculated. It can be found that the AF-DSDC machine is most cost-effective, and actually much better than the RF-DSPM machine.

9.10.4 Axial-Flux Flux-Switching DC Motor Drive

Similar to the previous case, an AF-FSDC machine is designed for EV in-wheel motor drive with the same specifications as listed in Table 9.5. Differing from the previous case, the topology of this AF-FSDC machine adopts a single-stator double-rotor structure rather than a double-stator single-rotor one, as shown in Figure 9.38. It installs with both the armature winding and DC field winding on both sides of the stator, and both windings adopt the concentrated winding arrangement (Lee *et al.*, 2014a). On the basis of this structure, the machine is directly mounted with the tire to achieve the in-wheel direct drive. The corresponding design equations are the same as the radial-flux FSDC machine as given by Eq. (9.10). Thus, when the pole arrangements of this AF-FSDC machine are selected as $m=3$, $i=2$, and $j=1$, it ends up with $N_s = 12$ and $N_r = 10$. That is, it is a three-phase AF-FSDC machine with 12 poles on both sides of the stator and 10 poles in each rotor. Because of the inherent pole split of each stator pole of the FSDC machine, the 12 stator poles can be considered as the 24 equivalent stator poles. The key design data are summarized in Table 9.9.

As the two sets of windings are installed with their magnetic axes perpendicular to each other, the magnetic flux generated by the DC field winding in the stator is divided into two equal paths as shown in Figure 9.39. Hence, the attractive force between the stator and rotor 1 can be canceled out with the attractive force between the stator and rotor 2, thus solving the key drawback of those single-stator single-rotor axial-flux machines.

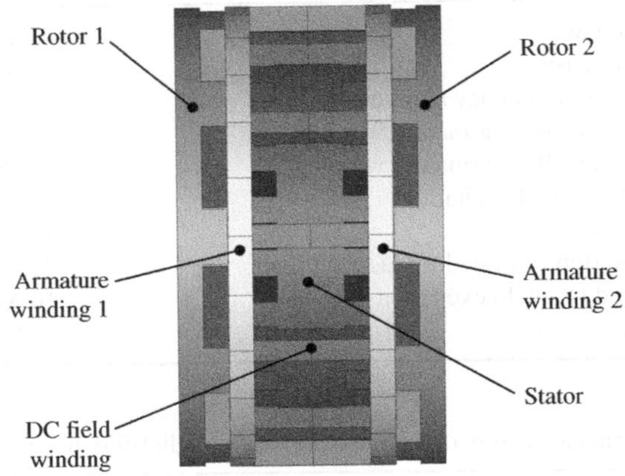

(a)

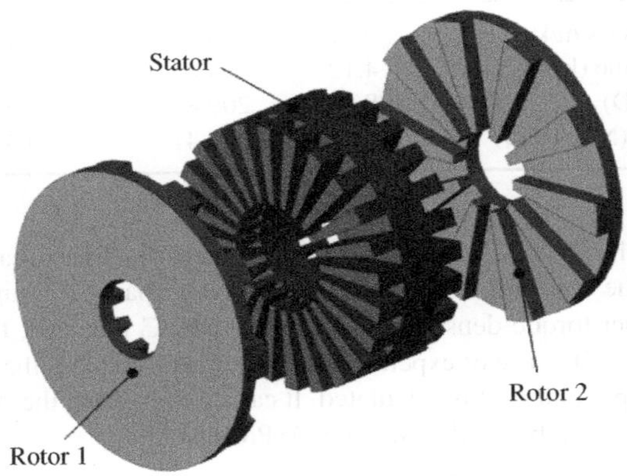

(b)

Figure 9.38 Axial-flux FSDC machine: (a) topology and (b) structure

Table 9.9 Key design data of AF-FSDC machine

Radial outside diameter	381 mm
Radial inside diameter	100 mm
Axial stack length	195 mm
Length of each air-gap	0.5 mm
Number of poles on each stator face	12
Number of teeth per stator pole	2
Number of equivalent poles on each stator face	24
Number of poles on each rotor face	10
Number of armature phases	3
Number of turns per armature coil	50
Number of turns per field coil	50

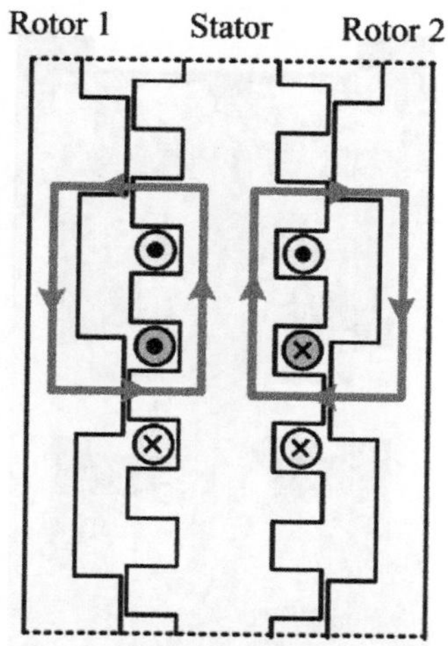

Figure 9.39 Magnetic flux paths of DC field in AF-FSDC machine

This AF-FSDC machine adopts the same power electronics as the radial-flux FSDC machine. That is, the full-bridge inverter is used to feed the armature windings, and the full-bridge DC-DC converter is employed to control the DC field windings. Similar to the radial-flux FSDC machine, the AF-FSDC machine can operate at the BLDC mode or BLAC mode. Similar to the AF-DSDC machine, this AF-FSDC machine can also offer the fault-tolerant operation when the DC field winding is under open-circuit fault. That is, during the malfunction of DC field excitation, the AF-FSDC machine can operate at the BLDC mode with the unipolar conduction scheme. Hence, under this fault-tolerant operation, it enables the FSDC motor drive operating alike an SR motor drive. The resulting torque level can be retained, but suffering from degraded efficiency and high torque ripple.

On the basis of three-dimensional finite element analysis, the no-load magnetic field distribution of this AF-FSDC machine is depicted in Figure 9.40. It confirms that the magnetic flux in the sandwiched stator is divided into two equal paths passing through the two rotors, leading to cancel out the attractive forces of the stator exerted on the two rotors. Since the flux densities at both air-gaps are almost the same, the two rotors can develop almost the same torque.

The flux linkage waveforms of the AF-FSDC machine under the rated DC field excitation of 5 A/mm^2 are shown in Figure 9.41. It can be seen that the machine offers the bipolar flux linkage, hence confirming the flux-switching feature. Meanwhile, it can be found that the machine offers balanced three-phase flux linkage patterns, confirming that the pole-pair arrangement and winding configurations are proper. The torque waveform of this machine under the rated armature current of 5 A/mm^2 and DC field excitation of 5 A/mm^2 is shown in Figure 9.42. It can be found that the developed torque can achieve an average value of 151.3 N m, which can well fulfill the desired torque specification. When the DC field excitation is temporarily increased to 10 A/mm^2, the developed torque can achieve an average value of 192.5 N m as shown in Figure 9.43, which is particularly useful for EV launch or overtaking.

Because of the merit of flexible DC field excitation, the torque–speed capability of this AF-FSDC machine with DC flux weakening for constant-power operation is shown in Figure 9.44. It confirms that the machine can maintain constant-power operation over a very wide speed range. Particularly, with the DC field excitation is weakened to 1 A/mm^2, the operating speed is 1183 rpm, which can fully cover the desired requirement of speed range.

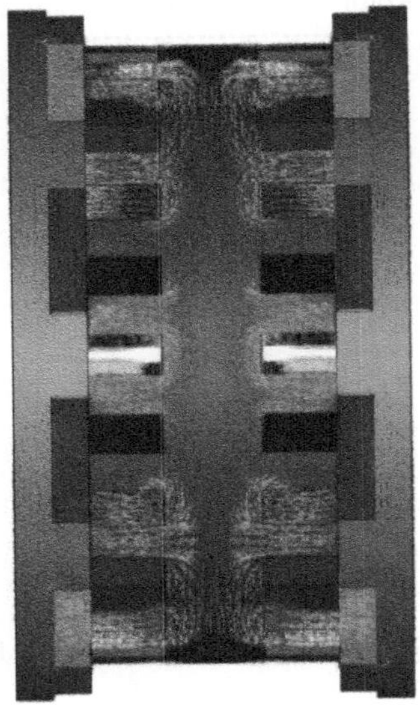

Figure 9.40 No-load magnetic field distribution of AF-FSDC machine

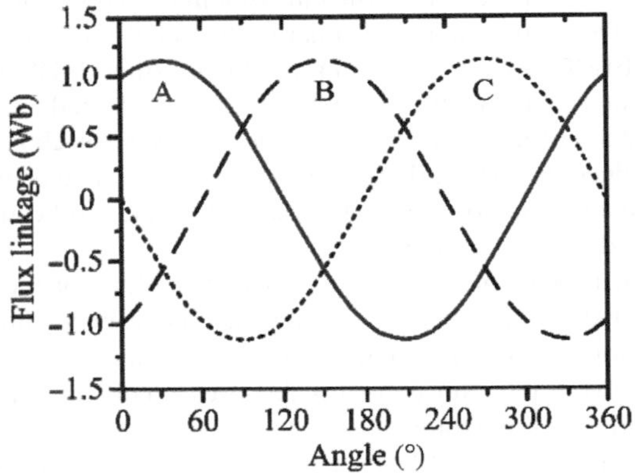

Figure 9.41 Flux linkage waveforms of AF-FSDC machine under rated DC field excitation

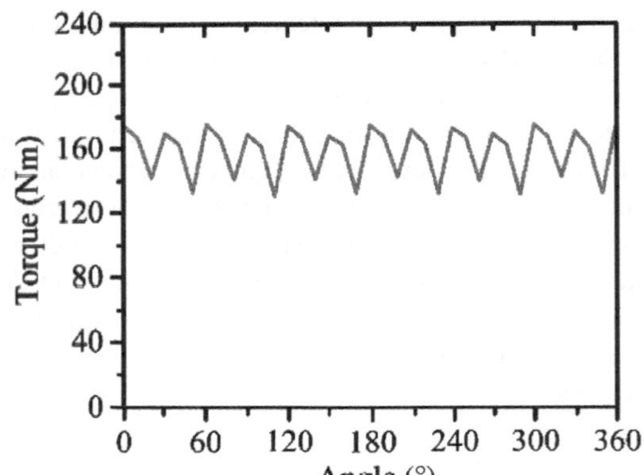

Figure 9.42 Torque waveform of AF-FSDC machine under rated DC field excitation

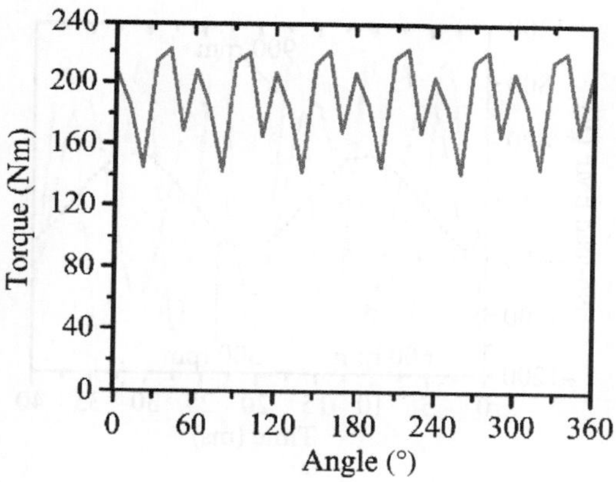

Figure 9.43 Torque waveform of AF-FSDC machine under strengthened DC field excitation

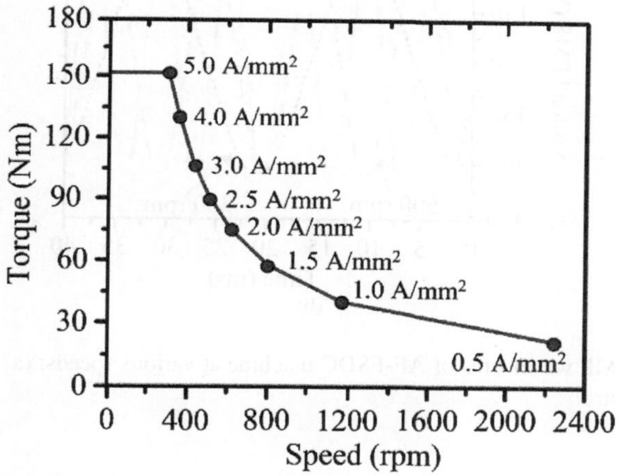

Figure 9.44 Torque–speed capability of AF-FSDC machine

In order to assess the merit of flux control of this AF-FSDC machine, the no-load EMF waveforms at various speeds with and without employing DC field control are shown in Figure 9.45. It can be found that the amplitudes of the output voltage vary with the operating speeds in the absence of DC field control, whereas the output voltage amplitudes can be kept constant over a wide speed range in the presence of DC field control. This feature is useful to effectively charge the battery during regenerative braking over a wide range of speeds.

The overall performances of the AF-FSDC machine are summarized in Table 9.10. In summary, the machine can offer high torque density and high power density while having flexible flux controllability for temporary torque boosting for EV launch or overtaking, wide speed range for constant-power operation, and constant output voltage regulation during EV regenerative braking.

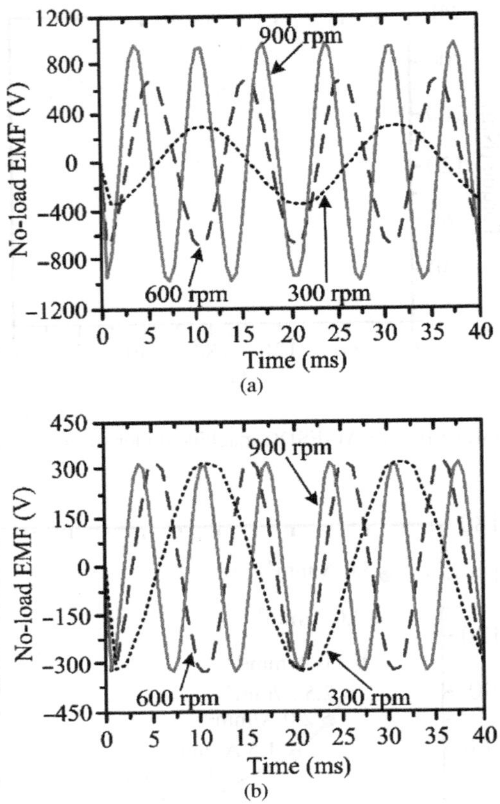

Figure 9.45 No-load EMF waveforms of AF-FSDC machine at various speeds: (a) without DC field control and (b) with DC field control

Table 9.10 Overall performances of AF-FSDC machine

Rated power	4.7 kW
Based speed	300 rpm
No-load EMF	316 V
Rated DC field excitation	5 A/mm^2
Rated torque	151.3 N m
Torque ripple at rated torque	28.4%
Boosted DC field excitation	10 A/mm^2
Boosted torque	192.5 N m

9.11 Potential Applications of Advanced Magnetless Motor Drives in EVs

Because of excellent performances, the PM motor drives play the major role for high-performance EV propulsion. Meanwhile, because of the high price and fluctuating supply of PM materials, the advanced magnetless motor drives will play the major role for well-performed EV propulsion desiring high cost-effectiveness.

Table 9.11 Comparison of potentiality of advanced magnetless motor drives

	SynR	DSDC	FSDC	VR	DFVR	AFM
Power density	Fair	Good	High	Low	Fair	Superb
Torque density	Fair	Good	High	High	High	Superb
Efficiency	Fair	High	High	Fair	High	High
Power factor	Low	Fair	Fair	Low	Fair	Fair
Controllability	Fair	High	High	Fair	High	High
Robustness	Low	Fair	Fair	High	Fair	Low
Manufacture	Hard	Fair	Fair	Easy	Fair	Hard
Material cost	Fair	Fair	Fair	Low	Fair	Fair
Maturity	High	Fair	Fair	High	Low	Low

As discussed earlier, there are six major types of advanced magnetless motor drives for EV propulsion: SynR, DSDC, FSDC, VR, DFVR, and AFM. It has to be noted that the AFM motor drives refer to the axial-flux versions of those viable RF magnetless motor drives such as the DSDC and FSDC motor drives. These six major types of advanced magnetless motor drives are compared in terms of their power density, torque density, efficiency, power factor, controllability, robustness, manufacturability, cost, and maturity. Table 9.11 summarizes the corresponding comparison.

In terms of power density and torque density, the AFM motor drives are the best because they can fully utilize the core material for torque production. The FSDC motor drive exhibits higher power and torque densities than the DSDC motor drive as it can offer bipolar flux linkage that can enhance the utilization of core material. Both the VR and DFVR motor drives possess the magnetic gearing effect, thus inherently offering high torque density for low-speed operation. Those doubly-fed magnetless motor drives with the capability of field control, namely the DSDC, FSDC, DFVR, and AFM, can achieve higher efficiency than those with singly-fed motor drives. As all magnetless machines inherently exhibit high inductance, the corresponding power factor is not high. Particularly, those singly-fed magnetless motor drives, SynR and VR, suffer from low power factor. Concerning controllability, those doubly-fed magnetless motor drives, DSDC, FSDC, DFVR, and AFM, are better than the singly-fed counterparts.

In terms of robustness and manufacturability, the SynR and AFM motor drives are nonperformers since the SynR machine relies on using the axially laminated iron core while the AFM machine desires the double-stator or double-rotor structure. On the contrary, the VR motor drive offers the highest simplicity, leading to be very robust and easy manufacturing. Also, the simple structure and singly-fed nature of the VR motor drive enables it to have the merit of lowest material cost.

Concerning maturity, the SynR and VR motor drives are relatively most mature because they have been developed for many decades. Then, the DSDC and FSDC motor drives are quite mature, which have been developed for over a decade. They are considered as the most prominent magnetless alternatives as compared with their PM counterparts, namely the DSPM and FSPM motor drives. As the DFVR motor drive is recently derived from the VR motor drive, it is immature. Meanwhile, since the AFM machines are recently derived from those viable RF counterparts, their corresponding maturity is low.

References

Agarlita, S.-C., Boldea, I. and Blaabjerg, F. (2012) High-frequency-injection-assisted 'active-flux'-based sensorless vector control of reluctance synchronous motors, with experiments from zero speed. *IEEE Transactions on Industry Applications*, **48**, 1931–1939.

Cruickshank, A.J.O., Anderson, A.F. and Menzies, R.W. (1971) Theory and performance of reluctance motors with axially laminated anisotropic rotors. *Proceedings of the IEEE*, **118**, 887–894.

Fan, Y. and Chau, K.T. (2008) Design, modeling, and analysis of a brushless doubly fed doubly salient machine for electric vehicles. *IEEE Transactions on Industry Applications*, **44**, 727–734.

Fukami, T., Matsuura, Y., Shima, K. et al. (2010) Development of a low-speed multi-pole synchronous machine with a field winding on the stator side. Proceedings of International Conference on Electrical Machines, pp. 1–6.

Kostko, J.K. (1923) Polyphase reaction synchronous motors. *Journal of American Institute of Electrical Engineers*, **42**, 1162–1168.

Lawrenson, P.J. and Gupta, S.K. (1967) Developments in the performance and theory of segmental-rotor reluctance motors. *Proceedings of the IEEE*, **114**, 645–653.

Lee, C.H. (1963) Vernier motor and its design. Proceedings of IEEE Winter General Meeting, pp. 343–349.

Lee, C.H.T., Chau, K.T., Chan, C.C. et al. (2014a) Development of axial-field flux-switching DC in-wheel motor drive for electric vehicles. Proceedings of International Conference on Electrical Engineering, pp. 1–6.

Lee, C.H.T., Chau, K.T., Liu, C. and Lin, F. (2014b) Design and analysis of a magnetless flux-switching DC-excited machine for wind power generation. *Journal of International Council on Electrical Engineering*, **4**, 80–87.

Lee, C.H.T., Liu, C. and Chau, K.T. (2014) A magnetless axial-flux machine for range-extended electric vehicles. *Energies*, **7**, 1483–1499.

Lee, C.H.T., Chau, K.T., Liu, C. *et al.* (2013) Quantitative comparison and analysis of magnetless machines with reluctance topologies. *IEEE Transactions on Magnetics*, **49**, 3969–3972.

Lee, H.-D., Kang, S.-J. and Sul, S.-K. (1999) Efficiency-optimized direct torque control of synchronous reluctance motor using feedback linearization. *IEEE Transactions on Industrial Electronics*, **46**, 192–198.

Matsuo, T. and Lipo, T.A. (1994) Rotor design optimization of synchronous reluctance machine. *IEEE Transaction on Energy Conversion*, **9**, 359–365.

Morales-Caporal, R. and Pacas, M. (2008) Encoderless predictive direct torque control for synchronous reluctance machines at very low and zero speed. *IEEE Transactions on Industrial Electronics*, **55**, 4408–4416.

Mukherji, K.C. and Tustin, A. (1974) Vernier reluctance motor. *Proceedings of the IEEE*, **121**, 965–974.

Profumo, F., Zhang, Z. and Tenconi, A. (1997) Axial flux machines drives: a new viable solution for electric cars. *IEEE Transactions on Industrial Electronics*, **44**, 29–45.

Supermagnete (2014) What is the Development of Neodymium Magnet Prices? http://www.supermagnete.de/eng/faq/price (accessed 1 May 2014).

Taibi, S., Tounzi, A. and Piriou, F. (2006) Study of a stator current excited vernier reluctance machine. *IEEE Transactions on Energy Conversion*, **21**, 823–831.

Tang, Y., Ilhan, E., Paulides, J.J.H., and Lomonova, E.A. (2013) Design considerations of flux-switching machines with permanent magnet or DC excitation. Proceedings of 15th European Conference on Power Electronics and Applications, pp. 1–10.

Tang, Y., Paulides, J.J.H., Motoasca, T.E. and Lomonova, E.A. (2012) Flux-switching machine with DC excitation. *IEEE Transactions on Magnetics*, **48**, 3583–3586.

Tounzi, A., Ramdane, B., and Zaim, M.E. (2008) Study of a rotor current excited vernier reluctance machine. Proceedings of 18th International Conference on Electrical Machines, pp. 1–6.

Xu, L., Xu, X., Lipo, T.A. and Novotny, D.W. (1991) Vector control of a synchronous reluctance motor including saturation and iron loss. *IEEE Transactions on Industry Applications*, **27**, 977–985.

10

Integrated-Starter-Generator Systems

The integrated-starter-generator (ISG) is an important device for hybrid electric vehicles (HEVs), which replaces the conventional starter motor and the generator (also called alternator or dynamo). Its other common name is the integrated-starter-alternator (ISA). Apart from cranking the engine and generating electricity to charge the battery, the ISG provides three unique functions for micro and mild HEVs: idle stop-start, regenerative braking, and power assistance.

In this chapter, various ISG systems, including their system configurations, machine structures, and operation modes, are presented, with emphasis on their applications to micro and mild HEVs. The corresponding design criteria, design examples, and application examples are also discussed.

10.1 Classification of HEVs

There are two major classifications of HEVs, one is based on the power flow and another is based on the hybridization level. The former is more academic in nature and can provide more technical information, whereas the latter is more straightforward and can be easily realized by the general public.

On the basis of the flows of electrical and mechanical powers, HEVs used to have two basic configurations – the series hybrid and parallel hybrid. With the introduction of some HEVs having the characteristics of both the series and parallel hybrids, the basic configurations were extended to three kinds – the series hybrid, parallel hybrid, and series-parallel hybrid. Moreover, with the advent of some HEVs with dual-axle propulsion that cannot be represented by these three kinds, the basic configurations have been extended to four kinds (Chau and Wong, 2002; Ehsani *et al.*, 2005):

- Series hybrid
- Parallel hybrid
- Series-parallel hybrid
- Complex hybrid

Figure 10.1 shows the four basic configurations of HEVs based on this classification, where B is the battery, D is the differential, E is the engine, F is the fuel tank, G is the generator, M is the motor, P is the power converter, T is the transmission, W is the wheel, the solid line represents the electrical link, the double line represents the mechanical link, and the dotted line represents the hydraulic link. Both the electrical and mechanical links are bidirectional in nature, and the hydraulic link is unidirectional. It can be observed that the key feature of the series hybrid is to couple the engine with the generator to produce electricity for pure electric propulsion, whereas the key feature of the parallel hybrid is to couple both the engine and electric motor with the transmission to propel the driving wheels. The series-parallel hybrid

Electric Vehicle Machines and Drives: Design, Analysis and Application, First Edition. K. T. Chau.

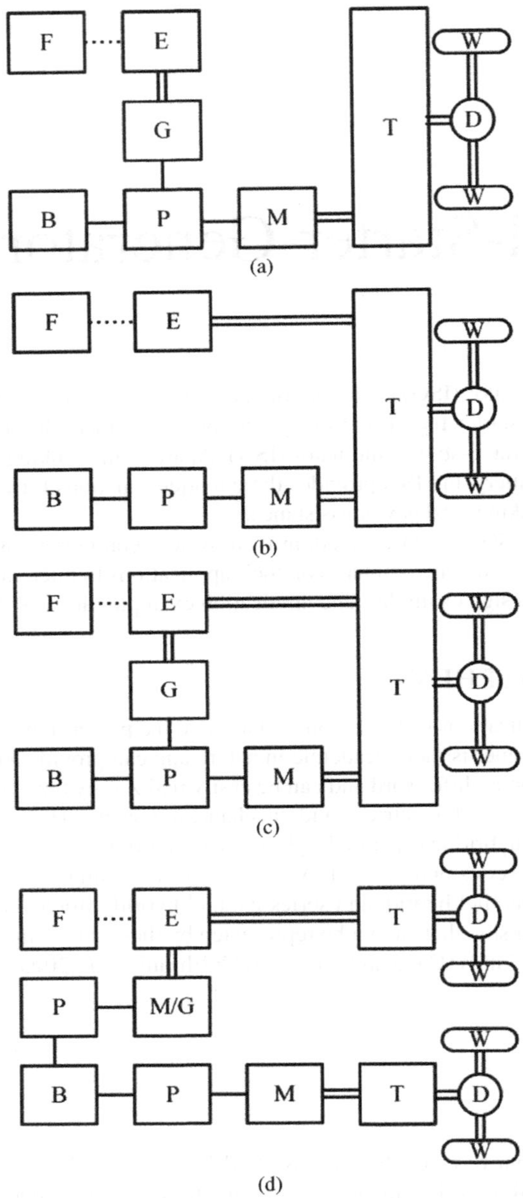

Figure 10.1 Classification of HEVs based on power flow: (a) series hybrid, (b) parallel hybrid, (c) series-parallel hybrid, and (d) complex hybrid

combines the features of both the series and parallel hybrids. The complex hybrid further combines the features of pure electric propulsion and hybrid propulsion.

The series hybrid is the simplest kind of HEVs. Since the engine is directly coupled with the generator, the mechanical output power of the engine is totally converted into electricity by using the generator. The converted electricity in turn charges the battery or feeds the electric motor to propel the driving wheels, depending on the load condition. Because of electrical wiring, it has the definite advantage of high flexibility to locate the engine-generator set. Since the engine can operate within a narrow range of speeds, the efficiency of the engine can readily be maximized. Although the series hybrid has the above-mentioned merits, it suffers from using three propulsion devices, namely the engine, generator, and electric motor, which need to be sized for the maximum output power. Unless it is designed to serve for short-trip operation such as commuting to work and shopping, the corresponding engine-generator set can adopt a lower power rating. In fact, most series hybrids that are used today are shorter-distance medium-duty vehicles such as the transit bus.

Differing from the series hybrid, the parallel hybrid allows both the engine and electric motor delivering power in parallel to propel the driving wheels. Since the engine and electric motor are coupled to the driveline of the wheels, the propulsion power can be provided by the engine alone, electric motor alone, or both. The electric motor can serve as a generator to charge the battery during regenerative braking or the light load condition. Better than the series hybrid, the parallel hybrid needs only two propulsion devices, namely the engine and electric motor. In addition, both the engine and electric motor can be sized smaller than those of the series hybrid because they share the maximum output power. Even for long-trip operation, only the engine needs to be rated for the maximum output power while the electric motor may still be sized to about one-half of the maximum level.

In the series-parallel hybrid, the configuration incorporates the features of both the series and parallel hybrids, but involving an additional mechanical link as compared with the series hybrid, or an additional generator as compared with the parallel hybrid. Although it possesses the advantageous features of both the series and parallel hybrids, the series-parallel hybrid is more complicated and costly. Nevertheless, with the ever-increasing demand on performance and flexibility, it is becoming accepted.

As reflected by its name, the complex hybrid involves a complex configuration that cannot be classified into the above-mentioned three kinds. It seems to be similar to the series–parallel hybrid in terms of the number of propulsion devices. However, there is a fundamental difference that the electric machine coupled with the engine can serve as both the electric motor and generator for bidirectional power flow in the complex hybrid, whereas the electric machine coupled with the engine works as only the generator for unidirectional power flow in the series–parallel hybrid. Most importantly, the three propulsion devices can be arranged to provide electric propulsion for one axle and hybrid propulsion for another axle, which cannot be offered by the series–parallel hybrid. A unique feature of this dual-axle propulsion system is the capability of axle balancing. For instance, when the wheels of the hybrid propulsion axle slip, the corresponding electric machine works as the generator to absorb the change of engine output power; through the battery, this power difference is then fed into the electric motor to drive the wheels of the electric propulsion axle. Of course, the regenerative braking of this configuration is better than the others because the electric machines in both axles serve to simultaneously recover the braking energy of all wheels to charge the battery.

On the basis of the hybridization level between the engine and electric motor, HEVs have been classified as (Ebron and Cregar, 2005; Chau and Chan, 2007)

- Micro hybrid
- Mild hybrid
- Full hybrid

Recently, this classification has been further extended to include the latest types of HEVs:

- Plug-in hybrid electric vehicle (PHEV)
- Range-extended electric vehicle (REV)

Figure 10.2 depicts their classification in terms of the energy source and propulsion device, whereas all kinds of hybrids lie between the internal combustion engine vehicle (ICEV) and the pure electric vehicle (PEV).

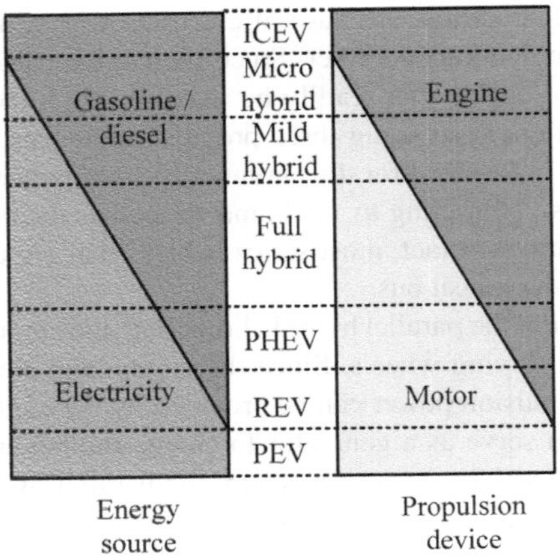

Figure 10.2 Classification of HEVs based on hybridization level

For the micro hybrid, the conventional starter motor is eliminated while the conventional generator is replaced by the ISG. This ISG is typically 3–5 kW with the system voltage of 14–42 V. Instead of propelling the vehicle, it offers two important hybrid features. One is to shut down the engine when the vehicle is at rest, the so-called idle stop-start feature, hence improving the fuel economy by 10–15% in urban driving environment. The micro hybrid also controls the generator in such a way that it produces electricity to charge the battery primarily during deceleration and braking, hence providing a mild amount of regenerative braking. The corresponding battery voltage ranges from 12 to 36 V. The Citroën C3, currently being sold in Europe, is a micro hybrid.

For the mild hybrid, the ISG is generally placed between the engine and the transmission. This ISG is typically 7–15 kW with the system voltage of 100–150 V. It can provide the hybrid features of idle stop-start and full regenerative braking. Also, as the ISG assists the engine to propel the vehicle, it can allow for a downsized engine. Since the rotor of the ISG replaces the engine flywheel in many designs, both the engine and ISG share the same shaft so that this hybrid cannot offer the electric launch feature, namely accelerating under electric power only. The battery voltage ranges from 36 to 144 V. The Honda Insight was probably the first commercially available mild hybrid.

Differing from the micro and mild hybrids, the full hybrid can run based on the engine alone, the electric motor alone, or a combination of both. Instead of using the ISG system, it adopts the electric variable transmission (EVT), also called electronic-continuously variable transmission, system to perform power splitting and power regulation. Therefore, it can offer all hybrid features, including the electric launch, idle stop-start, regenerative braking, and downsized engine, which enable an improvement of 30% in fuel economy. Typically, the corresponding electric motor is 50–60 kW with the system voltage of 500–600 V, and the battery capacity is 1–2 kWh with the nominal voltage of 200–300 V. The Toyota Prius is a full hybrid that is probably the most successful production HEV. Meanwhile, a derivative of this full hybrid is the power hybrid. The only difference is that the power hybrid does not have the feature of downsized engine. It installs the same size engine as that of a conventional ICEV so that the resulting hybrid propulsion power can provide more torque and better acceleration performance than the ICEV. A representative of this power hybrid is the Lexus RX450h.

Among the three types of hybrids, they need to carry various price premiums (a few hundred to several thousand dollars) in order to obtain 10–30% improvement in fuel economy. As depicted in Figure 10.3, the micro hybrid takes the advantage of cost-effectiveness in fuel economy, the full hybrid enjoys the highest fuel economy, and the mild hybrid offers a compromise. Anyway, each of them has its own niche of application.

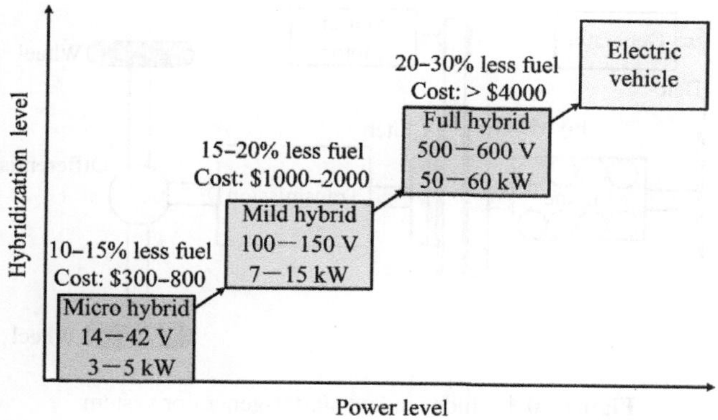

Figure 10.3 Comparison of price premium and fuel economy of various types of hybrids

For the PHEV, it provides all features of the full hybrid, while having an additional feature of plug-in rechargeable. In general, it is equipped with more battery packs than the full hybrid, typically 4–5 kWh with the nominal voltage of 200–300 V. Thus, it can offer a longer electric range and reduce the requirement for gasoline refueling. This plug-in hybrid can operate at the charge-depleting mode (the pure electric vehicle (EV) mode) to provide a decent electric range, and the blended mode in which most of the propulsion power is supplied by the battery but supplemented by the engine to maximize the fuel economy. Typically, the electric motor is about 60 kW with the system voltage of about 650 V. The corresponding electric motor and battery voltage ratings are typically 30–50 kW and 400–500 V, respectively. The Toyota Prius Plug-in Hybrid Vehicle (PHV) is a midsize plug-in hybrid, and has been commercially available since 2012.

For the REV, it is sometimes classified into the PHEV rather than as an individual type. It provides all features of the plug-in hybrid, and the unique feature of the series hybrid. Compared with the PHEV, it usually installs a smaller engine but a larger battery bank. As reflected by its name, it prefers to work as a PEV until the battery capacity drops to a predefined threshold from full charge. Then, it works as a series hybrid so that the engine drives the generator to keep the battery at the minimum capacity level and provide power to the electric motor. At higher speeds and loads, the REV operates as the plug-in hybrid or full hybrid. In order to provide the desired electric range, the battery requirement is typically over 16 kWh with the nominal voltage of 350–390 V. The corresponding electric motor needs to achieve 110 kW to enable the pure EV mode of operation. The General Motors (GM) Chevrolet Volt is a representative REV that has been on mass production since 2011.

In the following sections, the ISG systems will be dealt with emphasis on the micro and mild hybrids. The EVT systems for the full hybrid, PHEV, and REV, though including all features of ISG systems, will be discussed in next few chapters.

10.2 ISG System Configurations

In conventional automobiles, the starter motor and generator are separately coupled with the engine. That is, the conventional starter motor is coupled with the flywheel and located at the rear of the engine, whereas the conventional generator is located at the front of engine as shown in Figure 10.4, hence providing high starting torque for cold cranking and generating electricity for battery charging, respectively. This arrangement takes the advantage of simplicity, but suffers from poor utilization of both machines and thus resulting in heavy weight and bulky size. In order to incorporate both functions in a single unit, the development of ISG systems is a natural choice (Viorel *et al.*, 2004; Walker *et al.*, 2004).

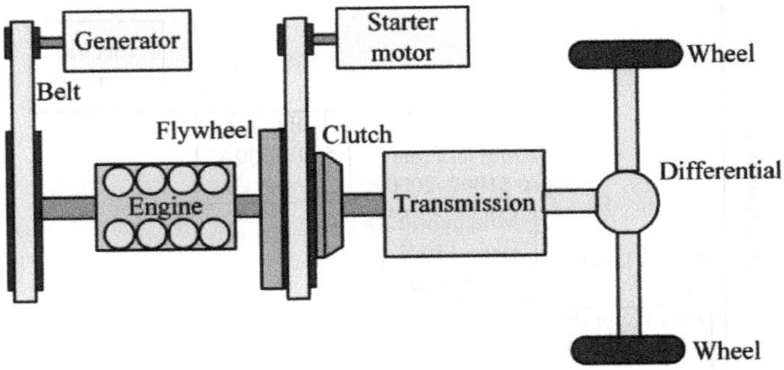

Figure 10.4 Independent starter-generator system

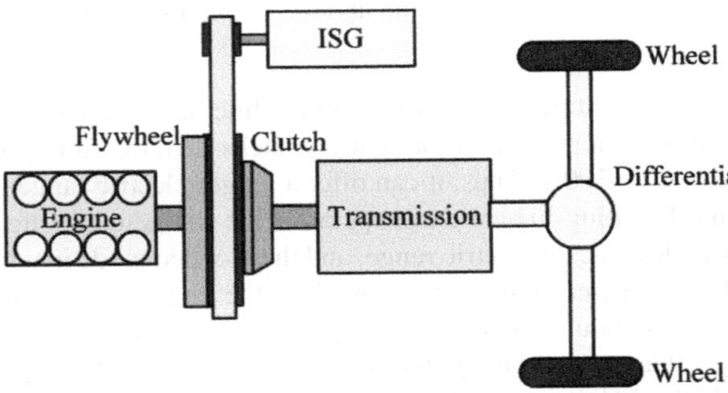

Figure 10.5 Belt-driven ISG system

For the micro and mild hybrids, the ISG replaces both the conventional starter motor and generator. It acts as a bidirectional electromechanical device, converting electrical energy to mechanical energy and vice versa. When operating at the motoring mode, it can crank the engine almost soundlessly and much faster than the conventional starter motor. When operating at the generation mode, it can generate electricity with much higher efficiency than the conventional generator.

By mounting the ISG in the conventional location of the starter motor or generator to couple with the flywheel and engine via the belt, it is so-called the belt-driven ISG system as shown in Figure 10.5. This configuration is simple and inexpensive. No significant modifications are required for the conventional vehicle chassis to accommodate this belt-driven ISG system. This allows the hybrids to be produced on the same assembly line as the conventional vehicles, resulting in substantial cost savings and enabling the automakers to flexibly adjust the production.

By directly mounting the ISG onto the crankshaft, the ISG rotor can serve as the flywheel inertia so that the aforementioned flywheel coupled with the engine can be eliminated. As shown in Figure 10.6, this configuration can offer direct control for both cranking and generation, and can avoid the belt-driven mechanism, hence reducing the construction cost and complexity. However, the corresponding ISG generally needs additional weight of copper and iron, and can withstand severe vibration of the crankshaft. Greater changes are required for the conventional vehicle chassis to accommodate this crankshaft ISG system.

10.3 ISG Machines

The electric machine is the core of the ISG system (Cai, 2004). Because of the fundamental drawback of wear and tear of carbon brushes, the DC machine is no longer acceptable for modern ISG systems

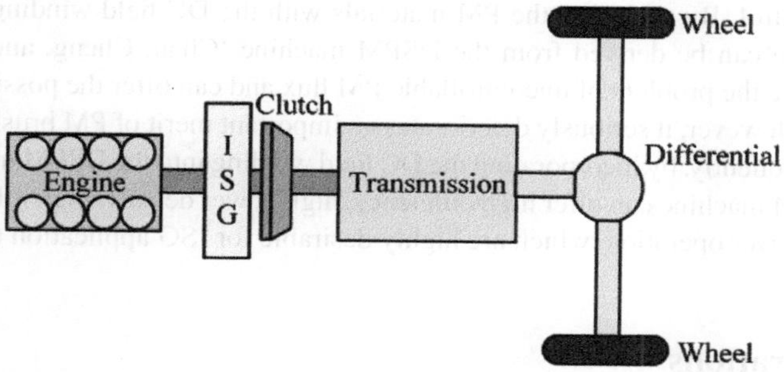

Figure 10.6 Crankshaft ISG system

even though it has been widely used as the starter motor for automobiles. Among those viable brushless machines, the induction machine and permanent-magnet (PM) brushless machines, including the PM brushless AC machine or commonly called the PM synchronous machine and the PM brushless DC machines, are widely used for modern ISG systems.

The induction machine and PM synchronous machine are currently the main competitors for application to ISG systems. The induction machine exhibits the advantages of low cost and high robustness, but needs complicated control to provide wide speed generation. It suffers from the drawbacks of small air-gap, limited efficiency, and cooling problem.

The PM synchronous machine offers the advantages of high efficiency and high power density, but suffers from high PM cost and limited constant-power operation range. Because of the independent field excitation in the rotor, the PM synchronous machine can be designed with a larger air-gap than the induction machine, which particularly facilitates mounting its rotor on the crankshaft where there are severe vibrations and oscillations. However, since the PM rotor is coupled with the engine and thus exposed to high temperatures, the thermal instability of PM material is a major concern.

Compared with the PM synchronous machine, the PM brushless DC machine can produce higher torque density. Its drawbacks are basically the same, namely the high PM cost, limited constant-power operation range, and thermal instability of PM rotor. Meanwhile, this PM brushless DC machine has a higher cogging torque and larger torque ripple than the PM synchronous machine, which are the main concern for ISG application.

The switched reluctance (SR) machine takes the definite advantages of low cost and high robustness, but has the drawbacks of relatively lower efficiency and lower power density than the PM brushless machines. As the rotor has no excitation, there is no problem of thermal instability. The corresponding low inertia can enable fast response. Because of its robust rotor, the SR machine is particularly attractive for the crankshaft ISG system (Viorel *et al.*, 2004).

In order to strengthen the motor torque to assist the engine for fast acceleration and widen the constant-power operation range, the PM synchronous machine can incorporate the concept of double-stator single-rotor arrangement (Niu, Chau, and Jiang, 2008b; Niu, Chau, and Yu, 2009). The resulting double-stator PM synchronous machine can inherit the merits of high efficiency and high power density from the conventional PM synchronous machine while achieving high torque density and wide constant-power operation range because of the feature of double stators. That is, both stators can interact simultaneously with the same rotor to develop the high torque, while two sets of stator windings can be electronically switched in such a way that the overall induced electromotive force (EMF) can be reduced at high speeds to achieve wide constant-power operation range.

By incorporating the advantageous features of both the PM brushless DC machine and SR machine, the doubly-salient permanent-magnet (DSPM) machine offers the merits of high efficiency, high power density, high robustness, and low inertia (Cheng *et al.*, 2003; Chau *et al.*, 2005). But, it still has the drawback of inflexible flux control. By replacing the PM materials with the DC field winding, the doubly-salient DC (DSDC) machine can be derived from the DSPM machine (Chau, Cheng, and Chan, 2002). This arrangement can solve the problem of uncontrollable PM flux and can offer the possibility to online optimize the efficiency. However, it seriously deteriorates an important merit of PM brushless machines, high power density. Consequently, by incorporating the DC field winding into the DSPM machine, the resulting hybrid-excited DSPM machine can offer high efficiency, high power density, high robustness, low inertia, and wide constant-power operation, which are highly desirable for ISG application (Chau *et al.*, 2006).

10.4 ISG Operations

In general, the ISG system possesses five main operations, two of them are inherited from the conventional vehicles – cranking and electricity generation, and the others are unique for the hybrids – idle stop-start, regenerative braking, and power assistance.

10.4.1 Cranking

The ISG works as a starter motor spinning the engine to about 200 rpm in order to initiate the engine operation under its own power. The cranking time is of very short duration, typically less than 2 seconds. When adopting a 12-V battery, the cranking currents are 150–185 and 185–200 A for the four-cylinder and eight-cylinder engines, respectively.

10.4.2 Electricity Generation

The ISG serves as a generator that converts the mechanical energy of the spinning engine into electrical energy for battery charging. When adopting a 12-V battery, the charging currents range from a few to a hundred amperes, depending on the battery type, state-of-charge (SOC), and ambient temperature as well as the maximum output current of the ISG and the speed of the engine. Typically, when adopting the lead-acid battery or lithium-ion battery, the battery charging is accomplished in three stages: bulk, absorption, and float. At the bulk stage, the charging current is only limited by the maximum output current of the ISG so that the battery is rapidly charged at constant current. At the absorption stage, the battery is charged at the constant voltage while the charging current is decreasing until the battery is fully charged. At the float stage, a constant voltage is applied to the battery to compensate the battery self-discharge so that it will maintain the full charge indefinitely.

10.4.3 Idle Stop-Start

In the idle stop-start operation mode, the ISG allows the engine to turn off and hence save fuel at stops, and instantly restart the engine upon pressing of the accelerator pedal. Hence, it is also called the auto start-stop mode of operation. This feature alone is estimated to enable the hybrids having 10–15% gain in fuel economy over the conventional vehicles especially during urban driving. In addition to fuel saving, this operation mode can significantly reduce the roadside emissions, which are particularly annoying and detrimental to our health during traffic-light signaling and traffic jams.

Differing from normal engine cranking, the ISG can instantly restart the engine for idle stop-start, typically less than 300 ms. There are various versions of idle stop-start mode of operation. For instance, when the hybrid is approaching to stop in front of the traffic light, the idle stop-start process is scheduled as follows:

- In the case of automatic transmission, once the vehicle stops completely, the engine will be shut down automatically after a few seconds. In the case of manual transmission, after a complete stop, if the shift is in the "Neutral" position and the clutch pedal is released, the engine will then be shut down.
- After turning off the engine, the indicator light located in the instrument cluster will blink to signify that the engine is off.
- In the case of automatic transmission, once depressing the accelerator pedal, the engine will be restarted instantly. In the case of manual transmission, if the clutch pedal is pressed and the shift is in the "Drive" or "Reverse" position, the engine will then be restarted.
- After restarting, the indicator light will turn off. The driver essentially should not notice the idle stop-start process.

It should be noted that a large number of cars, including both hybrids and nonhybrids, in Europe now feature the idle stop-start as a standard. Although drivers do have concerns on how many restarts that the ISG system can handle, more than one million restarts can generally be guaranteed.

10.4.4 Regenerative Braking

In the regenerative braking mode of HEVs, the ISG works as a generator that captures the vehicle kinetic energy and turns it into electricity to recharge the battery. That is, when releasing the accelerator pedal or pressing the brake pedal, the regenerative braking torque is produced to slow down the vehicle. If the desired braking torque is larger than the regenerative braking torque, the difference will be provided by the mechanical brake. It should be noted that the regenerative braking can only slow down the vehicle, and the mechanical braking is always necessary to fully stop the vehicle.

There are two standard regenerative braking strategies: the maximum-power regeneration and good-pedal-feel regeneration. The former can produce the largest amount of regenerative braking power, while the latter can emulate the comfortable feeling of mechanical braking.

Firstly, the control strategy for maximizing the regenerative braking power (Jung *et al.*, 2010) is depicted in Figure 10.7. After collecting all necessary input data from the brake pedal, battery, motor,

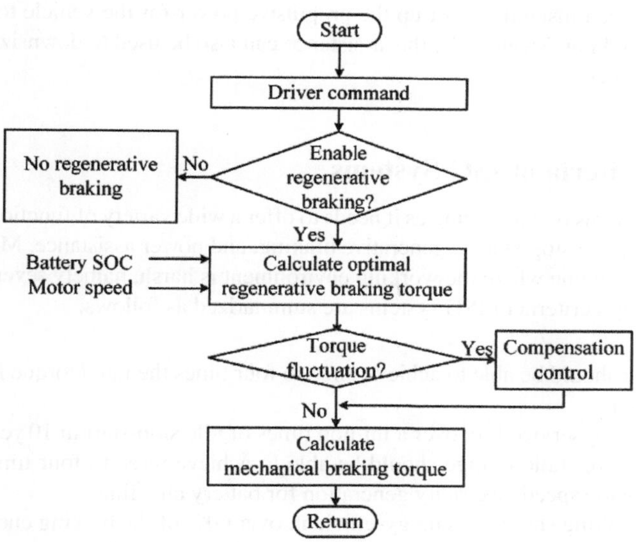

Figure 10.7 Regenerative braking control strategy

and transmission, the controller decides whether regenerative braking should be performed. For instance, if the battery SOC is too high, the regenerative braking will not be activated; otherwise, it may damage the battery. If the regenerative braking goes ahead, the controller will calculate the optimal regenerative braking according to the battery SOC and motor speed. In case of torque fluctuation during the regenerative braking process, the time-delay compensation control will be activated to minimize the fluctuation. Consequently, the difference between the brake command and the regenerative braking torque will be equal to the mechanical braking torque. These two braking torques are modified at each time step in response to the changing input data. In this control strategy, the regenerative braking torque provides an abnormal pedal feel. When the need for mechanical braking arises, there is a drastic change in the brake master cylinder pressure, which may mislead the driver's brake command. Thus, this maximum-power regenerative braking may create a serious safety problem.

Secondly, the use of good-pedal-feel regenerative braking can solve the abnormal pedal sensation (Bracken and Selker, 2013). In this control strategy, the regenerative braking is applied only after the wheel cylinder pressures become stable. Thus, the brake master cylinder pressure will not change significantly. However, this good-pedal-feel regeneration strategy suffers from about 10% drop in regenerative power as compared with the maximum-power regeneration strategy. Nevertheless, it is unacceptable to trade power regeneration for safety.

The coordinated regenerative braking control strategy, which combines the merits of maximum-power regeneration and good-pedal-feel regeneration, is attractive (Zhang *et al.*, 2012). In this strategy, the regenerative braking is combined with the hydraulic braking to emulate the normal braking feel. That is, the pressure in the wheel cylinder is gradually increased while the regenerative braking torque provides the difference to meet the brake command. If the brake command is above the threshold, the regenerative braking is maximized and the hydraulic braking fluid is pumped into the accumulator. It will be pumped back if the brake command exceeds the maximum regenerative braking torque. Hence, it can maximize the power regeneration while maintaining a uniform pedal feel.

10.4.5 Power Assistance

The ISG works as a motor that can provide additional power to assist the engine for fast acceleration. For instance, the ISG can instantly boost up the propulsive power for the vehicle to perform launching, hill-climbing, or overtaking. Meanwhile, this assistance can also be used to downsize the engine without reducing the performance.

10.5 Design Criteria of ISG Systems

The design of ISG systems is challenging as it needs to offer a wide variety of functions, namely cranking, electricity generation, idle stop-start, regenerative braking, and power assistance. Meanwhile, it needs to closely work with the engine where the working environment is harsh, namely severe vibration and high temperature. The design criteria of ISG systems are summarized as follows:

- The starting torque should be able to achieve three to four times the rated torque in order to satisfy the cranking requirement.
- It needs to have a long service life, over a million times of idle stop-start in 10 years.
- The constant-power operation range should be able to achieve three to four times the base speed in order to facilitate wide speed electricity generation for battery charging.
- The regenerative braking should be energy-efficient, over 60% of the braking energy.
- It should be able to offer high torque at speeds below the base speed for power assistance.
- It should be able to offer high efficiency of over 85%, covering wide ranges of torques and speeds.

- It should be able to withstand working vibrations up to 20 g.
- It should be able to work well over a wide range of temperatures, from −30 to 120 °C.
- It should offer good serviceability and zero maintenance for ISG machines.
- The system cost should be acceptable.

The ISG system is mainly composed of two components – the ISG machine and ISG converter. Between them, the ISG machine plays the key role to satisfy the above-mentioned criteria, while the ISG converter serves to regulate the bidirectional power flow within the ISG system. Thus, the design criteria of the ISG machine essentially coincide with the system design criteria, while following additional design criteria of individual machines.

10.6 Design Examples of ISG Systems

There are various ISG systems when adopting various ISG machines. Two recently developed ISG systems, namely based on the double-stator PM synchronous machine and hybrid-excited DSPM machine, are elaborated.

10.6.1 Double-Stator PM Synchronous Machine-Based ISG System

The double-stator PM synchronous machine is one of the most promising candidates to work as an ISG for HEVs, especially the mild hybrids (Niu, Chau, and Jiang, 2008a). The configuration and structure of this ISG machine are shown in Figure 10.8 in which there are two concentric stators with two sets of three-phase armature windings wounded on 24 slots, and one cup-rotor with 22 pieces of PMs separately mounted on the inner and outer surfaces. Two winding arrangements, three-phase single-layer and three-phase double-layer, can be adopted in both the inner and outer stators as shown in Figure 10.9. In general, the double-layer winding arrangement can produce more sinusoidal back EMF waveforms than the single-layer winding arrangement.

This double-stator PM synchronous machine possesses the following distinct features for ISG operation:

- The use of double stators functions to significantly increase the torque density for cranking, and provides versatile connection modes for generation over a wide speed range.
- The rotor core is designed like a cup with PMs. This cup-rotor can effectively shorten the magnetic circuit path and hence reduce the iron core, thus improving the torque density.
- Because the permeability of PMs is similar to that of air, the mutual inductance between the inner and outer stators is negligible, hence improving the controllability.
- Since the coil span of stator windings is designed to have one slot pitch, the phase flux paths are essentially independent. In this sense, the mutual inductance of phase windings is negligible, hence further improving the controllability.
- Since the machine has 24 slots in each stator and 22 poles totally, the slot pitch is 11/12 of the pole pitch. This fractional number of slots per pole per phase can significantly reduce the cogging torque that usually occurs in PM synchronous machines.
- The multipole structure can shorten the magnetic circuit path, while the one-slot-pitch coil span winding arrangement can shorten the end-winding, thus improving the utilization of both iron and copper materials, and hence the torque density.

The double-stator PM synchronous machine can be considered as a combination of two machines: a single-stator inner-rotor machine and a single-stator outer-rotor machine (Wang *et al.*, 2011). These two machines have essentially the same armature diameter and two rotors are fused to form a single rotor,

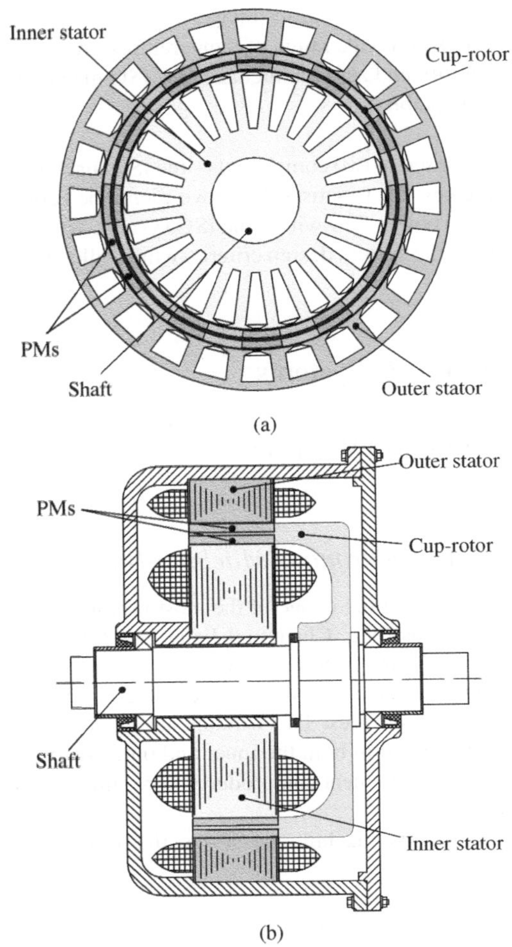

Figure 10.8 Double-stator PM synchronous machine: (a) configuration and (b) structure

hence reducing the yoke of rotors. The main magnetic circuits of this double-stator PM synchronous machine as well as the traditional single-stator inner-rotor PM synchronous machine and the traditional single-stator outer-rotor PM synchronous machine are depicted in Figure 10.10, where F_1 is the inner-rotor PM magnetomotive force (MMF), F_2 is the outer-rotor PM MMF, R_1 is the magnetic reluctance of the outer-stator iron core, R_1' is the magnetic reluctance of the inner-stator iron core, R_2 is the magnetic reluctance of the outer air-gap and PM, R_2' is the magnetic reluctance of the inner air-gap and PM, R_3 is the magnetic reluctance of the radial rotor iron core, and R_4 is the magnetic reluctance of the rotor yoke in the single-stator machine. It can be found that the double-stator configuration can effectively shorten the magnetic circuit path and hence improve the utilization of iron material.

The principle of operation of this double-stator PM synchronous machine is similar to that of the conventional PM synchronous machine, while providing additional operation modes. When the ISG is required to provide high starting torque for cranking, the double-stator windings are arranged in such a way that the same phases are connected in series. When the ISG is required to provide wide constant-power operation range, namely the requirement of constant output voltage over a wide range of speeds for battery charging, the double-stator windings are flexibly connected to vary the resultant EMF.

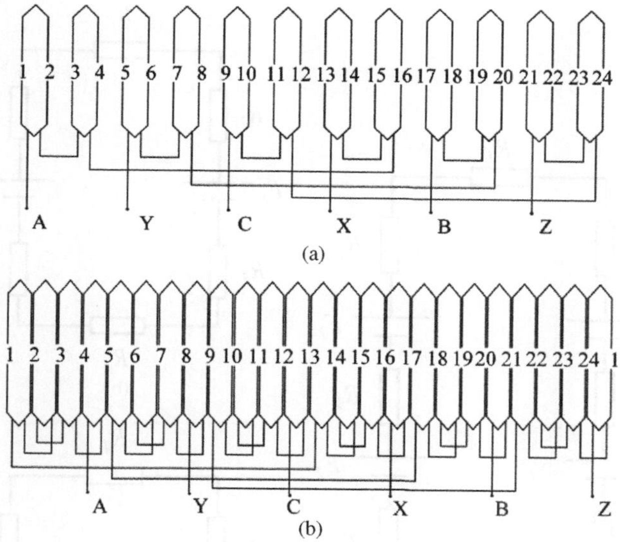

Figure 10.9 Winding arrangements of double-stator PM synchronous machine: (a) single-layer winding and (b) double-layer winding

When there is a spatial angle of 20° between the two stators, it yields six possible ways to connect the two sets of three-phase windings. As shown in Figure 10.11, different composite vectors can be generated by using different winding connections. When the machine speed ω is below its base speed ω_b, the same phases (for instance, E_A and E_a) are connected in series to form E_1, the so-called Mode 1 of operation. Between ω_b and $1.2\omega_b$, the adjacent antiphases are connected in series to form E_2, the so-called Mode 2. Similarly, between $1.2\omega_b$ and $1.5\omega_b$, E_3 is formed for Mode 3. Then, between $1.5\omega_b$ and $2\omega_b$, the adjacent phases are connected in series to form E_4 for Mode 4. Similarly, between $2\omega_b$ and $3\omega_b$, E_5 is formed in Mode 5. Finally, between $3\omega_b$ and $6\omega_b$, the same antiphases are connected in series to form E_6 for Mode 6. On the basis of the corresponding vector diagrams, it can be deduced that $E_1 \approx 6E_6$, $E_2 \approx 5E_6$, $E_3 \approx 4E_6$, $E_4 \approx 3E_6$, and $E_5 \approx 2E_6$, indicating that the generated EMF can be discretely tuned in terms of E_6. For further fine tuning or with speed beyond $6\omega_b$, the conventional flux-weakening control can be employed to adjust the output voltage.

In order to enable flexible connections between the two sets of three-phase armature windings, a matrix of power switches can be utilized as shown in Figure 10.12. The switching combinations of the aforementioned six modes of operation are tabulated in Table 10.1. For instance, the switches S_{X1}, S_{Y2}, and S_{Z3} are turned on while the others are off when the machine operates in the Mode 1 with $\omega < \omega_b$. With a very wide range of speeds, the machine can experience all six modes of operation.

The key design data of this double-stator PM synchronous machine are listed in Table 10.2. The corresponding generated output voltage waveforms at the six modes of operation are shown in Figure 10.13. It can be found that the magnitude of these output voltages can be maintained almost constant over a speed range of six times the base speed, from 800 to 4800 rpm.

10.6.2 Hybrid-Excited Doubly-Salient PM Machine-Based ISG System

The hybrid-excited DSPM machine is also one of the most promising candidates to work as an ISG for the mild hybrids (Liu, Chau, and Jiang, 2010). The configuration and structure of this machine are shown in Figure 10.14, which is an outer-rotor machine. It has a doubly-salient structure with 36 salient poles in the

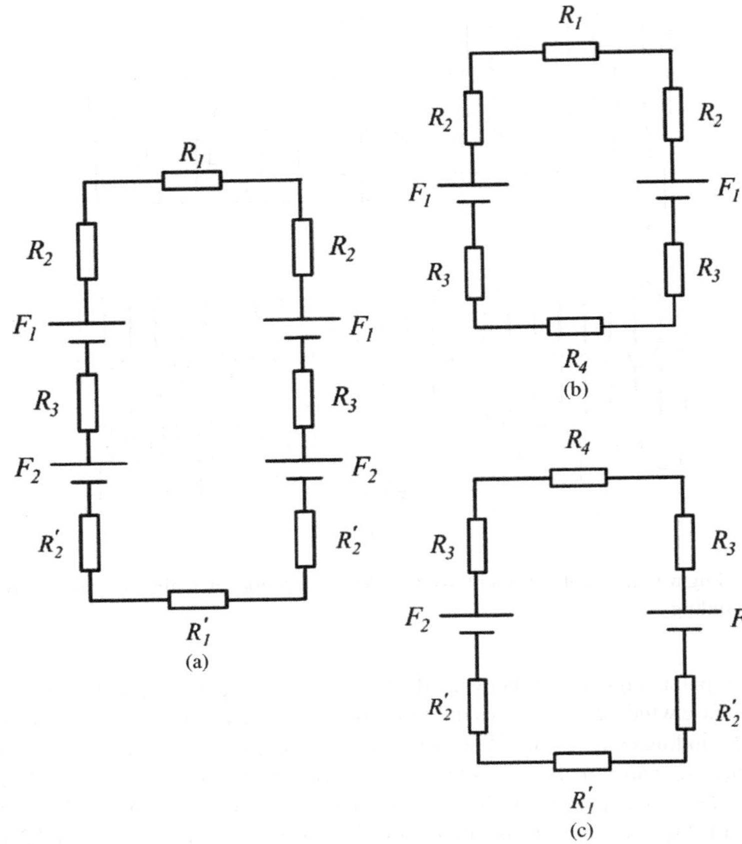

Figure 10.10 Magnetic circuits of PM synchronous machines: (a) double-stator, (b) single-stator inner-rotor, and (c) single-stator outer-rotor

stator and 24 salient poles in the rotor. It incorporates the armature winding, PMs, and DC field winding in the stator, while the rotor is simply salient iron core. Meanwhile, the stator adopts a two-layer structure: the outer-layer stator accommodates the armature winding and the inner-layer stator accommodates both the PMs and DC field winding. The armature winding employs a fractional-slot concentrated winding arrangement, whereas the DC field winding uses a simple DC winding arrangement. Moreover, there is a pair of iron bridges in shunt with each PM pole. With this unique structure, this ISG system offers the following distinct advantages:

- By tuning the magnitude and direction of the DC field current, the machine can flexibly regulate the air-gap flux. Hence, with flux strengthening, it can instantaneously provide high starting torque for cranking and boosting torque for hill-climbing or overtaking. In addition, with online flux control, the machine can maintain constant generated voltage for battery charging over wide ranges of speeds and loads.
- The outer-rotor topology can enable full utilization of the space of inner stator (the part beneath the armature winding) to accommodate both the PMs and DC field winding, hence improving the power density. Also, since both the PMs and DC field winding are embraced by the rotor, the problem of flux leakage can be minimized.

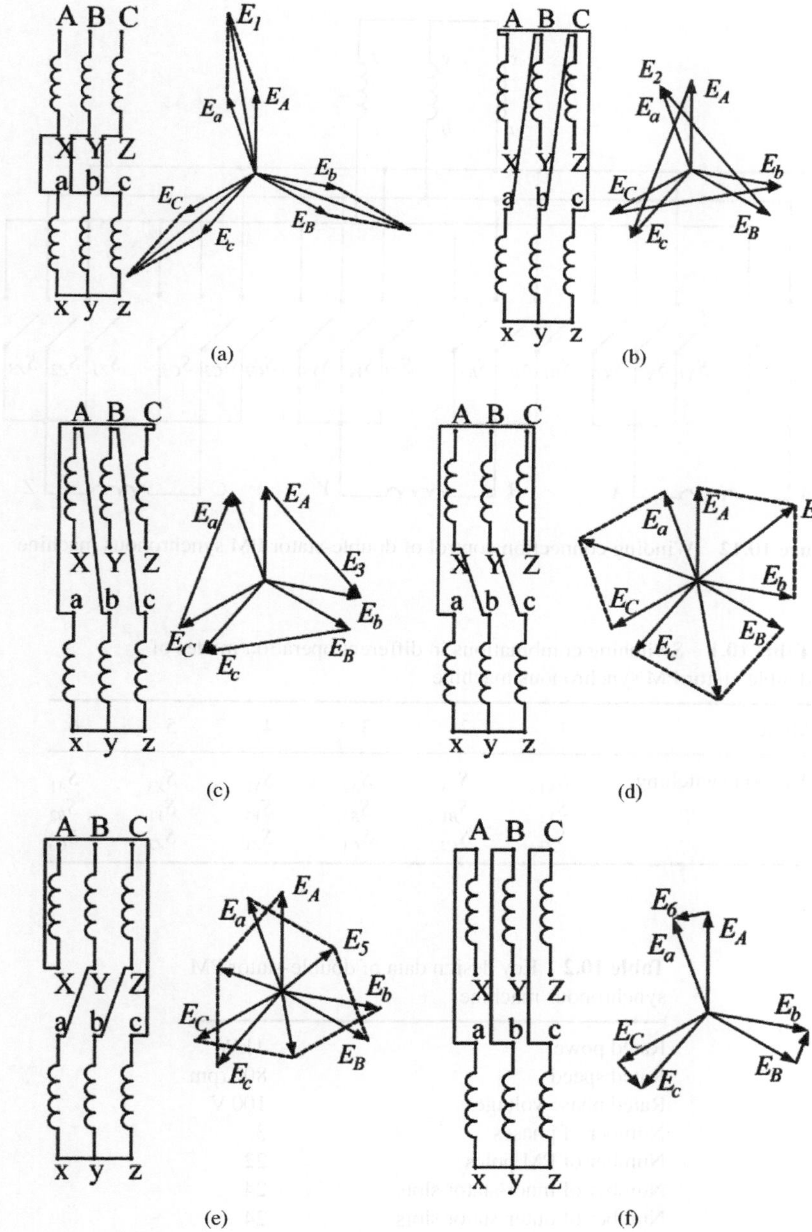

Figure 10.11 Winding connections and vector diagrams of double-stator PM synchronous machine: (a) Mode 1, (b) Mode 2, (c) Mode 3, (d) Mode 4, (e) Mode 5, and (f) Mode 6

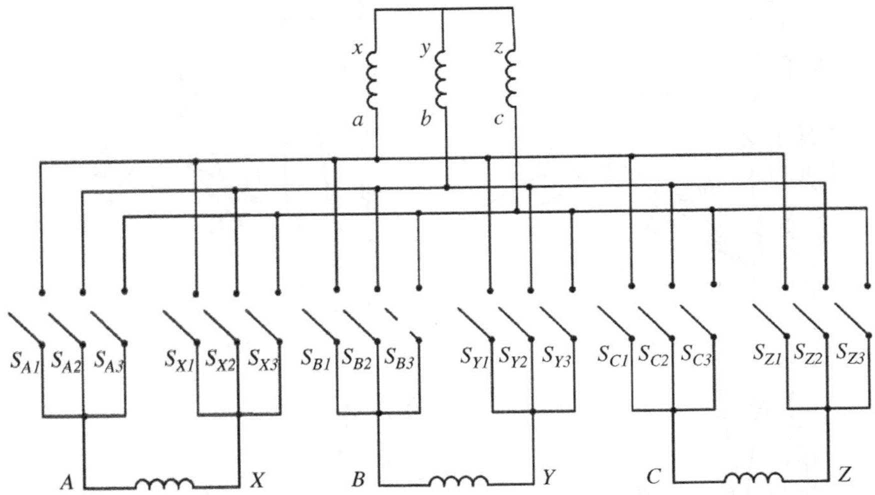

Figure 10.12 Winding connection control of double-stator PM synchronous machine

Table 10.1 Switching combinations in different operation modes of double-stator PM synchronous machine

Mode	1	2	3	4	5	6
Turn-on switching	S_{X1} S_{Y2} S_{Z3}	S_{A3} S_{B1} S_{C2}	S_{A2} S_{B3} S_{C1}	S_{X2} S_{Y3} S_{Z1}	S_{X3} S_{Y1} S_{Z2}	S_{A1} S_{B2} S_{C3}

Table 10.2 Key design data of double-stator PM synchronous machine

Rated power	4 kW
Rated speed	800 rpm
Rated phase voltage	100 V
Number of phases	3
Number of PM poles	22
Number of inner-stator slots	24
Number of outer-stator slots	24
Inner-stator inside diameter	92 mm
Inner-stator outside diameter	165 mm
Outer-stator inside diameter	191 mm
Outer-stator outside diameter	245 mm
Stack length	50 mm
Air-gap length	0.6 mm

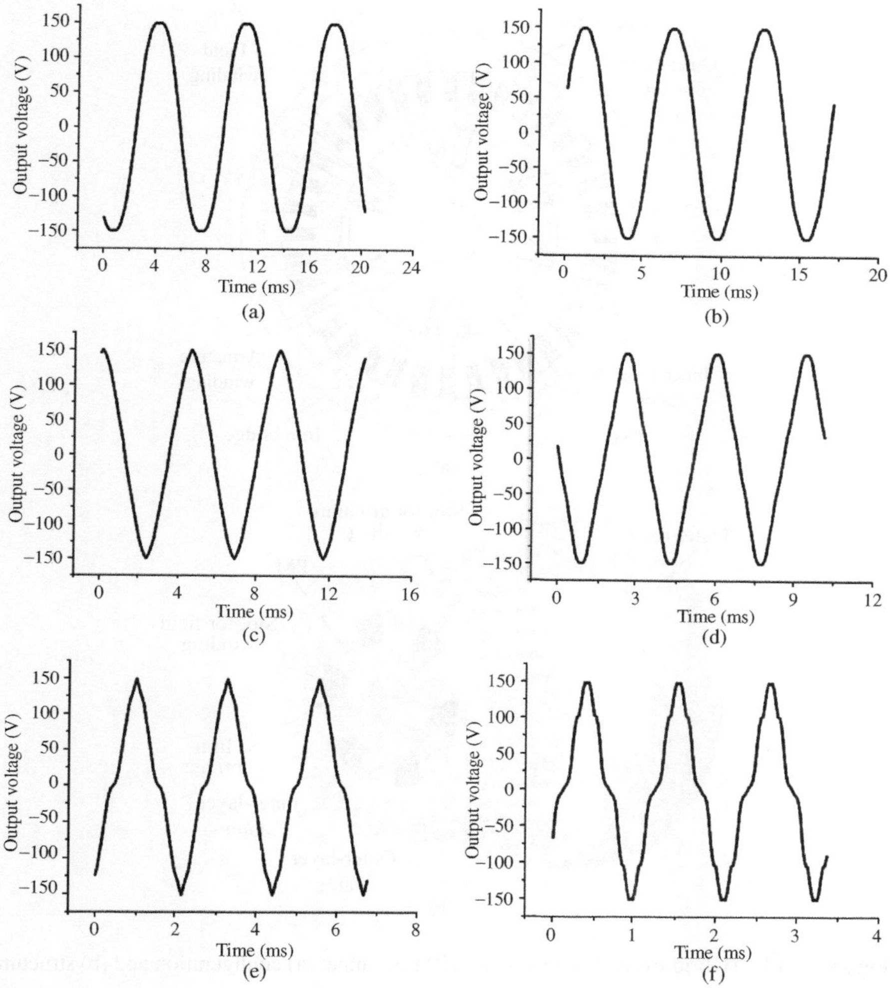

Figure 10.13 Generated output voltage of double-stator PM synchronous machine in different modes at different speeds: (a) 800 rpm, (b) 960 rpm, (c) 1200 rpm, (d) 1600 rpm, (e) 2400 rpm, and (f) 4800 rpm

- Since the rotor does not involve any windings or PMs, it can provide high mechanical integrity, which is essential to handle the high starting torque during cranking.
- The iron bridge functions to suppress the PM flux leakage via the inner-stator iron core, and amplify the effect of flux weakening when the field MMF is opposing the PM flux.
- The fractional-slot armature winding can shorten the magnetic flux path and the span of end-windings, which leads to reduce the use of both iron and copper materials, thus further improving the torque density. In addition, this multipole concentrated winding arrangement can significantly reduce the cogging torque, which usually occurs in conventional PM brushless machines.

To illustrate the ability and the range of controllable flux, an equivalent magnetic circuit of the hybrid-excited DSPM machine at the no-load condition is shown in Figure 10.15, where R_{PM} is the PM reluctance, R_b is the iron-bridge reluctance, R_g is the air-gap reluctance, F_{DC} is the DC field winding

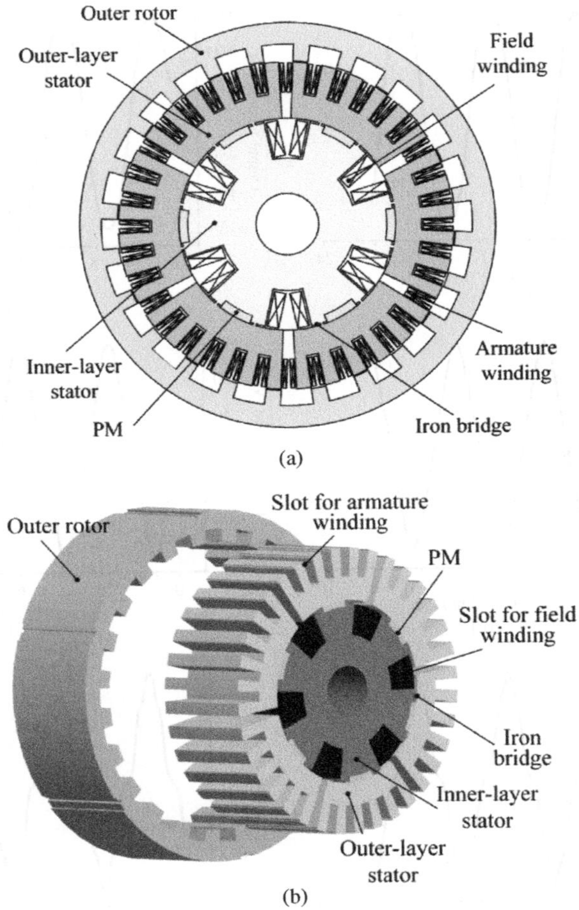

Figure 10.14 Hybrid-excited doubly-salient PM machine: (a) configuration and (b) structure

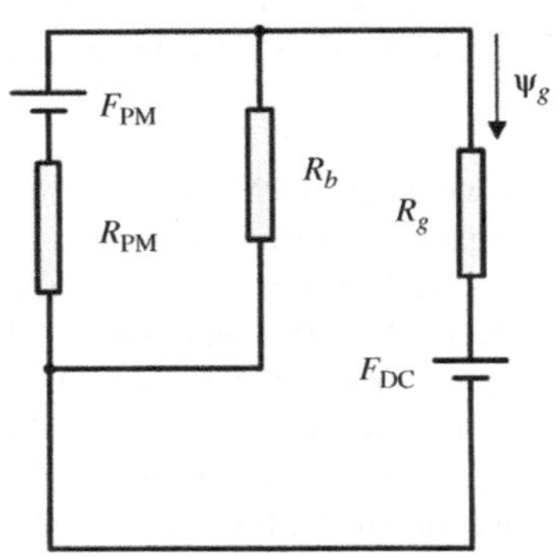

Figure 10.15 Magnetic circuit of hybrid-excited DSPM machine

MMF, and F_{PM} is the PM MMF. Hence, it yields two design equations:

$$\frac{F_{DC+}}{F_{PM}} = \left(\frac{\psi_{g+}}{\psi_{g0}} - 1\right) / \left(\frac{R_{PM}}{R_b} + 1\right) \tag{10.1}$$

$$\frac{F_{DC-}}{F_{PM}} = \left(1 - \frac{\psi_{g-}}{\psi_{g0}}\right) / \left(\frac{R_{PM}}{R_b} + 1\right) \tag{10.2}$$

where ψ_{g0} is the air-gap flux at $F_{DC} = 0$; F_{DC-} and ψ_{g-} are under flux weakening; and F_{DC+} and ψ_{g+} are under flux strengthening. Given the air-gap flux ranges of $(\psi_{g+}/\psi_{g0}) = 3$ and $(\psi_{g-}/\psi_{g0}) = 1/3$, the relationship between (R_{PM}/R_b), (F_{DC-}/F_{PM}), and (F_{DC+}/F_{PM}) can be obtained from Eqs. (10.1) and (10.2) as listed in Table 10.3. When selecting $(R_{PM}/R_b) = 7$, it results in $(F_{DC-} = F_{PM}/12)$ and $(F_{DC+} = F_{PM}/4)$, indicating that the DC field winding excitation needs only 8.3% of the PM excitation for flux weakening and 25% of the PM excitation for flux strengthening to achieve nine times change of air-gap flux.

Figure 10.16 illustrates the operating waveforms of this hybrid-excited DSPM machine. It is similar to those of the DSPM machine except that the air-gap flux linkages ψ_A, ψ_B, and ψ_C are controllable, which are composed of the PM flux linkage ψ_{PM} component and the DC field flux linkage ψ_{DC} component. Consequently, the electromagnetic torque T_e of this machine consists of three components: the PM torque component T_{PM}, which is due to the interaction between the PM flux linkage and the armature phase current i; the DC field torque component T_{DC}, which is due to the interaction between the DC field flux linkage and the armature phase current; and the reluctance torque component T_r, which is due to the variation of the winding inductance L. Mathematically, it is given by

$$T_e = T_{PM} + T_{DC} + T_r = i\frac{d\psi_{PM}}{dt} + i\frac{d\psi_{DC}}{dt} + \frac{1}{2}i^2\frac{dL}{dt} \tag{10.3}$$

The PM torque component actually dominates the torque production, whereas the reluctance torque component is minor and pulsating with zero average value. Thus, when the flux linkage increases with the rotor position, a positive armature current is applied, hence resulting in a positive torque. When the flux linkage decreases, a negative armature current is applied, also resulting in a positive torque. Meanwhile, the bidirectional DC field current is used to tune the DC field flux linkage and hence the DC field torque component according to various ISG operation modes.

Table 10.3 Selection of reluctance ratios of hybrid-excited DSPM machine

(R_{PM}/R_b)	(F_{DC-}/F_{PM})	(F_{DC+}/F_{PM})
10	2/33	2/11
9	1/15	1/5
8	2/27	2/9
7	1/12	1/4
6	2/21	2/7
5	1/9	1/3
4	2/15	2/5
3	1/6	1/2
2	2/9	2/3
1	1/3	1

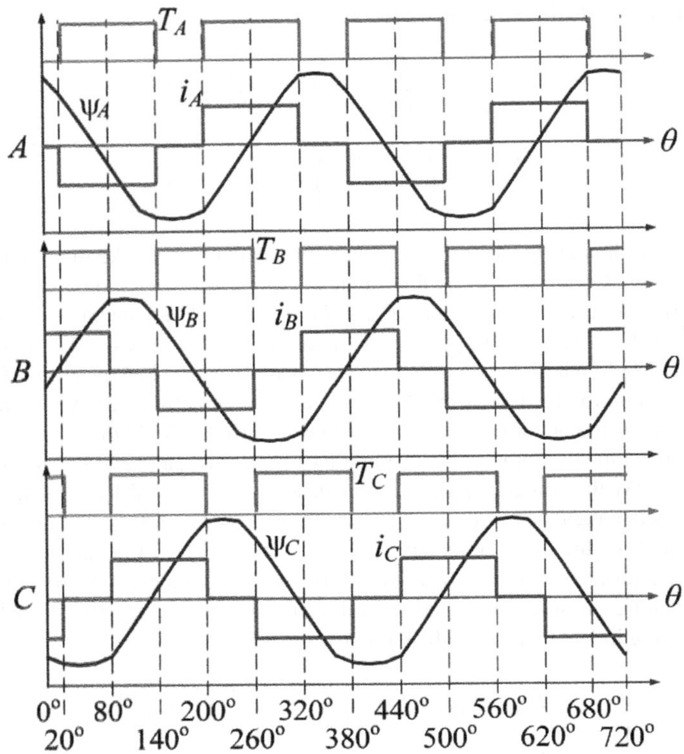

Figure 10.16 Operating waveforms of hybrid-excited DSPM machine

According to different ISG operation modes, the hybrid-excited DSPM machine works in the following ways:

- When the ISG works at the cranking mode, this machine acts as a motor. A temporary positive DC field current is applied to instantaneously strengthen the air-gap flux, hence achieving a high starting torque for engine cranking.
- When the ISG runs at the electricity generation mode, this machine serves as a generator. A positive or negative DC field current is applied to strengthen or weaken the air-gap flux, hence producing a constant output voltage for battery charging.
- When the ISG operates in the power assistance mode, this machine also acts as a motor. A short-term positive DC field current is applied to strengthen the air-gap flux, leading to providing a supplementary torque for boosting the driving torque.

By using finite element analysis, both the flux-strengthening and flux-weakening effects of this machine can be visualized as shown in Figure 10.17. It can be seen that when applying a positive DC field current, the PM flux is reinforced by the DC field flux, hence strengthening the air-gap flux. When applying a negative DC field current, the air-gap flux is effectively weakened. It can also be observed that the iron bridge can avoid PM flux leakage during flux strengthening while amplifying the effect of flux weakening when the DC field flux is opposing the PM flux.

The key design data of this hybrid-excited DSPM machine are listed in Table 10.4. The corresponding steady-state torque waveforms with and without flux strengthening are shown in Figure 10.18. It can

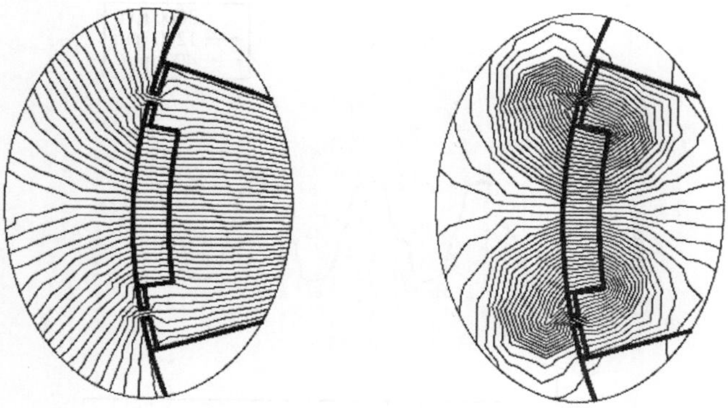

Figure 10.17 Magnetic field distributions of hybrid-excited DSPM machine: (a) positive DC field current and (b) negative DC field current

Table 10.4 Key design data of hybrid-excited DSPM machine

Rated power	2.5 kW
Rated speed	1200 rpm
Rated phase voltage	100 V
Number of phases	3
Number of stator poles	36
Number of rotor poles	24
Number of PM poles	6
Rotor outside diameter	270 mm
Rotor inside diameter	221.2 mm
Stator outside diameter	220 mm
Stator inside diameter	40 mm
Stack length	80 mm
Air-gap length	0.6 mm

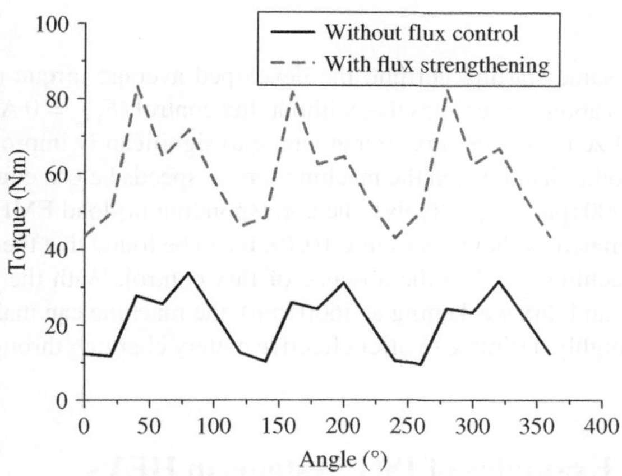

Figure 10.18 Steady-state torque waveforms of hybrid-excited DSPM machine with and without flux strengthening

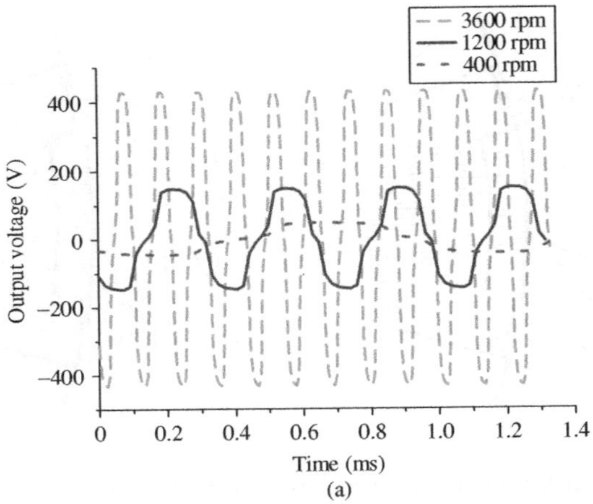

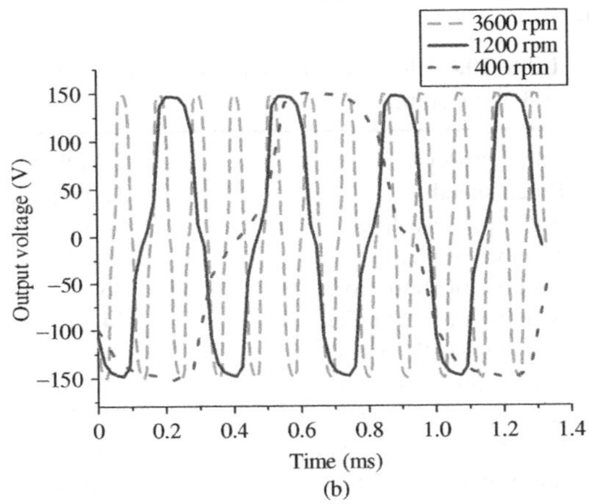

Figure 10.19 No-load EMF waveforms of hybrid-excited DSPM machine: (a) without flux control and (b) with flux control

be found that under the same starting current, the developed average torque under flux strengthening ($F_{DC} = +1000$ A-turns) is about three times that without flux control ($F_{DC} = 0$ A-turns). Hence, it verifies that this machine can utilize its unique flux strengthening to significantly improve the starting torque for engine cranking. On the other hand, when the machine runs at speeds below, equal to, and above the base speed (400, 1200, and 3600 rpm, respectively), the corresponding no-load EMF waveforms without and with flux control are simulated as shown in Figure 10.19. It can be found that the voltage amplitude varies significantly with the machine speed in the absence of flux control. With the use of flux control (flux strengthening at 400 rpm and flux weakening at 3600 rpm), the machine can maintain an almost constant output voltage, which is highly desirable to offer effective battery charging throughout a wide speed range of the machine.

10.7 Application Examples of ISG Systems in HEVs

Continental AG, a Germany-based automotive systems supplier, commenced the development of its integrated-starter-alternator-damper (ISAD) system in 1996. This ISAD system is an induction machine-based crankshaft ISG system in which the rotor serves as a flywheel mounted on the crankshaft between the engine and transmission (Hills, 2001). It not only offers the functions of the starter motor and generator, but also the damper function to stabilize the vibrations and oscillations of the whole drivetrain. In addition, it provides the functions of idle stop-start, regenerative braking, and power assistance, which are all essential for HEVs. The ISAD system went into series production in 2003, and was subsequently installed in both General Motors Company (GMC) Sierra and Chevrolet Silverado mild hybrids.

The ZF-Sachs' DynaStart is a representative crankshaft ISG system based on the PM synchronous machine. It is available with an internal or external rotor. Its design features are large diameter and short axial length, hence producing high torque and facilitating the requirements for powertrain integration. In contrast to conventional machines, the DynaStart adopts automatically wound coils on specially formed laminated cores. This design allows for cost-effective production. Its product range covers both passenger cars and commercial vehicles. The corresponding DynaStart SG325 can offer the maximum starting torque of 210 N m and maximum speed of 6500 rpm (ZF, 2014). This ISG system has been produced for standard applications since 2008. The first example for its production application in cars is the Mercedes-Benz S-400 hybrid.

Honda's in-house developed ISG technology is dubbed the integrated-motor-assist (IMA) system and was first introduced to the Insight hybrid in 1999. This IMA is based on the PM brushless DC machine that is mounted between the engine and transmission to act as a starter motor for cranking, a generator for regenerative braking, an engine balancer to counterbalance the vibrations in the engine, and an assistant traction motor to assist the engine whenever necessary. The IMA is designed with high torque characteristics at low speeds, which is particularly helpful to assist the engine for low-speed acceleration. It can contribute up to 10 kW at 1500 rpm and 78 N m at 1000 rpm to the powertrain (Honda, 2014). While the IMA can serve for the idle stop-start mode, a conventional 12-V starter motor is also present, which kicks in and starts the engine when the IMA battery temperature and SOC are below a predefined value. Currently, the IMA has been applied to different models of HEVs, including the Honda Insight, Honda Civic Hybrid, Honda Accord Hybrid, Honda CR-Z, and Acura ILX Hybrid.

10.8 Matured Technology for HEVs?

HEVs are on the transportation scene as an effective way to meet the increasing challenges of fuel economy and tailpipe emissions. Technical and business considerations continually drive HEVs to be improved day by day. In recent years, there has been a consensus that HEVs, especially the micro and mild hybrids, can present an economically viable solution for the manufacture of low-consumption low-exhaust vehicles in large production volumes. As the ISG system is the core technology for micro and mild hybrids, the corresponding research and development have been fast and encouraging. In fact, the induction machine or PM brushless machine-based ISG systems are becoming mature for mass production, which can further reduce the manufacturing cost. Nevertheless, the ISG system, especially the ISG machine technology, needs to be further developed to improve the starting torque, constant-power operation, motoring and braking efficiencies, robustness, and service life.

References

Bracken, J. and Selker, G. (2013) Regenerative Braking: Decelerating for the Future. University of Pittsburgh, Swanson School of Engineering, Session B7, 3077, pp. 1–8.

Cai, W. (2004) Comparison and review of electric machines for integrated starter alternator applications. Proceedings of IEEE Industry Applications Society Annual Meeting, pp. 386–393.

Chau, K.T. and Chan, C.C. (2007) Emerging energy-efficient technologies for hybrid electric vehicles. *Proceedings of the IEEE*, **95**, 821–835.

Chau, K.T., Cheng, M. and Chan, C.C. (2002) Nonlinear magnetic circuit analysis for a novel stator-doubly-fed doubly-salient machine. *IEEE Transactions on Magnetics*, **38**, 2382–2384.

Chau, K.T., Li, Y.B., Jiang, J.Z. and Liu, C. (2006) Design and analysis of a stator-doubly-fed doubly-salient permanent-magnet machine for automotive engines. *IEEE Transactions on Magnetics*, **42**, 3470–3472.

Chau, K.T., Sun, Q., Fan, Y. and Cheng, M. (2005) Torque ripple minimization of doubly salient permanent magnet motors. *IEEE Transactions on Energy Conversion*, **20**, 352–358.

Chau, K.T. and Wong, Y.S. (2002) Overview of power management in hybrid electric vehicles. *Energy Conversion and Management*, **43**, 1953–1968.

Cheng, M., Chau, K.T., Chan, C.C. and Sun, Q. (2003) Control and operation of a new 8/6-pole doubly salient permanent magnet motor drive. *IEEE Transactions on Industry Applications*, **39**, 1363–1371.

Ebron, A., and Cregar, R. (2005) Introducing Hybrid Technology. National Alternative Fuels Training Consortium eNews, July 1–5.

Ehsani, M., Gao, Y., Gay, S.E. and Emadi, A. (2005) *Modern Electric, Hybrid Electric, and Fuel Cell Vehicles: Fundamentals, Theory, and Design*, CRC Press, Boca Raton, FL.

Hills, A. (2001) Continental ISAD System Chosen by General Motors for Hybrid Truck Project, EV World, http://evworld.com/news.cfm?newsid=1160, (accessed September 2014).

Honda (2014) Insight IMA Technical Information and Emergency Handling Guide, http://www.techinfo.honda-eu.com/sv/sv/Images/104185.pdf Honda. (accessed September 2014).

Jung, K.H., Kim, D., Kim, H. and Hwang, S.-H. (2010) in *Analysis of a Regenerative Braking System for a Hybrid Electric Vehicle Using Electro-Mechanical Brakes, Urban Transport and Hybrid Vehicles* (ed S. Soylu), InTech.

Liu, C., Chau, K.T. and Jiang, J.Z. (2010) A permanent-magnet hybrid brushless integrated starter-generator for hybrid electric vehicles. *IEEE Transactions on Industrial Electronics*, **57**, 4055–4064.

Niu, S., Chau, K.T., and Jiang, J.Z. (2008a) A permanent-magnet double-stator integrated-starter-generator for hybrid electric vehicles. Proceedings of IEEE Vehicle Power and Propulsion Conference, H08359, pp. 1–6.

Niu, S., Chau, K.T. and Jiang, J.Z. (2008b) Analysis of eddy-current loss in a double-stator cup-rotor PM machine. *IEEE Transactions on Magnetics*, **44**, 4401–4404.

Niu, S., Chau, K.T. and Yu, C. (2009) Quantitative comparison of double-stator and traditional permanent magnet brushless machines. *Journal of Applied Physics*, **105**, 07F105, 1–07F105, 3.

Viorel, I.-A., Szabó, L., Löwenshein, L. and Şteţ, C. (2004) Integrated starter-generators for automotive applications. *Acta Electrotehnica*, **45**, 255–260.

Walker, A., Anpalahan, P., Coles, P. et al. (2004) Automotive integrated starter generator. Proceedings of International Conference on Power Electronics, Machines and Drives, pp. 46–48.

Wang, Y., Cheng, M., Chen, M. *et al.* (2011) Design of high-torque-density double-stator permanent magnet brushless motors. *IET Electric Power Applications*, **5**, 317–323.

ZF Friedrichshafen AG (2014) Electric Motors for Hybrid Drives for Passenger Cars and Commercial Vehicles, ZF Friedrichshafen AG, http://www.zf.com/media/media/en/document/corporate_2/downloads_1/flyer_and_brochures/cars_flyer/elektrischemaschinenfrhybridantriebefrpersonenkraftwagenundnutzfahrzeuge.pdf, (accessed September 2014).

Zhang, J., Lv, C., Gou, J. and Kong, D. (2012) Cooperative control of regenerative braking and hydraulic braking of an electrified passenger car. *Proceedings of the Institution of Mechanical Engineers, Part D: Journal of Automobile Engineering*, **226**, 1289–1302.

11

Planetary-Geared Electric Variable Transmission Systems

While the micro and mild hybrids take the advantages of using the integrated-starter-generator to offer some hybrid features, the full hybrid can provide all hybrid features and thereby achieve the highest fuel economy. The key technology of the full hybrid is the electric variable transmission (EVT) system, also called the electronic-continuously variable transmission system. Since the introduction of the first EVT system in 1997, there have been many derivatives developed by different automakers. Essentially, most of them are based on the use of planetary gearing to perform power-split, hence achieving EVT.

In this chapter, various planetary-geared electric variable transmission (PG EVT) systems for the full hybrids, including the system configurations and planetary gears as well as the input-split and compound-split planetary-geared (PG) types, are discussed. The corresponding design criteria, design example, and application examples are also presented.

11.1 System Configurations

The PG EVT system functions to offer electrically controllable power transfer from the engine to the wheels with continuously variable transmission – seamless torque transfer and smooth gear changing. This PG EVT system can generally be categorized into the input-split and compound-split types (Miller, 2006; Chau and Chan, 2007; Wang, Cheng, and Chau, 2009).

First, the general configuration of the input-split PG EVT system is shown in Figure 11.1, which is composed of a power-split device, two machines, an energy storage device, and a final drive. The key element is the power-split device, actually a planetary gear, which serves to split the engine power into two paths: one is a mechanical path from the engine to the wheels via the power-split device and final drive; another is an electrical path from the engine to the wheels via the power-split device, machine 1, machine 2, and final drive. An energy storage device is generally present between the machine 1 and machine 2 to serve as a power buffer, which can be a battery or an ultracapacitor. The battery takes the advantage of higher specific energy that facilitates larger energy storage capacity, whereas the ultracapacitor has the merit of higher specific power that favors handling of higher instantaneous power (Chau, Wong, and Chan, 2003). Two renowned input-split PG EVT systems are the Toyota and Ford systems.

Second, the general configuration of the compound-split PG EVT system is shown in Figure 11.2. Differing from the input-split counterpart, the compound-split configuration involves two power-split devices: one located at the input and another located at the output. Because of its four-port nature, it can offer two modes of operation under variable structure control. In general, two-mode operation can improve

Electric Vehicle Machines and Drives: Design, Analysis and Application, First Edition. K. T. Chau.

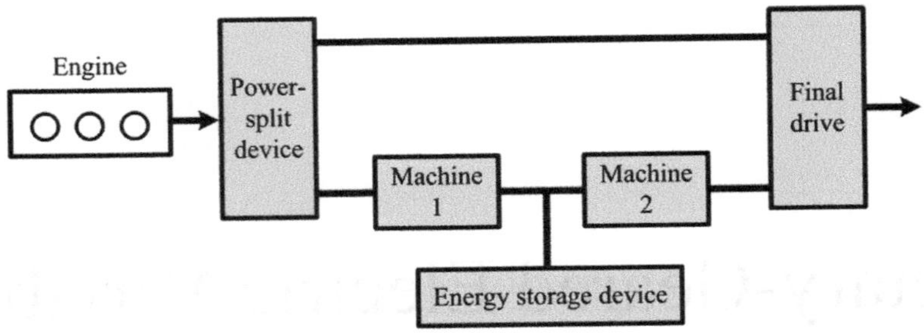

Figure 11.1 Input-split PG EVT system configuration

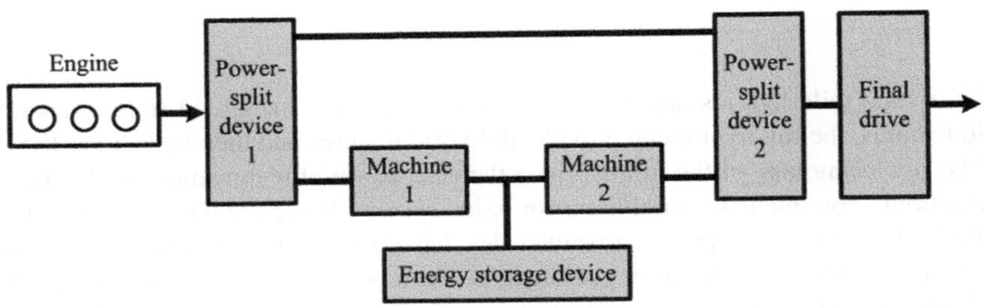

Figure 11.2 Compound-split PG EVT system configuration

the vehicular performance and fuel economy. Basically, in one mode, the compound-split system resembles the input-split system with output gearing. That is, the power-split device at the input serves for torque differential and the power-split device at the output serves as a torque multiplier. In another mode, the compound-split system operates both the input and output power-split devices serving for torque differential. There are three well-known compound-split PG-EVT systems: the GM-Allison (General Motors), Timken, and Renault systems.

11.2 Planetary Gears

The planetary gear, also called epicyclic gear, is the core of PG EVT systems. It consists of one or more planet gears revolving about a sun gear. The more the planets in the gear, the greater the load capability and the higher the torque density are resulted. The planet gears are mounted on a movable arm, called carrier, which can rotate relative to the sun gear. It also incorporates an outer ring gear to mesh with the planet gears. As shown in Figure 11.3, there are totally three shaft axes: the sun gear axis, ring gear axis, and planet carrier axis, all of them are usually coaxial. Among these three shafts, any one of them can be considered as the input or output. Thus, it can work as a two-input single-output or single-input two-output mechanical system.

Compared with the traditional parallel-axis gear, the planetary gear possesses the following advantages:

- Higher power density and higher torque density
- Higher power transmission efficiency
- Larger reduction ratio in a given volume
- Multiple kinematic combinations with coaxial shafting
- Better load distribution and greater stability

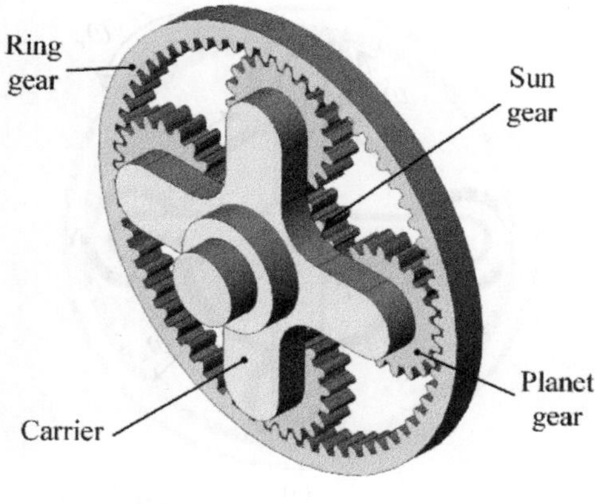

Figure 11.3 Planetary gear structure

However, the corresponding disadvantages are larger bearing loads and higher design complexity. Because of the unique feature of multiple kinematic combinations with coaxial shafting, it is the key device to perform the desired power-split for EVT.

A schematic of a planetary gear having four planet gears is shown in Figure 11.4, which is used to derive the speed relationship inside the planetary gear (Mi, Masrur, and Gao, 2011). First, the tangential speed v and angular speed ω of a gear are related by $v = r\omega$, where r is its radius. When two gears are coupled, they must have the same tangential speeds at the contact point. Hence, the speed relationship of two coupled gears is governed by

$$\frac{r_1}{r_2} = \frac{\omega_2}{\omega_1} \tag{11.1}$$

where ω_1 and ω_2 are their angular speeds and r_1 and r_2 are their radii. Meanwhile, the tooth pitch p and tooth number N of each gear are related by $2\pi r = pN$. When two gears are coupled, they must move with the same tooth pitches. Hence, the tooth relationship of two coupled gears is given by

$$\frac{r_1}{r_2} = \frac{N_1}{N_2} \tag{11.2}$$

where N_1 and N_2 are their tooth numbers.

For the planetary gear, there are two main points of contact: X and Y. The point X lies between a planet gear and the sun gear, whereas the point Y lies between that planet gear and the ring gear. Thus, the tangential speed at point X, v_X, can be written as

$$v_X = \omega_s r_s \tag{11.3}$$

$$v_X = \omega_p r_p - \omega_c r_c \tag{11.4}$$

where ω_s, ω_p, and ω_c are the angular speeds of the sun gear, planet gear, and carrier, respectively, and r_s, r_p, and r_c are their respective radii. Similarly, the tangential speed at point Y, v_Y, can be expressed as

$$v_Y = \omega_r r_r \tag{11.5}$$

$$v_Y = \omega_p r_p + \omega_c r_c \tag{11.6}$$

Eliminating v_X and v_Y can reduce Eqs. (11.3)–(11.6) as

$$\omega_s r_s = \omega_p r_p - \omega_c r_c \tag{11.7}$$

$$\omega_r r_r = \omega_p r_p + \omega_c r_c \tag{11.8}$$

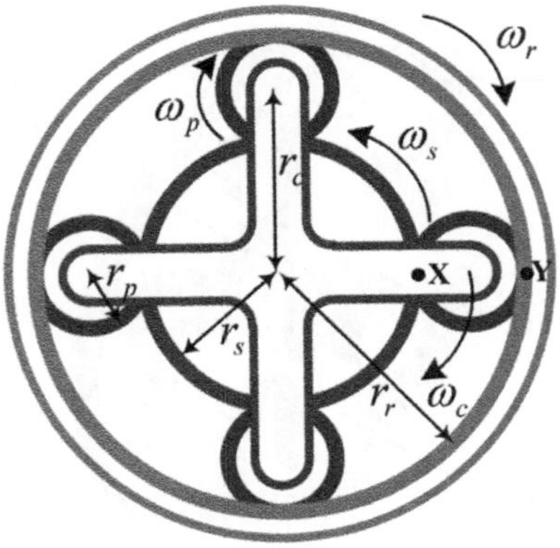

(a)

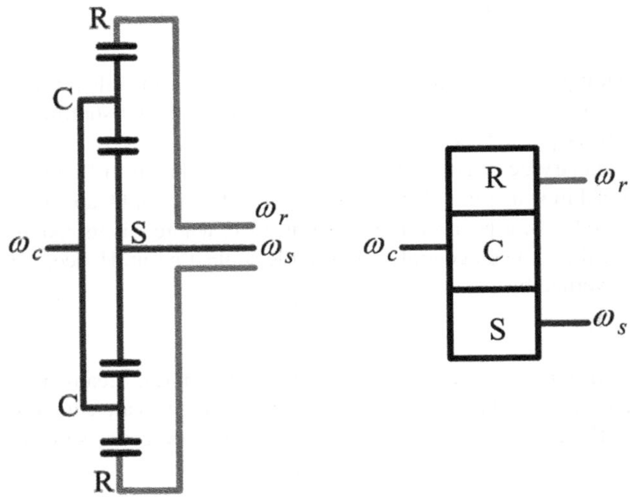

(b)

Figure 11.4 Planetary gear schematic: (a) model and (b) symbols

Then, further eliminating ω_p can deduce the following relationship:

$$\omega_r r_r = \omega_s r_s + 2\omega_c r_c \tag{11.9}$$

On the other hand, the radius relationships inside the planetary gear are given by

$$r_r = r_c + r_p \tag{11.10}$$

$$r_r = r_s + 2r_p \tag{11.11}$$

Eliminating r_p can deduce the following relationship:

$$r_c = (r_r + r_s)/2 \tag{11.12}$$

Substituting Eq. (11.12) into Eq. (11.9) results in

$$\omega_r r_r = \omega_c(r_s + r_r) + \omega_s r_s \tag{11.13}$$

In order to unify all rotational directions as positive for clockwise rotation and negative for anticlockwise rotation, Eq. (11.13) can be generally rewritten as

$$\omega_r r_r = \omega_c(r_s + r_r) - \omega_s r_s \tag{11.14}$$

On the basis of Eq. (11.2), the speed relationships of the planetary gear given by Eq. (11.14) can be obtained as

$$\omega_s = (1 + \rho)\,\omega_c - \rho\omega_r \tag{11.15}$$

where ρ is the planetary gear ratio defined as

$$\rho = \frac{N_r}{N_s} \tag{11.16}$$

11.3 Input-Split PG EVT Systems

There are two main input-split PG EVT systems: one was developed by Toyota for its Prius HEV (hybrid electric vehicle); another was developed by Ford for its Escape HEV. In order to properly assess their speed and torque relationships, their system dynamical equations need to be formulated. Detailed derivations of their system dynamics can be found in Miller and Everett (2005) and Miller (2010).

11.3.1 Toyota Hybrid System

The turning point of HEV development was the advent of Toyota Prius in 1997, which introduced the first PG EVT system (Sasaki, 1998). Figure 11.5 shows the basic configuration of Toyota hybrid system (THS). It is a well-known PG EVT system, which mainly consists of a planetary gear, a battery pack, the machine 1 (normally operating as a generator), the converter 1 (normally operating as a controlled rectifier), the machine 2 (normally operating as a motor), and the converter 2 (normally operating as an inverter). In this system, the planetary gear set plays a key role to split the engine power into two paths of power flow: mechanical and electrical. On the one hand, the mechanical power produced by the engine is transferred to the driveline via the ring gear. On the other hand, the sun gear is attached to the machine 1, which converts a portion of the engine power to electrical power so that it can be used to feed the machine

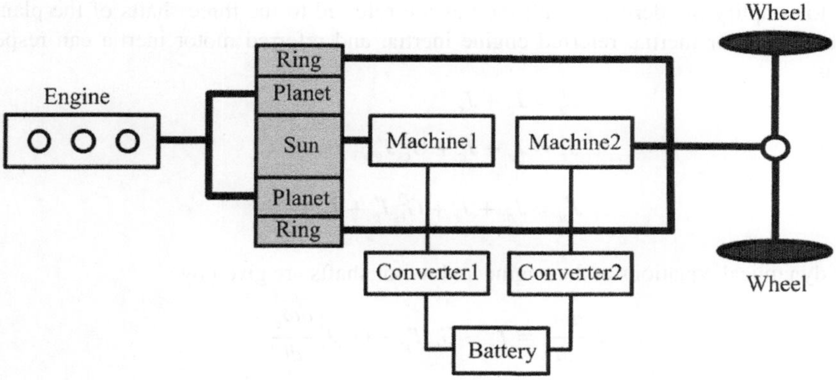

Figure 11.5 Toyota hybrid system

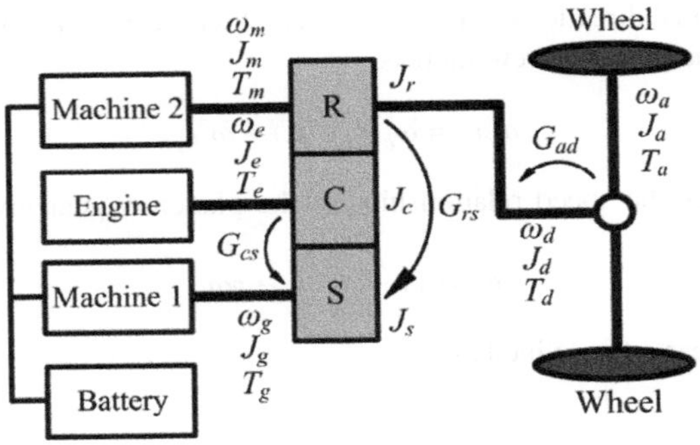

Figure 11.6 Toyota hybrid system model

2 to drive the driveline. The converter 1 and converter 2 work with the battery pack to buffer the electrical power transfer between the machine 1 and machine 2. If the desired vehicle power is greater than the engine power, the battery will release electrical power to supplement the engine power; otherwise, the battery will store the excessive engine power.

The model of the THS is shown in Figure 11.6, where the engine speed, inertia, and torque are denoted as ω_e, J_e, and T_e, respectively; the generator speed, inertia, and torque are denoted as ω_g, J_g, and T_g, respectively; the motor speed, inertia, and torque are denoted as ω_m, J_m, and T_m, respectively; the driveline speed, inertia, and torque are denoted as ω_d, J_d, and T_d, respectively; the axle speed, inertia, and torque are denoted as ω_a, J_a, and T_a, respectively; the ring gear, carrier, and sun gear inertias are denoted as J_r, J_c, and J_s, respectively; the carrier-to-sun and ring-to-sun gear ratios are denoted as G_{cs} and G_{rs}, respectively; and the axle-to-driveline gear ratio is denoted as G_{ad}. In the absence of clutch, the planetary gear runs with the vehicle in motion.

On the basis of Jourdain's principle, the torque relationships of the planetary gears can be described as

$$T_c = (1+\rho)T_s \tag{11.17}$$

$$T_c = \frac{(1+\rho)}{\rho}T_r \tag{11.18}$$

$$T_s = \frac{1}{\rho}T_r \tag{11.19}$$

where T_c, T_s, and T_r are the torques of the carrier, sun gear, and ring gear, respectively.

In order to simplify the derivation, all inertias are referred to the three shafts of the planetary gear. The referred generator inertia, referred engine inertia, and referred motor inertia can respectively be expressed as

$$J'_g = J_g + J_s \tag{11.20}$$

$$J'_e = J_e + J_c + G_{cs}^2 J'_g \tag{11.21}$$

$$J'_m = J_m + J_r + G_{rs}^2 J'_g + J_d + \frac{J_a}{G_{ad}^2} \tag{11.22}$$

Hence, the dynamical equations at the engine and motor shafts are given by

$$J'_e \frac{d\omega_e}{dt} = T_e - G_{cs}T_g - G_{cs}J'_g \frac{d\omega_g}{dt} \tag{11.23}$$

$$J'_m \frac{d\omega_m}{dt} = T_m - T_d - G_{rs}T_g - G_{rs}J'_g \frac{d\omega_g}{dt} \tag{11.24}$$

As the ring gear, carrier, and sun gear are directly coupled with the motor, engine, and generator, respectively, their speed relationships given by Eq. (11.15) can be rewritten as

$$\omega_g = (1+\rho)\,\omega_e - \rho\omega_m \tag{11.25}$$

which can also be expressed as

$$\omega_g = G_{cs}\omega_e + G_{rs}\omega_m \tag{11.26}$$

$$G_{cs} = 1+\rho \tag{11.27}$$

$$G_{rs} = -\rho \tag{11.28}$$

Substituting Eq. (11.26) into Eqs. (11.23) and (11.24), the dynamical equations at the engine and motor shafts are reduced as given by

$$J_{eo}\frac{d\omega_e}{dt} = T_e - G_{cs}T_g - J_{go}\frac{d\omega_m}{dt} \tag{11.29}$$

$$J_{mo}\frac{d\omega_m}{dt} = T_m - T_d - G_{rs}T_g - J_{go}\frac{d\omega_e}{dt} \tag{11.30}$$

where the equivalent lumped inertias are given by

$$J_{eo} = J'_e + G_{cs}^2 J'_g \tag{11.31}$$

$$J_{mo} = J'_m + G_{rs}^2 J'_g \tag{11.32}$$

$$J_{go} = G_{cs}G_{rs}J'_g \tag{11.33}$$

Hence, the generator and driveline torques can be obtained as

$$T_g = \frac{1}{G_{cs}}\left(T_e - J_{eo}\frac{d\omega_e}{dt} - J_{go}\frac{d\omega_m}{dt}\right) \tag{11.34}$$

$$T_d = T_m - \frac{G_{rs}}{G_{cs}}T_e + \left(\frac{G_{rs}}{G_{cs}}J_{eo} - J_{go}\right)\frac{d\omega_e}{dt} + \left(\frac{G_{rs}}{G_{cs}}J_{go} - J_{mo}\right)\frac{d\omega_m}{dt} \tag{11.35}$$

Equation (11.35) represents the final driving torque of this input-split PG EVT system and must be equal to the road load. At steady state, all the system dynamics become zero so that the steady-state generator torque and driveline torque are given by

$$T_g = \frac{1}{G_{cs}}T_e \tag{11.36}$$

$$T_d = T_m - \frac{G_{rs}}{G_{cs}}T_e \tag{11.37}$$

Substituting Eqs. (11.27) and (11.28) into Eqs. (11.36) and (11.37), the corresponding steady-state torques become

$$T_g = \frac{1}{(1+\rho)}T_e \tag{11.38}$$

$$T_d = T_m + \frac{\rho}{(1+\rho)}T_e \tag{11.39}$$

This input-split PG EVT mainly offers four operating modes: electric launch, cruising, acceleration, and regenerative braking:

- *Electric launch*: The engine is turned off to avoid operating at low efficiency. As depicted in Figure 11.7, the driveline torque is only contributed by the motor, which is powered by the battery. Hence, substituting $T_e = 0$ into Eq. (11.39) yields

$$T_d = T_m \tag{11.40}$$

- *Cruising*: Since the desired vehicle power is low, the engine power is split in such a way that the excessive engine power can be stored in the battery as depicted in Figure 11.8. The motor is turned off and the driveline torque is contributed by the engine via the ring gear only. Hence, substituting $T_m = 0$ in Eq. (11.39) yields

$$T_d = \frac{\rho}{(1+\rho)} T_e \tag{11.41}$$

- *Acceleration*: Since the desired vehicle power is high, the battery releases electrical power to supplement the engine power as depicted in Figure 11.9. Both the engine and motor contribute to the driveline torque as given by Eq. (11.39).
- *Regenerative braking*: The engine is turned off and the motor operates as a generator to produce a braking torque, which converts the braking energy into electrical energy to recharge the battery as depicted in Figure 11.10. Hence, substituting $T_e = 0$ in Eq. (11.39), the driveline torque becomes a negative motor torque as given by

$$T_d = -T_m \tag{11.42}$$

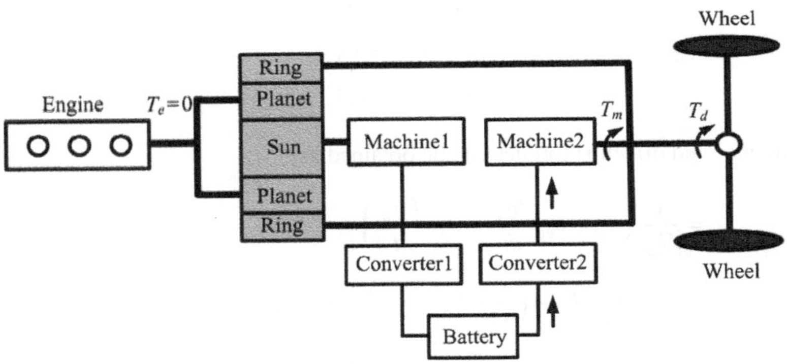

Figure 11.7 Electric launch mode

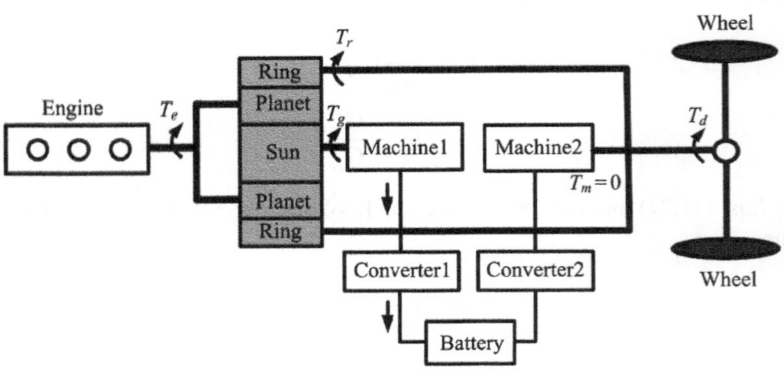

Figure 11.8 Cruising mode

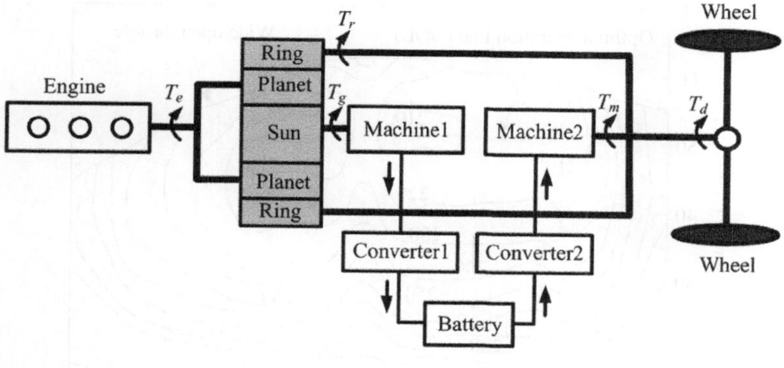

Figure 11.9 Acceleration mode

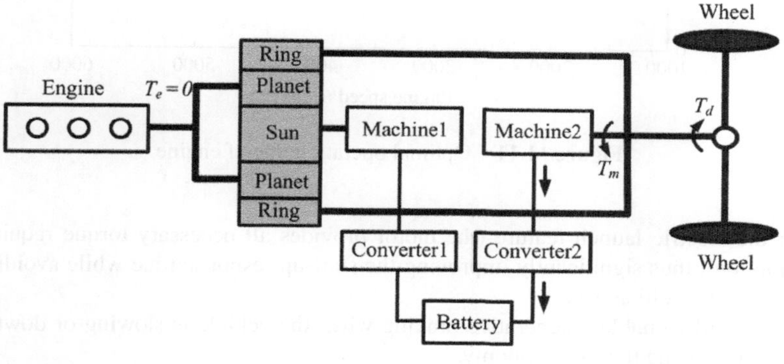

Figure 11.10 Regenerative braking mode

During normal operation with power-split, the control strategy is to enable the engine always working at the optimal operation line (OOL). Figure 11.11 shows a typical OOL of an engine in which the optimization is based on the minimum fuel consumption or maximum fuel economy. Given the vehicle driveline speed and demanded output power, the engine operating point (T_e, ω_e) can be deduced according to the OOL of its fuel consumption characteristics. On the basis of the driveline speed and engine speed, the required generator speed can be determined by using Eq. (11.25). This generator speed can be regulated by controlling its output power drawn by the associated converter. Then, the motor torque can be deduced by subtracting the engine torque at the ring gear from the demanded vehicle torque demand. By controlling the power taken by the generator and then feeding it back into the motor via the DC-link and battery, the engine speed can be maintained constant when the driveline speed is varying. Hence, a continuously variable ratio between the engine speed and wheel speed can be achieved.

The key advantages of this input-split PG EVT system for HEVs are summarized as follows:

- Because of the absence of clutches or shifting gears, it can significantly improve the transmission efficiency and reduce the overall size, hence increasing both the energy efficiency and power density.
- In the presence of continuously variable ratio between the engine speed and wheel speed, the engine can always operate at its most energy-efficient or OOL, hence resulting in considerable reduction of fuel consumption.
- The system can offer the idle stop-start feature (the engine is completely shut down when the vehicle is stopped and instantly cranked once the vehicle is launched), thus significantly reducing fuel consumption.

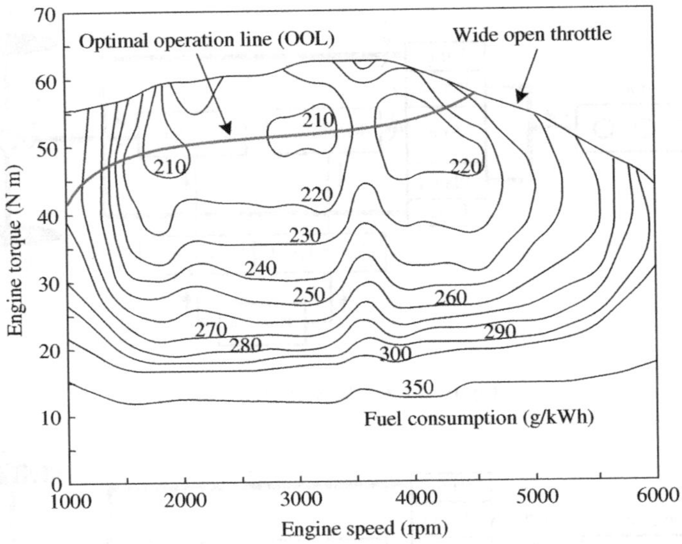

Figure 11.11 Optimal operation line of engine

- It can offer the electric launch feature (the motor provides all necessary torque required to put the vehicle in motion), thus significantly improving the start-up response time while avoiding the engine operating at very low efficiency.
- The system can fully enable regenerative braking when the vehicle is slowing or downhill coasting, hence further improving the fuel economy.
- The system facilitates full-throttle acceleration, where both the engine and motor operate simultaneously to provide the required power demand.

11.3.2 Ford Hybrid System

As shown in Figure 11.12, the Ford hybrid system (FHS) is another well-known input-split PG EVT. Similar to the THS, the FHS consists of a planetary gear, a motor, a generator, a battery pack, a controlled rectifier, and an inverter to perform the desired power-split. In addition, it incorporates output gearing between the motor and final drive. This output torque multiplier is generally in the form of spur gears.

The model of this FHS is shown in Figure 11.13 in which G_{12}, G_{13}, and G_{23} are the additional gear ratios of output gearing. Similar to the previous derivation, all inertias are first referred to the three shafts of the planetary gear. The referred generator inertia, referred engine inertia, and referred motor inertia are given by

$$J'_g = J_g + J_s \tag{11.43}$$

$$J'_e = J_e + J_c + G_{cs}^2 J'_g \tag{11.44}$$

$$J'_m = J_m + G_{13}^2 J_r + G_{13} G_{rs} J'_g + G_{12}^2 J_d + \frac{G_{12}}{G_{ad}} J_a \tag{11.45}$$

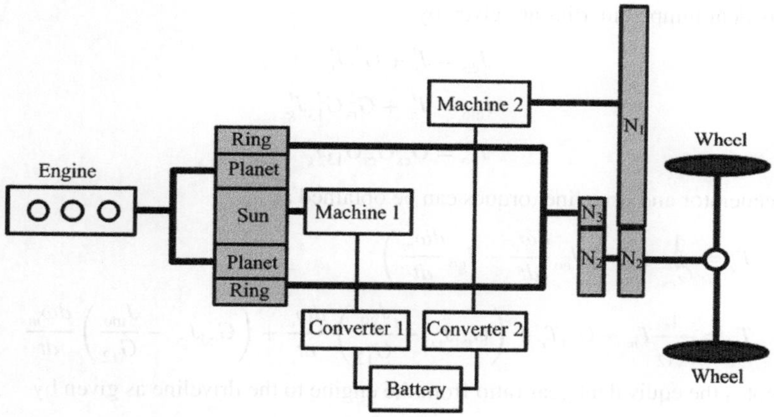

Figure 11.12 Ford hybrid system

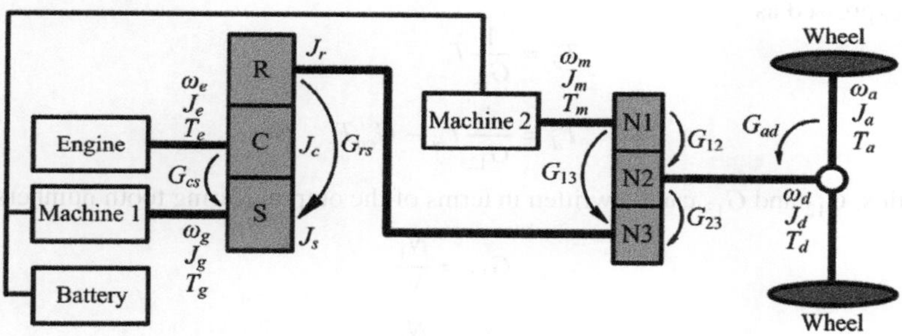

Figure 11.13 Ford hybrid system model

Consequently, the dynamical equations at the engine and motor shafts can be expressed as

$$J'_e \frac{d\omega_e}{dt} = T_e - G_{cs}T_g - G_{cs}J'_g \frac{d\omega_g}{dt} \tag{11.46}$$

$$J'_m \frac{d\omega_m}{dt} = T_m - G_{12}T_d - G_{rs}G_{13}T_g - G_{rs}G_{cs}G_{13}J'_g \frac{d\omega_g}{dt} \tag{11.47}$$

The generator speed can be expressed in terms of the engine speed and motor speed as given by

$$\omega_g = G_{cs}\omega_e + G_{rs}G_{13}\omega_m \tag{11.48}$$

Substituting Eq. (11.48) in Eqs. (11.46) and (11.47), the dynamical equations at the engine and motor shafts are rearranged as

$$J_{eo} \frac{d\omega_e}{dt} = T_e - G_{cs}T_g - J_{go} \frac{d\omega_m}{dt} \tag{11.49}$$

$$J_{mo} \frac{d\omega_m}{dt} = T_m - G_{12}T_d - G_{rs}G_{13}T_g - J_{go} \frac{d\omega_e}{dt} \tag{11.50}$$

where the equivalent lumped inertias are given by

$$J_{eo} = J'_e + G_{cs}^2 J'_g \tag{11.51}$$

$$J_{mo} = J'_m + G_{rs}^2 G_{13}^2 J'_g \tag{11.52}$$

$$J_{go} = G_{cs} G_{rs} G_{13} J'_g \tag{11.53}$$

Hence, the generator and driveline torques can be obtained as

$$T_g = \frac{1}{G_{cs}}\left(T_e - J_{eo}\frac{d\omega_e}{dt} - J_{go}\frac{d\omega_m}{dt}\right) \tag{11.54}$$

$$T_d = \frac{1}{G_{12}}T_m - G_{ed}T_e + \left(G_{ed}J_{eo} - \frac{J_{go}}{G_{12}}\right)\frac{d\omega_e}{dt} + \left(G_{ed}J_{go} - \frac{J_{mo}}{G_{12}}\right)\frac{d\omega_m}{dt} \tag{11.55}$$

where G_{ed} denotes the equivalent gear ratio from the engine to the driveline as given by

$$G_{ed} = \frac{G_{rs}G_{13}}{G_{cs}G_{12}} \tag{11.56}$$

At steady state, the system dynamics become zero. The steady-state generator torque and driveline torque can be expressed as

$$T_g = \frac{1}{G_{cs}}T_e \tag{11.57}$$

$$T_d = \frac{1}{G_{12}}T_m - G_{ed}T_e \tag{11.58}$$

The gear ratios, G_{12} and G_{13} can be written in terms of the output gearing tooth numbers as

$$G_{12} = \frac{N_1}{N_2} \tag{11.59}$$

$$G_{13} = \frac{N_1}{N_3} \tag{11.60}$$

Hence, substituting Eqs. (11.27) and (11.28) as well as Eqs. (11.59) and (11.60) into Eqs. (11.57) and (11.58), the corresponding steady-state torques become

$$T_g = \frac{1}{(1+\rho)}T_e \tag{11.61}$$

$$T_d = \frac{N_2}{N_1}T_m + \frac{\rho}{(1+\rho)}\frac{N_2}{N_3}T_e \tag{11.62}$$

Comparing Eqs. (11.61) and (11.62) with Eqs. (11.38) and (11.39), it illustrates that the FHS is very similar to the THS. Because of the same input-split feature, their generator torques are the same. However, because of the additional output gearing, the FHS can offer further mechanical advantage to the motor and engine torques, leading to have the effect of torque multiplier. Meanwhile, the corresponding operating modes are essentially the same as those of the THS.

11.4 Compound-Split PG EVT Systems

The compound-split PG EVT system is more complicated than the input-split system. The key is that it incorporates the feature of variable structure control. With the use of mechanical disconnect clutches, the system can offer various structures. The representative compound-split PG EVT system is the GM two-mode hybrid system. Consequently, there are various variants, such as the Renault and Timken systems.

11.4.1 GM Two-Mode Hybrid System

The GM two-mode hybrid system is illustrated in Figure 11.14. It is mainly composed of three clutches, two planetary gears, two machines, two converters, and a battery pack (Holmes and Schmidt, 2002). The corresponding equivalent model is depicted in Figure 11.15. The engine is coupled with the ring gear R_1 of the first planetary gear P_1 via the clutch CL_1. The machine M_1 normally operates as a generator, which is coupled with the sun gear S_1 of P_1. The carrier C_1 of P_1 is coupled with the carrier C_2 of the second planetary gear P_2, and both are linked up with the driveline. The machine M_2 normally operates as a motor, which is coupled with the sun gear S_2 of P_2. There are two clutches to enable the ring gear R_2 of P_2 connecting to M_1 (via the clutch CL_2) or grounded (via the clutch CL_3). By means of engaging or disengaging different clutch arrangements, the system can alter its structure to provide more flexibility for transmission control.

When the HEV operates at low-speed mode (dubbed the low-range mode) for urban driving, both the clutches CL_1 and CL_3 are engaged while the clutch CL_2 is disengaged. Thus, the first planetary gear is responsible for input power-split while the second planetary gear is in charge of output gearing, thus working alike the FHS. When the HEV operates at high-speed mode (dubbed the high-range mode) for highway cruising, both the clutches CL_1 and CL_2 are engaged while the clutch CL_3 is disengaged. Thus, both first and second planetary gears perform the power-split together, hence the compound-split. Neutral

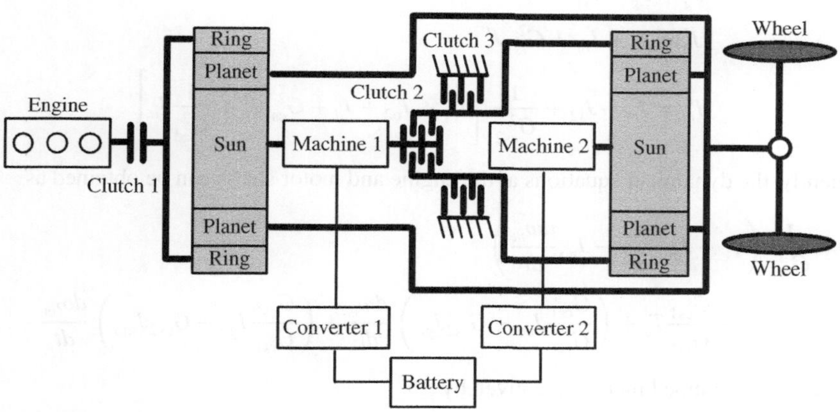

Figure 11.14 GM two-mode hybrid system

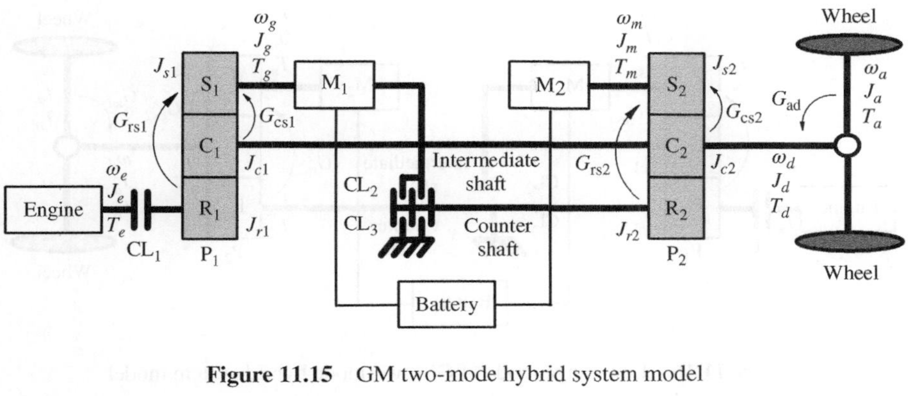

Figure 11.15 GM two-mode hybrid system model

occurs when both CL_2 and CL_3 are disengaged. Therefore, this two-mode PG EVT system can optimize the fuel economy for both low-speed and high-speed operations.

In order to properly assess their speed and torque relationships, their system dynamical equations under individual modes need to be formulated. Detailed derivations of their system dynamics can be found in Miller (2006, 2010).

11.4.1.1 Low-Range Mode

In this mode, the clutches CL_1 and CL_3 are engaged whereas CL_2 is disengaged. The first planetary gear works for power-split and the second planetary gear serves for output gearing. Figure 11.16 shows the equivalent model of this low-range mode, where the ring gear, carrier, and sun gear inertias of the first planetary gear are denoted as J_{r1}, J_{c1}, and J_{s1}, respectively; the ring gear, carrier, and sun gear inertias of the second planetary gear are denoted as J_{r2}, J_{c2}, and J_{s2}, respectively; the carrier-to-sun and ring-to-sun gear ratios of the first planetary gear are denoted as G_{cs1} and G_{rs1}, respectively; and the carrier-to-sun and ring-to-sun gear ratios of the second planetary gear are denoted as G_{cs2} and G_{rs2}, respectively.

Similar to the previous derivations, all inertias are first referred to the generator, engine, and motor shafts, which are given by

$$J'_g = J_g + J_{s1} \tag{11.63}$$

$$J'_e = J_e + J_{r1} + G_{\mathrm{rs1}}^2 J'_g \tag{11.64}$$

$$J'_m = J_m + J_{s2} + \frac{1}{G_{\mathrm{cs2}}^2}\left[J_{c1} + J_{c2} + J_d + G_{\mathrm{cs1}}^2 J'_g + \frac{1}{G_{\mathrm{ad}}^2} J_a\right] \tag{11.65}$$

Consequently, the dynamical equations at the engine and motor shafts can be obtained as

$$T_g = \frac{1}{G_{\mathrm{rs1}}}\left(T_e - J_{\mathrm{eo}}\frac{d\omega_e}{dt} - J_{\mathrm{go}}\frac{d\omega_m}{dt}\right) \tag{11.66}$$

$$T_d = G_{\mathrm{cs2}} T_m - \frac{G_{\mathrm{cs1}}}{G_{\mathrm{rs1}}} T_e + \left(\frac{G_{\mathrm{cs1}}}{G_{\mathrm{rs1}}} J_{\mathrm{eo}} - G_{\mathrm{cs2}} J_{\mathrm{go}}\right)\frac{d\omega_e}{dt} + \left(\frac{G_{\mathrm{cs1}}}{G_{\mathrm{rs1}}} J_{\mathrm{go}} - G_{\mathrm{cs2}} J_{\mathrm{mo}}\right)\frac{d\omega_m}{dt} \tag{11.67}$$

where the equivalent lumped inertias are given by

$$J_{\mathrm{eo}} = J'_e + G_{\mathrm{rs1}}^2 J'_g \tag{11.68}$$

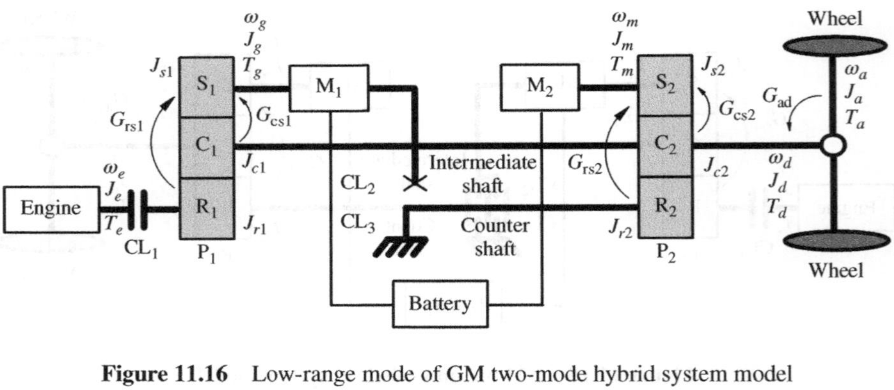

Figure 11.16 Low-range mode of GM two-mode hybrid system model

$$J_{\text{mo}} = J'_m + \frac{G^2_{\text{cs1}}}{G^2_{\text{cs2}}} J'_g \tag{11.69}$$

$$J_{\text{go}} = \frac{G_{\text{rs1}} G_{\text{cs1}}}{G_{\text{cs2}}} J'_g \tag{11.70}$$

At steady state, the corresponding generator torque and driveline torque can be expressed as

$$T_g = \frac{1}{G_{\text{rs1}}} T_e \tag{11.71}$$

$$T_d = G_{\text{cs2}} T_m - \frac{G_{\text{cs1}}}{G_{\text{rs1}}} T_e \tag{11.72}$$

According the speed relationships of a planetary gear given by Eqs. (11.26)–(11.28), the gear ratios G_{cs1} and G_{rs1} as well as G_{cs2} and G_{rs2} can be written as

$$G_{\text{cs1}} = 1 + \rho_1 \tag{11.73}$$

$$G_{\text{rs1}} = -\rho_1 \tag{11.74}$$

$$G_{\text{cs2}} = 1 + \rho_2 \tag{11.75}$$

$$G_{\text{rs2}} = -\rho_2 \tag{11.76}$$

where ρ_1 and ρ_2 are defined as the first and second planetary gear ratios, respectively. Hence, substituting Eqs. (11.73)–(11.76) into Eqs. (11.71) and (11.72), the corresponding steady-state torques become

$$T_g = -\frac{1}{\rho_1} T_e \tag{11.77}$$

$$T_d = (1 + \rho_2) T_m + \frac{(1 + \rho_1)}{\rho_1} T_e \tag{11.78}$$

which indicates that the driveline torque is a weighted combination of the engine and motor torques, and the weightings are based on the relevant planetary gear ratios. Also, since both ρ_1 and ρ_2 are normally greater than one, the driveline torque relies more from the motor, rather than from the engine.

11.4.1.2 High-Range Mode

In this mode, the clutches CL_1 and CL_2 are engaged while CL_3 is disengaged so that both the first and second planetary gears work for power-split. Figure 11.17 shows the equivalent model of this high-range mode. Similar to the previous derivations, the referred generator, engine, and motor inertias are given by

$$J'_g = J_g + J_{s1} \tag{11.79}$$

$$J'_e = J_e + J_{r1} + G^2_{\text{rs1}} J'_g \tag{11.80}$$

$$J'_m = J_m + J_{s2} + \frac{1}{G^2_{\text{cs2}}} \left[J_{c1} + J_{c2} + J_d + G^2_{\text{cs1}} J'_g + \frac{1}{G^2_{\text{ad}}} J_a \right] + \frac{1}{G^2_{\text{rs2}}} [J_{r2} + J'_g] \tag{11.81}$$

Consequently, the dynamical equations at the engine and motor shafts can be obtained as

$$T_g = \frac{1}{G_{\text{rs1}}} \left(T_e - J_{\text{eo}} \frac{d\omega_e}{dt} - J_{\text{go}} \frac{d\omega_m}{dt} \right) \tag{11.82}$$

$$T_d = \frac{1}{G_{\text{cs2}}} T_m - \frac{1}{G_{\text{rs1}} G_{\text{cs2}} G_{\text{rs2}}} T_e + \frac{1}{G_{\text{cs2}}} J'_{\text{eo}} \frac{d\omega_e}{dt} - \frac{1}{G_{\text{cs2}}} J_{mo} \frac{d\omega_m}{dt} \tag{11.83}$$

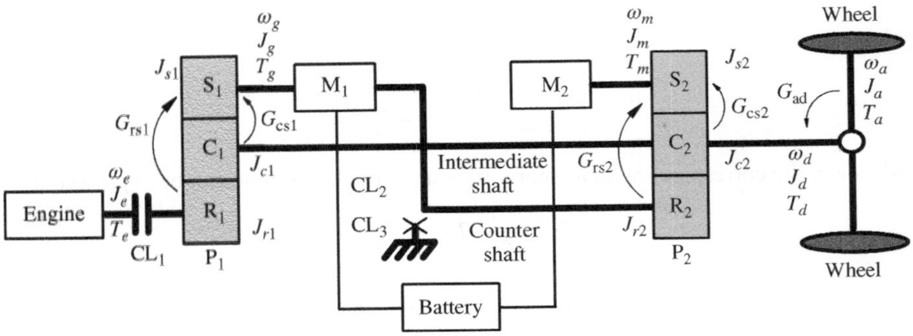

Figure 11.17 High-range mode of GM two-mode hybrid system model

where the equivalent lumped inertias are given by

$$J'_{eo} = \frac{1}{G_{rs1}G_{rs2}} J_{eo} - G_{rs1}G_{rs2}J'_g - G_{rs2}^2 J_{go} \tag{11.84}$$

$$J_{eo} = J'_e - \left(G_{rs1}^2 + \frac{G_{cs1}G_{rs2}G_{rs1}^2}{G_{cs2} + G_{cs1}G_{rs2}} \right) J'_g \tag{11.85}$$

$$J_{mo} = J'_m - \frac{1}{G_{cs1}G_{rs2}} J_{go} + \frac{G_{cs1}G_{rs2}}{G_{cs2} + G_{cs1}G_{rs2}} J'_g \tag{11.86}$$

$$J_{go} = \frac{G_{cs1}G_{rs1}}{G_{cs2} + G_{cs1}G_{rs2}} J'_g \tag{11.87}$$

At steady state, the corresponding generator torque and driveline torque can be expressed as

$$T_g = \frac{1}{G_{rs1}} T_e \tag{11.88}$$

$$T_d = \frac{1}{G_{cs2}} T_m - \frac{1}{G_{rs1}G_{cs2}G_{rs2}} T_e \tag{11.89}$$

Substituting Eqs. (11.73)–(11.76) into Eqs. (11.88) and (11.89), the corresponding steady-state torques become

$$T_g = -\frac{1}{\rho_1} T_e \tag{11.90}$$

$$T_d = \frac{1}{(1+\rho_2)} T_m - \frac{1}{\rho_1 \rho_2 (1+\rho_2)} T_e \tag{11.91}$$

which indicates that the driveline torque is also a weighted combination of the engine and motor torques. Comparing Eq. (11.91) and Eq. (11.78), the driveline torques in the low-range and high-range modes are very different. Particularly, the contribution of the engine torque to the driveline torque at the low-range mode depends only on the first planetary gear ratio whereas the relevant contribution at the high-range mode depends on both the first and second planetary gear ratios. In general, the engine is operated at constant speed to obtain the optimal fuel economy.

11.4.1.3 Electric Launch Mode

This mode is a kind of low-range mode with the clutches CL_1 and CL_3 engaged while CL_2 disengaged. In this mode, the engine is turned off to avoid operating at low efficiency. Hence, substituting $T_e = 0$ into Eq. (11.78), the driveline torque is given by

$$T_d = (1 + \rho_2)T_m \tag{11.92}$$

11.4.1.4 Regenerative Braking Mode

In this mode, the CL_1 and CL_2 are disengaged while only CL_3 is engaged. Both the engine and machine 1 (normally operating as a generator) are turned off. The machine 2 (normally operating as a motor) serves as a generator to produce a braking torque, which converts the braking energy into electrical energy to recharge the battery. Hence, substituting $T_e = 0$ in Eq. (11.78), the driveline torque becomes a negative motor torque as given by

$$T_d = -(1 + \rho_2)T_m \tag{11.93}$$

11.4.1.5 Mode Transition

Differing from the aforementioned THS and FHS, this GM two-mode hybrid system involves different physical structures during mode transition, especially between the low-range and high-range modes. Thus, the mode transition needs to be performed at proper conditions so that the mechanical disturbance can be minimized.

As depicted in Figure 11.18, the vehicle starts from operating at the electric launch mode with the engine off. When the motor itself is unable to fulfill the vehicle power demand, the engine is turned on to provide additional output power. Thus, the vehicle switches from the electric launch mode to the low-range mode. The transition between the low-range and high-range modes is mainly governed by the vehicle speed. Meanwhile, when the sun gear speed of the first planetary gear and the ring gear speed of the second planetary gear are the same, the transition between the low-range and high-range modes can take place without mechanical interruption. Whenever the brake pedal is pressed, there is a transition from the operation mode to the regenerative braking mode.

11.4.2 Renault Hybrid System

Figure 11.19 shows a schematic of the Renault compound-split PG EVT system, which is also called the infinitely variable transmission system (Villeneuve, 2004). It consists of two machines and two planetary

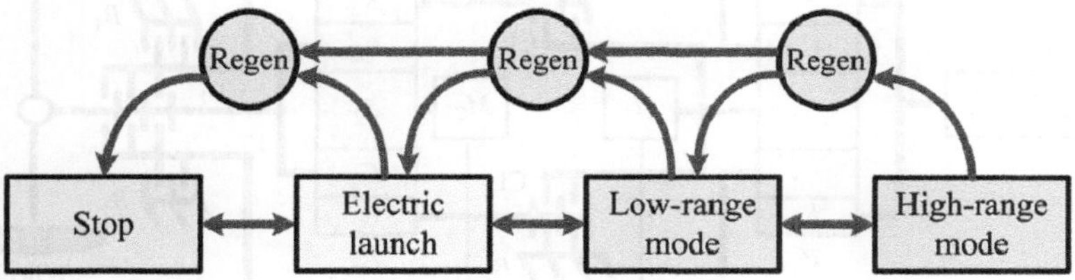

Figure 11.18 Mode transition of GM two-mode hybrid system

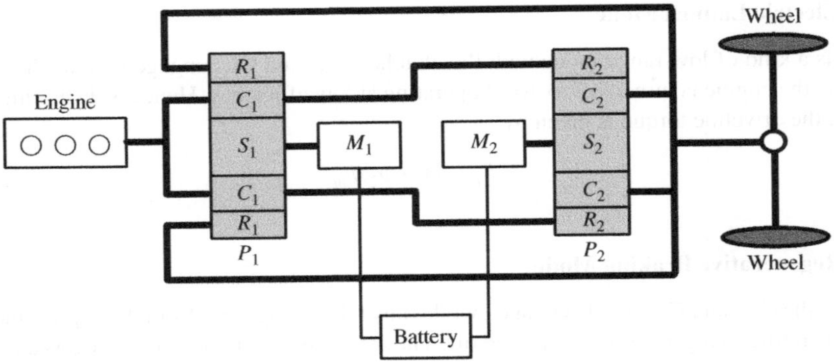

Figure 11.19 Renault hybrid system model

gears, but no clutches. The machine M_1 is coupled with the sun gear S_1 of the first planetary gear P_1 while the machine M_2 is coupled with the sun gear S_2 of the second planetary gear P_2. The engine is coupled with the carrier C_1 of P_1 and the ring gear R_2 of P_2. The carrier C_2 of P_2 and the ring gear R_1 of P_1 are linked up with the driveline. Since there are no clutches, the system involves only one physical structure that can avoid mechanical disturbances due to mode changes.

There is a full four-port arrangement: two mechanical ports and two electromechanical ports. By controlling the two machines, at a given input speed (engine speed), the output speed (vehicle speed) can vary "infinitely," namely continuously from reverse to forward speed passing through the zero speed without using any clutch or equivalent device.

11.4.3 Timken Hybrid System

The Timken hybrid system, also called the electromechanical infinitely variable transmission system, is shown in Figure 11.20 (Ai, Mohr, and Anderson, 2004). This two-mode compound-split system consists of two machines, two planetary gears, two clutches, and two brakes. The machine M_1 is coupled with the sun gear S_1 of the first planetary gear P_1, and can also be coupled with the ring gear R_2 of the second planetary gear P_2 via the clutch CL_2. The machine M_2 is coupled with the sun gear S_2 of P_2. The engine

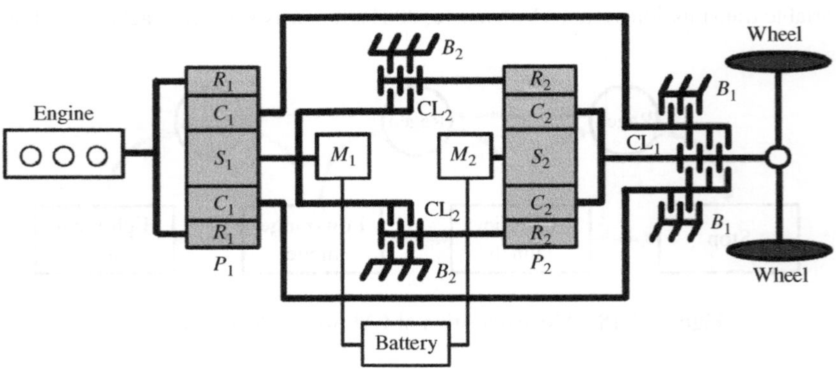

Figure 11.20 Timken hybrid system model

is coupled with the ring gear R_1 of P_1. The carrier C_2 of P_2 is coupled with the driveline, while the carrier C_1 of P_1 can also be coupled with the driveline via the clutch CL_1. The carrier C_1 and sun gear S_1 can be locked by the brakes B_1 and B_2, respectively.

By properly engaging the two clutches and two brakes, the Timken hybrid system can operate in various modes (Mi, Masrur, and Gao, 2011):

- *Electric launch*: Both the clutches CL_1 and CL_2 are disengaged, whereas both brakes B_1 and B_2 are activated. That is, the brake B_2 locks the ring gear R_2 of P_2 so that the machine M_2 delivers the torque to the sun gear S_2 via the carrier C_2 of P_2; meanwhile, the brake B_1 locks the carrier C_1 of P_1 so that the engine is stalled. Thus, the driveline torque is solely contributed by M_2.
- *Low-range mode*: The clutch CL_1 is engaged and the brake B_2 is activated, whereas the clutch CL_2 and brake B_1 are open. Thus, the system becomes the same as the low-range mode of the GM two-mode hybrid system.
- *High-range mode*: Both the clutches CL_1 and CL_2 are engaged, whereas both brakes B_1 and B_2 are deactivated. Thus, the system becomes the same as the high-range mode of the GM two-mode hybrid system.
- *Series mode*: Both the clutches CL_1 and CL_2 are engaged, while both brakes B_1 and B_2 are activated. Thus, the engine delivers power to the machine M_1 (operating as a generator) through the sun gear S_1 of P_1 with its carrier C_1 locked. Then, the generated power by M_1 feeds the machine M_2 (operating as a motor) to drive the sun gear S_2 of P_2 with its ring gear R_2 locked.

11.5 Design Criteria of PG EVT Systems

The design of PG EVT systems is challenging as the system needs to offer all hybrid features of a full hybrid, namely the electric launch, cranking, electricity generation, idle stop-start, regenerative braking, and EVT operation. All system devices need to closely collaborate with the engine, and work under harsh environment, such as large vibrations and high temperatures. The design goals of EVT systems are stringent, which are summarized as follows:

- Various EVT systems, such as the input-split and compound-split, are preferable for various types of HEVs. In general, the input-split types are particularly attractive for passenger cars or sport utility vehicles, whereas the compound-split types are preferable for vans or trucks.
- The starting torque of the motor should be up to four times the rated torque to enable electric launch. The corresponding torque density should be maximized to reduce the weight and size.
- The constant-power operation range of the motor should be up to four times the base speed to enable wide-speed operation. The corresponding power density should be maximized to reduce the weight and size.
- The output power of the generator should be sufficiently high to perform power-split of the engine output.
- The operating speed of the generator should be sufficiently wide to enable effective electricity generation for battery charging.
- The output torque of the generator, which operates as a starter motor, should be high enough to perform cold cranking.
- The planetary gear should be durable enough to perform power-split of the engine output, and form two paths of power flow.
- The engine should be durable enough to allow for frequent idle stop-start up to a million cycles.
- The motor should be able to offer efficiency higher than 85% over wide torque-speed regions.
- The motor should be able to perform regenerative braking with efficiency over 60%.
- The system should be able to withstand severe vibrations up to 20 g and extreme temperatures from −30 to 120 °C.

- The system should have high reliability and good serviceability. Particularly, the motor and generator should be maintenance-free.
- The system should offer excellent fuel economy to justify the increased complexity and additional cost.

11.6 Design Example of PG EVT Systems

As the input-split PG EVT system is the most fundamental type of PG EVT systems, it is selected for exemplification. Also, as the permanent magnet (PM) synchronous machine takes the definite advantages of high efficiency, high power density, and low torque ripple, it is selected as the machine topology for the input-split PG EVT system.

11.6.1 PM Synchronous PG EVT System Configuration

The system configuration has been discussed as depicted in Figure 11.5 in which both the machine 1 and machine 2 adopt the PM synchronous machine topology. In this system, the planetary gear plays a key role to perform power-split of the engine into two paths: one path is mechanical power to the driveline via the ring gear and another path is electrical power to the driveline via the machine 1 (normally operating as a generator), converter 1 (normally operating as a controlled rectifier), battery, converter 2 (normally operating as an inverter), and machine 2 (normally operating as a motor). The key of this design is the integration of the planetary gear and two machines. Since all these devices have their individual rotational shafts, the system arrangement shown in Figure 11.5 is not practical to perform such integration.

A practical system arrangement to integrate the planetary gear and two machines is depicted in Figure 11.21 in which the planetary gear is sandwiched between the two machines in such a way that the engine shaft coupled with the carrier of the planetary gear via the central core of the machine 1 and the ring gear of the planetary gear coupled with the rotor of the machine 2 are connected to the driveline

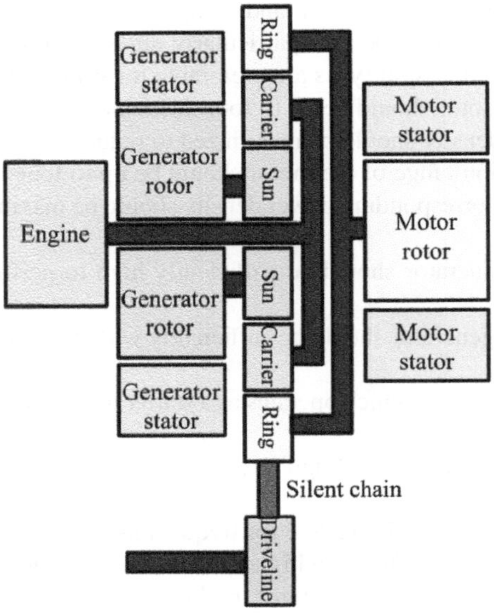

Figure 11.21 Practical arrangement of input-split PG EVT system

via a silent chain. This silent chain, having teeth on its links that mesh with teeth in the sprockets of the driveline, can perform effective torque transmission silently and smoothly.

In order to fulfill the propulsion requirement, the two PM synchronous machines (one mainly for motoring and one mainly for generation) need to be properly designed while the planetary gear and battery need to be properly sized. On the basis of the requirements of a typical passenger HEV, the specifications of the input-split PG EVT system are summarized in Table 11.1. The corresponding specifications of the two PM synchronous machines are listed in Table 11.2.

11.6.2 PM Synchronous Machine Design

Aiming to offer high torque density and high power density, the two PM synchronous machines preferably adopt the flux-focusing arrangement for the PM rotor. Rather than using the interior-circumferential PM rotor that has a significant flux leakage, both PM synchronous machines utilize the interior V-shaped PM rotor as shown in Figure 11.22, which is similar to that adopted by Prius (Staunton *et al.*, 2006; Burress *et al.*, 2011). This interior V-shaped PM rotor lies between the interior-radial PM rotor and interior-circumferential PM rotor, while offering the flexibility to adjust the V-angle in such a way that the degree of flux-focusing and the saliency ratio of d–q inductances can be optimized.

Because of the difference between d-axis and q-axis inductances, this machine topology has two torque components: PM torque and reluctance torque. In general, the PM torque component is larger than the

Table 11.1 Specifications of input-split PG EVT

Engine peak power	65 kW
Motor peak power	60 kW
Generator peak power	35 kW
Battery voltage	200 V
Number of ring gear teeth	78
Number of sun gear teeth	30
Number of planet gear teeth	23
Motor-to-wheel speed ratio	4

Table 11.2 Specifications of input-split PG EVT machines

Motor	
DC voltage	500 V
Rated power	35 kW
Peak power	60 kW
Rated torque	120 N m
Peak torque	416 N m
Constant-torque operation	0–1400 rpm
Constant-power operation	1400–6000 rpm
Generator	
DC voltage	500 V
Rated power	10 kW
Peak power	35 kW
Maximum speed	10 000 rpm

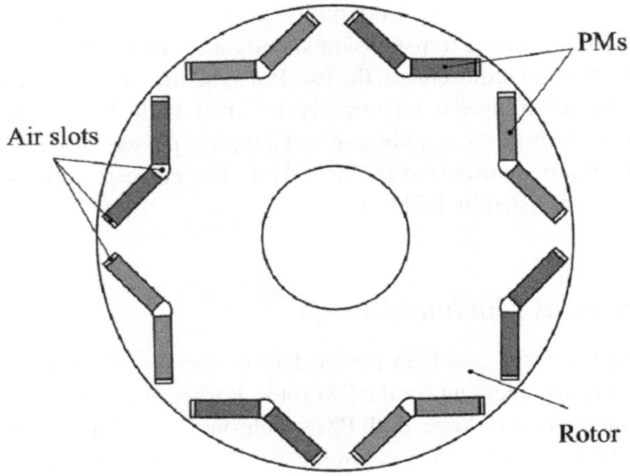

Figure 11.22 V-shaped PM rotor

reluctance torque component. Theoretically, the PM torque component can be maximized by applying the armature current with 0° electrical while the reluctance torque component can be maximized with the current angle of 45° electrical. Therefore the optimal current angle lies between them, depending on the working conditions, such as current magnitude and rotor speed.

On the basis of the specifications, the key design parameters and geometric dimensions of these two PM synchronous machines are listed in Table 11.3.

11.6.3 PM Synchronous Machine Analysis

By using finite element analysis, the electromagnetic field distributions of the two PM synchronous machines can readily be determined. Figure 11.23 shows a typical electromagnetic field distribution of

Table 11.3 Key parameters and dimensions of PM synchronous machines for input-split PG EVT system

	Motor	Generator
Number of phases	3	3
Number of poles	8	8
Stator outside diameter (mm)	280	240
Stator inside diameter (mm)	171.7	145
Stator stack length (mm)	50.8	83.6
Rotor outside diameter (mm)	170	143
Rotor inside diameter (mm)	54	86.4
Rotor stack length (mm)	50.8	83.6
Air-gap length (mm)	0.85	1.0
Number of stator slots	48	48
Number of stator turns per phase	88	72
PM type	Nd-Fe-B	Nd-Fe-B
PM dimensions (mm)	$20.25 \times 7 \times 50.8$	$19.1 \times 6.58 \times 83.6$

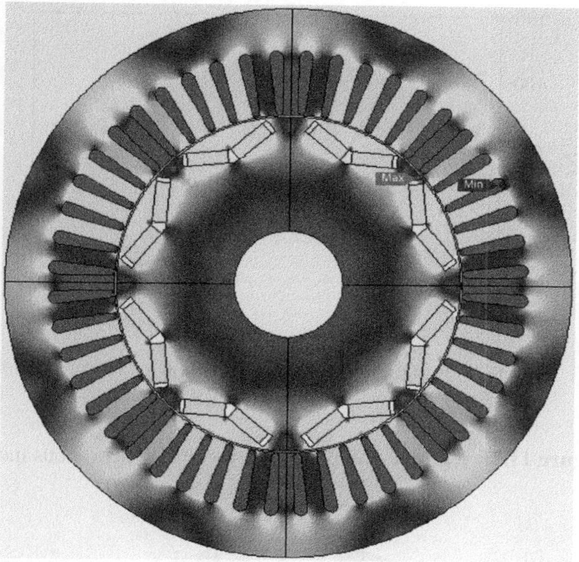

Figure 11.23 Magnetic field distribution of PM synchronous motor

the machine 2 (normally operating as a motor), which can assess the saturation level of the iron core and most importantly the air-gap flux density of the whole machine. The higher the air-gap flux density, the higher the developed torque per armature current.

In order to assess the torque capability of the PM synchronous motor, the developed torques are calculated by using finite element analysis at different rotor positions and different armature currents. The corresponding torque-angle characteristics are shown in Figure 11.24. It can be observed that the peak torque increases with the armature current. Consequently, the peak torque can achieve 416 N m at the peak current rating for speeds below the base speed of 1400 rpm. Beyond the base speed, the peak torque

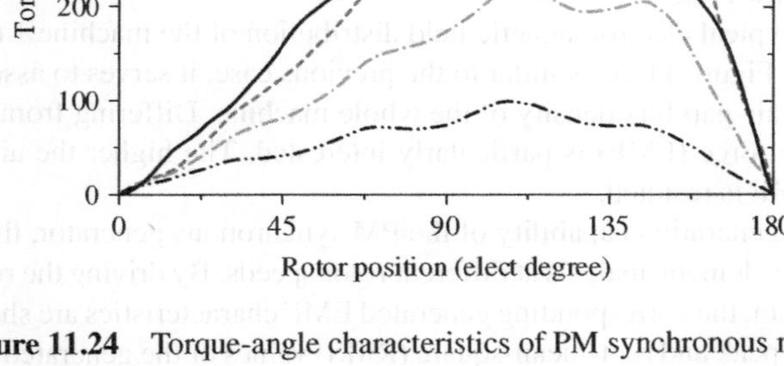

Figure 11.24 Torque-angle characteristics of PM synchronous motor

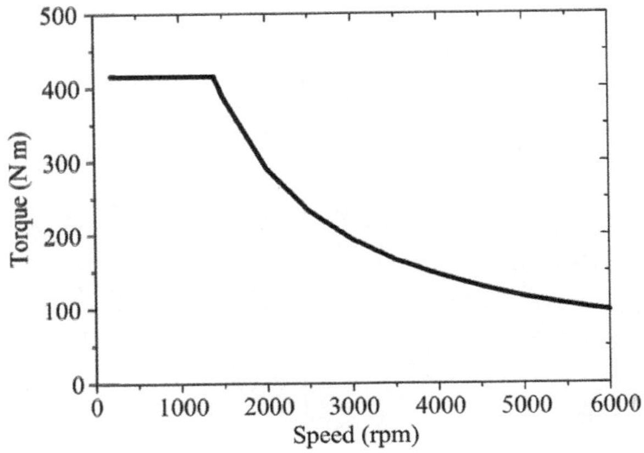

Figure 11.25 Torque-speed capability of PM synchronous motor

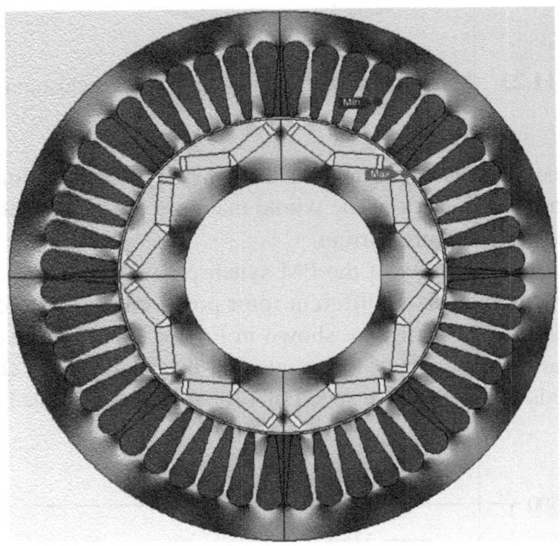

Figure 11.26 Magnetic field distribution of PM synchronous generator

decreases inversely with the speed for constant-power operation. The corresponding torque-speed capability is depicted in Figure 11.25.

On the other hand, a typical electromagnetic field distribution of the machine 1 (normally operating as a generator) is shown in Figure 11.26. Similar to the previous case, it serves to assess the saturation level of the iron core and the air-gap flux density of the whole machine. Differing from the previous case, the generated electromotive force (EMF) is particularly interested. The higher the air-gap flux density, the higher the generated EMF is resulted.

In order to assess the generation capability of the PM synchronous generator, the generated EMFs are calculated by using finite element analysis at different rotor speeds. By driving the rotor from 0 to the maximum speed of 10 000 rpm, the corresponding generated EMF characteristics are shown in Figure 11.27. It can be observed that the peak and root-mean-square (RMS) values of the generated EMF increase linearly

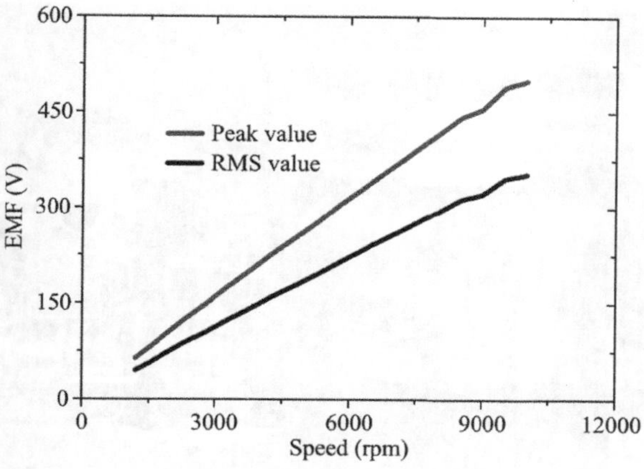

Figure 11.27 Generated EMF characteristics of PM synchronous generator

with the rotor speed. Consequently, the peak value can achieve 500 V at the maximum generator speed of 10 000 rpm.

11.7 Application Examples of PG EVT Systems in HEVs

A milestone of HEVs development is the Toyota Prius as shown in Figure 11.28. The first Prius was put on the market in 1997 and was the first mass-produced hybrid vehicle. Its input-split PG EVT system, also called the THS, was a technological breakthrough in the field of hybrid powertrains. The THS consisted of an engine, a generator, planetary gear, and a motor as shown in Figure 11.29. Its engine could offer 43 kW and 102 N m at 4000 rpm, and its motor could provide 305 N m at 0–940 rpm and 30 kW at 940–2000 rpm (Carfolio, 2014). The Prius family has been continually improved and extended such as the Prius plug-in

Figure 11.28 Toyota Prius (Source: Courtesy Wikimedia Commons, http://commons.wikimedia.org/wiki /File:1st_Toyota_Prius_--_01-13-2010.jpg)

Figure 11.29 Toyota hybrid system (Source: Courtesy Wikimedia Commons, http://commons.wikimedia.org/wiki/File:Toyota_1NZ-FXE_Engine_01.JPG)

Figure 11.30 Ford Escape Hybrid (Source: Courtesy Wikimedia Commons, http://en.wikipedia.org/wiki/File:2nd_Ford_Escape_Hybrid_--_04-29-2011_1.jpg)

hybrid, Prius v, and Prius c, aiming to deliver improved fuel economy through the development of new powertrains with advances in battery, motor, and engine technologies.

The Ford Escape Hybrid, released in 2004, was the first full hybrid sport utility vehicle (SUV) to hit the market as shown in Figure 11.30. It was equipped with an input-split PG EVT system, called the FHS. This FHS consisted of an engine, a generator, planetary gear, a motor, and a set of spur gears. The corresponding engine could offer 114 kW at 6000 rpm and 184 N m at 4500 rpm, and the motor could provide 70 kW at 5000 rpm (Automobile-catalog, 2014). The combined output power could achieve 132 kW.

Figure 11.31 Chevrolet Silverado Hybrid (Source: Courtesy Wikimedia Commons, http://en.wikipedia.org/wiki/File:Chevrolet_Silverado_Hybrid_--_01-07-2012.jpg)

GM launched the Silverado Hybrid in 2004. Starting from 2009, GM offered a second-generation Chevrolet Silverado Hybrid as shown in Figure 11.31. This hybrid truck was equipped with the GM two-mode hybrid system. Its 6-L V8 engine was rated at 248 kW at 5100 rpm and 498 N m at 4100 rpm (GM, 2014). Supplemented with two 60-kW electric machines, the combined output could achieve 282 kW.

11.8 Matured Technology for HEVs?

In recent years, the full hybrid has become one of the fastest growing segments of the automobile industry in the world. The Toyota Prius, Toyota Camry Hybrid, Ford Escape Hybrid, Ford Fusion Hybrid, GM Chevrolet Silverado Hybrid, General Motors Company (GMC) Sierra Hybrid, and Kia Optima Hybrid are commercial examples of this type of hybrids. They can provide all major hybrid features: the electric launch, engine cranking, electricity generation, idle stop-start, regenerative braking, and EVT operation.

As the EVT system is the core technology for the full hybrid, the corresponding research and development have been fast and encouraging. In fact, because of the popularity of the Toyota Prius, the PM synchronous input-split PG EVT system has become mature for mass production. Meanwhile, the compound-split PG EVT system is becoming mature, which can further reduce the manufacturing cost. Nevertheless, there are two technological shortcomings in the existing PG EVT system. First, while the PM synchronous machines can offer high efficiency and high power density, the PMs mounted on the rotor suffer from the difficulty in withstanding high centrifugal force, severe vibration, and high temperature for EVT operation. Second, while the planetary gear is the key element of the PG EVT system, this mechanical device inevitably suffers from wear and tear, gear noise, inefficient transmission, and need for regular lubrication. Thus, there are continual research and development to solve these shortcomings, such as using advanced magnetless machines for the PG EVT system or even replacing the mechanical planetary gear by a double-rotor machine or magnetic planetary gear.

References

Ai, X., Mohr, T., and Anderson, S. (2004) An electro-mechanical infinitely variable speed transmission. Proceedings of SAE World Congress, Paper No. 2004-01-0354.

Automobile-catalog (2014) 2012 Ford Escape Hybrid Limited AWD, Automobile-catalog.com, http://www.automobile-catalog.com/car/2012/1419305/ford_escape_hybrid_limited_awd.html (accessed September 2014).

Burress, T.A., Campbell, S.L., Coomer, C.L. *et al.* (2011) *Evaluation of the 2010 Toyota Prius Hybrid Synergy Drive System*, U.S. Department of Energy, Washington, DC.

Carfolio (2014) 1997 Toyota Prius Technical Specifications, Carfolio.com, http://www.carfolio.com/specifications/models/car/?car=230824, (accessed September 2014).

Chau, K.T. and Chan, C.C. (2007) Emerging energy-efficient technologies for hybrid electric vehicles. *Proceedings of the IEEE*, **95**, 821–835.

Chau, K.T., Wong, Y.S. and Chan, C.C. (2003) An overview of energy sources for electric vehicles. *Energy Conversion and Management*, **40**, 1021–1039.

General Motors (2014) Chevrolet Silverado Hybrid – 2012, http://media.cadillac.com/media/us/en/chevrolet/vehicles/silveradohybrid/2012.html (accessed September 2014).

Holmes, A.G. and Schmidt, M.R. (2002) Hybrid electric powertrain including a two-mode electrically variable transmission. U.S. Patent 6 478 705 B1.

Mi, C., Masrur, M.A. and Gao, D.W. (2011) *Hybrid Electric Vehicles: Principles and Applications with Practical Perspectives*, John Wiley & Sons, Ltd, Chichester.

Miller, J.M. (2006) Hybrid electric vehicle propulsion system architectures of the e-CVT type. *IEEE Transactions on Power Electronics*, **21**, 756–767.

Miller, J.M. (2010) *Propulsion Systems for Hybrid Vehicles*, IET, Stevenage.

Miller, J.M. and Everett, M. (2005) An assessment of ultracapacitors as the power cache in Toyota THS-II, GM-Allison AHS-2 and Ford FHS hybrid propulsion systems. Proceedings of IEEE Applied Power Electronics Conference and Exhibition, pp. 481–490.

Sasaki, S. (1998) Toyota's newly developed hybrid powertrain. Proceedings of International Symposium on Power Semiconductor Devices and ICs, pp. 17–22.

Staunton, R.H., Ayers, C.W., Marlino, L.D. *et al.* (2006) *Evaluation of 2004 Toyota Prius Hybrid Electric Drive System*, U.S. Department of Energy, Washington, DC.

Villeneuve, A. (2004) Dual mode electric infinitely variable transmission. Proceedings of SAE TOPTECH Meeting on Continuously Variable Transmission, 04CVT-19.

Wang, Y., Cheng, M. and Chau, K.T. (2009) Review of electronic-continuously variable transmission propulsion system for full hybrid electric vehicles. *Journal of Asian Electric Vehicles*, **7**, 1297–1302.

12

Double-Rotor Electric Variable Transmission Systems

As the electric variable transmission (EVT) system is the core technology for full hybrids, it is undergoing continual development. The existing EVT systems are all based on the use of planetary gearing, which inherently suffers from transmission loss, gear noise, and need for regular lubrication. In order to overcome these drawbacks, a new class of EVT systems has been developed. The key is to employ a double-rotor (DR) machine to perform the desired power-split.

In this chapter, various DR EVT systems, including the system configurations and DR machines as well as the basic and advanced DR types, are discussed. The corresponding design criteria, design example, and potential applications are also presented.

12.1 System Configurations

The planetary-geared (PG) EVT system, including the input-split and compound-split types, employs planetary gearing to perform the power-split of engine output. As the planetary gear is a kind of mechanical gears, it inevitably suffers from transmission loss, gear noise, and need for regular lubrication. In addition, backlash and misalignment problems are almost unavoidable. Instead of using a mechanical device, an electrical device is a good alternative to perform the power-split so that the problems associated with mechanical gearing can be solved fundamentally. The key is to employ a DR machine to perform the desired power-split.

Figure 12.1 describes the configuration of the PG EVT system, which consists of a planetary gear, a machine normally operating as a generator, another machine normally operating as a motor, a power converter normally operating as a controlled rectifier, another power converter normally operating as an inverter, and an energy storage device, normally a battery. When the engine drives the carrier of planet gears, the power is split into two paths: mechanical and electrical. The mechanical path from the engine to the driveline is via the ring gear of the planetary gear. The electrical path involves the sun gear of the planetary gear, generator, controlled rectifier, battery, inverter, and motor: the sun gear spins the generator rotor; the generator stator produces electricity; the controlled rectifier regulates the power flow between the generator stator and DC link; the battery serves as an energy buffer; the inverter regulates the power flow between the DC link and motor stator; and the motor rotor spins the driveline.

Figure 12.2 depicts the configuration of the DR EVT system. By allowing both the stator and rotor of the machine 1 (initially operating as a generator) to rotate freely, the machine 1 becomes a primitive DR machine, which can simultaneously offer the planetary gear function to split the engine power, and

Electric Vehicle Machines and Drives: Design, Analysis and Application, First Edition. K. T. Chau.

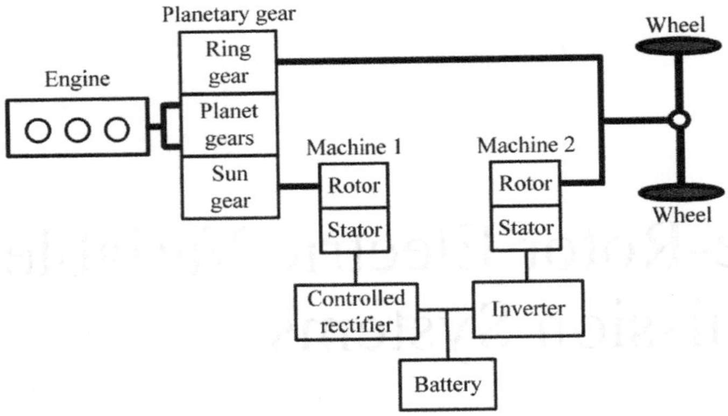

Figure 12.1 PG EVT system configuration

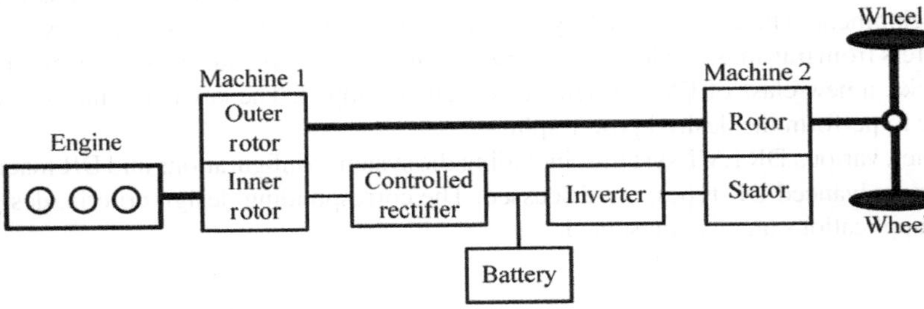

Figure 12.2 DR EVT system configuration using cascaded machines

the generator function to produce electricity. The machine 1 splits the engine power into two paths. One path is from the engine to the driveline via the inner and outer rotors of the machine 1 and then the rotor of the machine 2 (simply operating as a motor). Another path is via the inner rotor of the machine 1, controlled rectifier, battery, inverter, and machine 2: the engine spins the inner rotor; the inner rotor generates electricity; the controlled rectifier regulates the power flow between the inner rotor and DC link; the battery serves as an energy buffer; the inverter regulates the power flow between the DC link and motor stator; and the motor rotor spins the driveline. Therefore, the planetary gear and hence its problems can be completely eliminated.

Because of the nature of direct coupling between the outer rotor of the machine 1 and the rotor of the machine 2, they should be lumped together to form a single machine having a stator and two rotors – the so-called DR machine. As depicted in Figure 12.3, the system configuration is much simpler than that of

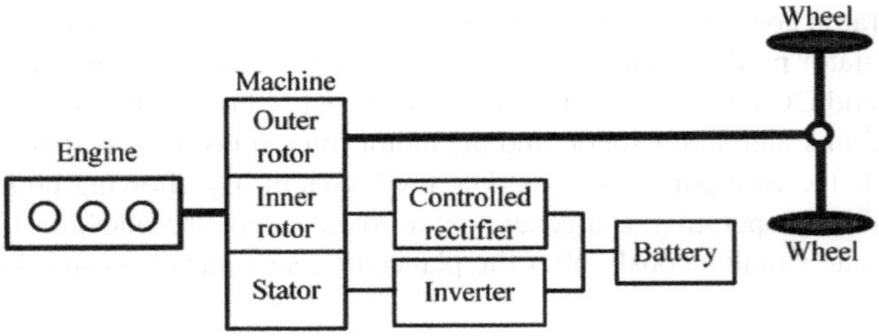

Figure 12.3 DR EVT system configuration using single machine

the PG EVT system. That is, when the engine spins the inner rotor, the engine power is split into two paths. One path goes through the inner rotor and outer rotor to the driveline line. Another path goes through the inner rotor, controlled rectifier, battery, inverter, stator, and outer rotor to the driveline. Thus, the outer rotor is the summing point of two driving torques.

12.2 Double-Rotor Machines

With the ever-increasing demand on electromechanical energy conversion, the development of electric machines has been extended from a traditional single-input single-output (SISO) system to a versatile multi-input multi-output (MIMO) system. As the MIMO electromechanical system involves more than two ports (an electrical port and a mechanical port), it is called the multi-port machine. When there are two mechanical ports, a multi-port machine is called the double-rotor machine.

12.2.1 Multi-port Machine Concept

A traditional electric machine, no matter what type is considered, is a kind of SISO electromechanical energy conversion system. In this regard, it can be described as a coupled electromagnetic field device with one electrical port and one mechanical port (Xu, 2006). As shown in Figure 12.4, the corresponding electrical port of a conventional electric machine is characterized by its input voltage and current while the corresponding mechanical port is characterized by its output torque and speed.

In order to improve the operating performance, such as control flexibility and efficiency optimization, some electric machines have incorporated two electrical inputs. For instance, the doubly-fed induction machine has a three-phase winding in the stator as one electrical input and a three-phase winding in the rotor as another electrical input so as to perform the slip energy recovery control. Also, the doubly-salient DC (DSDC) machine has the armature winding in the stator as one electrical input and the field winding in the stator as another electrical input so that it can achieve online efficiency optimization. As these machines have two electrical input ports and one mechanical output port, they can generally be classified as the dual-input single-output three-port machine as depicted in Figure 12.5.

Differing from the fast development of dual-input single-output three-port machines, the development of single-input dual-output three-port machines was slow. The major reason was simply due to the lack of applications desiring dual-output mechanical ports. With the ever-increasing applications of two rotating bodies, such as the two wheels of electric vehicles and the dual impellers of industrial mixers, the development of single-input dual-output three-port machines has been accelerated in recent years. As depicted in Figure 12.6, those single-input dual-output three-port machines have two shafts that are either laterally offset or concentrically arranged. Consequently, by adding one more electrical port to improve the operating performance, a dual-input dual-output four-port machine as shown in Figure 12.7 can readily be generated.

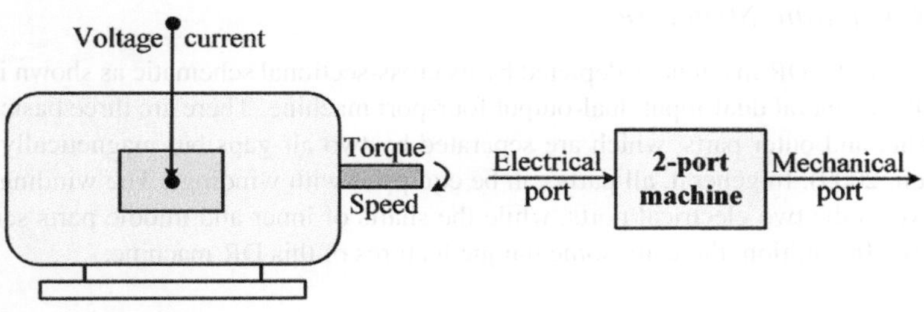

Figure 12.4 Single-input single-output two-port machine

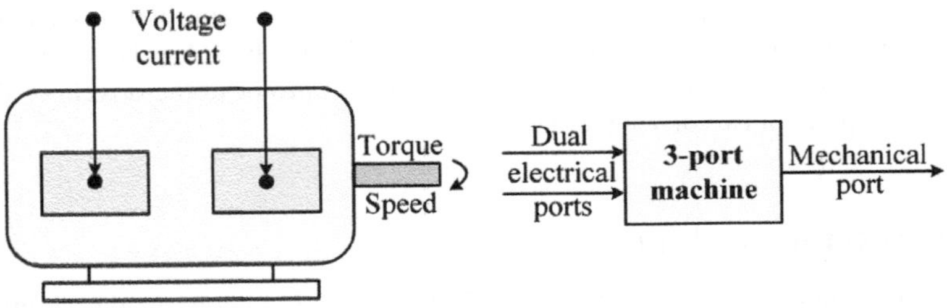

Figure 12.5 Dual-input single-output three-port machine

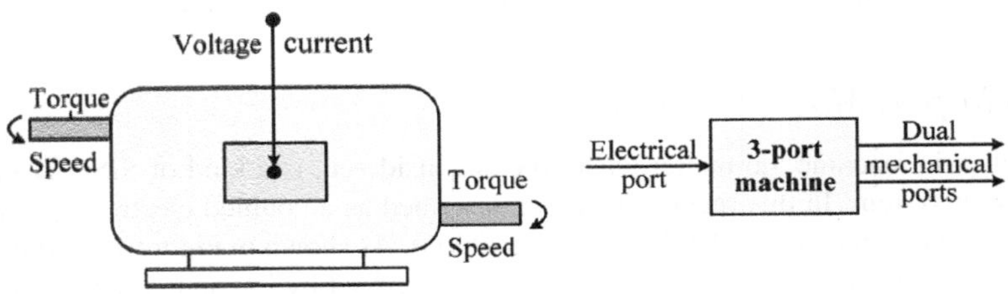

Figure 12.6 Single-input dual-output three-port machine

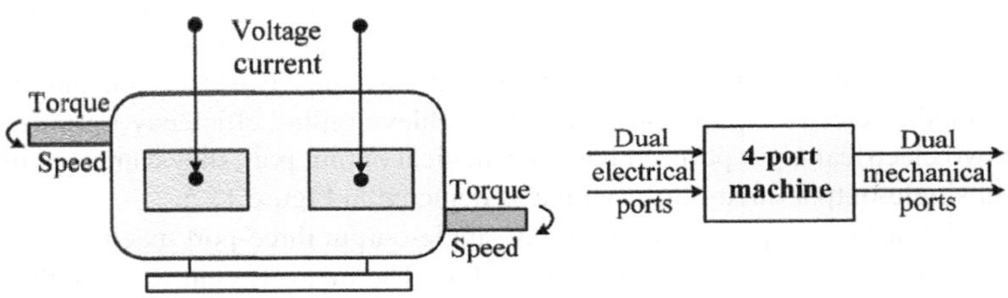

Figure 12.7 Dual-input dual-output four-port machine

Similar to all electromechanical devices, all ports are bidirectional. That is, the input and output ports are interchangeable, and both the electrical and mechanical ports allow for bidirectional power flow.

12.2.2 DR Machine Structure

A basic structure of the DR machine is depicted by its cross-sectional schematic as shown in Figure 12.8, which is actually a general dual-input dual-output four-port machine. There are three basic parts, namely the inner, middle, and outer parts, which are separated by two air-gaps but magnetically coupled (Xu, Zhang, and Wen, 2009). In general, all parts can be equipped with windings. The windings of inner and outer parts serve as the two electrical ports, while the shafts of inner and middle parts serve as the two mechanical ports. In addition, there are some unique features of this DR machine:

- The two mechanical ports can be freely assigned among the three parts, depending on which rotational parts are more favorable for specific application.

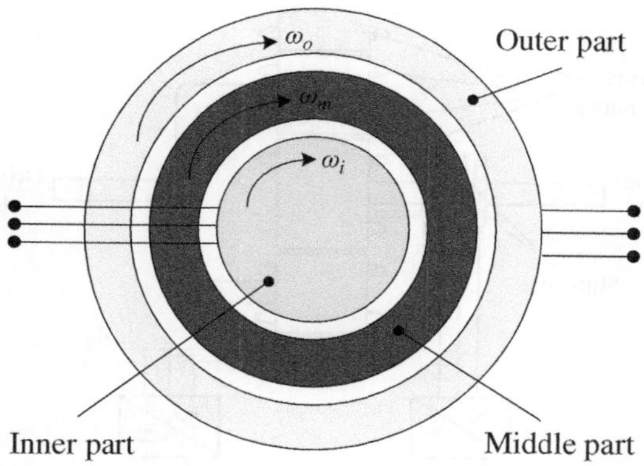

Figure 12.8 Basic structure of DR machine

- For the two electrical ports, one must be the winding terminals of the stationary part, while another is the winding terminals at one of the two rotational parts.
- The middle part is sandwiched by the inner and outer air-gaps so that the corresponding torque is the sum of the electromagnetic torque components produced at two air-gaps.
- The two mechanical ports can be operated as two driving bodies, two driven bodies, or one driving and one driven bodies.
- The two electrical ports can both accept and produce power, or one accepts and one produces power. They can also be connected together via two back-to-back power converters to enable internal power control.
- When the electrical port is located at the rotational part, a setup of slip-rings and brushes is necessary.

As mentioned earlier, the two mechanical ports can be freely assigned among the three parts. No matter how these parts are assigned, there are two possible operating modes, depending on the relative speeds of the two rotational parts (outer and inner rotors). For exemplification, the outer part is the stator, the middle part is the outer rotor, and the inner part is the inner rotor. First, when the two rotors are of the same speed, there is no relative speed and hence no electromechanical energy flow between them. The machine is essentially a traditional one with two shafts. Second, when the speed of inner rotor is different from that of outer rotor, there is a slip between them. The inner rotor needs to be excited with the slip frequency. The power flow through the inner rotor due to slip may be negative or positive, depending on whether the slip is positive or negative.

12.3 Basic Double-Rotor EVT Systems

A basic DR EVT system is mainly based on the induction DR machine. Although this DR machine is a single machine, it is usually split into two cascaded induction machines for the sake of explanation and modeling. Meanwhile, it helps elaborate various operation modes for the full hybrids.

12.3.1 DR EVT Structure

As shown in Figure 12.9, a basic DR EVT structure is composed of an induction DR machine, a power converter normally operating as a controlled rectifier, an energy storage device normally using a battery, and another power converter normally operating as an inverter (Hoeijmakers and Ferreira, 2006).

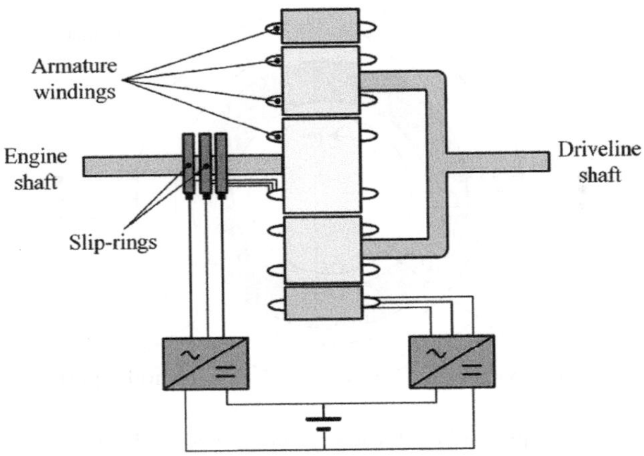

Figure 12.9 Induction DR EVT system

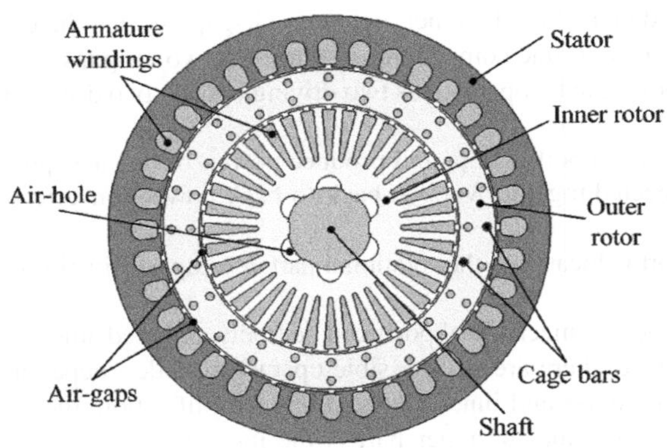

Figure 12.10 Induction DR machine

The induction DR machine has a stator, an outer rotor, and an inner rotor, which are all concentrically arranged as shown in Figure 12.10. The corresponding stator is usually wound with a three-phase distributed winding, which is similar to that of a traditional induction machine. The outer rotor is equipped with a squirrel cage shorted with two end-rings, which is similar to that of a conventional squirrel-cage induction machine. The inner rotor also has a three-phase distributed winding with its phase terminals separately connected to three slip-rings, which is similar to that of a normal slip-ring induction machine.

To facilitate the explanation and analysis of the induction DR EVT, the induction DR machine is split into two cascaded induction machines: machine 1 is a primitive DR machine and machine 2 is a traditional machine as shown in Figure 12.11. The outer rotor of the machine 1 is mechanically coupled with the rotor of the machine 2. It offers two paths of power flow. One path is the mechanical coupling between the outer rotor of the machine 1 and the rotor of the machine 2. Another path is electrical connection between the inner rotor of the machine 1 and the stator of the machine 2 via the back-to-back power converters and battery. Therefore, when the engine drives the machine 1, the machine 1 works as a generator while the machine 2 works as a motor.

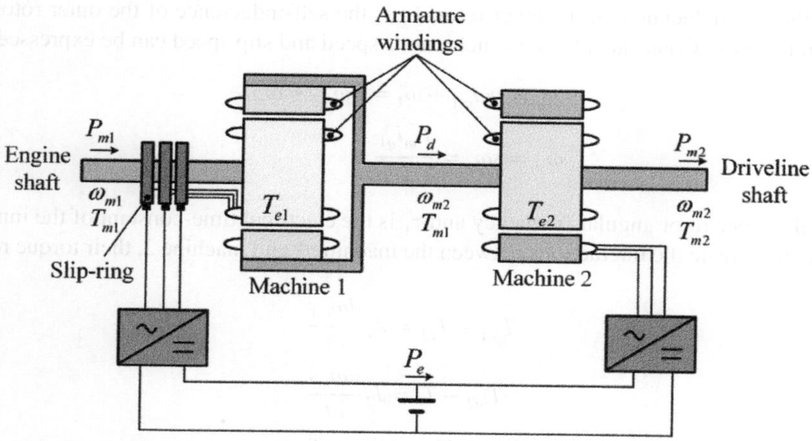

Figure 12.11 Induction DR EVT system model

12.3.2 DR EVT Modeling

The modeling of the machine 1 needs to take the effect of double rotors into account, whereas the modeling of the machine 2 is the same as that of a traditional induction machine (Cui, Cheng, and Chan, 2006).

Extending the coordinate transformation in standard induction machines, the flux linkage of the outer rotor of the machine is taken as the d-axis. Hence, the voltage equations of the machine 1 can be expressed as

$$u_{d1} = R_1 i_{d1} + \frac{d\psi_{d1}}{dt} - (\omega_s - p\omega_{m1})\psi_{q1} \quad (12.1)$$

$$u_{q1} = R_1 i_{q1} + \frac{d\psi_{q1}}{dt} + (\omega_s - p\omega_{m1})\psi_{d1} \quad (12.2)$$

$$0 = R_2 i_{d2} + \frac{d\psi_{d2}}{dt} \quad (12.3)$$

$$0 = R_2 i_{q2} + \omega_{\text{sl}}\psi_{d2} \quad (12.4)$$

where u_{d1} and u_{q1} are, respectively, the d-axis and q-axis phase voltages of the inner rotor, R_1 is the inner-rotor resistance, i_{d1} and i_{q1} are, respectively, the d-axis and q-axis phase currents of the inner rotor, ψ_{d1} and ψ_{q1} are, respectively, the d-axis and q-axis flux linkages of the inner rotor, ω_s is the synchronous speed, p is the number of pole-pairs, ω_{m1} is the inner-rotor mechanical speed, R_2 is the outer-rotor phase resistance, i_{d2} and i_{q2} are, respectively, the d-axis and q-axis phase currents of the outer rotor, ψ_{d2} is the d-axis flux linkage of the outer rotor, and ω_{sl} is the slip angular frequency. The corresponding flux linkages are given by

$$\psi_{d1} = L_1 i_{d1} + L_m i_{d2} \quad (12.5)$$

$$\psi_{q1} = L_1 i_{q1} + L_m i_{q2} \quad (12.6)$$

$$\psi_{d2} = L_m i_{d1} + L_2 i_{d2} \quad (12.7)$$

$$0 = L_m i_{q1} + L_2 i_{q2} \quad (12.8)$$

where L_1 is the self-inductance of the inner rotor, L_2 is the self-inductance of the outer rotor, and L_m is the mutual inductance. Consequently, the synchronous speed and slip speed can be expressed as

$$\omega_s = p\omega_{m1} + \omega_1 = p\omega_{m2} + \omega_{sl} \tag{12.9}$$

$$\omega_{sl} = s\omega_s = \frac{L_m i_{q1}}{\tau_1 \Psi_{d2}} \tag{12.10}$$

where ω_1 is the inner-rotor angular frequency and τ_1 is the electrical time-constant of the inner rotor.

In order to investigate the interactions between the machine 1 and machine 2, their torque relationships are written as

$$T_{m1} - T_{e1} = J_1 \frac{d\omega_{m1}}{dt} \tag{12.11}$$

$$T_{m2} - T_l = J_2 \frac{d\omega_{m2}}{dt} \tag{12.12}$$

$$T_{m2} = T_{e1} + T_{e2} \tag{12.13}$$

where T_{m1} and T_{m2} are the mechanical torques at the machine 1 and machine 2, respectively, J_1 and J_2 are the inertias of the shafts of the machine 1 and machine 2, respectively, T_{e1} and T_{e2} are the electromagnetic torques developed at the air-gaps of the machine 1 and machine 2, respectively, ω_{m1} and ω_{m2} are the mechanical speeds of the shafts of the machine 1 and machine 2, respectively, and T_l is the load torque. Hence, it yields

$$T_{m1} + T_{e2} = T_l + J_1 \frac{d\omega_{m1}}{dt} + J_2 \frac{d\omega_{m2}}{dt} \tag{12.14}$$

which indicates that the sum of the mechanical torque at the machine 1 and the electromagnetic torque developed at the air-gap of the machine 2 is equal to the sum of load torque and the accelerating torques of the inertias of the machine 1 and machine 2. Actually, the mechanical torque at the machine 1 is the engine torque, and the electromagnetic torque developed at the air-gap of the machine 2 is the motor torque of this EVT system.

12.3.3 DR EVT Operation

The EVT operation is based on the power-split of the engine output using the machine 1. That is, one part of the engine power passes from the inner rotor to the outer rotor of the machine 1 via electromagnetic field, and then drives the rotor of the machine 2. Another part of the engine power is converted into electrical power in the inner rotor of the machine 1, and then feeds the machine 2 via the slip-rings and two back-to-back power converters. Ignoring the power losses (Cheng *et al.*, 2007), their power relationships are given by

$$P_{m1} = T_{m1}\omega_{m1} = P_d + P_e \tag{12.15}$$

$$P_d = T_{e1}\omega_{m2} = T_{m1}\omega_{m2} \tag{12.16}$$

$$P_e = P_{m1} - P_d = T_{m1}(\omega_{m1} - \omega_{m2}) \tag{12.17}$$

where P_{m1} is the mechanical power of the shaft of the machine 1, P_d is the mechanical power directly transferred from the machine 1 to the machine 2, and P_e is the electrical power transferred from the machine 1 to the machine 2. The electrical power generated by the machine 1 feeds the machine 2, which in turn produces the torque given by

$$T_{e2} = \frac{P_e}{\omega_{m2}} = \frac{T_{m1}(\omega_{m1} - \omega_{m2})}{\omega_{m2}} \tag{12.18}$$

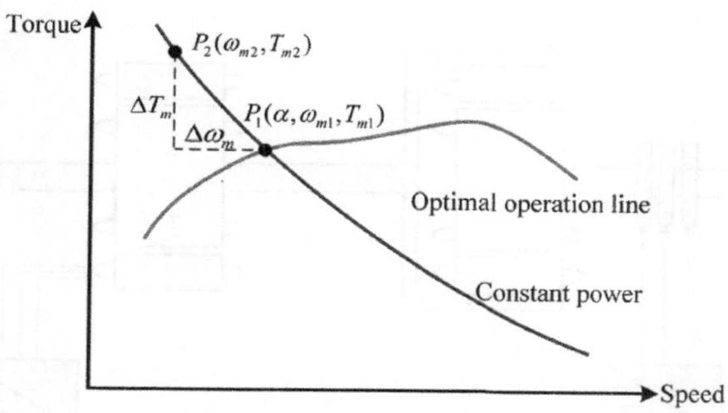

Figure 12.12 DR EVT control strategy

Hence, from Eqs. (12.16) and (12.18), the final driveline torque is given by

$$T_{m2} = T_{e1} + T_{e2} = T_{m1} + \frac{T_{m1}(\omega_{m1} - \omega_{m2})}{\omega_{m2}} = \frac{T_{m1}\omega_{m1}}{\omega_{m2}} = \lambda T_{m1} \quad (12.19)$$

where λ is the so-called EVT ratio, which is continuously variable.

Figure 12.12 shows a typical optimal operation line (OOL) of an engine. For a given throttle level α, which is specified by the accelerator pedal, the optimal engine torque and speed are governed by two functions:

$$T_{m1} = f_T(\alpha) \quad (12.20)$$

$$\omega_{m1} = f_\omega(\alpha) \quad (12.21)$$

When the vehicle operates at point P_2, where the required driveline torque and speed are respectively T_{m2} and ω_{m2}, point P_1 is the equivalent engine operating point on the OOL under the condition of constant power. The EVT controller commands the machine 1 to change the speed by $\Delta\omega_m$. Meanwhile, it commands the machine 2 to change the torque by ΔT_m. Hence, the engine can operate at its optimal condition while the vehicle can offer variable driveline torques and speeds.

Similar to other EVT systems, this DR EVT system can offer other operating modes on top of the aforementioned power-split or EVT mode, namely the electric launch, engine cranking, battery charging, and regenerative braking:

- In the electric launch mode, the engine does not work and the driveline torque is solely provided by the machine 2. The system power flow is depicted in Figure 12.13.
- In the engine cranking mode, the machine 1 works as a starter motor whereas the machine 2 does not work. Because of action and reaction, the direction of rotating field is opposite to the rotational direction of the inner rotor. As an engine usually has a specific rotational direction, special attention should be taken into account. Figure 12.14 shows the system power flow during cranking.
- In the battery charging mode, the inner rotor of the machine 1 generates electricity to charge the battery when the vehicle stops. At that time, the engine power is fully converted into electricity and stored in the battery. Figure 12.15 shows the system power flow during battery charging.
- In the regenerative braking mode, the machine 2 works as a generator to recover the braking energy. The corresponding power flow is shown in Figure 12.16.

12.4 Advanced Double-Rotor EVT Systems

Similar to the development of traditional machines, the induction DR machine is being challenged by other advanced DR machines, such as the permanent magnet (PM) synchronous DR machine and switched

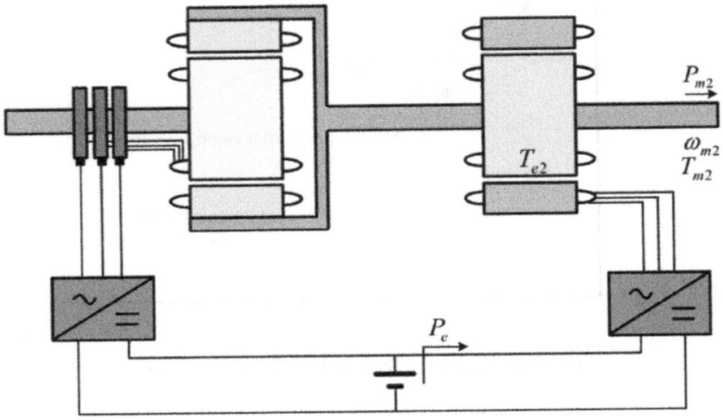

Figure 12.13 Power flow during electric launch

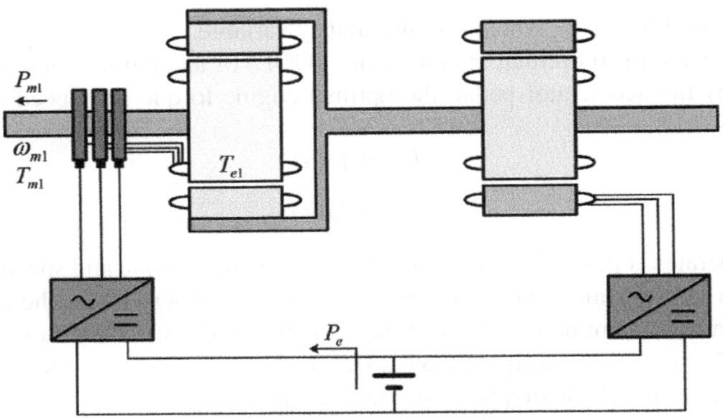

Figure 12.14 Power flow during cranking

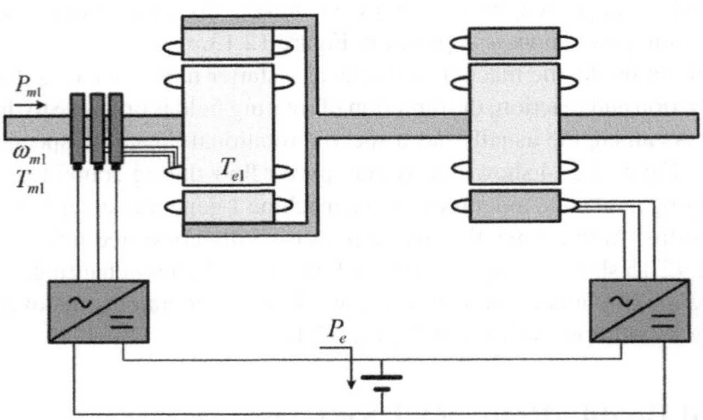

Figure 12.15 Power flow during battery charging

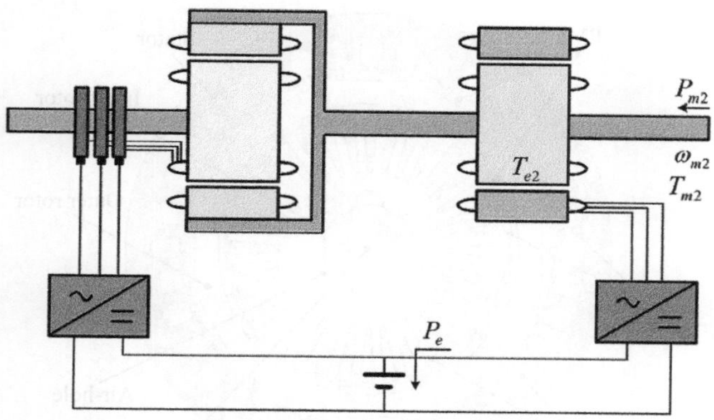

Figure 12.16 Power flow during regenerative braking

reluctance (SR) DR machine. In addition, there are axial-field DR machines and advanced magnetless DR machines that are promising for application to EVT systems.

12.4.1 PM DR EVT System

In order to improve the efficiency and power density, the DR machine can adopt the PM synchronous machine topology. Figure 12.17 shows the system configuration of this advanced PM DR EVT system. The corresponding machine topology is shown in Figure 12.18 in which there are the stator, outer rotor, and inner rotor.

Similar to a normal PM synchronous machine, the stator is equipped with the three-phase armature winding. In addition, similar to a normal outer-rotor PM synchronous machine, the inner rotor is equipped with the three-phase armature winding. There are two layers of PMs separately mounted on the inner and outer surfaces of the outer rotor. To ensure the mechanical integrity of the PMs during high-speed operation, the inner-layer PMs can be surface mounted on the outer-rotor inner surface,

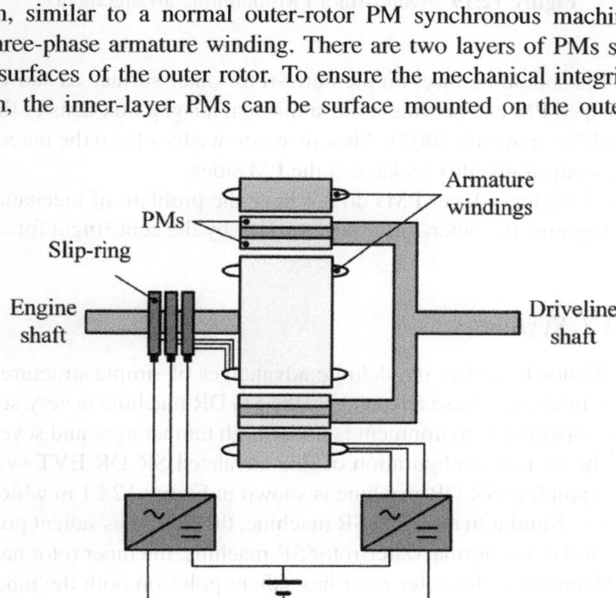

Figure 12.17 PM DR EVT system

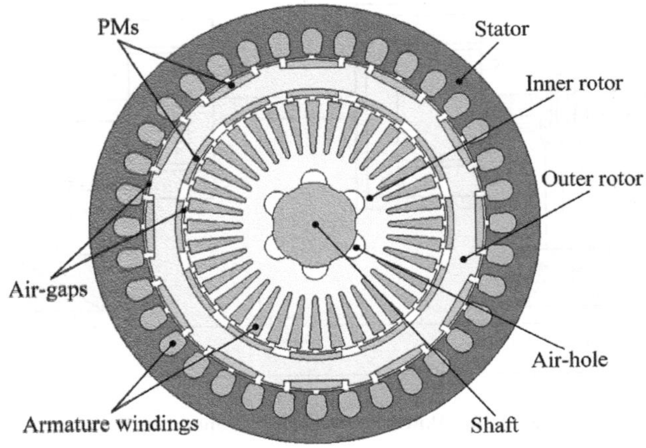

Figure 12.18 PM DR machine

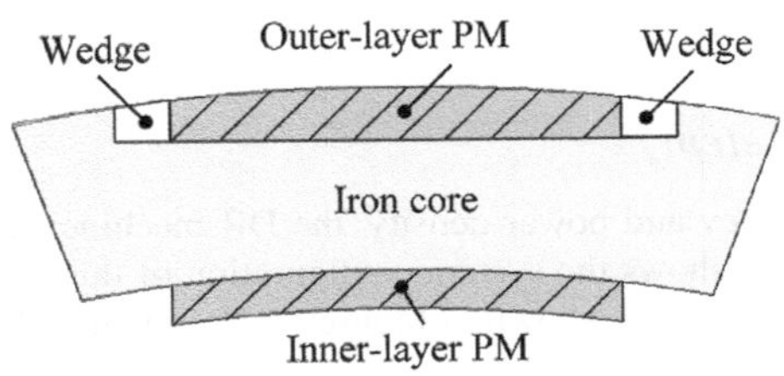

Figure 12.19 Outer-rotor PM mounting arrangement

while the outer-layer PMs should be inset on the outer-rotor outer surface as shown in Figure 12.19, in which the bread-loaf-shaped PM can provide more sinusoidal air-gap flux density than that with constant thickness (Nordlund and Sadarangani, 2002). Also, there are wedges from the outer-layer PM to the iron at the sides to avoid the undesirable flux leakage at the PM sides.

It should be noted that the inner-layer PMs do not have the problem of mechanical integrity because they are simply pressed against the outer-rotor inner surface by the centrifugal force.

12.4.2 SR DR EVT System

The SR machine is well known to offer the definite advantages of simple structure, high reliability, and low manufacturing cost. Inheriting these advantages, the SR DR machine is very suitable for application to DR EVT in which the operating environment is under high temperature and severe vibration.

Figure 12.20 shows the system configuration of this advanced SR DR EVT system (Cui, Yuan, and Wang, 2008). The corresponding SR DR machine is shown in Figure 12.21 in which there are the stator, outer rotor, and inner rotor. Similar to a normal SR machine, the stator has salient poles wound with phase windings. In addition, similar to a normal outer-rotor SR machine, the inner rotor has salient poles wound with phase windings. Meanwhile, the outer rotor has salient poles on both the inner and outer surfaces, but without any windings. Similar to other DR machines, the inner and outer rotors can be considered

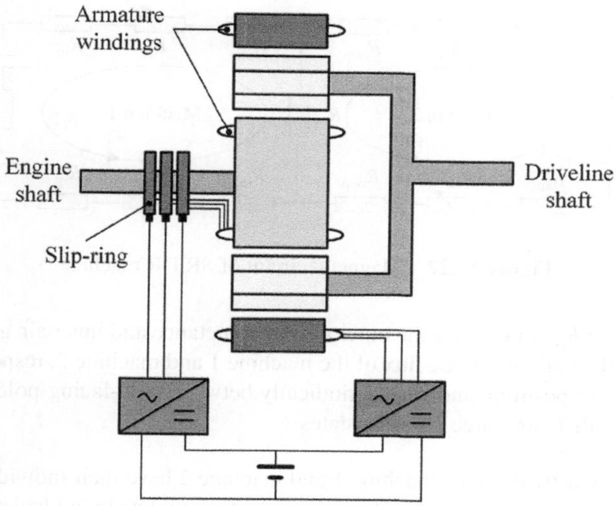

Figure 12.20 SR DR EVT system

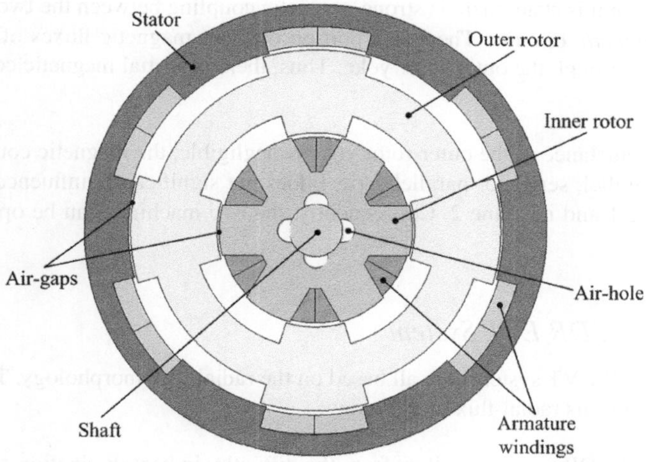

Figure 12.21 SR DR machine

as the machine 1, while the stator and outer rotor are treated as the machine 2. Both the machine 1 and machine 2 adopt the same pole arrangement, typically 6/4-pole or 8/6-pole.

Since the inner and outer air-gaps of the SR DR machine are small, there may be magnetic coupling between the inner and outer electromagnetic fields. Figure 12.22 shows its magnetic circuit, where F_s and F_s' are the magnetomotive forces (MMFs) of the stator coil per pole; F_i and F_i' are the MMFs of the inner-rotor coil per pole; R_{sy} and R_{sp} are the stator yoke reluctance and stator pole reluctance, respectively; R_{iy} and R_{ip} are the inner-rotor yoke reluctance and inner-rotor pole reluctance, respectively; R_{oy}, R_{oop}, and R_{oip} are the outer-rotor yoke reluctance, outer-rotor outer pole reluctance and outer-rotor inner pole

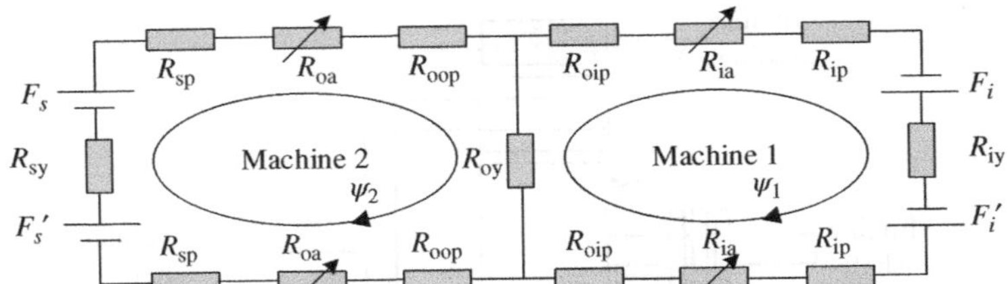

Figure 12.22 Magnetic circuit of SR DR machine

reluctance, respectively; R_{oa} and R_{ia} are the outer air-gap reluctance and inner air-gap reluctance, respectively; ψ_1 and ψ_2 are the main magnetic flux of the machine 1 and machine 2, respectively. Both R_{oa} and R_{ia} change with the rotor position, and vary periodically between pole-facing-pole and pole-facing-slot. It can be observed that there are three possible states:

- *Parallel magnetic circuits*: Both the machine 1 and machine 2 have their individual magnetic circuits, and their main magnetic fluxes pass the outer-rotor yoke to complete individual magnetic circuit loops. There is only mechanical coupling, but no magnetic coupling, between the two machines.
- *Series magnetic circuits*: Neither the machine 1 nor the machine 2 have their main magnetic fluxes passing the outer-rotor yoke to form a closed magnetic circuit. Hence, their main magnetic fluxes are virtually the same. In this state, there is strong magnetic coupling between the two machines.
- *Parallel-series magnetic circuits*: There is a portion of main magnetic fluxes of the machine 1 and machine 2 passing through the outer-rotor yoke. Thus, there is partial magnetic coupling between the two machines.

Provided that the reluctance of the outer-rotor yoke is negligible, the magnetic coupling under various magnetic circuits (parallel, series, or parallel-series) does not significantly influence the main magnetic fluxes of the machine 1 and machine 2. Consequently, the two machines can be operated separately or simultaneously.

12.4.3 Axial-Flux DR EVT System

The aforementioned DR EVT systems are all based on the radial-flux morphology. There are two fundamental shortcomings of this radial-flux morphology:

- The inner rotor of the DR machine suffers from the difficulty in heat dissipation or cooling.
- The magnetic loading of the DR machine is inherently limited by the tooth size and shape.

By converting the radial-flux morphology into the axial-flux morphology (Lee, Liu, and Chau, 2014), the resulting axial-flux DR EVT systems can solve the above-mentioned problems. Figure 12.23 shows the configuration of an axial-flux PM DR EVT in which the left rotor is equivalent to the inner rotor and the right rotor is equivalent to the outer rotor of a radial-flux counterpart (Eriksson and Sadarangani, 2002). The left rotor is wound with the three-phase armature winding, and is equipped with three slip-rings. The right rotor is mounted with two layers of PMs, and is mechanically coupled with the driveline. The stator is also wound with the three-phase armature winding, which is fed by the generated electricity from the left rotor via two back-to-back power converters and the battery. The principle of operation of this axial-flux PM DR EVT is exactly the same as its radial-flux counterpart.

It should be noted that this axial-flux PM DR machine takes the definite advantages of higher torque density and higher power density than its radial-flux counterpart. In addition, differing from the inner and outer rotors of the radial-flux DR machine, both the left and right rotors of the axial-flux DR machine can

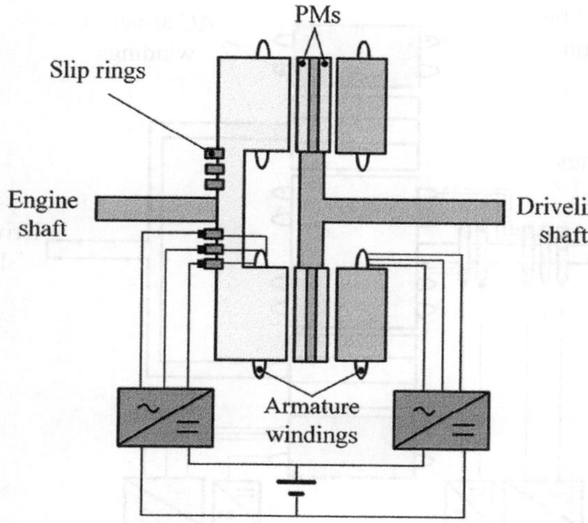

Figure 12.23 Axial-flux PM DR EVT system

have better exposure to surroundings, which greatly facilitates the heat dissipation and cooling arrangement. Nevertheless, the axial-flux machine generally has a different space requirement and suffers from a mounting problem in bearings due to the attractive forces among the stator and two rotors. Additional rotors or stators can alleviate such problem, but significantly increase the system complexity.

12.4.4 Advanced Magnetless DR EVT System

In recent years, many advanced magnetless machines have been proposed, such as the DSDC machine (Fan and Chau, 2008), flux-switching DC (FSDC) machine (Tang *et al.*, 2012), vernier reluctance (VR) machine (Lee, 1963), and vernier reluctance DC (VRDC) machine (Taibi, Tounzi, and Piriou, 2006). Among them, the DSDC machine is becoming mature for EV and hybrid electric vehicle (HEV) applications. The DSDC machine takes the definite advantages of higher torque density and higher power density than the SR machine while retaining the corresponding merits of simple structure, high reliability, and low manufacturing cost (Lee *et al.*, 2013). Increasingly, because of the controllable DC field current, the DSDC machine can online control the air-gap flux to facilitate various HEV operations, such as flux strengthening for electric launch or engine cranking, flux weakening for high-speed constant-power operation, and flux tuning for efficiency optimization. Thus, the DSDC DR machine is promising for application to DR EVT.

Figure 12.24 shows the system configuration of an advanced magnetless DR EVT system, namely the DSDC DR EVT system. Differing from the SR DR EVT system, there are two additional electrical input ports in the DSDC DR machine to allow for individual excitations of two DC field windings. Thus, the DSDC DR machine can be considered as a six-port device, in which there are four electrical ports and two mechanical ports. Among the four electrical ports, two are the AC armature windings, one located in the stator and another located in the inner rotor; meanwhile, other two electrical ports are the DC field windings, again one located in the stator and another located in the inner rotor. Among the two mechanical ports, one is the inner rotor coupled with the engine while another is the outer rotor connected to the driveline.

The structure of the DSDC DR machine is shown in Figure 12.25, which consists of a stator, an outer rotor, and an inner rotor. Similar to the SR DR machine, the stator has salient poles wound with phase windings while the inner rotor also has salient poles wound with phase windings. Differing from the SR DR machine, the stator and inner rotor have equipped with individual DC field windings. The outer rotor

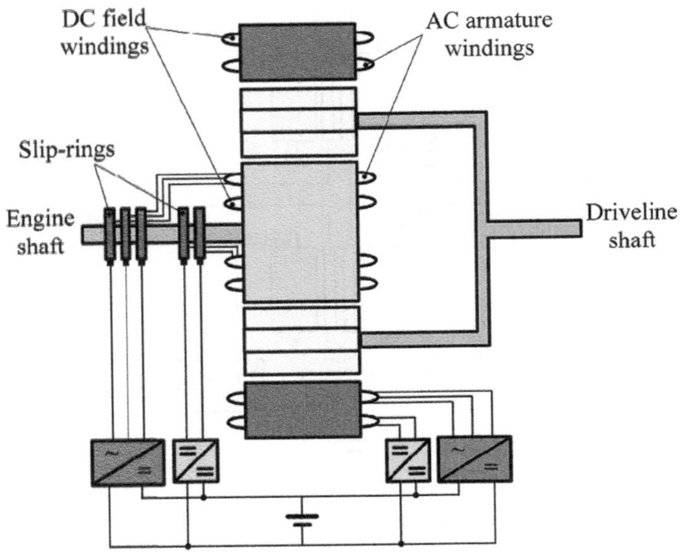

Figure 12.24 DSDC DR EVT system

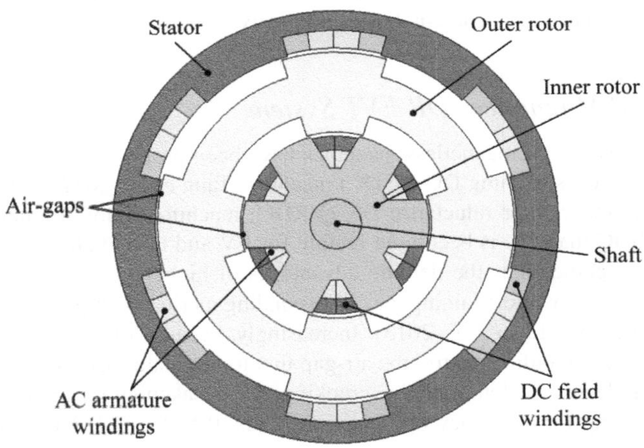

Figure 12.25 DSDC DR machine structure

has salient poles on both the inner and outer surfaces. Similar to other DR machines, the inner and outer rotors can be considered as the machine 1, while the stator and outer rotor are treated as the machine 2. Both the machine 1 and machine 2 adopt the same pole arrangement, such as 6/4-pole, 8/6-pole, or 12/8-pole.

While appreciating the advantages of the DSDC DR machine over the SR DR machine, there is a key drawback that the DSDC DR machine needs an additional pair of brushes and slip-rings to enable feeding the DC current into the inner rotor. Thus, this DSDC DR machine preferably adopts the three-phase arrangement, such as 6/4-pole or 12/8-pole, rather than the four-phase arrangement so that the number of brushes and slip-rings can be minimized.

Similar to the SR DR machine, the DSDC DR machine involves small inner and outer air-gaps, which may cause magnetic coupling between the inner and outer electromagnetic fields. Thus, the outer-rotor yoke needs to be properly designed. Provided that the reluctance of the outer-rotor yoke is low, the magnetic coupling does not significantly influence the main magnetic fluxes of the machine 1 and machine 2. Consequently, the two machines can be operated separately or simultaneously.

12.5 Design Criteria of DR EVT Systems

The DR EVT system is composed of three key components: the DR machine to perform the desired power-split as well as to provide motoring and generating, the back-to-back converters to provide controlled rectification and inversion for power flow regulation, and the battery pack for energy storage and buffering. Among them, as reflected by their specific names, such as the induction DR EVT, PM DR EVT, or SR DR EVT, the key component of DR EVT systems is the DR machine.

The system design criteria of these DR EVT systems for the full hybrids, no matter what kinds of DR machines adopted, are summarized as follows:

- The desired driveline torque and power for various vehicular operations should be satisfied by the summation of torques and powers provided by the engine and motor.
- The inner rotor (or the left rotor in case of the axial-flux morphology) of the DR machine should be able to split the engine power from no load to full load. Particularly, the slip-rings of the inner rotor (or left rotor) and the associated brushes must be able to withstand the split power and the maximum rotational speed.
- The power-split capability of the DR machine should enable the engine working on the OOL throughout the whole operating range.
- The power-handling capacity of the two back-to-back converters should be able to fully extract the split power and fully fulfill the power requirement for motoring.
- The battery capacity or state-of-charge (SOC) should enable the system to absorb the split power and also the regenerative braking energy.

Being the key component of DR EVT systems, the DR machine needs to be designed with special considerations. Magnetically, the outer-rotor yoke (or right-rotor yoke) should be properly designed in such a way that the corresponding reluctance is low enough; otherwise, the two machines cannot be independently operated and their operations will affect one another. Thermally, the inner rotor inherently suffers from the problem of heat dissipation, which needs special cooling facilities; in contrast, the left rotor of the axial-flux morphology will have better heat dissipation. Structurally, the whole DR machine involves two air-gaps with two rotating bodies spinning at different speeds, which desires robust structures and high-precision manufacturing. In particular, the slip-rings and the associated brushes degrade the overall mechanical integrity. Therefore, after the initialization of machine parameters according to the design criteria of relevant machines, the detail design has to adopt iterative multi-physics finite element analysis in which the magnetic field, thermal field, and structural calculations are iteratively carried out to satisfy the desired performances.

12.6 Design Example of DR EVT Systems

By adopting different types of DR machines, different types of DR EVT systems can be resulted. As mentioned earlier, the most basic and relatively mature one is the induction DR EVT system, whereas the advanced ones include the PM, SR, axial-flux, and magnetless counterparts. As the advanced magnetless machines are becoming attractive for HEV applications, the DSDC DR EVT system is adopted for design exemplification.

12.6.1 DSDC DR EVT System Configuration

The configuration of the DSDC EVT system has been depicted in Figure 12.24. The structure of the DSDC DR machine has also been shown in Figure 12.25 in which the inner rotor and outer rotor are considered as the machine 1, while the stator and outer rotor are treated as the machine 2. The system splits the engine power into two paths as depicted in Figure 12.26. One path directly feeds the driveline via the inner rotor and outer rotor, while another goes through the inner rotor, controlled rectifier, battery, inverter, stator, and outer rotor. These two main power flow paths are essentially bidirectional. Meanwhile, both the stator and inner rotor are fed by additional DC field currents via two separate DC-DC converters. As these DC field currents are for air-gap excitations, the corresponding power flows are unidirectional. By properly controlling the controlled rectifier and inverter as well as the two DC-DC converters, this DSDC DR EVT can offer the following features:

- The engine can always work on the desired OOL for the EVT operation mode. Other modes of operation, including the idle stop-start, electric launch, and regenerative braking, can also be provided.
- The flexibility of DC field current control and hence the flux control at both the inner and outer air-gaps can greatly facilitate the flux strengthening to boost up the torque for engine cranking and electric launch as well as the flux weakening to enable wide-range constant output voltage for battery charging. Additionally, these air-gap flux densities can be online tuned to achieve efficiency optimization under various operations.
- The drawbacks of mechanical gearing such as the transmission loss, gear noise, and need for regular lubrication can be totally eliminated. In addition, as the double-rotor machine performs both power-split and power summation, it essentially combines the functions of planetary gear, generator, and motor of the PG EVT system so that this DR EVT system has significantly smaller size and lighter weight than its PG EVT counterpart.

12.6.2 DSDC DR Machine Design

The core component of this DSDC DR EVT system is the DSDC DR machine. On the basis of the requirements of a typical passenger HEV, the specifications of the DSDC DR machine are listed in Table 12.1 in which the inner and outer rotors are considered as the machine 1, while the stator and outer rotor are considered as the machine 2. On the basis of the specifications, the DSDC DR machine is designed as shown in Figure 12.27, and its key parameters and geometric dimensions are listed in Table 12.2.

12.6.3 DSDC DR Machine Analysis

By using finite element analysis, the electromagnetic field distributions of the DSDC DR machine can readily be determined. Figure 12.28 shows the electromagnetic fields when the machine 1 (inner machine)

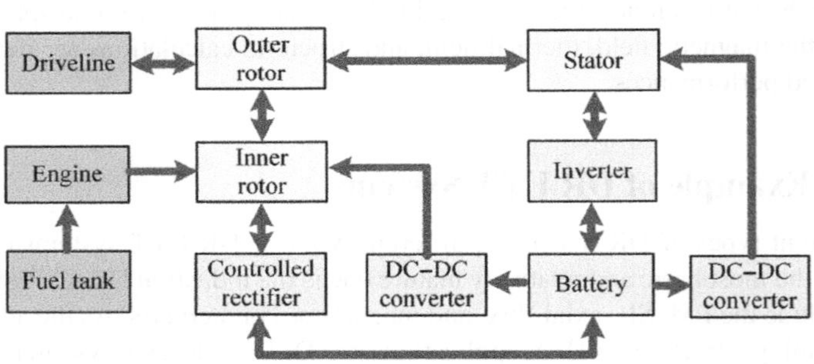

Figure 12.26 Power flow paths in DSDC DR EVT system

Table 12.1 Specifications of DSDC DR machine

Machine 1	
Rated power	8 kW
Maximum speed	10 000 rpm
Machine 2	
Rated power	12 kW
Rated torque	100 N m
Constant-torque operation	0–1200 rpm
Constant-power operation	1200–6000 rpm

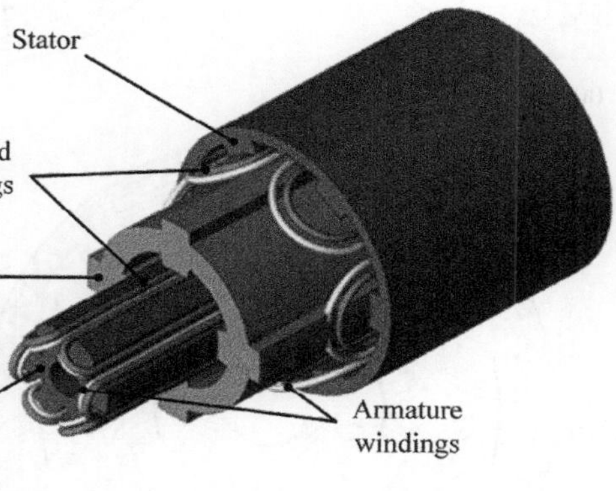

Figure 12.27 DSDC DR machine

Table 12.2 Key parameters and dimensions of DSDC DR machine

Number of phases in stator	3
Number of salient poles in stator	6
Number of stator armature turns per phase	15
Number of stator DC field turns per phase	20
Number of doubly-salient poles in outer rotor	4
Number of phases in inner rotor	3
Number of salient poles in inner rotor	6
Number of inner-rotor armature turns per phase	8
Number of inner-rotor DC field turns per phase	12
Stator outside diameter	300 mm
Stator inside diameter	241 mm
Outer-rotor outside diameter	240 mm
Outer-rotor inside diameter	120 mm
Inner-rotor outside diameter	119 mm
Inner-rotor inside diameter	40 mm
Outer air-gap length	0.5 mm
Inner air-gap length	0.5 mm
Effective axial length	350 mm

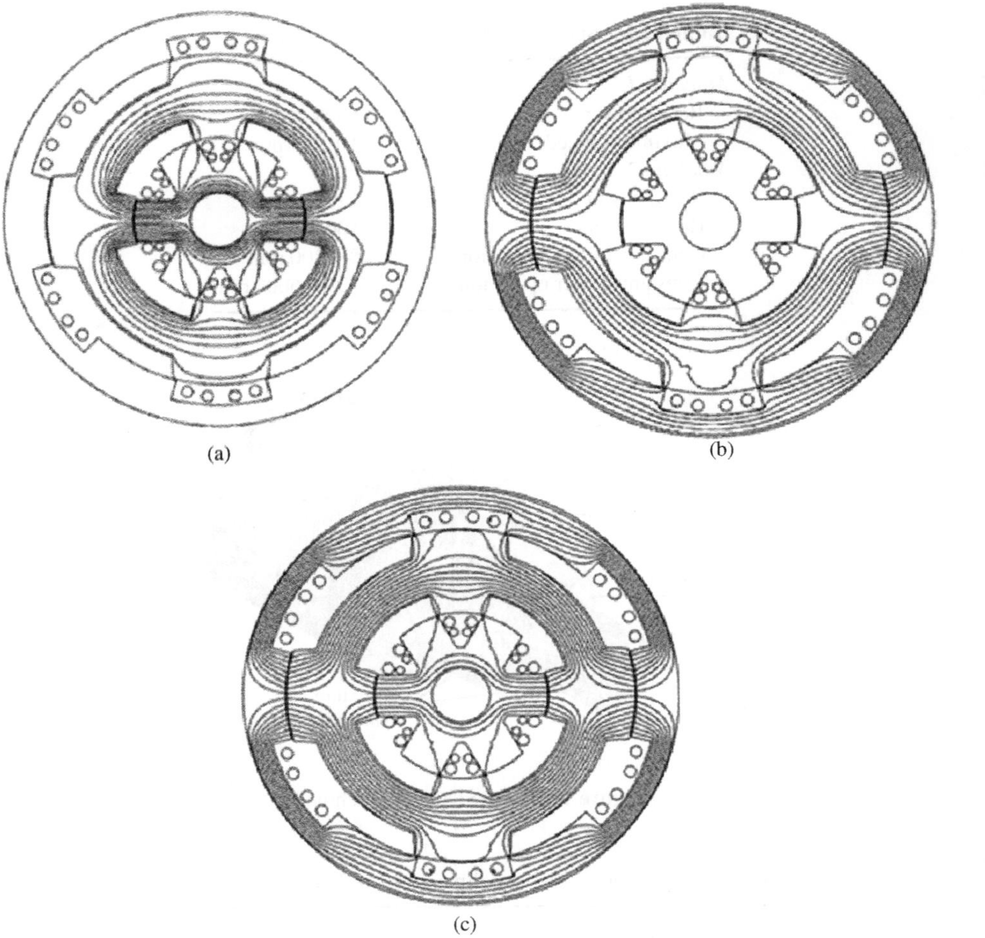

Figure 12.28 Electromagnetic field distributions of DSDC DR machine: (a) machine 1, (b) machine 2, and (c) both machine 1 and machine 2

and machine 2 (outer machine) run separately and simultaneously under the rated DC field current of 5 A/mm^2. It can be observed that both the machine 1 and machine 2 operate similarly with the magnetic flux lines focusing on the relevant pole tips. Since the DR machine, including the machine 1 and machine 2, is designed in such a way that the outer-rotor yoke is thick enough, technically low reluctance enough, it can be seen that there is no noticeable electromagnetic coupling between the machine 1 and machine 2. Thus, they can independently work at different modes of operation.

As mentioned earlier, the machine 1 and machine 2 of the DSDC DR machine can operate independently without affecting one another. To analyze the torque capability of the machine 2, the armature winding and DC field winding of the stator are fed with various currents while the inner rotor is unexcited and kept stationary. Figure 12.29a and b shows the outer-rotor torque-angle characteristics when the armature current is varied and the DC field current is kept at the rated value of 5 A/mm^2, and when the DC field current is varied and the armature current is kept at the rated value of 5 A/mm^2, respectively. It can be

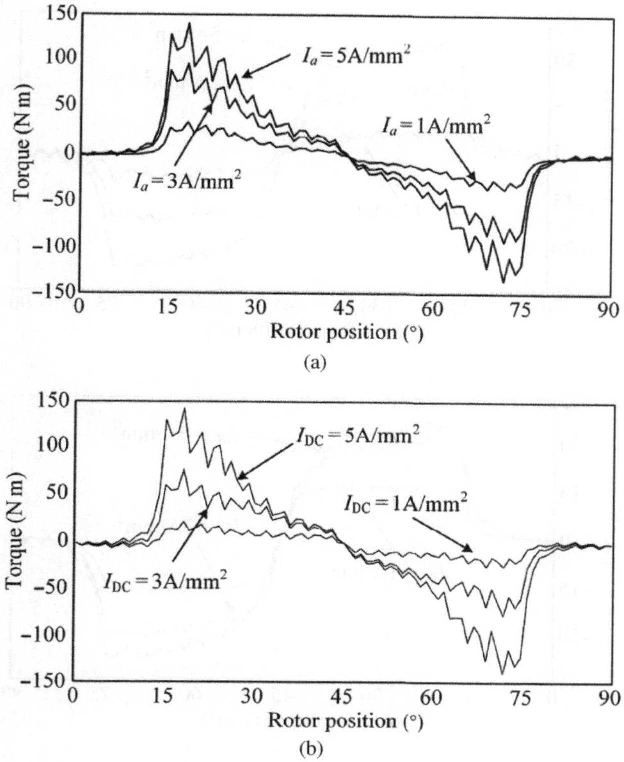

Figure 12.29 Torque-angle characteristics of machine 2: (a) various armature currents and (b) various DC field currents

observed that when the DC field current is kept at the rated value, the average torque at the rated armature current can reach the desired value of 100 N m. Meanwhile, it can also be observed that the DC field current can effectively adjust the developed torque, which is very useful to temporarily boost up the starting torque for engine cranking or electric launch.

Similarly, in order to assess the torque capability of the machine 1, the armature winding and DC field winding of the inner rotor are fed with various currents while the stator is unexcited and the outer rotor is kept stationary. Figure 12.30a and b shows the inner-rotor torque-angle characteristics when the armature current is varied and the DC field current is kept at the rated value of 5 A/mm^2 and when the DC field current is varied and the armature current is kept at the rated value of 5 A/mm^2, respectively. It can be found that when the DC field current is kept at the rated value, the average torque at the rated armature current can reach 30 N m. Similarly, the DC field current can help adjust the developed torque. It should be noted that the machine 1 normally operates as a generator, rather than a motor, to help split the engine power.

Since the inner rotor of the DR machine is directly coupled with the engine, the machine 1 normally operates as a generator. To evaluate the generation characteristics of the machine 1, the inner rotor is driven by the engine while the stator is unexcited and the outer rotor is kept stationary. Figure 12.31a and b shows the generated electromotive forces (EMFs) when the engine speed varies from zero to the maximum speed while the DC field current is fixed at the rated value of 5 A/mm^2 and when the DC field current is varied while the engine speed is kept at 5000 rpm, respectively. It can be found that when the DC

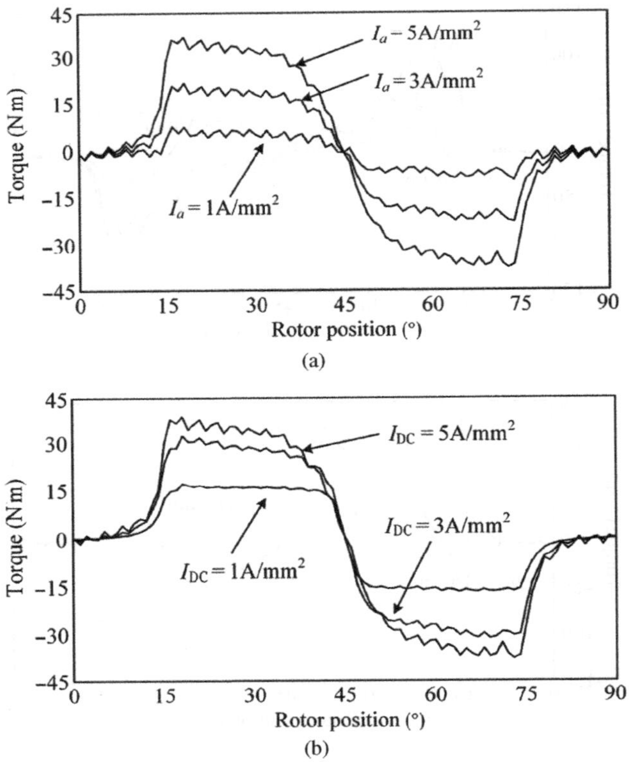

Figure 12.30 Torque-angle characteristics of machine 1: (a) various armature currents and (b) various DC field currents

field current is kept at the rated value, the generated EMF can reach 280 V at 5000 rpm. Meanwhile, the variation of DC field current can help adjust the generated EMF linearly, which is very useful to maintain the generated EMF for effective battery charging over a wide speed range.

12.7 Potential Applications of DR EVT Systems in HEVs

The aforementioned DR EVT systems, no matter what types of electric machines adopted, all rely on using double rotors to split the engine power in which the electrical power generated by the inner rotor is controlled via the slip-rings and brushes. The slip-rings and brushes inevitably cause reliability concern and need regular maintenance. These drawbacks cannot enable the DR EVT system really outperforming its PG EVT counterpart for application to HEVs. In fact, they have their individual advantages and disadvantages.

In recent years, there have been some attempts to get rid of these slip-rings and brushes, such as the use of an additional stator to capture the split power from the inner rotor being driven by the engine. This additional stator can be cascaded with the original stator along the driveline (Zheng *et al.*, 2013) or located inside the inner rotor (Niu, Ho, and Fu, 2013). These kinds of double-stator double-rotor arrangements will definitely make the system very complicated, deteriorating their realistic application to HEVs.

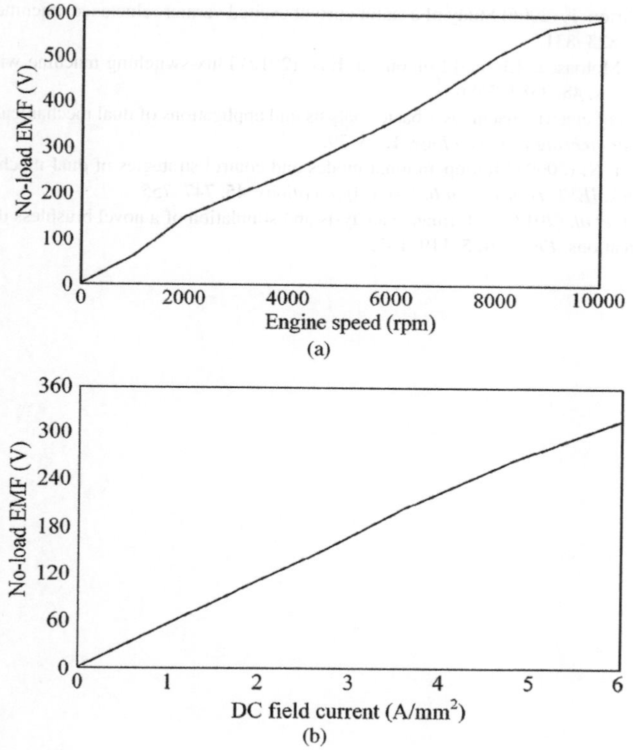

Figure 12.31 Generated EMF characteristics: (a) various engine speeds and (b) various DC field currents

References

Cheng, Y., Cui, S., Song, L. and Chan, C.C. (2007) The study of the operation modes and control strategies of an advanced electromechanical converter for automobiles. *IEEE Transactions on Magnetics*, **43**, 430–433.

Cui, S., Cheng, Y., and Chan, C.C. (2006) A basic study of electrical variable transmission and its application in hybrid electric vehicle. Proceedings of IEEE Vehicle Power and Propulsion Conference, pp. 1–4.

Cui, S., Yuan, Y., and Wang, T. (2008) Research on switched reluctance double-rotor motor used for hybrid electric vehicle. Proceedings of International Conference on Electrical Machines and Systems, pp. 3393–3396.

Eriksson, S. and Sadarangani, C. (2002) A four-quadrant HEV drive system. Proceedings of IEEE Vehicular Technology Conference, pp. 1510–1514.

Fan, Y. and Chau, K.T. (2008) Design, modeling, and analysis of a brushless doubly fed doubly salient machine for electric vehicles. *IEEE Transactions on Industry Applications*, **44**, 727–734.

Hoeijmakers, M. and Ferreira, J. (2006) The electric variable transmission. *IEEE Transaction Industry Applications*, **42**, 1092–1100.

Lee, C.H. (1963) Vernier motor and its design. Proceedings of IEEE Winter General Meeting, pp. 343–349.

Lee, C.H.T., Chau, K.T., Liu, C. *et al.* (2013) Quantitative comparison and analysis of magnetless machines with reluctance topologies. *IEEE Transactions on Magnetics*, **49**, 3969–3972.

Lee, C.H.T., Liu, C. and Chau, K.T. (2014) A magnetless axial-flux machine for range-extended electric vehicles. *Energies*, **7**, 1483–1499.

Niu, S., Ho, S.L. and Fu, W.N. (2013) A novel double-stator double-rotor brushless electrical continuously variable transmission system. *IEEE Transactions on Magnetics*, **49**, 3909–3912.

Nordlund, E. and Sadarangani, C. (2002) The four-quadrant energy transducer. Proceedings of IEEE Industry Applications Conference, pp. 390–397.

Taibi, S., Tounzi, A. and Piriou, F. (2006) Study of a stator current excited vernier reluctance machine. *IEEE Transactions on Energy Conversion*, **21**, 823–831.

Tang, Y., Paulides, J.J.H., Motoasca, T.E. and Lomonova, E.A. (2012) Flux-switching machine with DC excitation. *IEEE Transactions on Magnetics*, **48**, 3583–3586.

Xu, L. (2006) A new breed of electric machines—basic analysis and applications of dual mechanical port electric machines. *Journal of Electrical Engineering & Technology*, **1**, 73–79.

Xu, L., Zhang, Y. and Wen, X. (2009) Multioperational modes and control strategies of dual-mechanical-port machine for hybrid electrical vehicles. *IEEE Transaction Industry Applications*, **45**, 747–755.

Zheng, P., Wu, Q., Zhao, J. *et al.* (2013) Performance analysis and simulation of a novel brushless double rotor machine for power-split HEV applications. *Energies*, **5**, 119–137.

13

Magnetic-Geared Electric Variable Transmission Systems

The key of electric variable transmission (EVT) systems for hybrid electric vehicles (HEVs) is the power-split device. The planetary gear is the most commonly adopted power-split device. However, it suffers from the problems of transmission loss, gear noise, and need for regular lubrication. The double-rotor (DR) machine is a newly introduced gearless power-split device. However, it requires brushes and slip-rings, which are less reliable and incur need for regular maintenance. In order to overcome these drawbacks, a new class of magnetic-geared electric variable transmission (MG EVT) systems has been developed based on contactless magnetic gears, which can offer the definite advantage of brushless and pseudo-gearless power-split operation.

In this chapter, the MG EVT systems for HEVs, including the system configurations and multi-port magnetic gears as well as the magnetic planetary-geared (MPG) and magnetic concentric-geared (MCG) types, are discussed. The corresponding design criteria, design example, and potential applications are also presented.

13.1 System Configurations

The planetary-geared electric variable transmission (PG EVT) system makes use of planetary gearing to split the engine power into two parts: one part is mechanically coupled with the driveline via the ring gear; another part is electrically connected to the driveline via the sun gear, generator, controlled rectifier, battery, inverter, and motor. The system is depicted in Figure 13.1. Because of the use of planetary gear, it inevitably suffers from transmission loss, gear noise, and need for regular lubrication as well as backlash and misalignment. In addition, the system involving physical combination of a planetary gear and two machines is heavy and bulky.

The DR EVT system utilizes an electric machine with two rotors to split the engine power into two parts: one part is to mechanically propel the driveline via the outer rotor of the DR machine; another part is to electrically feed the driveline via the inner rotor of the DR machine, controlled rectifier, battery, inverter, and stator of the DR machine. The corresponding system is depicted in Figure 13.2. Even though the planetary gear and hence the associated problems can be completely eliminated, it suffers from other problems because of the use of brushes and slip-rings. That is, the brushes desire regular maintenance while the slip-rings degrade the system reliability. Nevertheless, as compared with the PG EVT system, the DR EVT system facilitates the system integration, leading to use a single electric machine to perform the desired power-split and then the power summation along the driveline.

Electric Vehicle Machines and Drives: Design, Analysis and Application, First Edition. K. T. Chau.

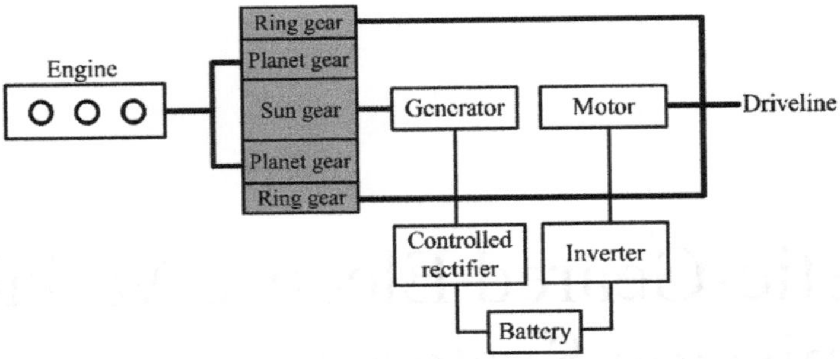

Figure 13.1 Planetary-geared EVT system configuration

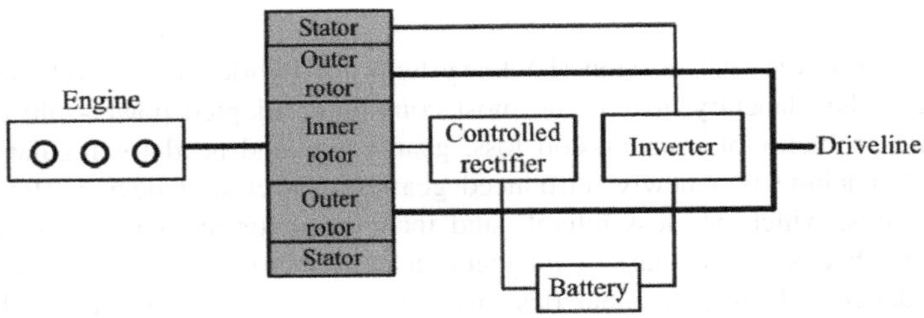

Figure 13.2 Double-rotor EVT system configuration

In recent years, the development of magnetic gears has been very promising. Compared with the mechanical gears, the magnetic gears have the advantages of high transmission efficiency, high reliability, no maintenance, and inherent overload protection capability. With the use of rare-earth permanent magnets (PMs), the magnetic gears can achieve a torque density over 100 kN/m^3, which is comparable with the mechanical counterparts. Therefore, by using a magnetic gear that is composed of the PM inner rotor, modulation ring, and PM outer rotor to supersede the planetary gear of the PG EVT system, the resulting MG EVT system can offer the definite advantage of brushless and pseudo-gearless power-split operation. Hence, the mechanical gear problems of PG EVT and the brush and slip-ring problems of DR EVT can be fundamentally solved. The system configuration of this MG EVT system is depicted in Figure 13.3.

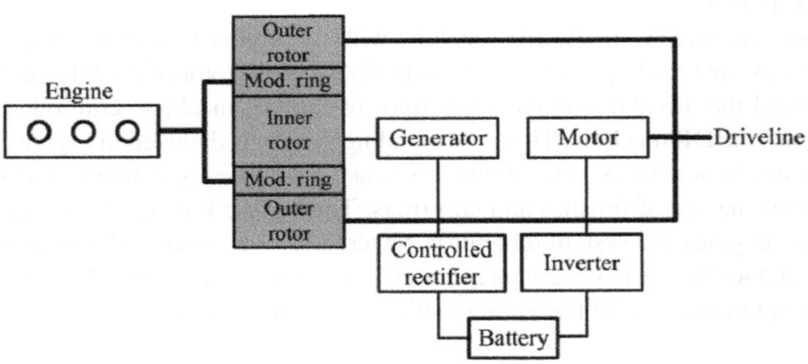

Figure 13.3 Magnetic-geared EVT system configuration

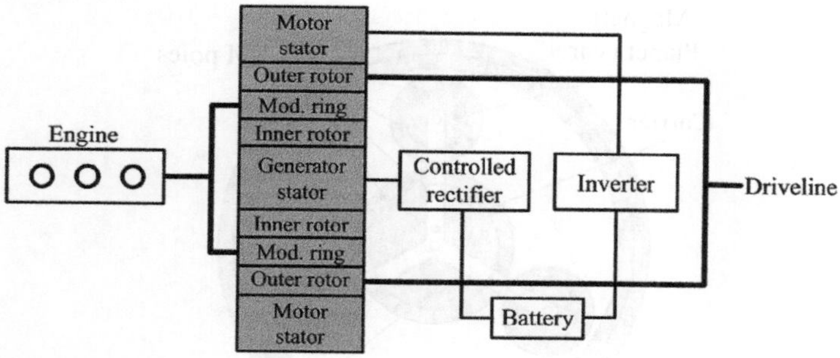

Figure 13.4 Integrated magnetic-geared EVT system configuration

Furthermore, the magnetic gear can readily be fused with an electric machine to form a magnetic-geared (MG) machine. Thus, it is possible to further integrate the magnetic gear with the two machines to form a single electric machine to perform the desired power-split and power summation. The system configuration of this integrated MG EVT system is depicted in Figure 13.4.

13.2 Multi-port Magnetic Gears

In order to perform the power-split, the magnetic gear needs to offer multiple mechanical ports. For MG EVT application, the magnetic gear should have an input mechanical port coupled with the engine and two output mechanical ports to divert the engine power. One of the output mechanical ports can also be integrated with an electric machine to become an electrical port. There are two major types of multi-port magnetic gears: magnetic planetary gear and magnetic concentric gear.

13.2.1 Magnetic Planetary Gears

By applying the concept of magnetic gearing to a mechanical planetary gear, the resulting magnetic planetary gear can inherit the multi-port feature from the mechanical planetary gear as well as the advantages of low transmission loss, silent operation, and no maintenance from the magnetic gear (Huang *et al.*, 2008).

Figure 13.5 shows the schematic of a magnetic planetary gear in which there are the magnetic sun gear, magnetic planet gears, carrier, and magnetic ring gear. The torque transmission between those magnetic gears is based on rare-earth PMs, rather than mechanical gear teeth. Because of the finite number of PM pole-pairs, which is usually much lower than the possible number of mechanical gear teeth, the design of magnetic planetary gears should be careful. In general, the more the number of magnetic planet gears, the higher the torque density of the whole magnetic planetary gear can be achieved. Hence, it can achieve the torque density up to about 100 kN m/m^3.

The geometric model of this magnetic planetary gear is shown in Figure 13.6. Since the pole pitch of each magnetic gear within the magnetic planetary gear must be the same, it yields

$$R_s\theta_s = R_p\theta_p = R_r\theta_r \tag{13.1}$$

where R_s, R_p, and R_r are the radii of the magnetic sun gear, magnetic planet gear, and magnetic ring gear, respectively, and θ_s, θ_p, and θ_r are the angular pole pitches of the magnetic sun gear, magnetic planet gear, and magnetic ring gear, respectively. These angular pole pitches can be written as $\theta_s = 2\pi/P_s$, $\theta_p = 2\pi/P_p$, and $\theta_r = 2\pi/P_r$, where P_s, P_p, and P_r are the numbers of pole-pairs of the

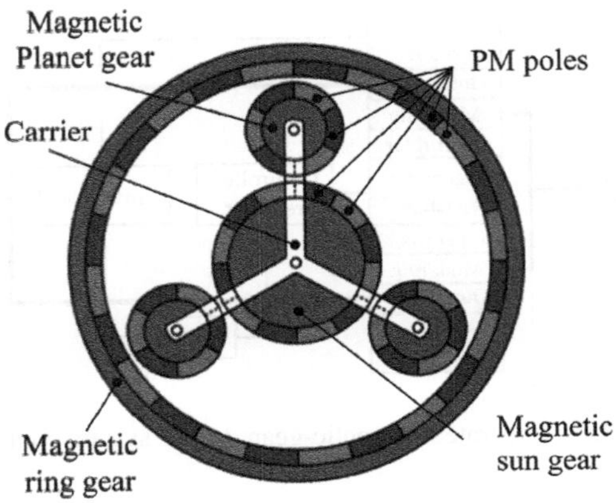

Figure 13.5 Magnetic planetary gear

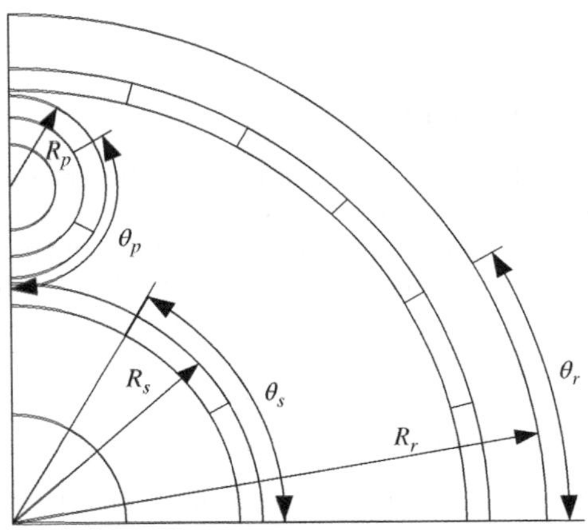

Figure 13.6 Geometric model of magnetic planetary gear

magnetic sun gear, magnetic planet gear, and magnetic ring gear, respectively. Meanwhile, their radii are related by

$$R_r = 2R_p + R_s \tag{13.2}$$

By using Eqs. (13.1) and (13.2), their pole-pairs are related by

$$P_r = 2P_p + P_s \tag{13.3}$$

Once the pole-pair numbers of these magnetic gears are set, the corresponding angular pole pitches can be determined accordingly. Then, the gear radii can be deduced, which should include an allowance for the air-gap between the magnetic gears. That is, the radii should include half the air-gap length between the magnetic gears.

Since the number of magnetic planet gears is the key parameter to govern the torque density of the whole magnetic planetary gear, there is a procedure (Huang *et al.*, 2008) to determine the maximum number of magnetic planet gears:

- After setting the pole-pair ratio between the magnetic sun and ring gears, it is simplified to deduce the natural numbers a and b as governed by

$$\frac{P_s}{P_r} = \frac{a}{b} \quad a, b \in N \tag{13.4}$$

- Having deduced a and b, the possible maximum number N'_p of magnetic planet gears is given by

$$N'_p = \frac{2P_r}{b} \tag{13.5}$$

- Then, the actual maximum number N_p of magnetic planet gears is determined by

$$N_p = \begin{cases} N'_p & \text{if } a \text{ and } b \text{ are odd} \\ N'_p/2 & \text{if } a \text{ or } b \text{ is even} \end{cases} \tag{13.6}$$

While increasing the number of magnetic planet gears can improve the overall torque density, it may not be necessary to install the maximum number of magnetic planet gears because there is a drawback that the cogging torque will also be increased.

On the basis of the same derivation as the mechanical planetary gear, the speed relationship of the magnetic planetary gear can be written as

$$\omega_s = (1 + \rho)\,\omega_c - \rho\omega_r \tag{13.7}$$

$$\rho = \frac{P_r}{P_s} \tag{13.8}$$

where ω_s, ω_c, and ω_r are the angular speeds of the magnetic sun gear, carrier, and magnetic ring gear, respectively, and ρ is defined as the magnetic planetary gear ratio. This relationship is essentially the same as that of the mechanical planetary gear except that the mechanical gear teeth are replaced by the magnetic gear PM poles.

Consequently, this magnetic planetary gear can directly replace the mechanical planetary gear to generate the MG EVT system. That is, the magnetic sun gear is coupled with an electric machine, which is normally operating as a generator; the magnetic ring gear is coupled with another machine, which is normally operating as a motor; and the carrier is driven by an engine. The operating principle and control strategy are the same as the PG EVT system.

13.2.2 Magnetic Concentric Gears

The invention of magnetic coaxial gear is a milestone for the development of magnetic gears (Atallah, Calverley, and Howe, 2004). This magnetic coaxial gear can offer much higher torque density than the magnetic spur gear (Jian *et al.*, 2009). The key characteristic of this coaxial configuration is that all PMs are involved in the torque transmission at the same time, hence achieving a high torque density up to 150 kN m/m^3. This magnetic coaxial gear is depicted in Figure 13.7 in which the inner and outer parts

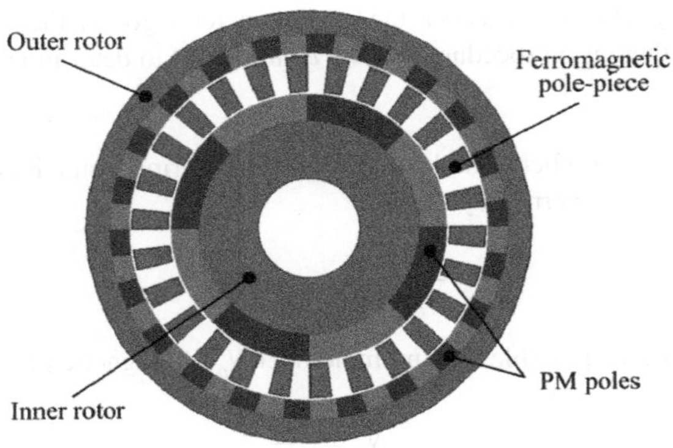

Figure 13.7 Magnetic coaxial gear

are mounted with PM pieces contributing to different numbers of pole-pairs, and the middle part is made of ferromagnetic segments, which serves to modulate the magnetic fluxes between the two PM parts. In general, this magnetic coaxial gear has two possible operations (Frank and Toliyat, 2009):

- The middle part is arranged as a stationary modulation ring so that the inner and outer parts are two PM rotors that rotate in opposite directions as shown in Figure 13.8a. This is actually the most common operation of the magnetic coaxial gear. The gear ratio G of inner-rotor to outer-rotor speeds is given by

$$G = \frac{P_o}{P_i} = \frac{N_s - P_i}{P_i} \tag{13.9}$$

  where P_i and P_o are the numbers of PM pole-pairs in the inner rotor and outer rotor, respectively, and N_s is the number of ferromagnetic segments in the modulation ring.
- The outer part is kept stationary so that the inner PM rotor and the modulation ring rotate in the same direction as shown in Figure 13.8b. The gear ratio G' of inner-rotor to modulation-ring speeds is given by

$$G' = \frac{N_s}{P_i} = \frac{P_i + P_o}{P_i} \tag{13.10}$$

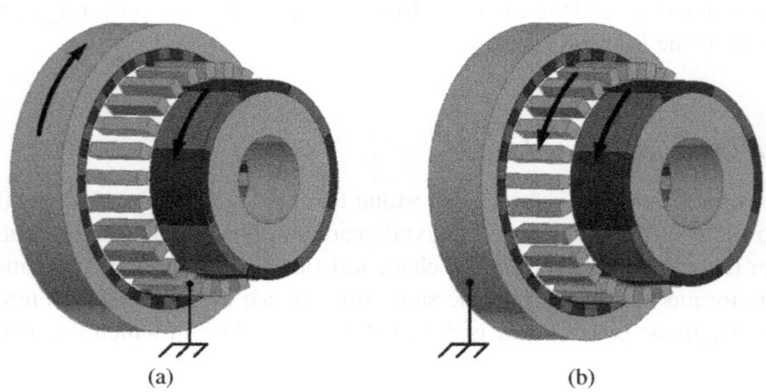

Figure 13.8 Magnetic coaxial gear operations: (a) modulation ring fixed and (b) outer rotor fixed

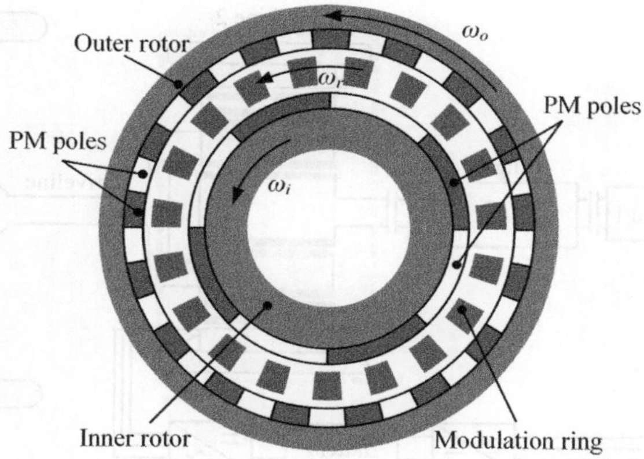

Figure 13.9 Magnetic concentric gear

Since all three parts of the magnetic coaxial gear are all rotatable, it can be considered as a multi-port magnetic gear. In order to distinguish it from the conventional magnetic coaxial gear, where the ferromagnetic ring is usually stationary, this multi-port magnetic gear is called the magnetic concentric gear. Figure 13.9 shows the schematic of the magnetic concentric gear where all parts are rotary and their speeds defined with the same direction are governed by

$$\omega_i + G\omega_o - (1 + G)\omega_r = 0 \tag{13.11}$$

where ω_i is the inner-rotor speed, ω_o is the outer-rotor speed, and ω_r is modulation-ring speed. When ignoring the transmission loss within the magnetic concentric gear, the torques are related by

$$T_i\omega_i + T_o\omega_o + T_r\omega_r = 0 \tag{13.12}$$

$$T_i + T_o + T_r = 0 \tag{13.13}$$

where T_i is the inner-rotor torque, T_o is the outer-rotor torque, and T_r is the modulation-ring torque.

13.3 Magnetic Planetary-Geared EVT System

By replacing the mechanical planetary gear with the magnetic planetary gear, the MPG EVT can readily be deduced. Since the magnetic planetary gear is cylindrical and coaxial, it can readily be integrated into the PM brushless machine. Figure 13.10 shows the MPG EVT system for the full hybrids in which the magnetic planetary gear is combined with the PM brushless machine to form a single machine (Zhu *et al.*, 2012), the machine 2. It can be observed that the engine is coupled with the rotor of the machine 1 and the magnetic sun gear of the machine 2. The machine 1 normally operates as a generator, while the magnetic planet gears of the machine 2 is attached to the driveline via the carrier. Meanwhile, the magnetic ring gear of the machine 2 is controlled by the stator winding. Thus, the engine power is split into two parts: one part is converted by the machine 1 into electricity, which in turn feeds the stator of the machine 2 via two back-to-back converters and the battery; another part is to directly feed the driveline via the magnetic planetary gear. Thus, the engine can always work at its optimal operation line under varying road load.

Figure 13.11 shows the structure of MPG PM brushless machine in which the magnetic planetary gear is artfully integrated into the PM brushless machine to achieve a compact structure. It consists of a stator wound with armature winding and various rotatable parts: the compound rotor, magnetic sun gear, carrier,

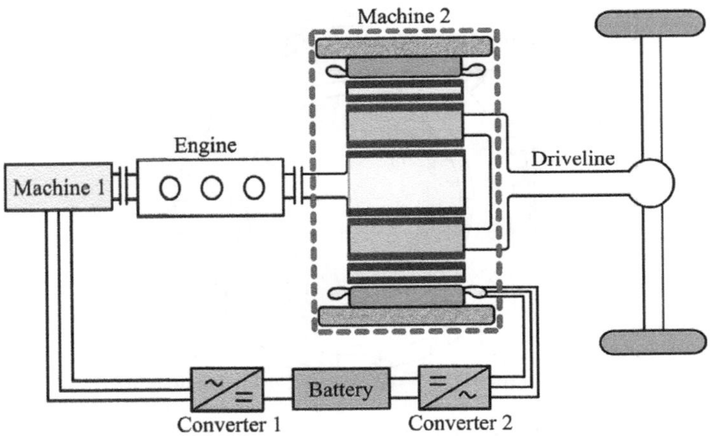

Figure 13.10 Magnetic planetary-geared EVT system

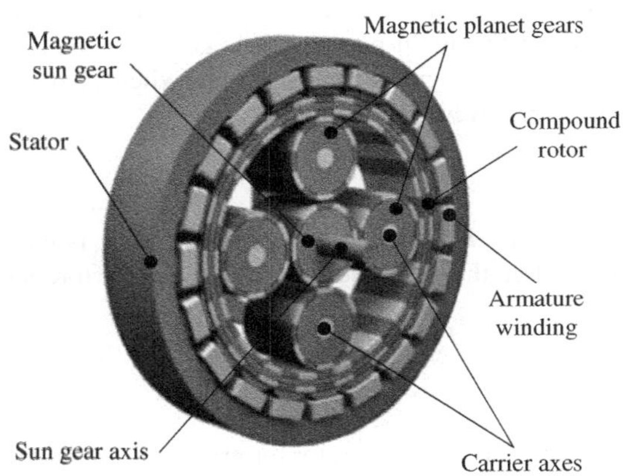

Figure 13.11 Magnetic planetary-geared PM brushless machine

and magnetic planet gears. The compound rotor, which has PM pieces on both the inner and outer surfaces, simultaneously serves as the rotor of the PM brushless machine and magnetic ring gear. The carrier is coupled with four magnetic planet gears so that the magnetic planet gears can revolute around the magnetic sun gear at the same time while the carrier serves as a mechanical output port. The stator develops torque on the compound rotor in the same way as the conventional PM brushless machine.

By using Eqs. (13.7) and (13.8), the carrier speed of the MPG PM machine can be expressed in terms of the magnetic sun gear speed and magnetic ring gear speed as given by

$$\omega_c = \frac{P_s}{P_s + P_r}\omega_s + \frac{P_r}{P_s + P_r}\omega_r \tag{13.14}$$

It can also be written in terms of the magnetic planetary gear ratio:

$$\omega_c = \frac{1}{1+\rho}\omega_s + \frac{\rho}{1+\rho}\omega_r \tag{13.15}$$

Since the magnetic sun gear speed can be electrically controlled by the machine 1 (generator) and the magnetic ring gear speed can also be electrically controlled by the machine 2 (motor), the carrier speed and hence the driveline speed can be flexibly controlled for different modes of operation:

- In the power-split mode, the torque and speed of the magnetic sun gear and the magnetic ring gear are in opposite directions, hence both the engine and PM brushless machine simultaneously deliver power to the carrier and hence the driveline. Their speed relationship is governed by Eq. (13.15). This mode is particularly useful for the vehicle to achieve the desired acceleration and hill-climbing capability.
- In the electric launch mode, the engine is shut down and the vehicle is solely driven by the PM brushless machine. Thus, substituting $\omega_s = 0$ in Eq. (13.15), the driveline speed (carrier speed) becomes directly proportional to the composite rotor speed (magnetic ring gear speed) as given by

$$\omega_c = \frac{\rho}{1+\rho}\omega_r \tag{13.16}$$

  which indicates that the driveline speed is slightly lower than the composite rotor speed and hence the driveline torque is somewhat amplified.
- In the cruising mode, the PM brushless machine is off and the vehicle is solely driven by the engine operating at a constant speed. Thus, substituting $\omega_r = 0$ in Eq. (13.15), the driveline speed (carrier speed) becomes directly proportional to the engine speed (magnetic sun gear speed) as given by

$$\omega_c = \frac{1}{1+\rho}\omega_s \tag{13.17}$$

  which indicates that the driveline speed is significantly lower than the engine speed and hence the driveline torque is greatly amplified.
- In the regenerative braking mode, the engine is shut down and the PM brushless machine operates as a generator to produce a braking torque, which in turn reduces the vehicle speed. Thus, the braking energy is converted into electricity by the PM brushless machine to charge the battery.

13.4 Magnetic Concentric-Geared EVT System

By replacing the mechanical planetary gear with the magnetic concentric gear, the MCG EVT system can be deduced. Serving as a power-split device, the magnetic concentric gear possesses three rotational mechanical ports. Thus, how to connect the engine to the input mechanical port as well as two electric machines to the output mechanical ports is essential for this EVT system. Figure 13.12 describes two possible configurations of this MCG EVT system: Type I and Type II (Jian and Chau, 2009a). In Type I, the PM inner rotor serves as an input port, which is coupled with the engine; the PM outer rotor is an output port, which is connected to the machine 2 and driveline shaft; and the modulation ring serves as another output port, which is connected to the machine 1. On the other hand, in Type II, the modulation ring is the input port, which is coupled with the engine; the PM outer rotor is an output port, which is connected to the machine 2 and driveline shaft; and the PM inner rotor is another output port, which is connected to the machine 1.

The system dynamical equations of the outer-rotor shaft for both Type I and Type II are the same as given by

$$T_o = T_{om} + T_{od} \tag{13.18}$$

$$T_{od} - T_d = J_d \frac{d\omega_o}{dt} \tag{13.19}$$

$$T_{om} + T_{m2} = J_{m2} \frac{d\omega_o}{dt} \tag{13.20}$$

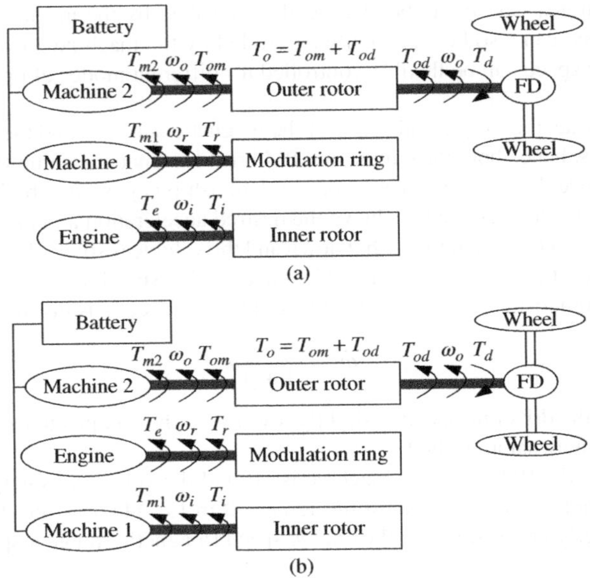

Figure 13.12 Possible configurations of magnetic concentric-geared EVT system: (a) Type I and (b) Type II

where T_{om} and T_{od} are the components of T_o on the machine 2 shaft and driveline shaft, respectively; T_{m2} and T_d are the output torque of machine 2 and load torque of driveline, respectively; and J_d and J_{m2} are the inertias of driveline shaft and machine 2 shaft, respectively.

The system dynamical equations of the modulation ring and inner-rotor shafts for Type I are given by

$$T_r + T_{m1} = J_{m1}\frac{d\omega_r}{dt} \tag{13.21}$$

$$T_e + T_i = J_e\frac{d\omega_i}{dt} \tag{13.22}$$

where T_{m1} and T_e are the output torques of the machine 1 and engine, respectively and J_{m1} and J_e are the inertias of machine 1 shaft and engine shaft, respectively. On the other hand, the system dynamical equations of the modulation ring and inner-rotor shafts for Type II are given by

$$T_e + T_r = J_e\frac{d\omega_r}{dt} \tag{13.23}$$

$$T_i + T_{m1} = J_{m1}\frac{d\omega_i}{dt} \tag{13.24}$$

For each type of configurations, the directions of the engine shaft and driveline shaft can be designed with the same direction or opposite directions. There are four possible cases:

- *Case IA*: Type I is adopted, and the directions of the engine shaft (ω_i) and driveline shaft (ω_o) are designed with the same direction.
- *Case IB*: Type I is adopted, and the directions of the engine shaft (ω_i) and driveline shaft (ω_o) are designed with the opposite directions.

- *Case IIA*: Type II is adopted, and the directions of the engine shaft (ω_r) and driveline shaft (ω_o) are designed with the same direction.
- *Case IIB*: Type II is adopted, and the directions of the engine shaft (ω_r) and driveline shaft (ω_o) are designed with the opposite directions.

As mentioned above, the goal of EVT is to satisfy the demands of the driveline and keep the engine working on the optimal operation line. For simplicity, an optimal operation point (OOP) is chosen for the engine and different working points of the driveline are investigated. Figure 13.13 marks the selected OOP (3000 rpm, 50 N m) for the engine and six typical working points for the driveline: W_1 (500 rpm, 320 N m), W_2 (700 rpm, 210 N m), W_3 (900 rpm, 190 N m), W_4 (1100 rpm, 130 N m), W_5 (1300 rpm, 100 N m), and W_6 (1500 rpm, 80 N m), while the corresponding powers are also plotted as a bar chart. It can be seen that the battery will assist the engine to supply power to the driveline for the working points W_1 and W_3, whereas the engine will charge the battery with the redundant power for the working points W_2, W_4, W_5, and W_6.

By using Eqs. (13.11)–(13.13) and (13.18)–(13.24), the operation points of the two machines can be calculated under the OOP of the engine and different working points of the driveline. Consequently, the results with different gear ratios of inner-rotor to outer-rotor speeds (G = 1, 2, and 3) are illustrated in Figures 13.14–13.16, respectively. It can be found that in both the Case IA and Case IB, the machine 1 works at low speeds and high torques. Thus, this machine 1 has to adopt the low-speed high-torque design which causes bulky machine size and high current rating, leading to be unattractive for HEV application. In contrast, in both the Case IIA and Case IIB, the machine 1 works at the low-torque and high-speed ranges. This machine 1 can adopt the high-speed low-torque design to reduce the weight and minimize the size, which are highly desirable for HEV application. Comparing the two cases of Type II, it can be observed that the Case IIA is more preferable, since the machine 2 in the Case IIA has lower operating torque than that in the Case IIB, and the operating speed of the machine 1 in the Case IIA is more suitable than that in the Case IIB. Referring to the gear ratio, the larger the value of G, the lower the operating torque and higher the operating speed of the machine 1 are resulted.

Consequently, the MCG EVT system adopts the Type II configuration and selects the Case IIA arrangement. That is, the engine is coupled with the modulation ring, the inner rotor is connected to the machine 1, and the outer rotor is connected to the machine 2 and driveline shaft, while the engine and driveline shafts are with the same direction. As shown in Figure 13.17, the armature windings of the two machines are electrically connected via two back-to-back converters and a battery.

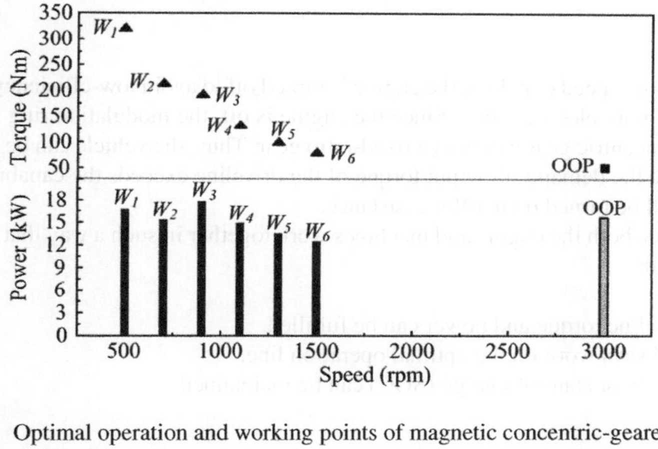

Figure 13.13 Optimal operation and working points of magnetic concentric-geared EVT system

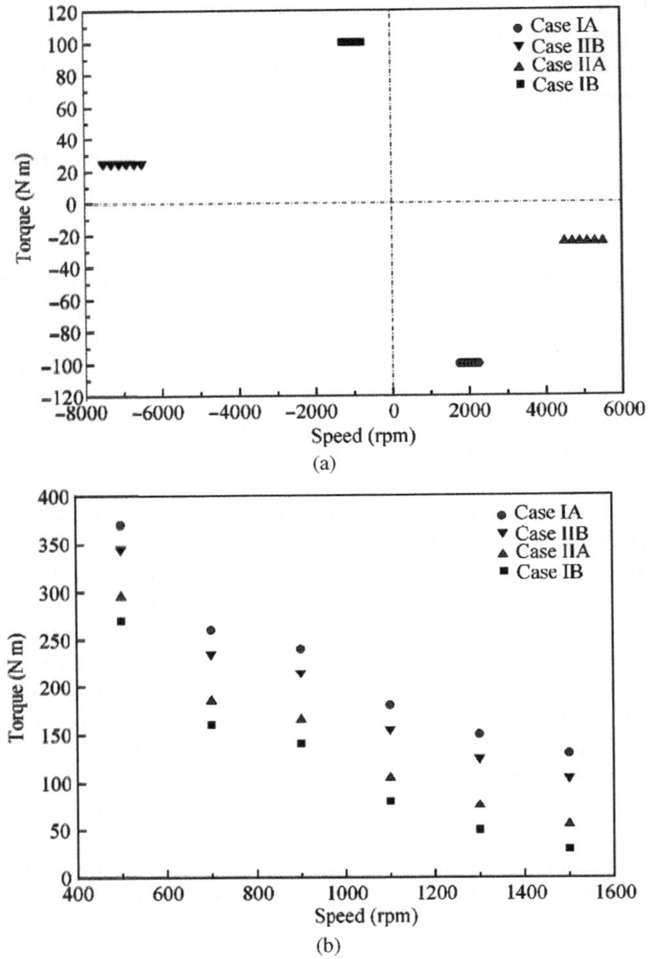

Figure 13.14 Operating points of magnetic concentric-geared EVT system under $G = 1$: (a) machine 1 and (b) machine 2

For launching or low-speed crawling, the engine is turned off to avoid low-efficiency operation, and the vehicle works at the pure electric mode. Since the engine is off, the modulation ring is always kept still, and the magnetic concentric gear works as a fixed-ratio gear. Thus, the vehicle can be solely propelled by the machine 1. Once the demand of output torque of the driveline exceeds the capability of the machine 1, the machine 2 will be turned on to offer assistance.

For EVT operation, both the engine and machines work together in such a way that the following three goals can be satisfied:

- The required driveline torque and power can be fulfilled.
- The engine can always work on the optimal operation line.
- The battery capacity or state-of-charge (SOC) can be maintained.

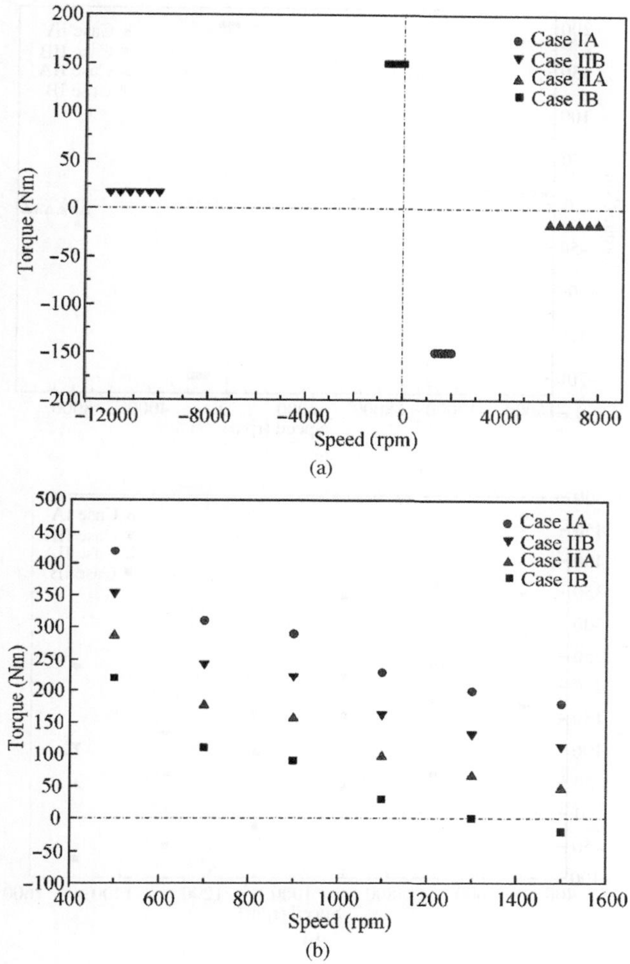

Figure 13.15 Operating points of magnetic concentric-geared EVT system under $G = 2$: (a) machine 1 (b) machine 2

Figure 13.18 shows the control block diagram of this MCG EVT system. Firstly, the torque and power demands of the driveline are calculated according to the accelerator command, brake command, and vehicle speed. Then, the throttle angle reference, α^* of the engine is deduced from the predefined optimal operation line. The adopted throttle angle is governed by the SOC of the battery. If the SOC level is lower than its rated value, the adopted throttle angle should be increased by $\Delta\alpha$, and the redundant power produced by the engine will be used to charge the battery; otherwise, if the SOC level is higher than its rated value, the adopted throttle angle should be decreased by $\Delta\alpha$, and the battery will discharge power to assist the engine. Next, the working points of the engine and the two machines can be determined. Finally, the resulting reference signals are fed into the machine controller to generate proper switching signals for the relevant converters.

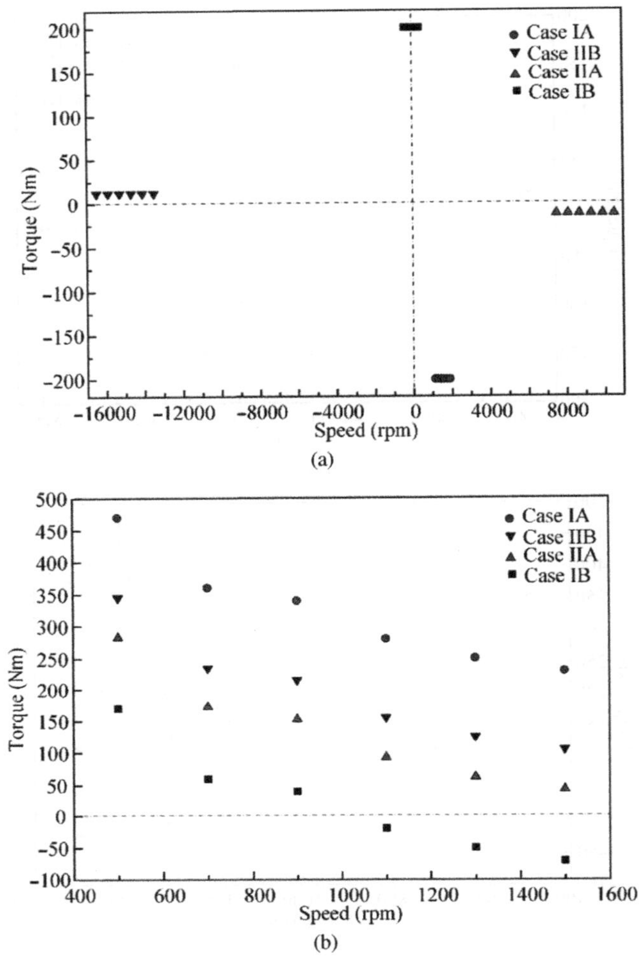

Figure 13.16 Operating points of magnetic concentric-geared EVT system under $G=3$: (a) machine 1 (b) machine 2

13.5 Design Criteria of MG EVT Systems

The MG EVT systems, including the MPG and MCG types, are very similar to the conventional PG EVT system in terms of system components. That is, they consist of a multi-port gear to perform power-split, two electric machines to provide motoring and generating, two back-to-back converters to provide controlled rectification and inversion for power flow regulation, and a battery pack for energy storage and buffering. The system design criteria of these MG EVT systems can be summarized as follows:

- The required driveline torque and power through the whole operation should be satisfied by the summation of torques and powers provided by the engine and motor.
- The magnetic gear should be able to split the engine power from no load to full load. In particular, the torque transmission capability of the magnetic gear must be able to withstand the maximum engine torque; otherwise, the phenomenon of magnetic torque slipping (Chau *et al.*, 2008) will be resulted.

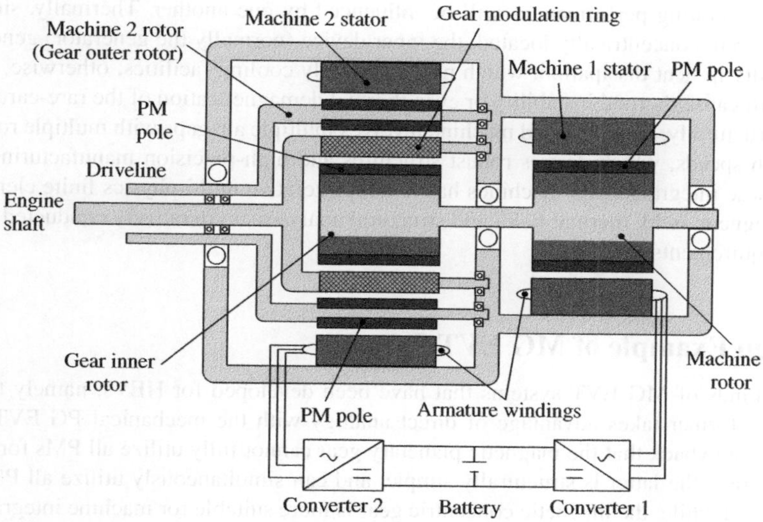

Figure 13.17 Structure of magnetic concentric-geared EVT system

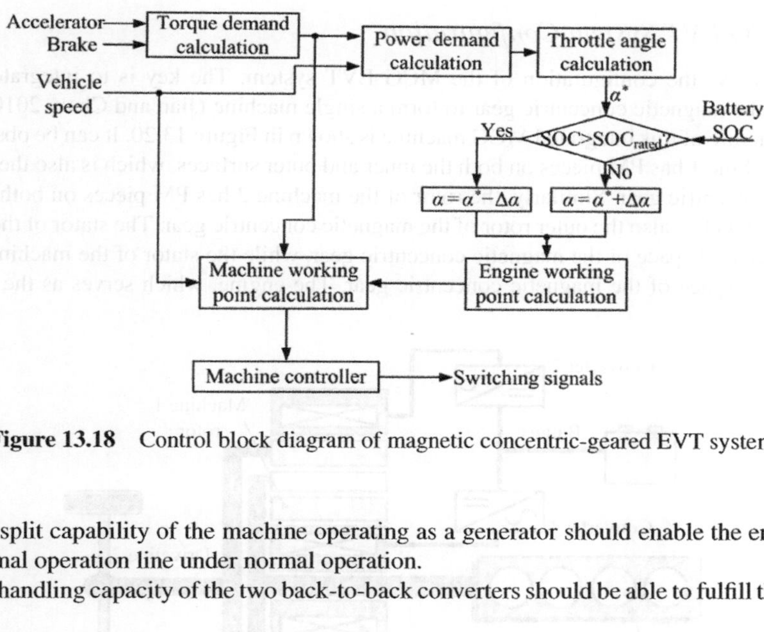

Figure 13.18 Control block diagram of magnetic concentric-geared EVT system

- The power-split capability of the machine operating as a generator should enable the engine working on the optimal operation line under normal operation.
- The power-handling capacity of the two back-to-back converters should be able to fulfill the power-split capability.
- The battery capacity or SOC should enable the system to absorb the excessive energy from the power-split process and also the recovered energy from regenerative braking.

While most of the system components of the MG EVT systems, including the magnetic gears, machines, and converters can be designed based on their individual design criteria, it is challenging to integrate the magnetic gear with the two machines to form a single MG machine. Magnetically, the magnetic fluxes of the three magnetic devices (magnetic gear, motor, and generator) should be decoupled as far as possible;

otherwise, their operating performances will be influenced by one another. Thermally, since the three magnetic devices are concentrically located, the inner device (normally the generator) generally suffers from the difficulty in heat dissipation, which needs specially cooling facilities; otherwise, the temperature increase will cause thermal instability or even thermal demagnetization of the rare-earth PMs (Chen *et al.*, 2012). Structurally, the integrated machine involves multiple air-gaps with multiple rotating bodies spinning at high speeds, which desires robust structures and high-precision manufacturing. Therefore, the design of these integrated MG machines has to adopt iterative multi-physics finite element analysis in which the magnetic field, thermal field, and structural analyses are iteratively conducted to satisfy the performance requirements.

13.6 Design Example of MG EVT Systems

There are two kinds of MG EVT systems that have been developed for HEVs, namely the MPG and MCG ones. The former takes advantage of direct analogy with the mechanical PG EVT system, but suffers from the drawback that the magnetic planetary gear cannot fully utilize all PMs for torque transmission. In contrast, the latter is structurally simple, and can simultaneously utilize all PMs for torque transmission. Meanwhile, the magnetic concentric gear is more suitable for machine integration since all its rotating parts are with the concentric axis. Thus, the integrated MCG EVT system is adopted for design exemplification.

13.6.1 MCG EVT System Configuration

Figure 13.19 shows the configuration of the MCG EVT system. The key is to integrate two electric machines into the magnetic concentric gear to form a single machine (Jian and Chau, 2010b; Jian *et al.*, 2011). The structure of this integrated MCG machine is shown in Figure 13.20. It can be observed that the rotor of the machine 1 has PM pieces on both the inner and outer surfaces, which is also the inner rotor of the magnetic concentric gear. Similarly, the rotor of the machine 2 has PM pieces on both the inner and outer surfaces, which is also the outer rotor of the magnetic concentric gear. The stator of the machine 1 is located in the central space of the magnetic concentric gear, while the stator of the machine 2 is situated at the peripheral space of the magnetic concentric gear. The engine, which serves as the system input,

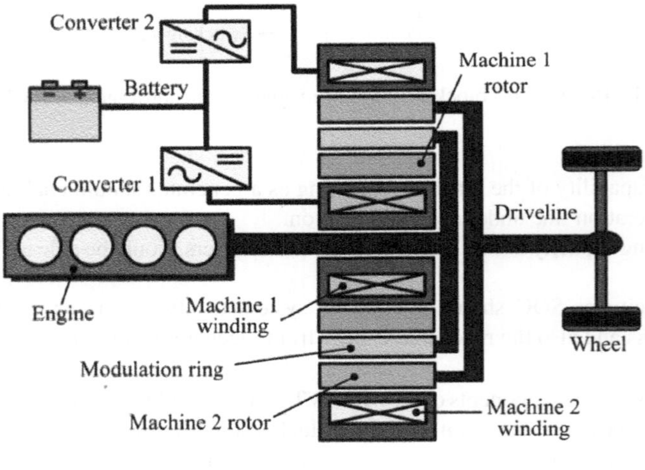

Figure 13.19 Integrated MCG EVT system

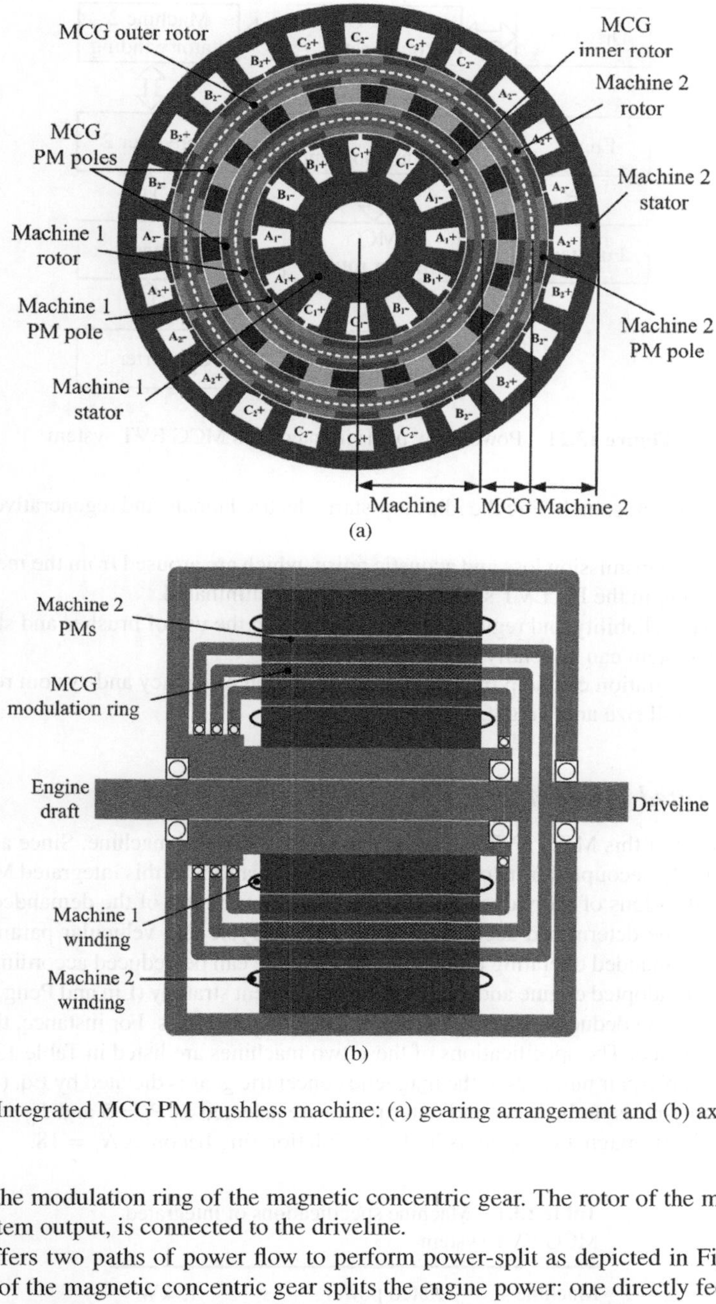

Figure 13.20 Integrated MCG PM brushless machine: (a) gearing arrangement and (b) axial arrangement

is coupled with the modulation ring of the magnetic concentric gear. The rotor of the machine 2, which serves as the system output, is connected to the driveline.

The system offers two paths of power flow to perform power-split as depicted in Figure 13.21. The modulation ring of the magnetic concentric gear splits the engine power: one directly feeds the driveline via the outer rotor, while another goes through the inner rotor, stator winding of the machine 1, converter 1, battery, converter 2, stator winding of the machine 2, and outer rotor. By properly adjusting the switching modes of the two power converters, this MCG EVT can offer the following advantageous features:

- Seamless match between the vehicle road load and engine optimal operating regime can be achieved, resulting in a considerable reduction of the fuel consumption and greenhouse gas emission.

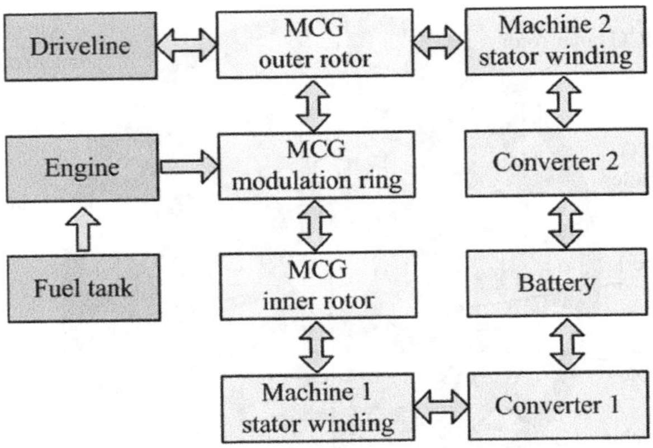

Figure 13.21 Power flow paths in integrated MCG EVT system

- All other hybrid features, including the idle stop-start, electric launch, and regenerative braking, can be retained.
- The drawbacks of transmission loss and acoustic noise, which are aroused from the mechanical contact mechanism existing in the PG EVT system, can be totally eliminated.
- The drawbacks of reliability and regular maintenance due to the use of brushes and slip-rings existing in the DR EVT system can be totally eliminated.
- High extent of integration can help improve the transmission efficiency and system reliability, as well as reduce the overall size and weight.

13.6.2 Integrated MCG Machine Design

The core component of this MCG EVT system is the integrated MCG machine. Since all three magnetic devices are essentially decoupled from one another, the specifications of this integrated MCG machine are based on the specifications of individual devices. First, the distribution of the demanded working points of the driveline can be determined according to the driving cycle and vehicular parameters. Then, the distribution of the demanded operating points of each machine can be deduced according to the fuel consumption map of the adopted engine and the energy management strategy (Liu and Peng, 2008). Different magnetic gear ratios can deduce different specifications of the machines. For instance, the magnetic gear ratio, $G = 2.6$ is selected. The specifications of these two machines are listed in Table 13.1.

The selection of pole-pair numbers of the magnetic concentric gear is dictated by Eq. (13.9). Hence, the numbers of PM pole-pairs in the inner and outer rotors are selected as $P_i = 5$ and $P_o = 13$, respectively, and the number of ferromagnetic segments in the modulation ring becomes $N_s = 18$.

Table 13.1 Machine specifications of integrated MCG EVT system

Machine 1	Rated power	15 kW
	Rated phase voltage	100 V
	Rated speed	2500 rpm
Machine 2	Rated power	30 kW
	Rated phase voltage	100 V
	Rated speed	950 rpm

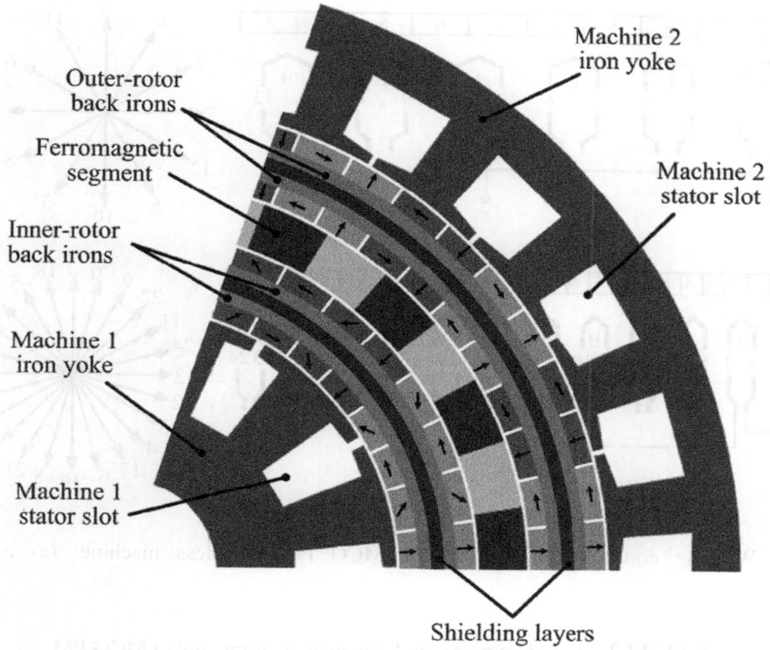

Figure 13.22 Halbach arrays in integrated MCG PM brushless machine

In this integrated machine, the electromagnetic fields of the magnetic concentric gear and the two machines should be decoupled as far as possible because it may adversely affect the system performances. Thus, the concept of Halbach arrays is employed (Jian and Chau, 2009b; 2010a). As shown in Figure 13.22, each PM pole in the inner rotor is constituted by four PM segments while each PM pole in the outer rotor is constituted by two PM segments. The corresponding directions of magnetization are denoted by the arrows. Since the Halbach arrays have the advantage of self-shielding capability, the back irons of both rotors can be very thin, hence saving the iron material and reducing the size. Meanwhile, two nonmagnetic shielding layers are separately laid within the two rotors in order to further enhance the decoupling effect. Moreover, the Halbach arrays can offer more sinusoidal and stronger air-gap flux density, hence reducing the cogging torque and improving the torque transmission capability of the magnetic concentric gear, respectively.

For the stators of the two machines, the fractional-slot concentrated windings are adopted (El-Refaie, 2010). Figure 13.23 depicts the armature winding connections of the two machines in which each armature coil encircles a single stator tooth. That is, the numbers of slots in the stators of the machine 1 and machine 2 are 12 and 24, respectively. The corresponding numbers of slots per pole per phase are 2/5 and 4/13, respectively. These fractional-slot concentrated windings offer some distinct advantages over the traditional distributed winding as follows:

- It can reduce the end-windings, hence saving the copper material and copper loss in the end region.
- It can shorten the machine axial length, hence improving the overall power density.
- It can ease the machine manufacture, hence saving the manufacturing cost.
- It can facilitate the use of higher slot fill factors.
- It can increase the *d*-axis inductance of the armature winding, hence widening the operating speed range (El-Refaie and Jahns, 2005).

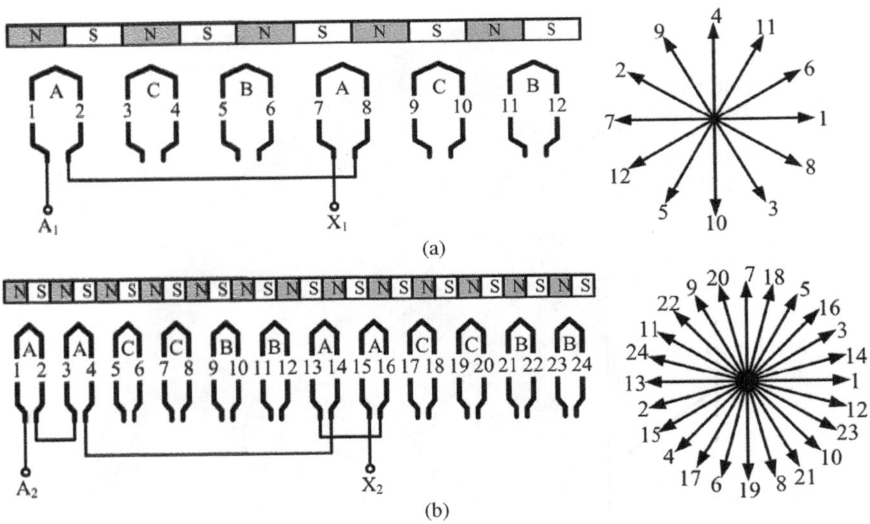

Figure 13.23 Winding connections in integrated MCG PM brushless machine: (a) machine 1 and (b) machine 2

Table 13.2 Key parameters and dimensions of integrated MCG PM brushless machine

Machine 1	No. of phases	3
	No. of stator slots	12
	No. of pole-pairs	5
	Inside radius of stator	30 mm
	Outside radius of stator	86.5 mm
Machine 2	No. of phases	3
	No. of stator slots	24
	No. of pole-pairs	13
	Inside radius of stator	153.5 mm
	Outside radius of stator	195 mm
Modulation ring	No. of ferromagnetic segments	18
	Thickness of modulation ring	15 mm
	Length of air-gaps	1 mm
	Thickness of PMs	6 mm
	Thickness of shielding layers	4 mm
	Thickness of back iron of rotors	4 mm
	Effective axial length	200 mm

Consequently, the whole MCG PM brushless machine, including the magnetic concentric gear and two machines, is designed. The corresponding key dimensions and parameters are listed in Table 13.2.

13.6.3 Integrated MCG Machine Analysis

The electromagnetic characteristics of this integrated MCG PM brushless machine can readily be assessed by using finite element analysis. First, the corresponding electromagnetic field distributions under no-load and full-load conditions are depicted in Figure 13.24. It can be seen that there is no noticeable flux line

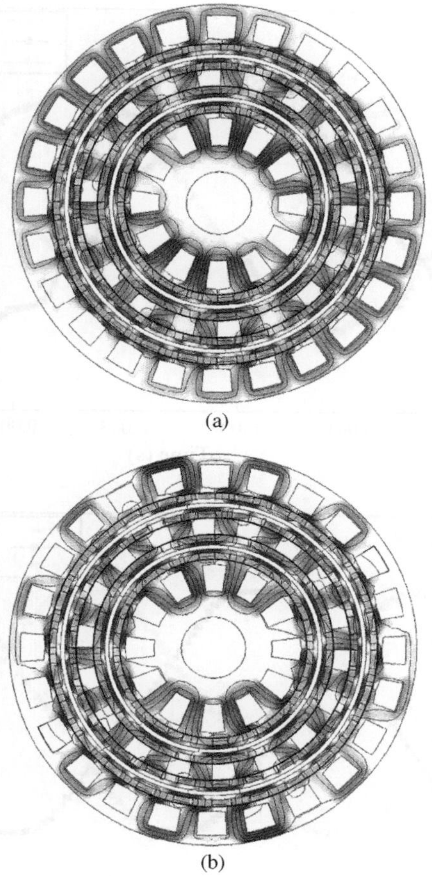

Figure 13.24 Electromagnetic field distributions in integrated MCG PM brushless machine: (a) no load and (b) full load

passing through the shielding layers, which illustrates that the Halbach arrays and shielding layers can offer the desired decoupling effect between the magnetic concentric gear and two machines. This can significantly improve the system controllability. Moreover, because of this decoupling nature, the magnetic concentric gear and two machines can be individually analyzed in spite of their integration into a single machine.

On the basis of the electromagnetic field distributions, the no-load electromotive force (EMF) can readily be deduced. Figure 13.25 shows the no-load EMF waveforms of the two machines when their rotors operate at their rated speeds. Moreover, the corresponding harmonic spectra are shown in Figure 13.26. It can be found that the harmonic contaminations in the two machines are actually quite small, which confirms that the Halbach-array PMs can improve the sinusoidal distribution of magnetic fields.

The torque transmission capability of the magnetic concentric gear can be calculated by rotating the inner rotor step by step while keeping the outer rotor and modulation ring standstill. Figure 13.27 shows the resulting torque-angle curves. It can be found that they vary sinusoidally in which the pull-out torques of the inner rotor, outer rotor, and modulation ring are 227, 590, and 816 N m, respectively. Thus, it can be deduced that the ratio of pull-out torques of the two rotors (T_o/T_i) is 2.6, which well agrees with the gear ratio of 2.6. In addition, the above-mentioned three pull-out torques agree well with Eq. (13.13) as the curves of the two rotors are out of phase with the curve of the modulation ring.

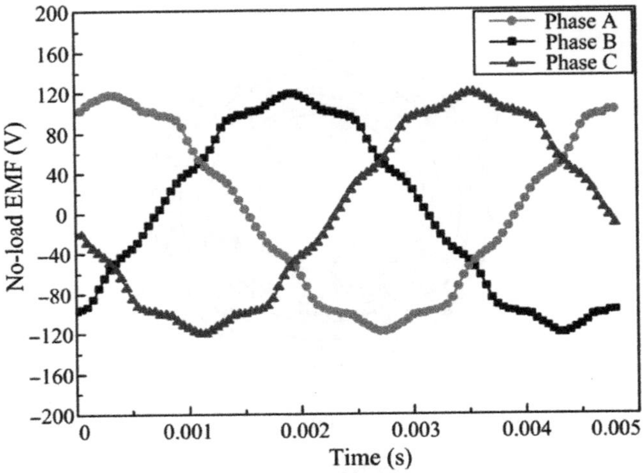

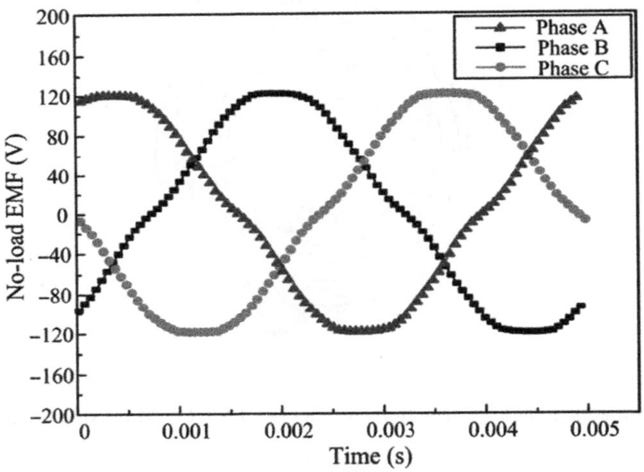

Figure 13.25 No-load EMF waveforms at rated speeds in integrated MCG PM brushless machine: (a) machine 1 and (b) machine 2

The cogging torques of the two machines are also calculated by using the finite element method. By keeping the two rotors standstill and rotating the modulation ring, the cogging torques aroused by the iron segments of the modulation ring are obtained as shown in Figure 13.28a. Then, by rotating the two rotors and the modulation ring simultaneously, the cogging torques aroused by the teeth of the stators are obtained as shown in Figure 13.28b. It can be observed that the cogging torques are insignificant as compared with the output torques.

Finally, some key parameters of the two machines are calculated as listed in Table 13.3. Since all PM poles are surface-mounted, the d-axis inductances L_d and q-axis inductances L_q of the two machines are almost the same. Figure 13.29 shows both the current and voltage constraints for operation of the two machines. The radius of the current limit circle R_I equals the rated current I_{rated}, while the radius of the voltage limit circle R_U is proportional to the voltage limit U_{lim} and inversely proportional to the rotational speed ω_r. The stator current i_s has to be located in the overlapping area of both the voltage and current limit circles. When the speed is lower than the base speed, the stator current is aligned along the q-axis to

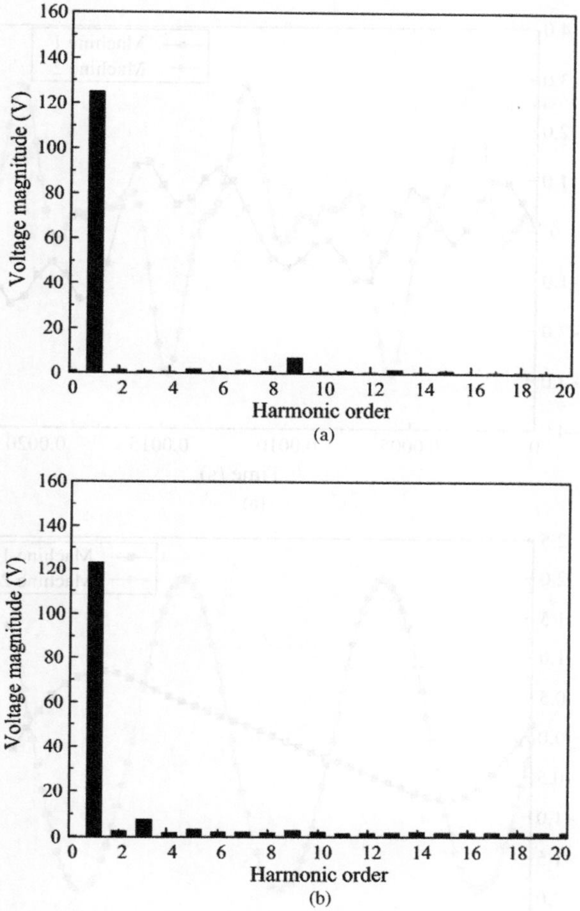

Figure 13.26 No-load EMF harmonic spectra at rated speeds in integrated MCG PM brushless machine: (a) machine 1 and (b) machine 2

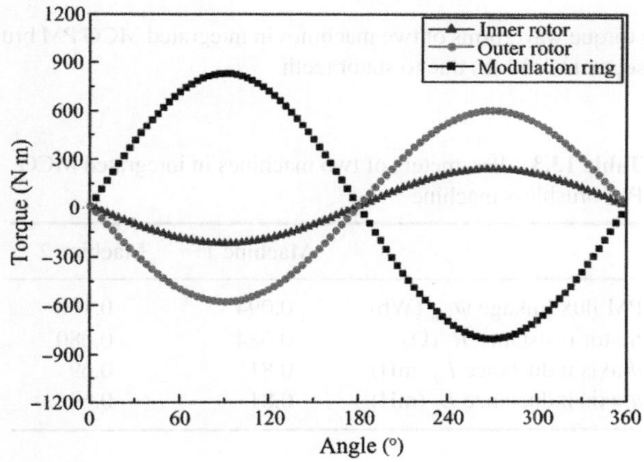

Figure 13.27 Torque-angle curves of magnetic concentric gear in integrated MCG PM brushless machine

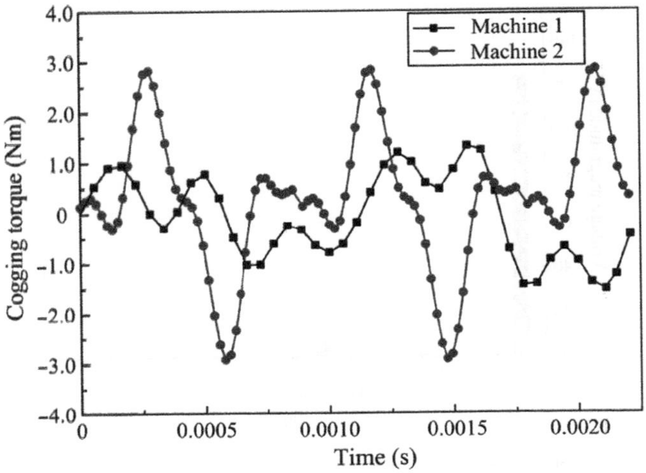

(a)

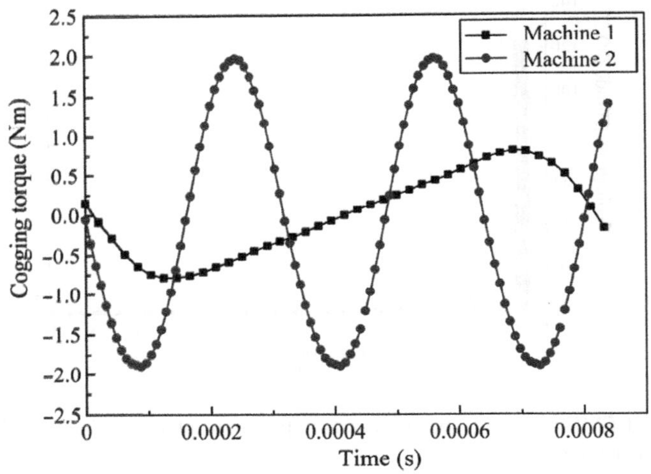

(b)

Figure 13.28 Cogging torque waveforms of two machines in integrated MCG PM brushless machine: (a) due to modulation ring iron segments and (b) due to stator teeth

Table 13.3 Parameters of two machines in integrated MCG PM brushless machine

	Machine 1	Machine 2
PM flux linkage ψ_{pm} (Wb)	0.094	0.108
Stator resistance R_s (Ω)	0.084	0.080
d-axis inductance L_d (mH)	0.81	0.69
q-axis inductance L_q (mH)	0.81	0.69

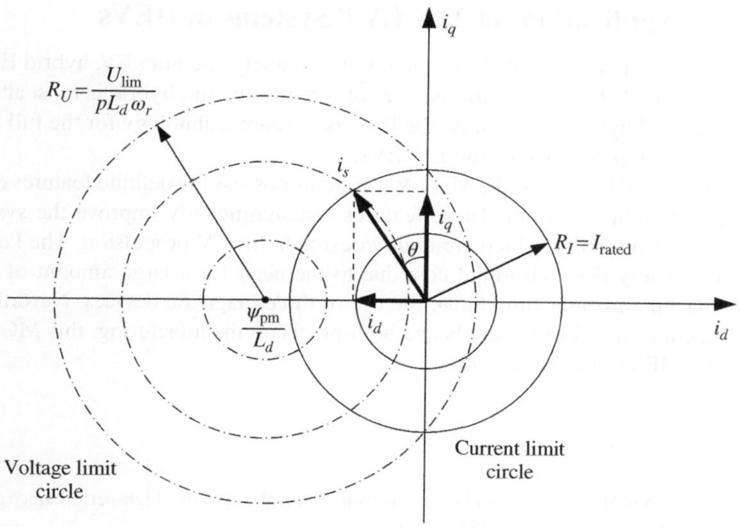

Figure 13.29 Current and voltage constraints for operation in integrated MCG PM brushless machine

offer the maximum torque per ampere, namely the constant-torque operation. Once the speed exceeds the base speed, the stator current deviates from the q-axis, and the corresponding i_d weakens the stator flux linkage to ensure that the back EMF is lower than the voltage limit, namely, the flux-weakening operation (Sun *et al.*, 2010). Figure 13.30 shows the resulting torque–speed capabilities of the two machines.

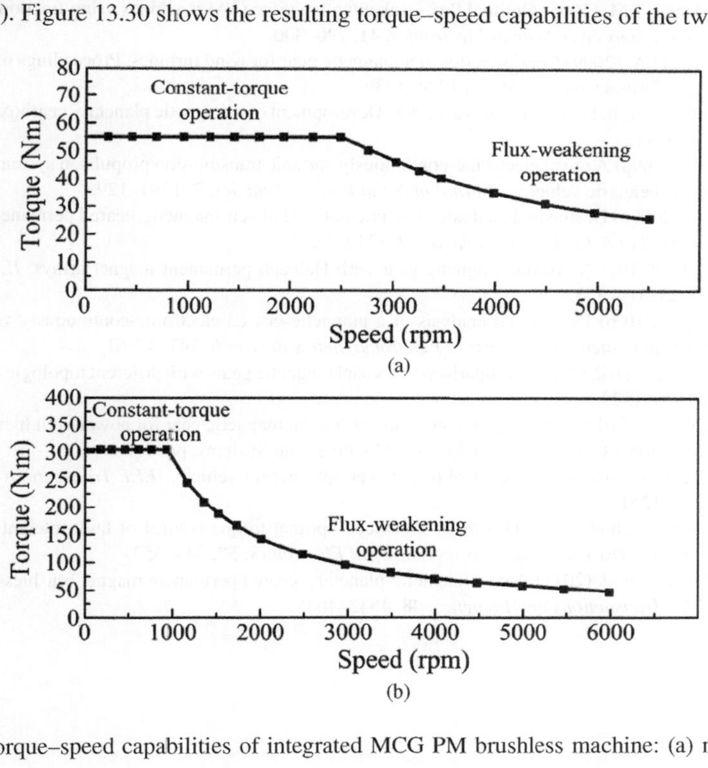

Figure 13.30 Torque–speed capabilities of integrated MCG PM brushless machine: (a) machine 1 and (b) machine 2

13.7 Potential Applications of MG EVT Systems in HEVs

In the foreseeable future, it is anticipated that the EV trio, namely, the pure EV, hybrid EV, and fuel-cell EV, will coexist in the market. Among various types of hybrids, the full hybrid is most attractive because it can offer all the desired hybrid features. As the EVT is the core technology for the full hybrid, the MG EVT systems exhibit high potentials for future HEVs.

Compared with other EVT systems, the MG EVT systems possess the definite features of brushless and pseudo-gearless power-split operation. These features can significantly improve the system efficiency, power density, and controllability, which are highly desirable for EV propulsion. The key challenges of this technology are mainly the high initial cost due to the need for a large amount of rare-earth PMs and the complex machine manufacture due to the use of three magnetic devices. Nevertheless, with the continual development of new PM materials and high-precision manufacturing, this MG EVT system is highly promising for HEVs in the near future.

References

Atallah, K., Calverley, S.D. and Howe, D. (2004) Design, analysis and realization of a high-performance magnetic gear. *IEE Proceedings – Electric Power Applications*, **151**, 135–143.

Chau, K.T., Zhang, D., Jiang, J.Z. and Jian, L. (2008) Transient analysis of coaxial magnetic gears using finite element comodeling. *Journal of Applied Physics*, **103**, 07F101, 1–07F101, 3.

Chen, M., Chau, K.T., Li, W. and Liu, C. (2012) Development of non-rare-earth magnetic gears for electric vehicles. *Journal of Asian Electric Vehicles*, **10**, 1607–1613.

El-Refaie, A.M. (2010) Fractional-slot concentrated-windings synchronous permanent magnet machines: opportunities and challenges. *IEEE Transactions on Industrial Electronics*, **57**, 107–121.

El-Refaie, A.M. and Jahns, T.M. (2005) Optimal flux weakening in surface PM machines using fractional-slot concentrated windings. *IEEE Transactions on Industrial Electronics*, **41**, 790–800.

Frank, N.W. and Toliyat, H.A. (2009) Gearing ratios of a magnetic gear for wind turbines. Proceedings of IEEE International Electric Machines and Drives Conference, pp. 1224–1230.

Huang, C.C., Tsai, M.C., Dorrell, D.G. and Lin, B.J. (2008) Development of a magnetic planetary gearbox. *IEEE Transactions on Magnetics*, **44**, 403–412.

Jian, L. and Chau, K.T. (2009a) A novel electronic-continuously variable transmission propulsion system using coaxial magnetic gearing for hybrid electric vehicles. *Journal of Asian Electric Vehicles*, **7**, 1291–1296.

Jian, L. and Chau, K.T. (2009b) Design and analysis of an integrated Halbach-magnetic-geared permanent-magnet motor for electric vehicles. *Journal of Asian Electric Vehicles*, **7**, 1213–1219.

Jian, L. and Chau, K.T. (2010a) A coaxial magnetic gear with Halbach permanent magnet arrays. *IEEE Transactions on Energy Conversion*, **25**, 319–328.

Jian, L. and Chau, K.T. (2010b) Design and analysis of a magnetic-geared electronic-continuously variable transmission system using finite element method. *Progress in Electromagnetics Research*, **107**, 47–61.

Jian, L., Chau, K.T., Gong, Y. *et al.* (2009) Comparison of coaxial magnetic gears with different topologies. *IEEE Transactions on Magnetics*, **45**, 4526–4529.

Jian, L., Xu, G., Wu, Y. et al. (2011) A novel power-train using coaxial magnetic gear for power-split hybrid electric vehicles. Proceedings of International Conference on Electrical Machines and Systems, pp. 1–6.

Liu, J. and Peng, H. (2008) Modeling and control of a power-split hybrid vehicle. *IEEE Transactions on Control Systems Technology*, **57**, 1242–1251.

Sun, Z., Wang, J., Jewell, G. and Howe, D. (2010) Enhanced optimal torque control of fault-tolerant PM machine under flux-weakening operation. *IEEE Transactions on Industrial Electronics*, **57**, 344–353.

Zhu, X., Chen, L., Quan, L. *et al.* (2012) A new magnetic-planetary-geared permanent magnet brushless machine for hybrid electric vehicle. *IEEE Transactions on Magnetics*, **48**, 4642–4645.

Index

120° conduction, 76, 84–5, 87, 92–3, 217, 259, 262
180° conduction, 84, 86–7, 92–3, 263

AC
link, 50
machine, 20, 34, 108, 117, 126
motor, 38, 124, 255
acceleration, 5, 67, 294, 297, 300, 313, 322–4, 375
accelerator, 298–9, 351, 379
accidental demagnetization, 165–6, 181, 193, 255
acoustic noise, 6, 27, 47, 108, 110, 117, 131, 135–7, 144, 195, 384
active regulation, 14, 16
actively clamped, 121
adaptive control, 57, 239
adhesive, 73, 214
advance angle control (AAC), 124–5, 129, 133, 139–41, 143
AF-DSDC
machine, 276–83, 285
motor, 272, 276
AF-FSDC
machine, 283–8
motor, 272, 283
aggregator, 14
air-gap flux density, 33, 41, 74–5, 94, 137, 153, 163, 165, 169, 171, 173, 178–81, 202–3, 217, 235–6, 241, 243–4, 246–7, 260, 278, 280, 337–8, 354, 385
air-gap length, 12, 33, 35, 62, 64, 73, 95, 97, 101, 133–4, 137–8, 142, 158–9, 175, 178, 183, 214, 240, 243, 247, 274, 276, 282, 306, 311, 336, 361, 370
air-gap permeance, 206–7, 234, 265–8
alternating current (*see* AC)
angle position control (APC), 157, 260
antiphase, 303
armature circuit, 20–23, 34
armature conductor, 34–5
armature reaction, 72, 93, 161, 165, 181, 217
armature resistance, 24, 29, 78, 237–8
armature slot, 34–5
armature voltage control, 29–30, 37
armature winding, 20, 34, 41, 72–3, 75–7, 84, 94, 96, 101, 148, 157, 159, 161–2, 165–7, 177–9, 181–3, 215, 217, 221, 228–36, 240, 242–3, 245–6, 251, 254–62, 264–9, 274, 276–7, 283–5, 301, 303–4, 307, 345, 353, 356–7, 362–3, 373–4, 377, 381, 385
artificial intelligent control, 93, 212
asymmetric bridge converter, 118–19
automatic transmission, 299
auxiliary resonant commutated pole inverter, 50
auxiliary resonant snubber inverter, 50–51
auxiliary switch, 27, 50, 120–122
axial length, 22, 41, 94, 112, 141, 209, 214, 217, 222, 243, 247, 269, 313, 361, 385–6
axial-flux DSDC (*see* AF-DSDC)
axial-flux FSDC (*see* AF-FSDC)
axial-flux machine, 254, 269, 272, 277–8, 283, 357
axial-flux morphology, 8, 253, 269, 272, 282, 356, 359
axially laminated, 256, 289
axle-to-driveline, 320

back iron, 149, 385–6
back-to-back converter, 5, 359, 373, 377, 380–381
backward magnetization, 172–4
base speed, 29–30, 52, 64, 88, 90–93, 124–5, 157, 283, 300, 303, 312, 333, 337, 388, 391
battery(ies), 4–5, 9–10, 13–16, 37, 61, 227
capacity, 4, 14, 294, 359, 378, 381
charging, 4–6, 11–12, 14, 165, 282, 295, 298, 300, 302, 304, 310, 312, 333, 351–2, 360, 364
cost, 4
lead-acid, 9, 37, 298
lithium-ion, 9, 67, 103, 144, 298
nickel-cadmium, 9, 37
nickel-metal hydride, 9
self-discharge, 298
swapping, 4, 6
zinc/air, 9
battery electric vehicle (BEV), 4
belt, 296
B–H loop, 167

Electric Vehicle Machines and Drives: Design, Analysis and Application, First Edition. K. T. Chau.

bidirectional power, 27–8, 211, 293, 301, 346
biomass, 2
bipolar flux, 158, 260–261, 285, 289
bipolar-magnetized, 203
BLAC mode, 83, 94, 153, 159, 161, 218, 221, 262–3, 285
BLDC mode, 95, 150, 159, 161, 176, 218, 221, 262–3, 277, 285
blended mode, 5, 295
boosting, 51–2, 278, 287, 304, 310
brake, 2, 299–300, 331–3, 379
braking energy, 31, 82–3, 293, 300, 322, 331, 351, 359, 375
braking torque, 299–300, 322, 331, 375
brushless AC (*see* BLAC mode)
brushless DC (*see* BLDC mode)

capacitive power transfer, 12
carbon brush, 7–9, 19–22, 34–5, 266, 296
carbon dioxide, 3
carbon monoxide, 3
carrier-to-sun, 320, 328
centrifugal force, 73, 110, 192, 202–3, 214, 341, 354
chaotic PWM, 47
charger, 4–5, 11–12, 16, 37
charging
  coordination, 14
  zone, 13
clutch, 299, 320, 323, 326–9, 331–3
coal, 2–3
coaxial shafting, 316–17
coenergy, 114, 152, 266
cogging torque, 101, 153, 158, 166, 218, 221, 224, 233, 240, 243, 246–8, 274–6, 297, 301, 307, 371, 385, 388
coil pitch, 242–3
coil span, 34, 76, 101, 166, 182, 235, 301
commutating frequency, 151, 259
commutator, 7, 19–22, 34–5, 75
  diameter, 34–5
  segment, 22, 34–5
commutatorless, 7, 39
complex permeance function, 206
compound-split, 315–16, 326–7, 331–3, 341, 343
compressed hydrogen gas, 6
concave stator pole, 158
concentrated winding, 75, 110, 112, 149, 157, 160, 241–3, 246, 262, 268, 276, 283, 304, 307, 385
constant-power operation, 35, 37, 52, 65, 72, 87–8, 91–3, 98, 103, 125, 139, 157, 165, 178, 239, 245–6, 248, 258, 260, 281, 285, 287, 297–8, 300, 302, 313, 333, 338, 357
constant-power region, 31, 52, 65
constant-torque operation, 35, 37, 64, 91, 98, 103, 124–5, 139, 157, 247–8, 391
constant-torque region, 30, 52, 65
controllability, 19, 28, 122–3, 149, 192–3, 255, 287, 289, 301, 387, 392
cooling, 33, 72–3, 136, 149, 192, 215, 228, 239–40, 297, 356–7, 359, 382
coordinate transformation, 54, 60, 84, 91, 349
copper loss, 45, 61, 72, 75, 77, 98, 102, 128, 178–9, 212, 241, 243, 246, 258, 269, 272, 385
copper material, 34, 75, 101, 243, 246, 259, 268–9, 301, 307, 385
core length, 33, 35, 62, 64, 97, 101, 132, 138, 142
core loss, 12, 45, 112, 167, 266, 273, 275
cost-effectiveness, 126, 224, 239, 253, 272, 283, 288, 294
cranking, 8, 165, 291, 295–6, 298, 300–302, 304, 307, 310, 312–13, 333, 341, 351–2, 357, 360, 363
crankshaft, 296–7, 313
creeping, 6
critical speed, 52–3, 125
crude oil, 2
cruising, 6, 87, 91, 125, 165, 178, 246, 258, 260, 322, 327, 375
current chopping control (CCC), 124, 129, 139, 141, 143, 157, 161, 183, 260
current commutation, 103, 119, 123
current constraint, 388, 391
current density, 34, 64, 95–6, 102, 135–6, 138, 142, 215, 233, 240, 278
current model, 91
current phasor, 186, 188–9
current ripple, 50, 83, 124
cycle life, 9, 14

d-axis, 73, 78–80, 84, 87–90, 93, 165, 212, 245, 256–7, 335, 349, 385, 388, 390
DC
  field control, 247–8, 282, 287
  field excitation, 162, 179, 258, 278, 281–3, 285–8
  field torque, 259, 263, 309
  link, 16, 50, 108, 120–121, 180, 323, 343–4
  motor, 19–20, 24–5, 27–39, 55
DC machine, 19–24, 33–5, 72, 74, 117, 296
  cumulative compound, 20–21, 24
  PM, 7, 21, 23, 35, 74, 81
  separately excited, 7, 20, 23, 28, 30–32, 35, 37–8, 55
  series, 7, 21, 23–4, 28–32, 35, 37, 117
  shunt, 7, 20, 23, 35
  wound-field, 21, 23
DC–DC converter, 19, 24, 27–9, 149, 263, 277, 285, 360
  multi-resonant, 27
  quasi-resonant, 27
  resonant, 27
  soft-switching, 27–8
  zero-transition, 27
decoupling condition, 57
decoupling effect, 385, 387
delta PWM, 47
detent torque, 274
DFVR
  machine, 8, 254, 266, 272
  motor, 266, 289
diesel, 1–2
direct current (*see* DC)
direct torque control (DTC), 51, 57–61, 72, 87, 159, 212, 239, 257
direct-axis (*see* *d*-axis)
distributed winding, 62, 75, 234, 241, 262, 266, 348, 385
distribution factor, 235
double-layer winding, 34, 62, 161, 301, 303
double-layer-stator topology, 166
double-rotor (*see* DR)
double-stator double-rotor, 364
double-stator PM brushless motor, 96
double-stator PM synchronous machine, 297, 301–3, 305–7
double-stator single-rotor, 269, 271, 276, 283, 297
doubly-fed motor, 255
doubly-fed vernier reluctance (*see* DFVR)
doubly-salient DC (*see* DSDC)
doubly-salient PM (*see* DSPM)
downsized engine, 4, 294

DR
EVT, 8, 343–4, 347–51, 353–60, 364, 368
machine, 8, 341, 343–8, 351, 353–63, 367
driveline torque, 321–2, 326, 329–31, 333, 351, 359, 375, 378, 380
DSDC
machine, 253, 257–62, 269, 272–83, 285, 298, 345, 357
motor, 257–8, 263, 272, 276–7, 289
DSPM
machine, 147–60, 162–3, 165–7, 169–71, 173–85, 187, 189–94, 257–9, 272, 281–3, 289, 297–8, 301, 303, 307–12
mode, 167, 169, 182–5, 191–2
motor, 149, 154–7, 159, 169, 177, 179–83, 189, 192–4, 258
dual control loops, 239
dual inverters, 228
dual-grid framework, 14–15
dual-motor propulsion, 239
dv/dt, 27, 50

eddy-current loss, 69, 158, 179, 215, 233
efficiency map, 100, 103, 131
efficiency optimization, 180, 212, 258, 345, 357, 360
efficiency-optimizing control, 165, 179–80, 212, 239, 257
electric launch, 4–6, 8, 294, 322, 324, 331, 333, 341, 351, 357, 360, 363, 375, 384
electric loading, 33, 62–3, 95, 112, 153, 233
electric propulsion, 8, 19–20, 37, 39–40, 51, 61, 70, 212, 291–3
electric range, 295
electric variable transmission (*see* EVT)
electric vehicles, 1
battery, 4
fuel-cell, 1, 3, 5–6, 8–10
hybrid, 1, 3–5, 7–8, 10, 14, 165, 291–5, 299, 301, 312–13, 319, 323, 327, 333, 335, 339, 341, 357, 359–60, 364, 367–77, 382, 392
plug-in hybrid, 1, 3, 5, 14, 293, 295
pure, 1, 3–5, 8–9, 11–14, 293, 295
range-extended, 1, 3, 5, 14, 293, 295
electrical link, 291
electrical port, 345–7, 357, 369
electricity generation, 4, 8, 298, 300, 310, 333, 341
electrocatalyst, 10
electromagnetic coupling, 362
electromagnetic field, 13, 33, 64–5, 96–7, 102, 138, 141–2, 205, 217–18, 278, 336, 338, 345, 350, 355, 359–60, 362, 385–7
electromagnetic interference (EMI), 12–13, 27, 50, 120, 124
electromechanical device, 296, 346
electromechanical infinitely variable transmission, 332
electromechanical port, 332
electromotive force (*see* EMF)
electronic differential, 68, 195, 227
electronic-continuously variable transmission (*see* EVT)
EMF
sinusoidal back EMF, 301
trapezoidal back EMF, 75, 183–4, 186
emission, 3–5, 298, 313, 383
end-winding, 75, 94, 101, 166, 214, 241, 243, 259, 268–9, 301, 307, 385
energy arbitrage, 14
energy buffer, 343–4
energy density, 6, 69–70
energy encryption, 13
energy resource, 2
energy service provider, 14
energy source, 1, 6, 9–10, 293
energy-conversion loop, 262
engine, 1, 4–5, 7–8, 10, 291–300, 310, 312–13, 315, 319–35, 339–41, 343–5, 348, 350–351, 357, 359–60, 363–5, 367, 369, 371, 373, 375–84
EVT, 333, 341, 350, 360, 378
extended mode, 5

far-field, 11
fault tolerance, 40, 166–7, 181
fault-tolerant control, 91, 93, 128, 183
faulty phase, 183, 191
ferromagnetic segment, 201, 206–7, 214–15, 217, 219, 222, 229, 372, 384, 386
Fiat Panda Elettra, 37
field weakening (*see* flux weakening)
field-oriented control (FOC), 51, 53–7, 60, 65, 72, 87–8, 91, 239, 245, 257
finite element analysis, 61, 64, 93, 96, 138, 141, 182, 192, 217, 221, 240–241, 243, 247, 272, 274–5, 278, 282, 285, 310, 336–8, 359–60, 382, 386
first-quadrant DC chopper, 24–5
five-leg inverter, 228, 238
flux barrier, 158
flux concentration, 74
flux focusing, 74, 160, 193, 203, 260, 335
flux leakage, 33, 74, 153, 158, 162, 178, 214, 304, 307, 310, 335, 354
flux modulation, 242, 247, 251
flux strengthening, 59, 149, 162, 164, 165, 178, 193, 260, 277, 304, 309–12, 357, 360
flux tuning, 260, 357
flux weakening, 29, 55, 87, 149, 214, 245, 260, 303, 357, 391
flux-controllable, 7, 148, 165, 242, 245–50
flux-mnemonic PM (*see* FMPM)
flux-modulation pole (FMP), 242–3, 245–7
flux-reversal DC (*see* FRDC machine)
flux-reversal PM (*see* FRPM)
flux-switching DC (*see* FSDC)
flux-switching PM (*see* FSPM)
flywheel, 9–10, 294–6, 313
FMPM
machine, 147–9, 165–6
motor, 192–4
force field, 61
Ford Escape Hybrid, 340–341
Ford Fusion Hybrid, 341
Ford hybrid system (FHS), 324–7, 331, 340
Ford Ranger, 67
forward magnetization, 173–4
four-quadrant DC chopper, 24–6
four-quadrant operation, 28, 82–3
four-quadrant zero-current transition DC–DC converter, 28–9
four-quadrant zero-voltage transition DC–DC converter, 28–9
four-wheel drive, 144, 228

fractional-slot, 166, 246, 304, 307, 385
FRDC machine, 7
friction and windage, 176, 179
FRPM
machine, 7, 147–8, 157–62, 165–6, 177, 192–4
motor, 157, 159, 161, 193–4
FSDC
machine, 7, 254, 260–263, 272, 274–6, 283–8, 357
motor, 260, 263, 272, 274, 283, 285, 289
FSPM
machine, 7, 148, 160–164, 260, 272
motor, 160–161, 193, 260, 263, 289
fuel cells, 5, 9–10
proton exchange membrane, 10
solid oxide, 10
solid polymer, 10
fuel consumption, 5, 323–4, 383–4
fuel economy, 4–5, 8, 294–5, 298, 313, 315–6, 323–4, 328, 330, 334, 340
fuel-cell electric vehicle (FEV), 1, 3, 5–6, 8–10
full throttle, 324
full-bridge converter, 260
fuzzy logic control, 212, 239

gasoline, 1–3, 295
gear inner rotor, 215, 217–19, 221–2, 224, 381
gear noise, 5, 8, 341, 343, 360, 367
gear outer rotor, 215, 217, 219, 221–2, 224, 381
gear ratio, 35, 64, 96, 137, 144, 200, 209, 217, 222, 242, 319–20, 324, 326, 328–30, 371–2, 374, 377, 384, 387
geothermal, 2
GM Chevrolet Silverado Hybrid, 341
GM two-mode hybrid system, 326–8, 330–331, 333, 341
GMC Conceptor Electric G-Van, 37
GMC Sierra Hybrid, 341
gridable, 1, 3–5, 14

Halbach array, 202, 218, 385, 387
half-bridge converter, 260
Hall sensor, 56, 239
hard switching, 28, 50, 120, 123–4
harmonic spectrum, 221, 223, 387, 389
harmonic torque, 236
harmonics, 47, 50, 80, 83, 210, 217, 235, 236, 256
H-bridge converter, 162, 246
healthy phase, 182–4, 186, 188, 191
heat tolerance, 108, 112
HE-DSPM
machine, 162–3, 165–6, 169, 177–80, 298, 301, 303, 307–12
motor, 177, 179–81
HEPM
machine, 147–9
motor, 192–4
high-precision manufacturing, 359, 382, 392
high-range mode, 327, 329–31, 333
hill-climbing, 6, 140, 165, 300, 304, 375
Honda Insight, 294, 313
hub motor (*see* in-wheel motor)
hybrid electric vehicle (HEV), 1, 3–5, 7–8, 10, 14, 165, 291–5, 299, 301, 312–13, 319, 323, 327, 333, 335, 339, 341, 357, 359–60, 364, 367–77, 382, 392
hybrid energy system, 10
hybrid excitation, 162–3, 165
hybrid propulsion, 8, 292–4
hybrid-excited DSPM (*see* HE-DSPM)
hybrid-excited PM (*see* HEPM)
hybridization, 1, 9, 291, 293
hybrids, 8, 291–6, 298–9, 301, 303, 313, 315, 341, 343, 347, 359, 373, 392
complex, 291–3
full, 1, 5, 293–5
micro, 1, 4, 8, 293–6, 313
mild, 1, 4, 8, 293–6, 301, 303, 313
parallel, 291–3
plug-in, 1, 5, 293, 295, 339–40
power, 294
series, 291–3, 295
series–parallel, 291–3
hydraulic braking, 300
hydraulic link, 291
hydrocarbon, 3
hydrogen, 1, 5–6
hydropower, 2
hysteresis current, 47–8, 83, 119, 124, 239
hysteresis loop, 167–8
hysteresis loss, 78, 179
hysteresis model, 167–8
hysteresis-current PWM, 47–8
idle stop-start, 4–5, 8, 291, 294, 298–300, 313, 323, 333, 341, 360, 384
IGBT, 28, 47, 122
immature, 194, 253, 289
incremental flux, 58
incremental inductance, 129–30
independent system operator, 14
induction motor, 39–40, 46–7, 50–53, 55–8, 60–68, 70, 72, 83, 87, 131, 133, 145, 253
inductive power transfer, 12
inertia, 87, 117, 211, 296–8, 320–321, 324, 326, 328–30, 350, 376
inner air-gap, 96, 206–10, 217–19, 221–3, 302, 347, 355–6, 359–61
inner-rotor structure, 110, 251
input-split, 315–16, 319, 321–4, 326, 333–6, 339–41, 343
inset-magnet, 231–2
instability, 24, 68–9, 149, 192, 255, 260, 297, 382
insulated gate bipolar transistor (*see* IGBT)
integrated-starter-alternator, 291, 312
integrated-starter-generator (*see* ISG)
intergrid, 14
interior-circumferential, 73–5, 94, 335
interior-radial, 73–5, 93, 96–7, 335
internal combustion engine (*see* engine)
internal combustion engine vehicle (ICEV), 1–5, 9, 11, 293–4
inter-turn, 189–92
intragrid, 14
in-wheel motor, 195, 211–12, 215–24, 227–8, 254, 269, 276, 283
iron bridge, 178, 304, 307, 310
iron loss, 34, 61, 133, 151, 178–9, 212, 214, 222, 240, 256, 259, 269
iron material, 34, 301–2, 307, 385
iron segment, 256, 388, 390
ISG
machine, 296, 301, 313
operation, 298, 301, 309–10
system, 8, 291, 294–301, 303–4, 312–13

Jourdain's principle, 320

Kia Optima Hybrid, 341

lamination material, 35, 64, 97, 101, 135, 138, 142
lap winding, 34
Laplacian and quasi-Poissonian equations, 205
limit circle, 388, 391
limp-home mode, 50
linear-flux morphology, 8
liquid fuel, 2
liquid hydrogen, 6
load angle, 237
load capability, 316
low inertia, 297–8
lower-deck inverter, 239
low-range mode, 327–8, 330–331, 333
lubrication, 5, 8, 195, 197, 224, 341, 343, 360, 367
lunar roving vehicle, 37

machine dimension, 33, 61, 64, 131, 136, 140, 158, 240, 272–3, 275
magnetic axial-flux gear, 201–2
magnetic coaxial gear, 195–6, 201–5, 207, 210–211, 215, 228–9, 371–3
magnetic concentric gear, 369, 371, 373, 375, 378, 382–7, 389
magnetic concentric-geared EVT (*see* MCG EVT)
magnetic coupling, 12, 199, 355–6, 359, 362
magnetic decoupling, 221
magnetic energy, 15, 119, 241, 266
magnetic gear, 195, 197–8, 200–201, 209, 211, 214–15, 217, 224, 227, 367–71, 373, 380–381, 384
  converted, 198, 200
  field-modulated, 198, 200
magnetic gearing effect, 231–3, 240, 246, 264, 289
magnetic gearing ratio, 264, 269
magnetic involute gear, 198–9
magnetic linear gear, 201
magnetic loading, 33, 62–3, 94–5, 132, 356
magnetic planetary gear, 195–6, 200, 341, 369–71, 373, 382
  carrier, 369, 371, 373–5
  magnetic planet gear, 200, 369–71, 373–4
  magnetic ring gear, 200, 369–71, 373, 375
  magnetic sun gear, 369–71, 373–5
magnetic planetary gear ratio, 371, 374
magnetic planetary-geared EVT (*see* MPG EVT)
magnetic resonant coupling, 12–13
magnetic spur gear, 195, 197, 199–200, 371
magnetic worm gear, 198–9
magnetic-geared (*see* MG)
magnetless machine, 7–8, 253–5, 269, 272, 283, 289, 341, 357, 359
magnetless motor, 105, 253–5, 269, 272, 288–9
maintenance, 8–9, 19, 22, 38–9, 144, 197, 224, 266, 301, 334, 364, 367–9, 384
manual transmission, 299
manufacturability, 8, 192, 194, 203, 205, 289
mass production, 295, 313, 341
maturity, 19, 38, 68, 91, 192–4, 203, 205, 213, 289
maximum power point tracking, 10
maximum torque per ampere, 91, 212, 391
MCG EVT, 375–84
mechanical coupling, 348, 356
mechanical differential, 195, 227
mechanical integrity, 72–4, 110, 162, 165, 202–3, 205, 214, 243, 255, 307, 353–4, 359
mechanical link, 291, 293
mechanical loss, 179
mechanical port, 332, 345–7, 357, 369, 375
mechanical stress, 202, 233
memory machine, 70, 165
metal hydride, 6, 9
metal-oxide-semiconductor field-effect transistor (*see* MOSFET),
MG
  EVT, 8–9, 367–9, 371, 380, 382, 392
  machine, 195–6, 203, 205, 211–13, 228–30, 369, 381–2
  motor, 195, 211–13, 215, 218–19, 221, 224
middle air-gap, 217, 219, 221
minimum permeance, 264
minimum reluctance, 112, 255
modulating function, 206, 243
modulation index, 239
modulation ring, 196, 201, 205–7, 214–15, 217, 219, 229, 368, 372–3, 375–8, 383–4, 387–8
MOSFET, 27, 122
motor starting, 14
move-and-charge, 6
MPG EVT, 373–4
multi-input multi-output (MIMO) system, 345
multiple-input converter, 10–11
multipole, 101, 231, 246, 301, 307
multi-port machine, 345
  four-port machine, 345–6
  three-port machine, 345–6
  two-port machine, 345
multi-port magnetic gear, 367, 369, 373
multi-tooth DSDC machine, 272–5
multi-tooth FSDC machine, 274–6
multi-tooth SR machine, 111, 272–6
mutual inductance, 55, 79, 81, 116, 211, 240, 301, 350

natural gas, 2, 6
natural operation, 125
near-field, 11–12
neighborhood electric vehicle, 37–8
nine-switch inverter, 228, 239
Nissan Leaf, 103, 105
nitrogen oxide, 3
nonmagnetic, 74, 214, 222, 385
nonrare-earth, 224
nuclear, 2
number of turns, 43, 45, 63–4, 94, 135, 178, 183, 186, 191, 235, 279, 284

observer, 129, 131, 169
oceanic, 2
oil, 2
open-circuit fault, 183, 186, 191
open-slot, 231
optimal operation line, 5, 323–4, 351, 373, 377–9, 381
optimal operation point, 377
optimal PWM, 47
outer air-gap, 96, 206–10, 217, 219, 221–2, 302, 347, 355–6, 359–61
outer-rotor structure, 110, 131, 246, 251, 272
output gearing, 316, 324, 326–8
overload capability, 6
overload protection, 368
overtaking, 6, 165, 178, 278, 285, 287, 300, 304

parallel magnetic circuit, 356
parallel resonant DC link, 50
parallel-magnet structure, 94
parallel-series magnetic circuit, 356
parasitic inductance, 50
park-and-charge, 12
particulate matter, 3
peak shaving, 14–15
permanent magnet (*see* PM)
petrol (*see* gasoline)
PG EVT, 8, 315–16, 319, 321–4, 326, 328, 331, 333–6, 343–5, 360, 364, 367–8, 371, 373–4, 380, 382, 384
phase inductance, 115–16, 119, 124–5, 129–30, 152, 231
phase-advance angle control, 91–3, 103
phase-decoupling, 182
phasor diagram, 188, 190, 237
PI controller, 87, 155–6, 239
PID regulator, 183
pitch factor, 235
planetary gear, 5, 64, 96, 98, 110, 137, 200, 315–20, 324, 327–35, 339–41, 343–4, 360, 367–9, 371, 373, 375
  carrier, 316–17, 320–321, 327–8, 332–4, 343
  planet gear, 316–17, 335, 343
  ring gear, 5, 96, 316–17, 319–23, 327–8, 331–5, 343, 367
  sun gear, 5, 96, 316–17, 319–21, 327–8, 331–5, 343, 367
planetary gear ratio, 319, 329–30
planetary-geared EVT (*see* PG EVT)
plug-in hybrid electric vehicle (PHEV), 1, 3, 5, 14, 293, 295
PM
  brushless machine, 7, 70, 72–3, 75, 77–8, 80, 94, 108, 110, 149, 206, 211, 215, 228–30, 240–244, 257, 297–8, 307, 313, 373–5, 383, 385–91
    BLAC, 72, 212, 297
    BLDC, 72, 74–8, 80–81, 84, 87, 94, 101, 161, 211, 241, 297, 313
    DC machine, 21, 23, 35, 74, 81
    interior-circumferential, 73–5, 94, 335
    interior-radial, 73–5, 93, 96–7, 335
    surface-inset, 73–5, 93
    surface-mounted, 72–5, 81, 90, 93, 101, 255, 353, 388
  brushless motor, 68–71, 82, 87, 93–6, 103–5, 131, 144, 147, 192, 212, 215, 224, 227, 253
    AC motor, 218
    DC motor, 82, 91–3, 96, 100–103, 105, 192, 215, 217–18
  immunity, 192–3
  material, 7, 68–71, 93, 105, 147, 161, 165–7, 170, 176–7, 193, 201, 211, 217, 222, 224, 234, 243, 247, 253, 255, 257–8, 260, 272, 283, 288, 297–8, 392
    aluminum-nickel-cobalt (Al-Ni-Co), 69, 165–8, 170–173, 182, 191
    ferrite, 69–71
    neodymium-iron-boron (Nd-Fe-B), 70–71, 165–6, 170–173, 176, 197, 201, 217, 222, 243, 247, 253, 336
    samarium-cobalt (Sm-Co), 70–71, 165–6
  properties, 70
    coercivity, 69–71, 93–4, 165–7, 172
    Curie temperature, 69–71
    energy product, 69–71, 170, 241
    remanence, 69–71, 74, 93, 95, 165, 167–8, 241
  synchronous machine, 72–80, 87–8, 93, 96, 161, 237, 239, 241, 251, 255, 257, 297, 301–3, 313, 334–6, 341, 353
    motor, 83–4, 87–91, 93–4, 96–100, 102–3, 105, 159, 192, 238–9, 255, 257, 337
  torque, 152, 157, 159, 161, 167, 176, 182, 184, 191, 309, 335–6
pole arc, 33, 35, 62, 76, 95, 115, 134, 138, 142, 150, 153, 158, 173, 175, 214, 219, 273, 275
pole height, 33, 35, 134–5, 173, 273, 275
pole pitch, 33, 62–3, 94, 115, 117, 134, 152–3, 157, 159, 231, 235–6, 301, 369–70
pole-pair arrangement, 258, 285
polyphase, 257, 266
position encoder, 56, 72, 87, 213
power assistance, 8, 291, 298, 300, 310, 313
power density, 6, 19, 50, 69, 123, 153, 228, 253, 297, 316, 353, 385
power factor, 8, 62–3, 158, 231, 237, 257, 266, 289
power ratio, 77
power regulation, 294
power summation, 360, 367, 369
power-handling capacity, 359, 381
power-split, 5, 105, 294, 315–17, 323–4, 327–9, 333–4, 343, 350–351, 359–60, 367–9, 375, 380–381, 383, 392
powertrain, 313, 339–40
pre-fault, 188–9
Preisach theory, 167
price premium, 294–5
proportional-integral (*see* PI controller)
proportional-integral-derivative (*see* PID regulator)
propulsion device, 1, 5, 293
pseudo-gearless, 9, 195, 367–8, 392
pull-out torque, 387
pulse-width modulation (*see* PWM)
pure electric mode, 5, 378
pure electric vehicle (PEV), 1, 3–5, 8–9, 11–14, 293, 295
PWM
  carrier, 91
  control, 82, 119, 122
  inverter, 39, 47, 57, 238

q-axis, 73, 78–80, 84, 87–90, 256–7, 335, 349, 388, 390–391
quadrant, 24–9, 31, 38, 60, 82–3, 167–8, 261
  first, 24–5
  four, 24–6, 28–9, 38, 82–3
  second, 24, 167–8
  two, 24–8, 31
quadrature-axis (*see* *q*-axis)
quasi-resonant DC link, 50

radial force, 136–7
radial-flux DSDC machine, 277, 282
radial-flux FSDC machine, 283, 285
radial-flux morphology, 8, 356

random PWM, 47
range per charge, 4, 11, 37, 82, 212, 227, 258
range-extended electric vehicle (REV), 1, 3, 5, 14, 293, 295
rare-earth, 68, 70, 72, 144, 197, 211, 224, 253, 368–9, 382, 392
reaction torque, 80, 152, 255
reactive regulation, 14, 16
recoil line, 165–6, 168
recoil permeability, 167
reduced-power region, 53
reformer, 5
refueling infrastructure, 5–6
regenerative braking, 2, 4–5, 8, 24–5, 27–8, 31–2, 227, 282, 287, 291, 293–4, 298–300, 313, 322–4, 331, 333, 341, 351, 353, 359–60, 375, 381, 384
regional transmission organization, 14
regular PWM, 47
relative permeability, 95, 167, 234
reliability, 6, 8, 27, 50–51, 72, 91, 123, 144, 181–2, 213, 239, 334, 354, 357, 364, 367–8, 384
reluctance torque, 73–4, 80, 114, 152, 155, 157, 159, 176, 182, 184, 255–6, 259, 263, 309, 335–6
remanence, 69–71, 74, 93, 95, 165, 167–8, 241
remediation, 183, 188, 191
Renault hybrid system, 331–2
renewable power, 3, 14
renewable transients, 14–15
resonant capacitor, 27, 121
resonant DC link, 50, 120–121
resonant inductor, 27, 50
resonant snubber capacitor, 50
resonant switch, 120
resonant transition, 50
ring-to-sun gear ratio, 320, 328
roadside emission, 3–5, 298
robustness, 6, 131, 147, 182, 192–4, 203, 231, 289, 297–8, 313
rotating field, 8, 42–4, 229, 240, 242–3, 246, 255, 264–6, 268, 351
rotor diameter, 33, 35, 63–4, 97, 112, 132–3, 135, 159, 259, 273, 275
rotor flux linkage, 53, 55–6, 58–9
rotor teeth pairing, 158
rotor-PM, 8, 147, 161, 165, 229–34, 242
saliency, 73, 256–7, 335
salient pole, 7, 110, 147, 149, 158, 166, 178, 182, 258, 303–4, 354, 357–8, 361
saturation, 35, 57, 64, 78, 96, 102, 112, 114–15, 117, 127, 131, 135, 138–9, 217, 233, 240, 256, 260, 273, 275, 337–8
scalar magnetic potential, 205
segmental rotor, 256
self-inductance, 81, 91–2, 167, 192, 211, 259, 263, 350
self-searching control, 179–80
self-shielding, 202, 218, 221–2, 385
self-tuning control, 212
sensing coil, 56, 93
sensorless control, 87, 90–91, 93, 124, 128–9, 212–13, 239, 257
series magnetic circuit, 356
series mode, 333
series resonant DC link, 50
series-magnet structure, 93–4
service life, 300, 313
shaft diameter, 95, 97, 101, 135
shielding layer, 385–7
shoot-through fault, 117
short-circuit fault, 181, 189, 191–2
short-pitch winding, 235
signal injection, 91, 129
silent chain, 335
single-input single-output (SISO) system, 345
single-layer winding, 34, 62, 161, 301, 303
single-stator double-rotor, 269, 271, 283
single-stator single-rotor, 269–70, 277, 283
singly-fed motor, 193, 289
sinusoidal back EMF, 301
sinusoidal PWM, 47
sinusoidal waveform, 78, 87
six-step inverter, 46
skewing, 137, 153, 158, 161, 173, 183
slip frequency, 347
slip ring, 8–9, 266, 347–8, 350, 356, 358–9, 364, 367–8, 384
slot height, 63–4, 240
slot pitch, 101, 166, 182, 242–3, 264–5, 301
slot width, 63–4
slot-fill factor, 64, 97, 101, 136, 138, 142, 215, 240, 385
slots per pole per phase, 101, 233, 235, 265, 301, 385
slotting effect, 206, 235
sluggish, 14, 53, 59, 61
Smart Fortwo, 103–4, 106
snubber, 50–51
soft switching, 27–8, 50–51, 119–20, 124
solar, 2, 10, 14
space harmonic, 242, 246
space-vector modulation, 83–4
space-vector PWM, 47, 49–50, 58, 87, 239
specific energy, 6, 9–10, 315
specific power, 9, 315
speed oscillation, 117
spinning reserve, 14
split capacitor, 154–5, 260
split-pole, 231, 242
squirrel-cage induction machine, 348
SR
  converter, 108, 117–24, 130, 133
    1.5m-switch, 118–19
    asymmetric bridge, 118–19
    C-dump, 119, 123
    R-dump, 119
    resonant DC-link, 120–121
    resonant switch, 120
    soft-switching, 119–20, 124
    zero-transition, 120, 122–3
  machine, 7, 108–17, 119, 128
  mode, 167, 169, 182–5, 191–2
  motor, 126, 132–4, 136, 144
stability, 70, 105, 147, 165–6, 227, 251, 316
stack length, 94, 153, 176, 178, 183, 214, 236, 274, 276, 279, 282, 284, 306, 311, 336
starter motor, 4, 291, 294–8, 313, 333, 351
start-up time, 99, 103
state-of-charge (SOC), 298, 300, 313, 359, 378–9, 381
stator diameter, 33, 35, 101, 133, 259, 273, 275
stator flux linkage, 53, 57–60, 131, 391
stator voltage vector, 58–61, 211
stator-PM, 7–8, 105, 147–9, 157, 160–162, 165–7, 173–4, 177, 192–4, 229, 231–3, 257
stator-PM machine, 7, 147–8, 157, 160–162, 165–7, 173–4, 177
  doubly-salient PM, 147–60, 162–3, 165–7, 169–71, 173–85, 187, 189–94, 257–9, 272, 281–3, 289, 297–8, 301, 303, 307–12
  flux-mnemonic PM, 147–9, 165–6

stator-PM machine (*continued*)
flux-reversal PM, 7, 147–8, 157–62, 165–6, 177, 192–4
flux-switching PM, 7, 148, 160–164, 260, 272
hybrid-excited PM, 147–9
stepwise control, 82
stray capacitance, 27, 50
stray loss, 179
stress tensor, 222
sulfur oxide, 3
supercapacitor (*see* ultracapacitor)
surface-inset, 73–5, 93
surface-magnet, 231–2
surface-mounted, 72–5, 81, 90, 93, 101, 255, 353, 388
switched reluctance (*see* SR)
switching frequency, 27, 50, 64, 83–4, 112, 114, 119–21, 133, 151
switching loss, 27–8, 46, 50, 83, 87, 119–22, 124, 151, 259
synchronism, 73, 82
synchronized resonant, 50
synchronous impedance, 237
synchronous inductance, 79
synchronous reactance, 237
synchronous reluctance (*see* SynR)
synchronous speed, 44, 56, 234, 265, 267–8, 349–50
SynR
machine, 253–7, 289
motor, 255, 257

tangential force, 136
tangentially magnetized, 203–4
temperature coefficient, 69–71
Tesla Roadster, 67–8
thermal dissipation, 178
thermal field, 359, 382
thermal instability, 68–9, 149, 192, 255, 260, 297, 382
thermal power, 2
Thevenin, 172–3
three-leg inverter, 228, 238–9
throttle angle, 379
Timken hybrid system, 332–3
tire rim, 215, 251
tooth width, 63, 95, 240
toroidal-field, 260–261
torque differential, 316
torque multiplier, 316, 324, 326
torque pulsation, 82
torque ripple, 21, 61, 87, 99, 103, 105, 108, 110, 112, 117, 119, 121, 124, 126, 128, 131, 133, 136–7, 140–142, 144, 155, 161, 183, 189, 191, 202–3, 222, 240, 243, 248, 263, 266, 273–7, 285, 297, 334
torque transmission capability, 380, 385, 387
torque-angle curve, 139, 141, 143, 222, 224, 387, 389
torque-ripple minimization (TRM), 124, 126, 128–9, 136
torque-sharing function (TSF), 126–9, 136
cosine, 127–8
cubic, 127
exponential, 127
linear, 126–8
nonlinear, 126, 128
torque-speed capability, 35–6, 65, 67, 98, 103, 125, 139–41, 144, 281, 285, 287, 338
torque-speed characteristic, 20, 23–4, 29, 31, 46, 52, 82
total harmonic distortion, 14
Toyota Camry Hybrid, 341
Toyota hybrid system (THS), 319–20, 324, 326, 331, 339–40
Toyota Prius, 14, 294–5, 319, 335, 339–41
transmission gear, 141, 227
transmission loss, 5, 8, 98, 112, 195, 227, 343, 360, 367, 369, 373, 384
transverse-flux morphology, 8
trapezoidal back EMF, 75, 183–4, 186
trapezoidal waveform, 82
two-arm modulation, 239
two-layer structure, 304
two-quadrant DC chopper, 24–6, 31
two-quadrant zero-current transition DC–DC converter, 27–8
two-quadrant zero-voltage transition DC–DC converter, 27–8
two-wheel drive, 228, 238

ultracapacitor, 4, 9–10, 315
ultrahigh-speed flywheel, 9–10
unipolar flux, 147
unipolar-magnetized, 203–4
upper-deck inverter, 239
utilization factor, 239

variable-voltage variable-frequency control (*see* VVVF control)
vector control, 56, 159, 161–2, 165
vehicle-to-grid, 6, 14–16
vehicle-to-vehicle, 14
vernier effect, 7, 264
vernier gauge, 231
vernier machine, 7
doubly-fed reluctance, 8, 254, 266, 272
PM, 7–8, 227–34, 237–51, 272
reluctance, 7–8, 228, 254, 264–6, 357
AC, 8, 266, 268
DC, 8, 266, 268–9, 357
vernier PM (*see* VPM)
vernier reluctance (*see* VR)
vernier reluctance AC (*see* VRAC machine)
vernier reluctance DC (*see* VRDC machine)
vibration, 56, 135–7, 144, 147, 193, 197, 296–7, 300–301, 313, 333, 341, 354
viscous friction, 87, 117, 211
voltage constraint, 388, 391
voltage deviation, 14
voltage regulation, 7, 287
voltage ripple, 178
voltage space vector, 47–8, 83
VPM
machine, 7–8, 227–34, 237–51, 272
motor, 227, 238–40, 242, 245, 251
VR
machine, 7–8, 228, 254, 264–6, 357
motor, 264, 289
VRAC machine, 8, 266, 268
VRDC machine, 8, 266, 268–9, 357
V-shaped PM rotor, 335–6
VVVF control, 51–3

wave winding, 34
wear and tear, 22, 296, 341
wind power, 2, 14, 200
winding factor, 62, 94, 282
wireless power transfer, 6, 13–14
wound-field, 21, 23, 256, 260, 261
wound-rotor induction machine, 39–40
wound-rotor synchronous machine, 73, 255

yoke thickness, 33, 35, 64, 95, 97, 134–6, 138, 142, 240, 273, 275

zero-current switching, 27, 50, 121
zero-current transition, 27–9, 122
zero-emission, 3
zero-transition, 27, 120, 122–4
zero-voltage switching, 27, 50
zero-voltage transition, 27–9, 122

Printed and bound by CPI Group (UK) Ltd, Croydon, CR0 4YY
16/09/2026
14963390-0003